Law of Sines

$$\frac{a}{\sin A} = \frac{b}{\sin B} = \frac{c}{\sin C}.$$

Law of Cosines

$$a^2 = b^2 + c^2 - 2bc \cos \theta,$$
$$b^2 = c^2 + a^2 - 2ca \cos \theta,$$
$$c^2 = a^2 + b^2 - 2ab \cos \theta.$$

Law of Tangents

$$\frac{a-b}{a+b} = \frac{\tan \frac{1}{2}(A-B)}{\tan \frac{1}{2}(A+B)}, \qquad \frac{b-c}{b+c} = \frac{\tan \frac{1}{2}(B-C)}{\tan \frac{1}{2}(B+C)},$$

$$\frac{c-a}{c+a} = \frac{\tan \frac{1}{2}(C-A)}{\tan \frac{1}{2}(C+A)}.$$

The Half-angle Formulas in Terms of the Sides of a Triangle

$$\tan \tfrac{1}{2}A = \frac{r}{s-a}, \qquad \tan \tfrac{1}{2}B = \frac{r}{s-b}, \qquad \tan \tfrac{1}{2}C = \frac{r}{s-c},$$

$$\sin \tfrac{1}{2}A = \sqrt{\frac{(s-b)(s-c)}{bc}}, \qquad \sin \tfrac{1}{2}B = \sqrt{\frac{(s-a)(s-c)}{ac}},$$

$$\sin \tfrac{1}{2}C = \sqrt{\frac{(s-a)(s-b)}{ab}},$$

where $\quad r = \sqrt{\dfrac{(s-a)(s-b)(s-c)}{s}} \quad$ and $\quad s = \tfrac{1}{2}(a+b+c).$

The Area of a Triangle

$$K = \tfrac{1}{2}ab \sin C.$$

$$K = \tfrac{1}{2}a^2 \frac{\sin B \sin C}{\sin A}.$$

$$K = \sqrt{s(s-a)(s-b)(s-c)},$$

where $\quad s = \dfrac{a+b+c}{2}.$

PLANE TRIGONOMETRY

a new approach

The Appleton-Century Mathematics Series

Raymond W. Brink and John M. H. Olmsted, Editors

PLANE TRIGONOMETRY

a new approach

C. L. Johnston

East Los Angeles College

APPLETON-CENTURY-CROFTS
EDUCATIONAL DIVISION
New York MEREDITH CORPORATION

Copyright © 1970 by

MEREDITH CORPORATION

All rights reserved

This book, or parts thereof, must not be used or reproduced in any manner without written permission. For information address the publisher, Appleton-Century-Crofts, Educational Division, Meredith Corporation, 440 Park Avenue South, New York, N.Y. 10016.

Library of Congress Card Number: 79-110245

ACKNOWLEDGMENTS

The following material has been taken from Raymond W. Brink, *Plane Trigonometry,* 3rd ed. (New York, Appleton-Century-Crofts, 1959): Page 61, proof, paragraph 1, from Brink, p. 68; Page 103, paragraph 2, 3, from Brink, p. 13; Page 170, Section 1003, 1. 1–13, from Brink, p. 131; Page 194, Section 1102, from Brink, p. 187; Page 222, paragraph 3, 4, from Brink, p. 199; Page 227, Section 1305, from Brink, p. 196.

The following material has been taken from C. L. Johnston, *Slide Rule,* 4th ed. (Iowa, Wm. C. Brown Co., 1967): Page 171, Ex. 1, 2, 3, from Johnston, p. 84; Page 178, Ex. 1, 2, 3, 4, from Johnston, p. 8.

740-1

PRINTED IN THE UNITED STATES OF AMERICA

390-48365-6

PREFACE

Students of Plane Trigonometry often encounter more difficulty in the study of the subject than is justified by the essential simplicity of the material. It is for this reason that in this book we avoid what appears to be one of the principal sources of their confusion.

In traditional textbooks on trigonometry, the student is introduced to all of the six basic trigonometric functions each immediately following the other. Then, very soon, he is required to give from memory, or to find from tables, the value of each of the six functions for an angle of any magnitude. For us who have lived with this subject for a long time, it requires no effort to determine both the correct algebraic sign and the numerical value, regardless of the value of the angle or real number on which the function depends. But experience shows that this is not such an easy matter for many students.

One of the principal innovations of this textbook is that here we take just one trigonometric function at a time and carry it through evaluations, trigonometric equations, applications to the right and the general triangles, and graphs. For example, by the time the student has lived with only the sine function through these applications, it has become thoroughly established in his mind and will not easily be forgotten. This is done with the sine function in Chapter 2, the cosine function in Chapter 3, and the tangent function in Chapter 4. Then, having this thorough knowledge of the sine, cosine, and tangent functions, the student has no difficulty extending his knowledge to include the definitions and applications of the reciprocal functions, which are treated in Chapter 5.

After studying the first five chapters a student should be able to (1) solve both right and general triangles and many basic problems involving vectors, (2) draw the graphs of all the trigonometric functions as well as composite functions, and (3) solve many trigonometric equations. In fact, for many students who are enrolled in some two-year curriculum such as drafting, mechanical engineering, or electronics, or some four-year curriculum such as medicine, dentistry, pharmacy, optometry, the first five chapters will give the necessary trigonometry. This does not mean that this textbook presents a "watered down" version of trigonometry. In fact I have used these first five chapters, experimentally, in classes for two semesters and found that one can ultimately cover the subject matter in less time by using this approach than can be done by the traditional method. This leaves more time for the subjects of the later chapters, such as inverse functions and complex numbers, which are often slighted at the end of the course. The entire book offers a complete course in Plane Trigonometry and is especially adaptable to courses of various lengths and purposes.

Modern notation has not been used, since many of our students do not have a background in modern mathematics and the use of modern notation would have required an additional introductory chapter and thus increased the length of the course. Since only a limited number of days are available for the course, the addition of new topics would leave less time for trigonometry. It is my opinion that to do a good job of teaching a complete course in plane trigonometry we must apply all of our time to that subject.

The unit circle has been used extensively as an aid in defining functions and developing formulas and identities.

An abundance of illustrative examples with solutions, two hundred seventy-five in number, anticipates the difficulties of the student and at the same time sets before him applications of basic principles and orderly solutions of exercises. The discussion of trigonometric graphs is more complete than in many other books on the subject. One hundred thirty-seven illustrative figures help to clarify the proofs, definitions, and examples. The thirteen hundred eighty-eight exercises allow a student to obtain practice on all parts of the theory by working either the odd-numbered or the even-numbered exercises. Answers to odd-numbered exercises appear at the end of the text. Answers to the even-numbered exercises are available in a separate pamphlet.

Basic trigonometric formulas are listed inside the front cover and the basic trigonometric identities inside the back cover so as to be immediately available when needed.

I am deeply indebted to Professor Raymond W. Brink, Consulting Editor of the Appleton-Century Mathematics Series, for his many suggestions and painstaking attention to my manuscript. The elegant proof of the Law of Cosines, the complete set of Tables, and more than a score of other passages in the text of this book were taken by permission from the *Third Edition* of Dr. Brink's *Plane Trigonometry*, Appleton-Century-Crofts, N.Y., 1959.

Whittier, California

C.L.J.

CONTENTS

PLANE TRIGONOMETRY

a new approach

Special formulas :

① sine / cosine function — pg. 47
② sine + cos = 1 unit circle function — pg 52
③ law of cosines — pg. 60

1

Introduction

101 TRIGONOMETRY

Trigonometry—a branch of mathematics that deals with the relationships between the angles and sides of triangles and the theory of the periodic functions connected with them—is a basic tool used in the development of mathematics and many sciences such as physics, engineering, astronomy, and the like.

102 RECTANGULAR COORDINATES

Although it is assumed that the student has had experience with the rectangular coordinate system, a brief review of the subject is given.

Two perpendicular lines are drawn meeting at O (Fig. 101). The point O is called the **origin**, the line OX the **x-axis**, and the line OY the **y-axis**. A convenient unit of length is used to mark off distances to the right and left and up and down from the origin O. Distances to the right are taken as positive values of x and distances to the left are taken as negative values of x. Positive values of y are measured upward and the negative values of y are measured downward.

The position of any point P on the xy-plane is determined by a pair of numbers called the **coordinates** of the point. The distance of P to the right or left of the origin is called the **abscissa** or **x-coordinate** of point P, and y, the vertical distance of P from the x-axis, is the **ordinate** or **y-coordinate** of point P. Point P is said to have the coordinates (x, y) and may be referred to as *the point (x, y)*.

1

Example 1.   Locate the point (2, 3).

Start at the origin and move two units to the right, then up three units (Fig. 101).

Example 2.   Locate the point (0, −2).

Start at the origin, but, because the first number is zero, do not move right or left.   The second number being negative directs us down two units (Fig. 101).

Example 3.   Locate the point (−3, 2).

Start at the origin and move left three units, then up two units (Fig. 101).

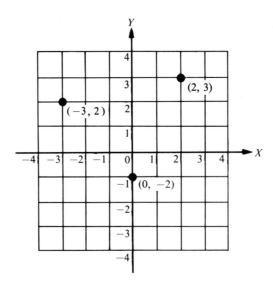

Figure 101

103   THE FORMATION OF ANGLES

A plane angle is formed if two half lines have the same end-point. This end-point is called the **vertex** of the angle and the two half lines are the **sides of the angle** (Fig. 102). We can think of the angle as being **generated** when a half line whose end-point is the vertex of the angle rotates in the plane about the vertex from the position of one side of the angle until it coincides with the

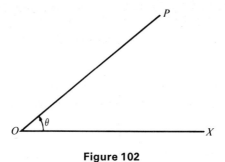

Figure 102

other side. Its initial position is the **initial side** and its final position is the **terminal side** of the angle.

Angle *XOP* is generated when a half line with the initial position *OX* rotates about *O* to the terminal position *OP*. We call *OX* the initial side and *OP* the terminal side of the angle. An angle is often denoted by a single Greek or italic letter (Fig. 102).

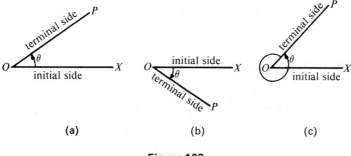

       (a)   (b)   (c)

Figure 103

An angle may be positive or negative and of any magnitude. Angles are said to be positive when the generating half line rotates counter-clockwise, and negative when the generating half line rotates clockwise (Fig. 103).

104   ANGLES IN STANDARD POSITION

An angle is said to be in **standard position,** with respect to the rectangular coordinate system, when its vertex is at the origin and its initial side extends along the positive *x*-axis.

The *xy*-plane is divided by the coordinate axes into four quadrants as shown in Figure 104. On some occasions we shall use the abbreviation Q I, Q II, Q III, and Q IV, to represent the respective quadrants.

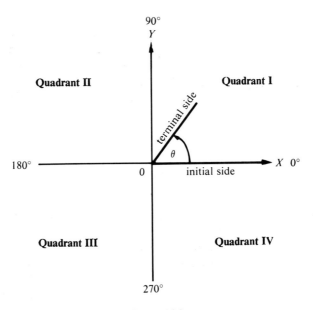

Figure 104

Angles in standard position are often spoken of as being in the quadrant which contains the terminal side (Fig. 105).

(a) 150° is an angle in the second quadrant;
(b) −50° is an angle in the fourth quadrant;
(c) 200° is an angle in the third quadrant.

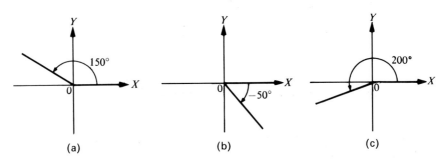

Figure 105

Coterminal Angles. *Angles are said to be* **coterminal** *if their terminal sides coincide when the angles are in standard position.* Examples of coterminal angles: 90° and −270°; 30° and 390°; −30° and 330°; etc. (Fig. 106). Any angle has an unlimited number of coterminal angles.

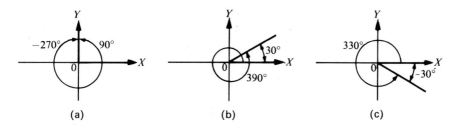

(a)   (b)   (c)

Figure 106

Quadrantal Angles. *Angles in standard position and having their terminal sides along a coordinate axis are called* **quadrantal angles.** The quadrantal angles include 0°, 90°, 180°, 270°, and their coterminal angles.

In addition to the *x*- and *y*-coordinates of the point *P* a third number, *r*, the distance of *P* from the origin, is often used. This distance is called the **radius vector** of *P*. In this course the radius vector, *r*, will always be considered positive or zero regardless of the signs of *x* and *y*. Referring to the right triangle *OAP* (Fig. 107) we have:

$$r^2 = x^2 + y^2,$$
$$r = \sqrt{x^2 + y^2}.$$

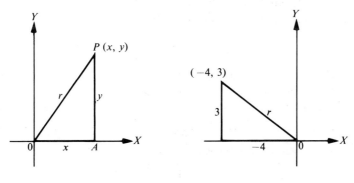

Figure 107   Figure 108

Example.   Find the distance from the origin to the point (−4, 3).

Solution.   Plot the point.
Then

$$r^2 = (-4)^2 + 3^2$$
$$= 16 + 9$$
$$= 25,$$
$$r = \sqrt{25} = 5.$$

105   DISTANCE BETWEEN TWO POINTS

Let two points, P_1 and P_2, have coordinates (x_1, y_1) and (x_2, y_2), and let d be the length of the line segment joining them (Fig. 109). These points can be located anywhere on the xy-plane. Draw a horizontal line through P_1 and a vertical line through P_2. These two lines intersect, forming a 90° angle at point $P_3(x_2, y_1)$. Draw the line segment P_1P_2 which represents the distance d between points P_1 and P_2. Then by the theorem of Pythagoras,

$$d^2 = (x_2 - x_1)^2 + (y_2 - y_1)^2,$$

(1)
$$\boldsymbol{d = \sqrt{(x_2 - x_1)^2 + (y_2 - y_1)^2}.}$$

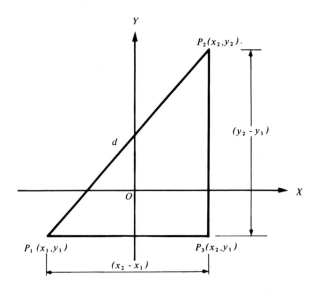

Figure 109

If the points P_1 and P_2 were interchanged the results would be the same. That is:

$$d = \sqrt{(x_2 - x_1)^2 + (y_2 - y_1)^2} = \sqrt{(x_1 - x_2)^2 + (y_1 - y_2)^2}.$$

Example 1.   Find the distance between the points $A(-3, -2)$ and $B(9, 3)$.

$$\begin{aligned} d &= \sqrt{(x_2 - x_1)^2 + (y_2 - y_1)^2} \\ &= \sqrt{[9 - (-3)]^2 + [3 - (-2)]^2} \\ &= \sqrt{(9 + 3)^2 + (3 + 2)^2} \\ &= \sqrt{144 + 25} = \sqrt{169} = 13. \end{aligned}$$

Example 2.   Find the distance between the points $A(-3, 1)$ and $B(0, -2)$.

Solution.   We use Formula (1) and Table VII to find the required distance.

$$\begin{aligned} d &= \sqrt{(x_2 - x_1)^2 + (y_2 - y_1)^2} \\ &= \sqrt{[0 - (-3)]^2 + [(-2) - (+1)]^2} \\ &= \sqrt{3^2 + (-3)^2} = \sqrt{18}. \end{aligned}$$

In order to find $\sqrt{18}$, let $18 = 10(1.8)$. Then on page *104* of Table VII, follow down the column under "n" to the number 1.80. Go across the line to the column under "$\sqrt{10n}$" and read $\sqrt{18} = 4.24264$ (to five decimal places). Therefore the required distance is 4.24264.

Example 3.   Use Table VII to find the value of each of the following expressions:

(a) $\sqrt{235}$.

Solution.   Set $235 = 100(2.35)$. Then

$$\begin{aligned} \sqrt{235} &= \sqrt{100(2.35)} \\ &= 10\sqrt{2.35}. \end{aligned}$$

To find $\sqrt{2.35}$ look on page *105* of Table VII. Follow down the column under "n" to the number 2.35. Go across the line to the column under "$\sqrt{n}$" and read $\sqrt{2.35} = 1.53297$. Therefore $10\sqrt{2.35} = 15.3297$ (to four decimal places).

(b) $\sqrt{0.0785}$.

Solution.   Set $0.0785 = 0.01(7.85)$. Then

$$\sqrt{0.0785} = \sqrt{0.01(7.85)}$$
$$= 0.1\sqrt{7.85} = 0.1(2.80179) = 0.280179$$

(to six decimal places).

106 RELATED ANGLES

Angles are usually designated by greek letters such as α (alpha), β (beta), γ (gamma), θ (theta), ϕ (phi), ω (omega), etc.

Let θ be any angle in standard position. *We define the **related angle** of θ as the angle, in the interval from $0°$ to $90°$ inclusive, formed by the terminal side of θ and the horizontal axis.*

Examples: (*See* Fig. 110.)
1. The related angle of $150°$ is $30°$.
2. The related angle of $210°$ is $30°$.
3. The related angle of $-150°$ is $30°$.

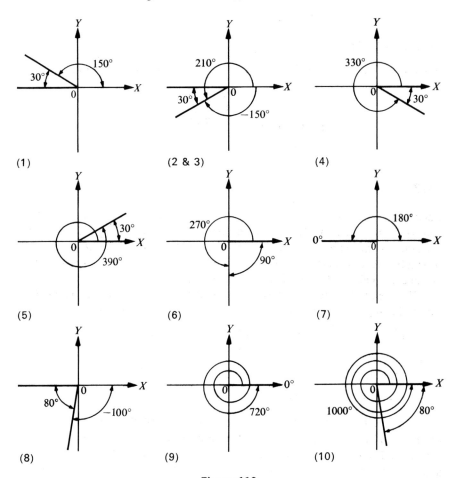

Figure 110

4. The related angle of 330° is 30°.
5. The related angle of 390° is 30°.
6. The related angle of 270° is 90°.
7. The related angle of 180° is 0°.
8. The related angle of $-100°$ is 80°.
9. The related angle of 720° is 0°.
10. The related angle of 1000° is 80°.

Example 11.   Find the related angle of 127° 34.2′.

Solution.   We subtract 127° 34.2′ from 180° in the following manner:

$$
\begin{array}{ll}
179°\ 60' & (=180°) \\
127°\ 34.2' & (-) \\
\hline
\text{Related angle} = \quad 52°\ 25.8'
\end{array}
$$

Example 12.   Find the related angle of 223° 17.8′.

Solution.   We subtract 180° from 223° 17.8′ in the following manner:

$$
\begin{array}{ll}
223°\ 17.8' & \\
180°\ \ 0.0' & (-) \\
\hline
\text{Related angle} = \quad 43°\ 17.8'
\end{array}
$$

107   THE 30°-60° RIGHT TRIANGLE

Each angle of an equilateral triangle is 60°. The bisector of any angle of this triangle is the perpendicular bisector of the side opposite the angle. (Consider, for example, the equilateral triangle of Figure 111 in which each side is 2 units in length.) It follows that in any 30°-60° right triangle the side opposite the 30° angle is half the length of the hypotenuse (Fig. 112).

We use the Pythagorean theorem to find the side opposite the 60° angle. The following solutions relate to parts *a*, *b*, and *c* of Figure 112.

$$
\begin{array}{lll}
a^2 + 1^2 = 2^2, & b^2 + 5^2 = 10^2, & c^2 + x^2 = (2x)^2, \\
a^2 + 1 = 4, & b^2 + 25 = 100, & c^2 + x^2 = 4x^2, \\
a^2 = 3, & b^2 \doteq 75, & c^2 = 3x^2, \\
a = \sqrt{3}. & b = \sqrt{(25)(3)} = 5\sqrt{3}. & c = \sqrt{3}x.
\end{array}
$$

The lengths of the respective sides of the 30-60° right triangle are in the ratio of 1 to 2 to $\sqrt{3}$, and the student should memorize this. Notice that the shortest side is opposite the smallest angle and the longest side is opposite the largest angle. We will have occasion to refer frequently to this triangle and to the ratios of the lengths of the respective sides (Fig. 113).

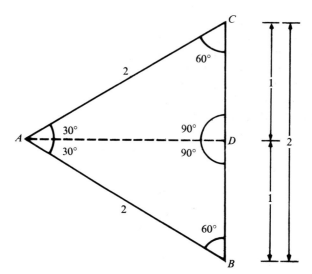

Figure 111

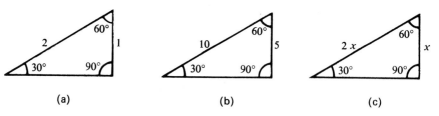

(a)   (b)   (c)

Figure 112

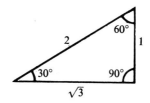

Figure 113

108   THE 45° RIGHT TRIANGLE

*An **acute** angle is a positive angle less than 90°.*

The sum of the two acute angles of a right triangle is 90°. If one of the two acute angles is 45° the other must also be 45°, making an isosceles right triangle.

We use the Pythagorean Theorem to find the hypotenuse of a 45° right triangle. We denote the length of the equal sides by s and that of the hypotenuse by h. First consider the 45° right triangle having equal sides 1 unit long (Fig. 114).

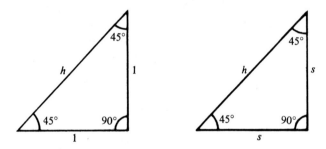

Figure 114

$$h^2 = 1^2 + 1^2 \qquad h^2 = s^2 + s^2$$
$$= 1 + 1 = 2, \qquad = 2s^2,$$
$$h = \sqrt{2}. \qquad h = \sqrt{2}s.$$

In the 45° right triangle the hypotenuse is always $\sqrt{2}$ times the length of one of the equal sides. The student should memorize the 1 to 1 to $\sqrt{2}$ ratios of the respective sides. These will be used many times in this course (Fig. 115).

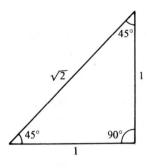

Figure 115

109 PYTHAGOREAN RIGHT TRIANGLES

Right triangles whose three sides are integers are often referred to as **Pythagorean right triangles**; for example, the 3, 4, 5 right triangle and the 5, 12, 13 right triangle are Pythagorean triangles (Fig. 116).

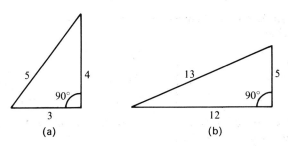

Figure 116

These triangles will be used on a number of occasions in this course. Note:

$$5^2 = 4^2 + 3^2, \qquad 13^2 = 12^2 + 5^2,$$
$$25 = 16 + 9 = 25. \qquad 169 = 144 + 25 = 169.$$

$(7, 24, 25)$

EXERCISES

Name the quadrant in which each angle in Exercises 1–12 terminates.

1. 100°.   **2.** 170°.   **3.** 200°.   **4.** 260°.

5. −10°   **6.** −70°.   **7.** −200°.   **8.** −250°.

9. 400°.   **10.** 440°.   **11.** 500°.   **12.** 530°.

Find the related angle for each angle in Exercises 13–48.

13. 100°.   **14.** 110°.   **15.** 210°.   **16.** 200°.

17. 265°.   **18.** 269°.   **19.** 275°.   **20.** 280°.

21. −140°.   **22.** 140°.   **23.** −160°.   **24.** −200°.

25. −210°.   **26.** −300°.   **27.** −280°.   **28.** 540°.

29. 360°.   **30.** −270°.   **31.** 450°.   **32.** 180°.

33. 720°. **34.** 900°. **35.** 530°. **36.** 630°.

37. 10° 15′. **38.** 20° 50′. **39.** −40° 10′. **40.** −60° 50′.

41. 140° 25′. **42.** 150° 35.′ **43.** 255° 10.3′. **44.** 235° 27.3′.

45. 300° 10.7′. **46.** 350° 50.6′. **47.** 151° 15.4′. **48.** 179° 25.9′.

Find the radius vector, r, for each of the points in Exercises 49–54.

49. (3, 4). **50.** (5, 12). **51.** (−5, 12). **52.** (−3, 4).

53. (5, 0). **54.** (0, −4).

In Exercises 55–60, use Formula (1) and Table VII to find the distance between the pairs of points.

55. (2, 1) and (6, 4). **56.** (−3, 0) and (0, 4).

57. (2, −3) and (5, 8). **58.** (2, −3) and (−5, 2).

59. (2, 3) and (−6, 1). **60.** (0, −5) and (−5, 0).

Find the length of the perimeters of the triangles whose vertices are at the points given in Exercises 61 and 62.

61. (0, 1), (4, 4), (6, 2). **62.** (10, 10), (−1, −2), (−4, 5).

Find the missing coordinate for each of the points in Exercises 63–70.

63. $x = 3, r = 5$, point is in Q I.

64. $x = 3, r = 5$, point is in Q IV.

65. $y = 5, r = 13$, point is in Q II.

66. $y = −5, r = 13$, point is in Q III.

67. $x = 1, r = 2$, point is in Q IV.

68. $x = 1, r = \sqrt{2}$, point is in Q I.

69. $y = −1, r = \sqrt{2}$, point is in Q III.

70. $y = 1, r = 2$, point is in Q II.

2

The sine function

201 THE DEFINITION OF A FUNCTION

By means of the relation

$$A = \pi r^2,$$

we can find the area of a circle when its radius is known. For example, when the radius is 3 inches the area is 9π, or approximately 28.3 square inches. The area, A, is said to be a **function** of the radius, r, because the value of A can be found when the value of r is known.

Definition. *A variable y is said to be a **function** of a variable x if x and y are so related that when a value is assigned to x, one and only one corresponding value is determined for y.*

202 THE SINE FUNCTION

Let θ be any angle. Place it in standard position (Fig. 201), and let P be a point on the terminal side of θ, other than the origin. Associated with P are three numbers; these are its x and y coordinates and its radius vector, r, the distance of P from the origin.

Six different fractions, or ratios, can be formed from the numbers x, y, and r. These ratios are called **trigonometric functions** of θ. In this chapter we shall learn a great deal about one of these ratios, namely y/r, which is called the **sine function.**

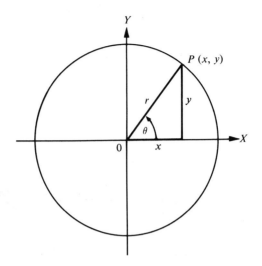

Figure 201

203   DEFINITION OF THE SINE OF AN ANGLE

1. Place an angle θ in standard position.

2. Let point P with coordinates (x, y) be any point, other than the origin, on the terminal side of angle θ.

3. Let r be the distance of P from the origin. This distance r will always be considered as a positive number no matter what position P may occupy on the coordinate plane.

Then the sine of θ (*written* $\sin \theta$) *is defined as the ratio of y to r, or*

$$\sin \theta = \frac{y}{r}.$$

Special case.   Let θ be an acute angle of a right triangle. Let this triangle be placed in the first quadrant with θ in standard position. Let the point (x, y) be the end of the hypotenuse in the first quadrant. With the triangle in this position y becomes the side of the triangle opposite θ and r becomes the hypotenuse of the triangle. Then, from the definition of the sine of an angle,

$$\sin \theta = \frac{y}{r} = \frac{\text{side opposite } \theta}{\text{hypotenuse}}.$$

Examples: (*See* Fig. 202.)

1. $\sin 30° = \frac{1}{2}.$

2.
$$\sin 60° = \frac{\sqrt{3}}{2} = \tfrac{1}{2}\sqrt{3}.$$

3.
$$\sin 45° = \frac{1}{\sqrt{2}} = \tfrac{1}{2}\sqrt{2}.$$

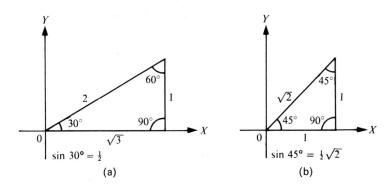

(a)   (b)

Figure 202

204 SIMILAR TRIANGLES

If two triangles have the three angles of one equal respectively to the three angles of the other, the triangles are similar. In plane geometry we learned that the ratios of corresponding sides of similar triangles are the same

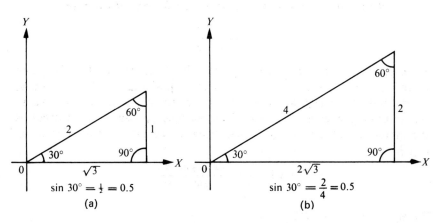

(a)   (b)

Figure 203

regardless of the size of the triangle. Therefore, corresponding angles of any two similar right triangles have equal sines. This is illustrated in Figure 203.

205   THE UNIT CIRCLE

The circle with radius $r = 1$ unit and with its center at the origin is called **the unit circle.** Because of the fact that corresponding sides of similar triangles have equal ratios, we can simplify the task of finding the sine of a given angle θ as follows (Fig. 204):

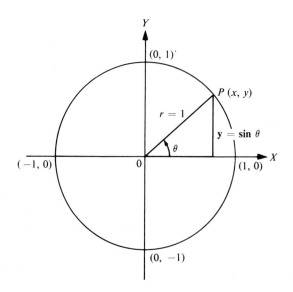

Figure 204

1. Place θ in standard position.
2. Use the origin as a center, and construct a circle with radius equal to one unit.
3. Let $P(x, y)$ be the point where the terminal side of θ meets the unit circle.
   Then $r = 1$, and

$$\sin \theta = \frac{y}{1} = y.$$

Important relationship: *In the unit circle* $\sin \theta = y$.

It follows that the sine of an angle is positive when y is positive and negative when y is negative. Therefore, *the sines of first and second quadrant angles are positive, and the sines of third and fourth quadrant angles are negative* (Fig. 205).

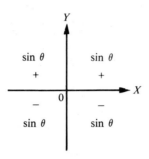

Figure 205

From Figure 204 we observe the sines of the quadrantal angles to be as follows:

$$\sin 0° = 0, \qquad \sin 180° = 0,$$
$$\sin 90° = 1, \qquad \sin 270° = -1.$$

Again by observing Figure 204, note that the sine of an angle is never greater than 1 nor less than -1.

206   THE NUMERICAL VALUE OF A NUMBER

Symbols that will be used in this section are:

$<$, read *is less than,*
$\leqq$, read *is less than or equal to,*
$>$, read *is greater than,*
$\geqq$, read *is greater than or equal to.*

*The **numerical value** or **absolute value** of a real number a is written $|a|$, and is equal to a if $a \geqq 0$ and to $-a$ if $a < 0$.*

Thus, for example, $|5| = 5$, since $5 > 0$. But $|-3| = 3$, for $-3 < 0$, and consequently $|-3| = -(-3) = 3$. These examples illustrate the fact that *the numerical value of a number is never negative.*

Example: $|-5| = |5| = 5$.

207 VALUES OF THE SINES OF CERTAIN SPECIAL ANGLES

Let θ be any angle in standard position, and let $P(x, y)$ be the point of intersection of the terminal side of θ with the unit circle. Recall that under these conditions the sine of θ has the same value as the y-coordinate of point P. Therefore, as θ varies from 0 to 90°, to 180°, to 270°, and finally to 360°, the sine of θ correspondingly varies from 0 to 1, to 0, to -1, and back to 0 (Fig. 206).

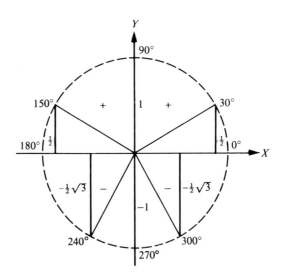

Figure 206

From Figure 206 we can find the values of certain special angles. For example, $\sin 30° = \frac{1}{2}$, $\sin 90° = 1$, $\sin 150° = \frac{1}{2}$, $\sin 180° = 0$, $\sin 240° = -\frac{1}{2}\sqrt{3}$, $\sin 270° = -1$, $\sin 300° = -\frac{1}{2}\sqrt{3}$, $\sin 360° = 0$.

The sine of any angle has the same numerical value as the sine of its related angle. The sine is positive when a point on the terminal side lies above the x-axis and negative when a point on the terminal side lies below the x-axis (Fig. 205).

To find the sine of an angle, place the angle in standard position and then determine its quadrant and the sine of its related angle.

Example 1. Find $\sin 150°$. (*See* Fig. 207a.)

Since 150° is in the second quadrant, $\sin 150°$ is positive. The related angle is 30°. Therefore, $\sin 150° = \sin 30° = \frac{1}{2}$.

Example 2.   Find sin 300°. (*See* Fig. 207b.)
Because 300° is in the fourth quadrant, sin 300° is negative. The related angle is 60°. Therefore, sin 300° = $-\sin 60° = -\frac{1}{2}\sqrt{3}$.

Example 3.   Find sin(−225°). (*See* Fig. 207c.)
Since −225° is in the second quadrant, sin(−225°) is positive. The related angle is 45°. Therefore, sin(−225°) = sin 45° = $\frac{1}{2}\sqrt{2}$.

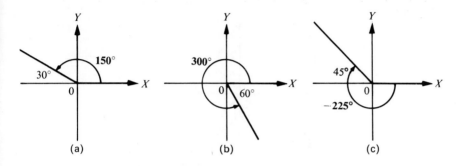

(a)   (b)   (c)

Figure 207

From elementary algebra we recall certain facts about *powers* and *roots*:

(1) $\left(\dfrac{a}{b}\right)^n = \dfrac{a^n}{b^n}$; $\begin{array}{l} n \text{ is a positive whole number.} \\ b \neq 0 \ (\neq \text{ is read } is\ not\ equal\ to). \end{array}$

(2) $\sqrt{a}\sqrt{b} = \sqrt{ab}$; $a > 0, b > 0.$

(3) $(\sqrt{a})^2 = \sqrt{a}\sqrt{a} = \sqrt{a^2} = a.$

(4) $(\frac{3}{2})^2 = \frac{9}{4}.$

(5) $\dfrac{1}{\sqrt{2}} = \dfrac{1}{\sqrt{2}} \cdot \dfrac{\sqrt{2}}{\sqrt{2}} = \dfrac{\sqrt{2}}{2} = \frac{1}{2}\sqrt{2}.$

Powers of sin θ are written:

$$(\sin \theta)^3 = \sin^3 \theta, \qquad (\sin \theta)^2 = \sin^2 \theta, \qquad (\sin \theta)^n = \sin^n \theta.$$

Example 4. $\sin^2 60° = (\sin 60°)^2 = (\frac{1}{2}\sqrt{3})^2 = \frac{3}{4}.$

Example 5. $\sin^3 210° = (\sin 210°)^3 = (-\frac{1}{2})^3 = -\frac{1}{8}.$

Example 6.   Let θ be an angle in standard position with its terminal side passing through the point (3, −4). Find sin θ.

Solution: Sketch the angle θ (Fig. 208).

Solve for r.

$$r = \sqrt{3^2 + (-4)^2}$$
$$= \sqrt{9 + 16}$$
$$= \sqrt{25} = 5.$$

Hence $\qquad\qquad \sin\theta = \dfrac{y}{r} = \dfrac{-4}{5} = -\dfrac{4}{5}.$

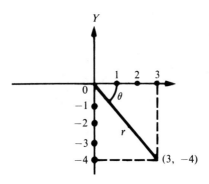

Figure 208

EXERCISES

In Exercises 1–20, find the value of each term, and then combine the terms.

1. $\sin 30° + \sin 180° + \sin 60°$.

2. $\sin 90° + \sin 150° + \sin 0°$.

3. $\sin(-90°) + \sin(-150°) + \sin(-30°)$.

4. $\sin 210° + \sin 270° + \sin 360°$.

5. $\sin(-30°) + \sin 720° + \sin 45°$.

6. $\sin(-90°) + \sin(-180°) + \sin(135°)$.

7. $\sin^2 45° + \sin^2 135° + \sin^2 180°$.

8. $\sin^2 60° + \sin^2 90° + \sin^2 150°$.

9. $2 \sin 150° + \sin^2(-60°) + \sin^2 90°$.

10. $\sin^2(-60°) + \sin 180° + \sin 210°$.

11. $\sin 330° + \sin(-150°) + \sin 150°$.

12. $\sin^3 30° + \sin^4 45° + \sin^2 135°$.

13. $\sin^3 210° + \sin^4 60° + \sin^2 225°$.

14. $\sin 390° + \sin 450° + \sin 540°$.

15. $\sin 720° + \sin(-270°) - \sin(-150°)$.

16. $\sin(-135°) - \sin^2(-60°) + \sin 360°$.

17. $\sin 0° + \sin 180° + \sin 360°$.

18. $\sin 720° + \sin(-180°) + \sin 540°$.

19. $2 \sin^2 60° + 4 \sin^2 45° + 6 \sin^2 135°$.

20. $4 \sin^2 120° + 2 \sin^2 150° + 6 \sin^2 210°$.

In each of the Exercises 21–28, let θ be an angle in standard position with its terminal side passing through the point given. Find $\sin \theta$.

21. $(3, 4)$.   **22.** $(5, 12)$.   **23.** $(-3, 4)$.   **24.** $(\sqrt{3}, 1)$.

25. $(-1, -1)$.   **26.** $(0, 3)$.   **27.** $(-2, 0)$.   **28.** $(1, -\sqrt{3})$.

208   PERIODIC FUNCTIONS

We have seen that if an angle is increased by 360° the value of the sine of the angle is unchanged.

Examples:
1. $\sin \ 0° = \sin 360°$
2. $\sin 30° = \sin 390°$
3. $\sin 60° = \sin 420°$.
That is, if θ is any angle, then

$$\sin(\theta + 360°) = \sin \theta, \text{ or } \sin(\theta \pm n \cdot 360°) = \sin \theta \ (n \text{ an integer}).$$

The values of the sine of an angle thus recur periodically, the period of recurrence being 360°

It will be seen later that all the trigonometric functions repeat their values in cycles and are therefore called **periodic functions.**

The periodic nature of the trigonometric functions makes them of great importance in the study of natural science. Many natural phenomena are of a periodic nature, as for example a vibrating string, a swinging pendulum, the rotation of the earth, the movements of the planets, and the wave motions in the transmission of sound, light, and electricity.

209   A TABLE OF SINES

To find sin θ when θ is known, and to find θ when sin θ is known, refer to Table IV.

For angles in the interval $0°$ to $45°$, use the degree and sine columns that read down from the top of the page. For angles in the interval $45°$ to $90°$, read up from the bottom of the page.

Example 1.   Evaluate sin $20°$.

Solution.   Locate $20°$ in the degree column, page *94*, then read the value of sin $20° = .3420$, to the right of $20°$ in the sine column.

Example 2.   Evaluate sin $74°$.

Solution.   Locate $74°$ in the degree column, page *95*, and read the value of sin $74° = .9613$ to the left in the column marked " Sine " at the bottom of the page.

Example 3.   Evaluate sin $200°$.

Solution: The related angle is $20°$. The angle $200°$ is in the third quadrant. Therefore,

$$\sin 200° = -\sin 20° = -.3420.$$

Example 4.   Find the acute angle θ when sin $\theta = .3256$.

Solution.   Look in the sine column until .3256 is found, on page *94*. Then read the value of $\theta = 19°\ 00'$, to the left in the degree column.

Example 5.   Find an angle θ in the second quadrant such that sin $\theta = .8988$.

Solution.   Search the sine columns until .8988 is located, on page *94*, in the column marked " Sine " at the bottom of the page. To the right in the degree column find $64°\ 00'$, which is the related angle of the desired angle, or angles, in the second quadrant (Fig. 209).

One value of θ is $180° - 64°\ 00' = 116°\ 00'$. There are, however, infinitely many values of θ in the second quadrant for which sin $\theta = .8988$. These values of θ can be given as follows: $\theta = 116°\ 00' + n \cdot 360°$ where n is any integer.

Example 6.   Find an angle θ in the third quadrant such that sin $\theta = -.3420$.

Solution.   Search the sine columns until, on page *94*, $+.3420$ is located in the column marked " Sine " at the top of the page. To the left in the degree

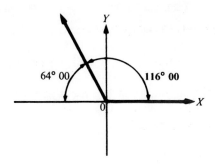

Figure 209

column find 20° 00′, which is the related angle of the desired angle, or angles, in the third quadrant (Fig. 210). Hence $\theta = 180° + 20° = 200°$. A more general answer would be: $\theta = 200° + n \cdot 360°$ where n is any integer.

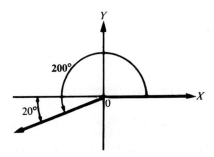

Figure 210

EXERCISES

In Exercises 1–21, using Table IV, find the value of the functions listed.

1. sin 10°.

2. sin 40°.

3. sin 160°.

4. sin 170°.

5. sin 220°.

6. sin 230°.

7. sin 280°.

8. sin 290°.

9. sin (−10°).

10. sin(−15°)

11. sin(−100°).

12. sin(−110°).

13. sin 20° 40′.

14. sin 30°50′.

15. sin 100°50′.

16. sin 100° 10′.

17. sin 265°40′.

18. sin 255°10′.

19. sin 760°.   **20.** sin 1000°.   **21.** sin 45°.

In Exercises 22–27, using Table IV, find the acute angle of the functions listed.

22. sin θ = .1219.   **23.** sin θ = .3907.   **24.** sin θ = .6157.

25. sin θ = .6988.   **26.** sin θ = .7660.   **27.** sin θ = .9356.

In Exercises 28–31, using Table IV, find the angle θ where 90° < θ < 180° and sin θ has the given value.

28. sin θ = .1045.   **29.** sin θ = .1736.

30. sin θ = .9886.   **31.** sin θ = .8307.

In Exercises 32–35, using Table IV, find the angle θ where 180° < θ < 270° and sin θ has the given value.

32. sin θ = −.4067.   **33.** sin θ = −.5446.

34. sin θ = − .9283.   **35.** sin θ = −.8675.

In Exercises 36–39, using Table IV, find the angle θ where 270° < θ < 360° and sin θ has the given value.

36. sin θ = −.4226.   **37.** sin θ = −.4384.

38. sin θ = −.9474.   **39.** sin θ = −.8897.

210   SIGNIFICANT DIGITS

Applied problems are usually stated in terms of data obtained by *measurement* or *observation*. The accuracy of such data is subject to the mechanical limitations of the instruments used.

When an astronomer states that it is 93,000,000 miles from the earth to the sun, or a machinist states that the diameter of a bar of metal is 1.287 inches, we appreciate that both numbers are only approximate. The entries in the tables of trigonometric functions are usually approximations of never-ending decimals.

Digit defined.   *A digit is any one of the ten Arabic numerals,* 0, 1, 2, 3, 4, 5, 6, 7, 8, 9, *so named from counting upon the fingers* (Latin *digitus*, finger). Thus 785 is a three-digit number; 80 is a two-digit number; etc.

In most computation we necessarily deal with approximations. Thus, if a four-place table gives sin 34° 00′ = .5592, we know merely that the exact value of sin 34° 00′ is at least as near to .5592 as it is to either .5591 or .5593, and therefore it lies in the range from .55915 to .55925. In this case we say that 2 is the **last significant digit.** *In general, the position of the last significant digit is determined such that the error made by using the approximate value instead of* *the exact value is not more than 5 units in the next place to the right of the last significant digit.*

The *significant digits* of a number written in decimal form are those digits that begin with the first one not zero and end with the last one which is definitely known. Thus, in the number 27.43, the significant digits are 2, 7, 4, 3; in the number 0.00412, the significant digits are 4, 1, 2; in the number 0.315, the significant digits are 3, 1, 5. In the number recorded as 3.510, the significant digits are 3, 5, 1, 0, the final 0 being significant since it is used to indicate a degree of accuracy. Zeros used to fix the position of the decimal point are not significant.

In a measurement recorded as 180,000 for example, it is impossible to tell which of the zeros are significant until further information is given concerning the known accuracy of the measurement. By expressing such numbers in a form called *scientific notation* (Section 212), or *standard notation*, the significant digits can be indicated. In the absence of this additional information concerning the position of the last significant digit, we shall assume that it is the last nonzero digit in a number. Thus, in the absence of other information, the last significant digit of 1800 is 8, that of 1492 is 2, and that of 0.00326 is 6.

211   APPROXIMATIONS

It is often desirable to express a number with fewer digits than were originally given. This process is called **rounding off the number.** We shall use the following rules whenever rounding off takes place:

(1) *If the part to be dropped amounts to less than 5 units in the first discarded place, keep the last retained digit unchanged.* (*See* Example 1.)

(2) *If the part to be dropped amounts to more than 5 units in the first discarded place, increase the last retained digit by one unit.* (*See* Example 2.)

(3) *If the part to be dropped is exactly 5, round off the preceding digit to an even number.* (*See* Examples 3 and 4.)

Examples:

1. $3.1416 = 3.14$ to three significant digits.
2. $3.1416 = 3.142$ to four significant digits.
3. $47.45 = 47.4$ to three significant digits.
4. $47.35 = 47.4$ to three significant digits.

The symbol $\approx$ is put between two expressions to indicate that the right-hand expression is an **approximation** to the left-hand expression. This symbol is read *is approximately equal to*. Thus, $\sqrt{3} \approx 1.73$; $\pi \approx 3.1416$; $\sin 21° \approx .3584$; etc.

In general, the result of a calculation is only as accurate as the least accurate number used in the calculation.

212   SCIENTIFIC NOTATION

Scientific notation (sometimes called *standard notation*) gives us a convenient way of writing numbers so as to display their degree of accuracy. Assuming that 40950 and 0.04095 have four-place accuracy, in scientific notation we should write them as

$$4.095 \times 10^4 \quad \text{and} \quad 4.095 \times 10^{-2}, \text{ respectively.}$$

It will be noticed that the factor 4.095 is the same for both numbers, and that it consists of the sequence of significant digits with the decimal point placed immediately after the first such digit.

A number is said to be expressed in scientific notation when it is represented as the product of an integral power of 10 and a number between 1 and 10 (or 1 itself).

To change a number to scientific notation, always place the first nonzero digit immediately to the left of the decimal point and then multiply the number so written by the appropriate power of ten.

Examples:

1. $830 = 8.3 \times 100 = $ . . . . 8.3×10^2.
2. $83 = 8.3 \times 10 = $ . . . . 8.3×10^1.
3. $8.3 = 8.3 \times 1 = $ . . . . 8.3×10^0.

(By definition $10^0 = 1$)

4. $0.83 = \dfrac{83}{100} = \dfrac{8.3}{10} = $ . . . . 8.3×10^{-1}.

5. $0.083 = \dfrac{83}{1000} = \dfrac{8.3}{100} = \dfrac{8.3}{10^2} = $ . . . . . 8.3×10^{-2}.

The number 1800, by our assumption in Section 210, is understood to be accurate to two significant digits, that is, to hundreds. If we wish to indicate that this number is accurate to units, then scientific notation can be used as follows: $1800 = 1.800 \times 10^3$.

213   APPLICATION OF THE SINE FUNCTION TO RIGHT TRIANGLES

In studying triangles it is common practice to denote the vertices and also the values of the corresponding angles by capital letters such as *A*, *B*, *C*. When this is done we denote the lengths of the opposite sides by the corresponding small letters such as *a*, *b*, *c* (Fig. 211).

In general we shall use a capital letter, such as *A*, to denote the point or

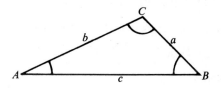

Figure 211

vertex as well as the angle. When there is any possibility of confusing two such meanings we shall denote the angle whose vertex is A by $\angle A$ or $\angle CAB$.

As we know, a triangle has six *parts*, three sides and three angles. To *solve* a triangle, when we are given some of its parts, means to find the remaining parts. However, in some of the problems the student will be asked to solve for only specified parts.

We can solve for the sides of a right triangle when the hypotenuse and an acute angle are known.

Example 1.   Solve the right triangle ABC, in which $c = 1.243$, $A = 32°\ 50'$, and $C = 90°\ 00'$ (Fig. 212). Here we have used the letter A to denote a *point*

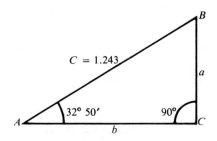

Figure 212

(namely one of the vertices of the triangle) and also an angle whose vertex is at that point.

Solution.   We use Table IV and the sine function.

$$\sin 32°\ 50' = \frac{a}{1.243},$$

$$a = (1.243)(\sin 32°\ 50')$$
$$= (1.243)(.5422) = .6740 \qquad \text{(to four significant digits).}$$
$$B = 90° - 32°\ 50' = 57°\ 10'.$$

$$\sin 57° \ 10' = \frac{b}{1.243},$$

$$b = (1.243)(\sin 57° \ 10')$$
$$= (1.243)(.8403) = 1.044 \qquad \text{(to four significant digits)}.$$

Example 2.   Solve the oblique triangle ABC for the altitude drawn from vertex C to side AB, given $A = 41° \ 20'$ and $AC = 1.065$.

Solution.   Let the altitude h meet side AB at D (Fig. 213). The altitude divides triangle ABC into two right triangles. In triangle ADC,

$$\sin 41° \ 20' = \frac{h}{1.065},$$

$$h = (1.065)(\sin 41° \ 20')$$
$$= (1.065)(.6604) = .7033 \ \text{(to four significant digits)}.$$

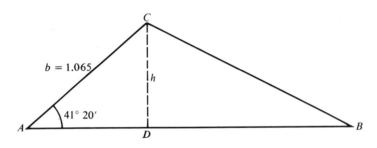

Figure 213

We can use the sine function to solve for the angles of a right triangle when one side and the hypotenuse are known.

Example 3.   Solve the right triangle ABC, in which $c = 10.00$, $a = 6.428$, and $C = 90° \ 00'$.

Solution.

$$\sin A = \frac{6.428}{10.00} = .6428;$$

$$A = 40° \ 00' \quad \text{(see page } 96 \text{ of Table IV)}.$$
$$B = 90° \ 00' - 40° \ 00' = 50° \ 00'.$$

Example 4.   The equal sides of an isosceles triangle are 20.00 units long, and the altitude is 14.94. Find the base angles of the triangle (Fig. 214).

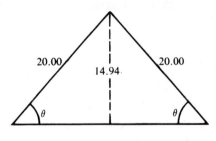

Figure 214

Solution.

$$\sin \theta = \frac{14.94}{20.00} = .7470;$$

$$\theta = 48° \; 20' \quad \text{(see page } 96 \text{ of Table IV).}$$

EXERCISES

State the number of significant digits in each of the numbers in Exercises 1–6.

1. 2,001. **2.** 0.002. **3.** 200.5.

4. 0.050. **5.** 40.050. **6.** 0.00505.

In Exercises 7–12 round off each of the approximate numbers to one less significant digit.

7. 12.24. **8.** 3.95. **9.** 4.05.

10. 8.666. **11.** 1,600. **12.** 24,000.

In Exercises 13–18 write each of the numbers in scientific notation.

13. 93,000,000. **14.** 0.000,006. **15.** 31.47.

16. 184.50. **17.** 0.015. **18.** 0.108.

In Exercises 19–24 write each of the numbers in a form to indicate definitely that the number is known to three significant figures.

19. 4,000. **20.** 20,000. **21.** 93,000,000.

22. 0.03200. **23.** 0.00150. **24.** 21.5.

In Exercises 25–28, use Table IV and solve each of the right triangles for side *a*. Angle $C = 90°$.

25. $A = 38° \; 10', c = 4.551.$ **26.** $A = 44° \; 40', c = 0.1852.$

27. $A = 51° \; 10', c = 2485.$ **28.** $A = 3° \; 40', c = 5.085.$

In Exercises 29–32, use Table IV and solve each of the right triangles for angle *A*. Angle $C = 90°$.

29. $a = 5.640,\ c = 30.00.$ **30.** $a = 2.060,\ c = 4.000.$

31. $a = 41.45,\ c = 50.00.$ **32.** $a = 22.47,\ c = 30.00.$

33. A 15-foot ladder just reaches the base of a second-storey window. The ladder makes an angle of 73° with the level paving below. Find the height of the window.

34. A surveyor measures the distance down a sloping surface from point *A* to point *C* to be 97.98 feet. The sloping surface makes an angle of 25° 30′ with the horizontal. Determine the difference in elevation of points *A* and *C*.

214   THE LAW OF SINES

*In **any** triangle the sides are proportional to the sines of the opposite angles;*

$$\frac{a}{\sin A} = \frac{b}{\sin B} = \frac{c}{\sin C}.$$

Proof.   Let *A*, *B*, and *C* be the angles of any triangle, and let *a*, *b*, and *c* be the opposite sides. We consider two triangles, one in which all angles are acute (Fig. 215a), and the other in which one angle, angle A, is obtuse (Fig. 215b).

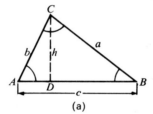
(a)
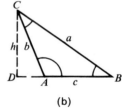
(b)

Figure 215

Draw the altitude *h* from vertex *C* to side *AB* or *AB* extended. In both right triangles thus formed,

$$\sin B = \frac{h}{a} \quad \text{and} \quad \sin A = \frac{h}{b}.$$

Therefore,

$$h = a \sin B \quad \text{and} \quad h = b \sin A.$$

Then by equating the two values of *h* we have

$$a \sin B \quad = \quad b \sin A.$$

Divide both sides of the equation by sin A sin B.

$$\frac{a \sin B}{\sin A \sin B} = \frac{b \sin A}{\sin A \sin B}.$$

This reduces the equation to

(1)
$$\frac{a}{\sin A} = \frac{b}{\sin B}.$$

By drawing the altitude from angle B we can show that

$$\sin C = \frac{h}{a} \quad \text{and} \quad \sin A = \frac{h}{c}.$$

Then

$$h = a \sin C \quad \text{and} \quad h = c \sin A,$$

and

$$a \sin C = c \sin A,$$

which yields

(2)
$$\frac{a}{\sin A} = \frac{c}{\sin C}.$$

Combining Equations (1) and (2) yields the *Law of Sines*:

$$\frac{a}{\sin A} = \frac{b}{\sin B} = \frac{c}{\sin C}.$$

The Law of Sines is used in solving any triangle when two opposite parts (angle A and side a, or angle B and side b, or angle C and side c) are given along with another known side or angle.

Example. Find side b of triangle ABC when $a = 7.05$, $A = 40°\ 10'$, and $B = 35°\ 10'$.

Solution.

$$\frac{b}{\sin B} = \frac{a}{\sin A},$$

$$\frac{b}{\sin 35°\ 10'} = \frac{7.05}{\sin 40°\ 10'},$$

$$b = \frac{(7.05) \sin 35°\ 10'}{\sin 40°\ 10'}$$

$$= \frac{(7.05)(.5760)}{.6450} = 6.30 \quad \text{(to three significant digits).}$$

EXERCISES

In Exercises 1–4, use Table IV and solve the oblique triangles for the missing sides.

1. $A = 52° \ 00'$, $B = 63° \ 00'$, $b = 1.15$.

2. $B = 43° \ 00'$, $C = 77° \ 10'$, $c = 5.61$.

3. $A = 38° \ 10'$, $C = 110° \ 20'$, $a = 2.10$.

4. $A = 41° \ 20'$, $B = 35° \ 10'$, $c = 1.20$.

5. A building along a line AB makes direct measurement of the line impossible. An offset point C is established, and, by measurement, it is found that $CB = 70.5$ feet, angle $CAB = 54°$, and angle $ABC = 47°$. Find AB.

6. Points A and B are on opposite banks of a river. Point C is 235 feet from A and on the same side of the river as A. By measurement, $\angle CAB = 105° \ 30'$ and $\angle ACB = 50° \ 40'$. Find the distance across the river between points A and B.

7. Points A and B are on the same side of a river and are 155 feet apart. Point C is located across the river in such a way that $\angle ABC = 77° \ 10'$ and $\angle BAC = 69° \ 20'$. Find the distance across the river between points B and C.

215   THE AMBIGUOUS CASE

To solve a triangle when given two sides and the angle opposite one of them, the following possibilities exist:

1. Only one triangle is possible.
2. Two different triangles are possible.
3. A triangle cannot be constructed using the given parts.

Let the given parts be a, b, and A. (Fig. 216.)

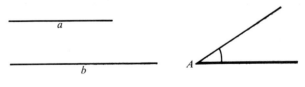

Figure 216

First, draw a freehand figure to help you visualize the problem (Fig. 217). Lay out the figure by always placing one side of the given angle (in this example, A) along a horizontal working line, L, as shown (Fig. 217). Of the

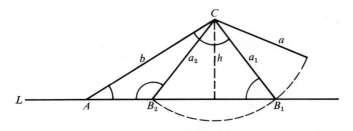

Figure 217

two given sides the one which is not opposite the given angle (*b* in this example) should be placed along the oblique side of the given angle. The other end of *b* determines the point of vertex *C*. Side *a* must begin at *C*. Next compute the perpendicular distance from *C* to line *L*. This perpendicular distance (*b* sin *A*) is the altitude *h* of the triangle as drawn from vertex *C*.

If *A* is an acute angle, there are the following possibilities:

1. If $a < h$, then no triangle is possible (Fig. 218a).
2. If $a = h$, then there is just one triangle—a right triangle (Fig. 218b).
3. If *a* is greater than *h* but less than *b* (that is, $h < a < b$), then there are two triangles—triangles AB_1C and AB_2C (Fig. 217).
4. If *a* is greater than *h* and also greater than *b* (that is, $h < b < a$), then there is only one triangle (Fig. 219).

IF $a < h \Rightarrow$ NO $\triangle$'s FORMED
IF $a = h \Rightarrow 1 \triangle = $ rt. $\triangle$
IF $a > b \& A > h \Rightarrow 1 \triangle$ FORMED
IF $h < a < b \Rightarrow 2 \triangle$'s FORMED

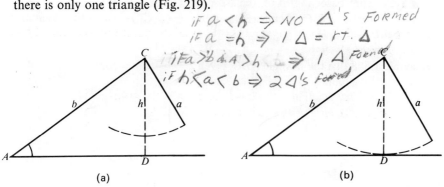

(a)   (b)

Figure 218

If *A* is a right angle or an obtuse angle, and if a triangle is possible, then one and only one triangle exists (Fig. 220).

After we have determined which of the possibilities is applicable, we can complete the solution using the Law of Sines.

If there are two triangles (*see* Fig. 217), then $\angle AB_1C$ and $\angle AB_2C$ are

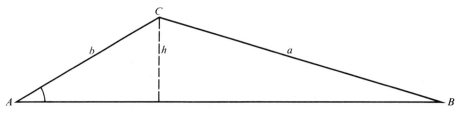

Figure 219

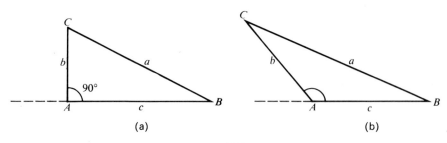

(a)   (b)

Figure 220

supplementary angles, that is, the sum of these angles is $180°$ ($\angle B_1 B_2 C =$ $\angle AB_1 C$, being base angles of the isosceles triangle $B_2 B_1 C$. $\angle B_1 B_2 C$ is a supplement of $\angle AB_2 C$; therefore its equal, $\angle AB_1 C$, is also supplementary to $\angle AB_2 C$).

Example 1. Solve the triangle ABC, given $a = 6.45$, $b = 10.0$, $A = 40°\ 10'$.

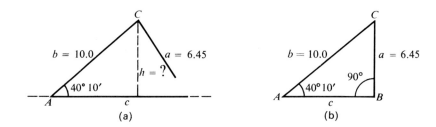

(a)   (b)

Figure 221

Solution. Make a freehand sketch of the figure involved in the problem, as shown in Figure 221a.

Compute the altitude h.

$$\sin A = \frac{h}{b},$$

$$h = b \sin A$$
$$= 10 \sin 40° \ 10'$$
$$= 10 \ (.6450) = 6.45 \text{ (to three significant figures)}.$$

Hence $h = a$, and therefore triangle ABC is a right triangle (Fig. 221b). Solve for angle C and side c.

$$C = 90° - 40° \ 10' = 49° \ 50'.$$

$$\sin C = \frac{c}{b},$$

$$c = b \sin C$$
$$= 10 \sin 49° \ 10'$$
$$= 10 \ (.7642) = 7.64 \text{ (to three significant figures)}.$$

Example 2.   Solve the triangle ABC, given $a = 6.00$, $b = 10.0$, $A = 30°$.

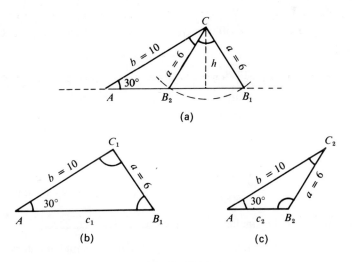

Figure 222

Solution.   Make a freehand sketch of the figure involved in the problem as shown in Figure 222a.

First compute the altitude h.

$$\sin A = \frac{h}{b},$$

$$h = b \sin A$$

$$= 10 \sin 30°$$

$$= 10 \, (.5000) = 5.$$

Hence $h < a < b$, therefore two different triangles are possible.

First we use the Law of Sines to solve the larger triangle AB_1C_1. (*See* Fig. 222b.)

$$\frac{\sin B_1}{b} = \frac{\sin A}{a},$$

$$\sin B_1 = \frac{b \sin A}{a}$$

$$= \frac{10 \sin 30°}{6} = .8333;$$

$$B_1 = 56° \ 30' \qquad \text{(to the nearest 10').}$$

$$A + B_1 + C_1 = 180°,$$

$$30° + 56° \ 30' + C_1 = 180°,$$

$$C_1 = 93° \ 30'.$$

$$\frac{c_1}{\sin C_1} = \frac{a}{\sin A},$$

$$c_1 = \frac{a \sin C_1}{\sin A}$$

$$= \frac{6 \sin 93° \ 30'}{\sin 30°}$$

$$= \frac{6 \, (.9981)}{.5000} = 11.98 \qquad \text{(to four significant figures).}$$

Next we solve the smaller triangle AB_2C_2. (*See* Fig. 222c.)

$$B_2 = 180° - B_1 \qquad (B_2 \text{ and } B_1 \text{ are supplementary} \\ \text{angles})$$

$$= 180° - 56° \ 30' = 123° \ 30'.$$

$$A + B_z + C_2 = 180°,$$

$$30° + 123° \ 30' + C_2 = 180°,$$

$$C_2 = 26° \ 30'.$$

$$\frac{c_2}{\sin C_2} = \frac{a}{\sin A},$$

$$c_2 = \frac{a \sin C_2}{\sin A}$$

$$= \frac{6 \sin 26° \ 30'}{\sin 30°}$$

$$= \frac{6 \,(.4462)}{.5000} = 5.354 \text{ (to four significant figures)}.$$

EXERCISES

In the following Exercises, find the number of possible triangles for the data, and complete the solution of all possible triangles (express angles to the nearest 10 minutes). Draw the figure approximately to scale.

1. $A = 41° \ 00'$, $a = 11.0$, $b = 10.0$.

2. $A = 51° \ 00'$, $a = 11.0$, $b = 10.0$.

3. $C = 25° \ 00'$, $c = 4.11$, $b = 10.0$.

4. $C = 28° \ 00'$, $c = 4.51$, $b = 10.0$.

5. $A = 43° \ 00'$, $a = 8.00$, $b = 10.0$.

6. $A = 37° \ 00'$, $a = 7.00$, $b = 10.0$.

7. $B = 28° \ 10'$, $a = 10.0$. $b = 4.72$.

8. $B = 35° \ 10'$, $a = 10.0$, $b = 5.76$.

9. $C = 63° \ 00'$, $a = 10.0$, $c = 9.00$.

10. $C = 67° \ 40'$, $a = 10.0$, $c = 9.50$.

216 TRIGONOMETRIC EQUATIONS

A **trigonometric equation** is an equality that involves one or more trigono-metric functions of an unknown angle, and which is true for certain particular values of the unknown angle but not for all values. Any value of the unknown angle for which the quality is true is called a **solution** of the trigonometric equation and is said to **satisfy** the equation. Since the same functions of coterminal angles are equal (and there are an unlimited number of angles coterminal with a given angle), a trigonometric equation, in general, has an unlimited number of solutions, provided that it does have a solution. In most cases we will limit our solutions to angles from 0° to 360°—we shall include 0° but not 360°. Our first examples are not of trigonometric equations but of algebraic equations whose methods of solution are often useful in solving trigonometric equations.

Example 1.   Solve $x^2 + x = 0$.

Solution.   Factor the left-hand member of the equation.

$$x(x + 1) = 0.$$

Set each factor equal to zero.
Thus

$$x = 0 \qquad \text{or} \qquad x + 1 = 0.$$

Therefore,

$$x = 0 \qquad \text{and} \qquad x = -1$$

are the solutions or roots of the given equation.

Example 2.   Solve $4x^2 - 1 = 0$.

Solution.

FIRST METHOD   SECOND METHOD

$$4x^2 - 1 = 0,$$ $$4x^2 - 1 = 0,$$

$$4x^2 = 1,$$ $$(2x + 1)(2x - 1) = 0,$$

$$x^2 = \tfrac{1}{4},$$ $$2x + 1 = 0$$ or $$2x - 1 = 0,$$

$$x = \pm\tfrac{1}{2}.$$ $$2x = -1$$ or $$2x = 1,$$

$$x = -\tfrac{1}{2} \quad \text{and} \quad x = \tfrac{1}{2}.$$

Example 3.   Solve the equation $\sin \theta = \tfrac{1}{2}$, for θ where $0° \leq \theta < 360°$.

Solution.   First, determine the related angle. The related angle is 30°, because $\sin 30° = \tfrac{1}{2}$. Second, determine the quadrants which apply. They are quadrants I and II because $\sin \theta$ is positive. Third, place the related angle in each of these quadrants. Thus 30° is the related angle of itself and of 150°. (Fig. 223.)
Therefore, the solutions are 30° and 150°.

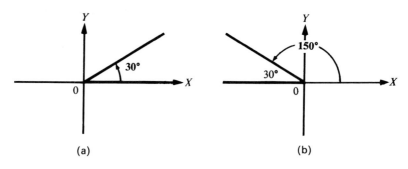

(a)   (b)

Figure 223

Example 4.   Solve the equation $\sin \theta = -\frac{1}{2}$, for θ where $0° \leq \theta < 360°$.

Solution.   The related angle is $30°$ and $\sin \theta$ is negative. Therefore, θ is in quadrant III or IV, and the solutions are $210°$ and $330°$. (Figure 224.)

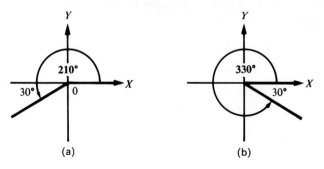

Figure 224

Example 5.   Solve the equation $\sin 2\theta = 0$, for θ where $0° \leq \theta < 360°$.

Solution.   Let $\alpha = 2\theta$.

Then,

$$\sin \alpha = 0,$$

and

$$\alpha = 0°, 180°, 360°, 540°, 720°, \text{etc.}$$

But $\alpha = 2\theta$, therefore,

$$2\theta = 0°, 180°, 360°, 540°, 720°, \text{etc.}$$

We were required to find θ in the range $0° \leq \theta < 360°$, therefore,

$$\theta = 0°, 90°, 180°, 270°.$$

Example 6.   Solve the equation $\sin \frac{1}{2}\theta = 1$, for θ where $0° \leq \theta < 360°$.

Solution.   Let $\alpha = \frac{1}{2}\theta$.

Then,

$$\sin \alpha = 1,$$

and

$$\alpha = 90°, 450°, \quad \text{etc.}$$

But $\alpha = \frac{1}{2}\theta$, therefore,

$$\frac{1}{2}\theta = 90°, 450°, \quad \text{etc., and}\quad \sin \frac{1}{2}\theta = 1 \qquad \text{if} \quad \theta = 180°, 450°, \text{etc.}$$

We were required to find θ in the range $0° \leqq \theta < 360°$. Therefore,

$$\theta = 180° \text{ is the only solution.}$$

Example 7.   Solve the equation $2 \sin^3 \theta + \sin^2 \theta - \sin \theta = 0$, for θ where $0° \leqq \theta < 360°$.

Solution.   Let $\sin \theta = x$, then the equation

$$2 \sin^3 \theta + \sin^2 \theta - \sin \theta = 0$$

takes the form

$$2x^3 + x^2 - x = 0,$$

which factors into

$$x(2x^2 + x - 1) = 0,$$

or

$$x(2x - 1)(x + 1) = 0.$$

Set each factor equal to zero.

Thus

$$x = 0 \qquad \text{or} \qquad 2x - 1 = 0 \qquad \text{or} \qquad x + 1 = 0.$$

Then

$$x = 0 \qquad\qquad\qquad x = \tfrac{1}{2} \qquad\qquad\qquad x = -1.$$

But

$$x = \sin \theta.$$

Therefore

$$\sin \theta = 0 \qquad\qquad \sin \theta = \tfrac{1}{2} \qquad\qquad \sin \theta = -1.$$

Hence

$$\theta = 0°, 180°, \qquad \theta = 30°, 150°, \qquad \theta = 270°$$

are the solutions or roots of the equation.

EXERCISES

Solve each of the following equations for all of the values of θ such that $0° \leqq \theta < 360°$.

1. $\sin \theta = 1$.   2. $\sin \theta = -1$.

3. $\sin \theta = 0$.   4. $\sin \theta = \tfrac{1}{2}$.

5. $\sin \theta = \dfrac{\sqrt{3}}{2} = \tfrac{1}{2}\sqrt{3}$.   6. $\sin \theta = -\dfrac{\sqrt{3}}{2} = -\tfrac{1}{2}\sqrt{3}$.

7. $\sin \theta = -\dfrac{1}{\sqrt{2}} = -\tfrac{1}{2}\sqrt{2}.$ **8.** $\sin \theta = 3.$

9. $\sin \theta = 2.$ **10.** $2 \sin \theta = 1.$

11. $2 \sin \theta = -1.$ **12.** $\sin^2 \theta = 1.$

13. $\sin^2 \theta + \sin \theta = 0.$ **14.** $\sin^2 \theta - \sin \theta = 0.$

15. $4 \sin^2 \theta - 1 = 0.$ **16.** $4 \sin^2 \theta - 3 = 0.$

17. $2 \sin^2 \theta + \sin \theta - 1 = 0.$ **18.** $\sin^2 \theta - 2 \sin \theta + 1 = 0.$

19. $\sin^2 \theta + 2 \sin \theta + 1 = 0.$ **20.** $\sin^3 \theta - \sin \theta = 0.$

21. $2 \sin^3 \theta - \sin^2 \theta - \sin \theta = 0.$ **22.** $\sin^2 \theta + \sin \theta - 2 = 0.$

23. $\sin 2\theta = 1.$ **24.** $\sin 2\theta = -1.$

25. $\sin^2 2\theta + \sin 2\theta = 0.$ **26.** $\sin^2 2\theta - \sin 2\theta = 0.$

27. $2 \sin^2 2\theta + \sin 2\theta - 1 = 0.$ **28.** $\sin^2 \tfrac{1}{2}\theta - 2 \sin \tfrac{1}{2}\theta + 1 = 0.$

217   GRAPH OF THE SINE FUNCTION

Here we discuss the graph of the fundamental sine function, $y = \sin \theta$. Graphs of the more general sine functions, $y = a \sin b\theta$, will be discussed in Chapter 9. The unit circle can be used to show the increasing and decreasing values of the sine function as the angle changes from $0°$ to $360°$. Let the variable angle θ be in standard position with respect to a coordinate system. It was shown in Section 203 that the value of $\sin \theta$ is the same as the value of y at the point where the terminal side of θ crosses the unit circle. We will show the relationship of θ and $\sin \theta$ by means of a graph.

Example 1.   Graph $y = \sin \theta.$

We take values of θ along the horizontal axis and the corresponding values of y on the vertical axis. On the θ-axis we choose a convenient length to represent the equal distances between $0°, 30°, 60°$, etc. Then by plotting many points of $y = \sin \theta$, we determine the shape of the curve. The coordinates of some of these points are given in the attached table.

θ	$0°$	$30°$	$60°$	$90°$	$120°$	$150°$	$180°$	$210°$	$240°$	$270°$	$300°$	$330°$	$360°$
y	0	$\tfrac{1}{2}$	$\tfrac{1}{2}\sqrt{3}$	1	$\tfrac{1}{2}\sqrt{3}$	$\tfrac{1}{2}$	0	$-\tfrac{1}{2}$	$-\tfrac{1}{2}\sqrt{3}$	-1	$-\tfrac{1}{2}\sqrt{3}$	$-\tfrac{1}{2}$	0

We use Figure 225 to emphasize that on the unit circle the value of $\sin \theta$ is the same as the value of y. In Figure 226 the sine function is carried through one complete cycle.

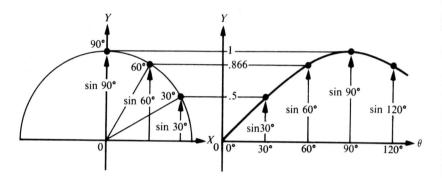

Figure 225

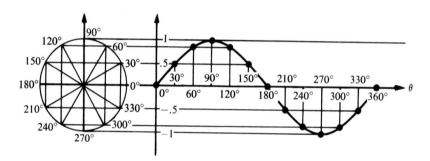

Figure 226

The curve shown in Figures 226 and 227 is called **the graph of the sine function** or **the sine wave curve**. Because of its wave form the sine curve is very important in the study of wave motion in electrical engineering and physics. The maximum distance of the curve from the θ-axis is called the **amplitude** of the curve. The amplitude of this curve is said to be 1, not ± 1. This curve repeats its cycle in an interval of $360°$ and is therefore said to have a **period** of $360°$. (Fig. 227.)

Example 2.   Graph $y = 2 \sin \theta$. (Fig. 228.)
The y values on the $y = 2 \sin \theta$ curve are double those of the $y = \sin \theta$ curve. Therefore, the $y = 2 \sin \theta$ curve has an amplitude of 2 and oscillates back and forth between $y = 2$ and $y = -2$. Both curves have a period of $360°$.

Example 3.   Graph $y = 2 + \sin \theta$. (Fig. 229.)
For each value of θ the corresponding value of y for this curve is 2 greater than the value of y for the standard sine curve, $y = \sin \theta$. Thus by the addition of the constant 2 the curve was **translated** upward 2 units. The period and amplitude remain unchanged.

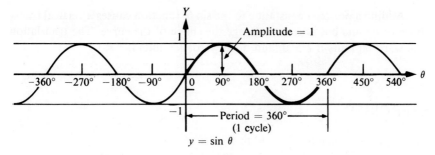

Figure 227

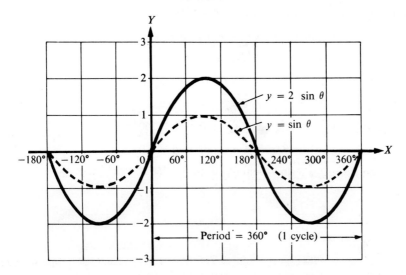

Figure 228

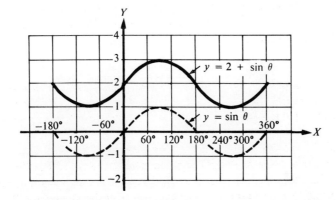

Figure 229

Adding a nonzero constant c to the sin θ function causes a vertical translation of c units but does not change the shape of the curve. The translation is upward $|c|$ units if $c > 0$, and downwards $|c|$ units if $c < 0$.

EXERCISES

1. Graph $y = \frac{1}{2} \sin \theta$ from $\theta = 0°$ to $\theta = 360°$.

2. Graph $y = 3 \sin \theta$ from $\theta = 0°$ to $\theta = 360°$.

3. Graph $y = -\sin \theta$ from $\theta = 0°$ to $\theta = 360°$.

4. Graph $y = 1 + \sin \theta$ from $\theta = 0°$ to $\theta = 360°$.

5. Graph $y = 1 - \sin \theta$ from $\theta = 0°$ to $\theta = 360°$.

6. Graph $y = -1 + \sin \theta$ from $\theta = 0°$ to $\theta = 360°$.

7. Graph $y = -2 - \sin \theta$ from $\theta = 0°$ to $\theta = 360°$.

8. Graph $y = 1 + 2 \sin \theta$ from $\theta = -180°$ to $\theta = 360°$.

9. Graph $y = 1 - 2 \sin \theta$ from $\theta = -180°$ to $\theta = 360°$.

10. Graph $y = -2 + 3 \sin \theta$ from $\theta = 0°$ to $\theta = 360°$.

$$\sin \theta = \cos (90 - \theta) \quad \text{rule}$$
$$\text{or}$$
$$\cos \theta = \sin (90 - \theta)$$

The cosine function

301 THE SINE FUNCTION REDEFINED

The student will recall (Section 102) that the distance of a point $P(x, y)$ to the right or left of the origin is called the *abscissa* or x-coordinate of point P, and that y, the vertical distance of P from the x-axis, is called the *ordinate* or y-coordinate of point P. The radius vector of P is the distance of P from the origin.

The definition of the sine of an angle, given in Section 203, can now be phrased as follows: Let θ be an angle placed in standard position with respect to a system of coordinates. Choose any point on the terminal side of θ other than the origin. Then *the sine of θ is the ratio of the ordinate to the radius vector of that point. (See* Fig. 301.)

$$\sin \theta = \frac{\text{ordinate}}{\text{radius vector}}.$$

302 DEFINITION OF THE COSINE OF AN ANGLE

1. *Place an angle θ in standard position.*

2. *Let point P, whose coordinates are (x, y), be any point on the terminal side of angle θ, other than the origin.*

3. *Let r be the distance of P from the origin. This distance r will always be considered as a positive number.*

47

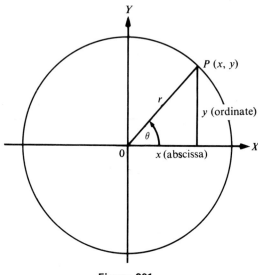

Figure 301

Then the cosine of θ (written cos θ*) is defined as the ratio of x to r:*

$$\cos \theta = \frac{x}{r},$$

or *the cosine of θ is the ratio of the abscissa to the radius vector of P:*

$$\cos \theta = \frac{\text{abscissa}}{\text{radius vector}}.$$

Special case.   Let θ be an acute angle of a right triangle. Let this triangle be placed in the first quadrant with θ in standard position. Let the point (x, y) be the end of the hypotenuse in the first quadrant. With the triangle in this position x becomes the side of the triangle adjacent to θ and r is the hypotenuse of the triangle. Then, from the definition of the cosine of an angle,

$$\cos \theta = \frac{x}{r} = \frac{\text{side adjacent to } \theta}{\text{hypotenuse}}.$$

Examples: (*See* Fig. 302.)

1. $\cos 30° = \dfrac{\sqrt{3}}{2} = \tfrac{1}{2}\sqrt{3}.$

2. $\cos 60° = \tfrac{1}{2}.$

3. $\cos 45° = \dfrac{1}{\sqrt{2}} = \tfrac{1}{2}\sqrt{2}.$

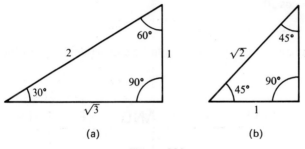

(a)   (b)

Figure 302

303 COMPLEMENTARY ANGLES

Two angles, A and B, are complementary, and each is the complement of the other, if their sum is equal to 90°. We shall adopt this definition even when one of the angles is negative or 0°. Thus 30° and 60°, 0° and 90°, 120° and $-30°$, are pairs of complementary angles.

Observe that the cosine of an angle is equal to the sine of its complement:

$$\cos 60° = \tfrac{1}{2} \quad\text{and}\quad \sin 30° = \tfrac{1}{2}.$$

Therefore, $\cos 60° = \sin 30°.$

Note in Figure 303 that

$$\cos A = \frac{b}{c} \quad\text{and}\quad \sin B = \frac{b}{c}.$$

Therefore, $\underline{\underline{\cos \theta = \sin (90° - \theta).}}$

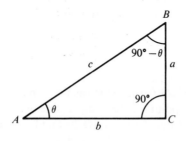

Figure 303

Examples:

1. $\cos 10° = \sin 80°.$

2. $\sin 40° = \cos 50°$.

3. $\sin 120° = \cos(-30°)$.

The sine and the cosine are said to be **cofunctions**, each of the other.

304   THE UNIT CIRCLE AND THE COSINE FUNCTION

Let θ be any angle in standard position, and let $P(x, y)$ be the point of intersection of the terminal side of θ with the unit circle. Then by definition,

$$\cos \theta = \frac{x}{1} = x.$$

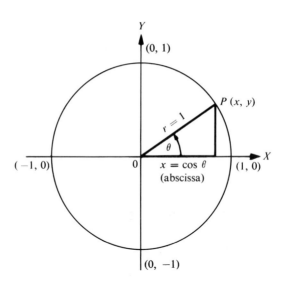

Figure 304

Clearly the sign of $\cos \theta$ is the same as the sign of x when θ is in standard position and (x, y) is a point on the terminal side of θ. Therefore, the cosine is positive in quadrants I and IV, negative in quadrants II and III. (Fig. 305.)

From Figure 304 we find the values of the cosines of quadrantal angles.

$$\cos 0° = 1, \qquad \cos 180° = -1,$$
$$\cos 90° = 0, \qquad \cos 270° = 0.$$

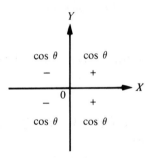

Figure 305

Again by observing Figure 304, note that the cosine of an angle is never greater than 1 or less than −1.

From Figure 306 we can find the values of certain special angles. For example, $\cos 60° = \frac{1}{2}$, $\cos 150° = -\frac{1}{2}\sqrt{3}$, $\cos 240° = -\frac{1}{2}$, $\cos 300° = \frac{1}{2}$.

The cosine of any angle has the same numerical value as the cosine of its

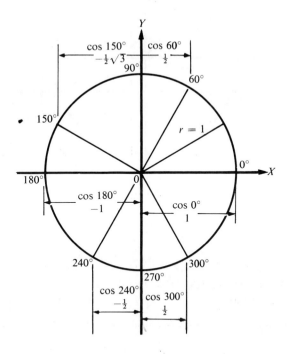

Figure 306

*related angle. The cosine is positive when a point on the terminal side lies to the
right of the y-axis and negative when a point on the terminal side lies to the left
of the y-axis.*

305 $\boxed{\sin^2\theta + \cos^2\theta = 1}$

Let θ be any angle in standard position, and let $P(x, y)$ be the point of intersection of the terminal side of θ with the unit circle. Then the coordinates of P
are $(\cos\theta, \sin\theta)$. From these coordinates of P we find $\mathbf{sin^2\,\theta + cos^2\,\theta = 1}$, by
the Pythagorean theorem. (Fig. 307.)

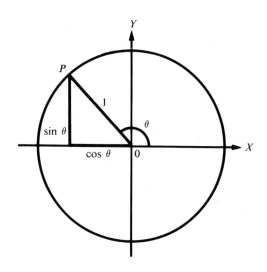

Figure 307

Examples:

1. $\sin^2 45° + \cos^2 45° = (\tfrac{1}{2}\sqrt{2})^2 + (\tfrac{1}{2}\sqrt{2})^2 = \tfrac{1}{2} + \tfrac{1}{2} = 1.$

2. $\sin^2 120° + \cos^2 120° = (\tfrac{1}{2}\sqrt{3})^2 + (-\tfrac{1}{2})^2 = \tfrac{3}{4} + \tfrac{1}{4} = 1.$

3. $\sin^2 180° + \cos^2 180° = (0)^2 + (-1)^2 = 1.$

4. $\sin^2 315° + \cos^2 315° = (-\tfrac{1}{2}\sqrt{2})^2 + (\tfrac{1}{2}\sqrt{2})^2 = \tfrac{1}{2} + \tfrac{1}{2} = 1.$

To find the cosine of an angle, place the angle in standard position, and
then determine its quadrant and related angle.

Example 5.   Find cos 120°.
   Since 120° is in the second quadrant, cos 120° is negative. The related angle is 60°. Therefore, cos 120° = −cos 60° = −½.

Example 6.   Find cos 300°.
   Because 300° is in the fourth quadrant, cos 300° is positive. The related angle is 60°. Therefore, cos 300° = cos 60° = ½.

Example 7.   Find cos(−450°).
   −450° is a quadrantal angle. The related angle is 90°. Therefore, cos(−450°) = cos 90° = 0.

EXERCISES

   In Exercises 1–20, find the value of each term and then combine the terms.

1. cos 0° + cos 90° + cos 180° + cos 270°.

2. cos 60° + cos² 45° + cos² 30° + cos 90°.

3. cos 120° + cos² 180° + cos 300°.

4. cos 135° + cos 225° + cos 330°.

5. sin² 30° + cos² 30°.

6. sin² 60° + cos² 60°.

7. sin² 135° + cos² 135°.

8. sin(−60°) + cos(−60°) + cos(−270°).

9. sin 180° + cos 180° + sin 60° + cos 30°.

10. cos(−30°) + sin(−30°) + cos(−300°).

11. cos 30° + sin 60°.

12. cos 60° + sin 30°.

13. cos 0° + sin 90°.

14. sin 0° + cos 90°.

15. cos 720° + cos(−720°) − cos(−150°).

16. cos(−135°) − sin(−60°) + cos 360°.

17. cos(−60°) + sin(−60°) + cos 120°.

18. cos 540° + cos 0° + cos 90°.

19. 2 cos² 30° + 4 cos² 45° + 6 cos² 135°.

20. 4 cos² 120° + 2 cos² 150° + 6 cos² 225°.

In each of the Exercises 21–26, let θ be an angle in standard position with its terminal side passing through the given point. Find both the sine and the cosine of θ.

21. (4, 3).   **22.** (12, 5).   **23.** (3, -4).

24. (1, $\sqrt{3}$).   **25.** (-1, $\sqrt{3}$).   **26.** (3, 0).

In Exercises 27–44 use Table IV and the given or related angle to find the value of each function.

27. cos 10°.   **28.** cos 40°.   **29.** cos 160°.

30. cos 170°.   **31.** cos 220°.   **32.** cos 230°.

33. cos 280°.   **34.** cos 290°.   **35.** cos(-10°).

36. cos(-15°).   **37.** cos(-100°).   **38.** cos(-110°).

39. cos 100° 50′.   **40.** cos 100° 10′.   **41.** cos 265° 10′.

42. cos 255° 10′.   **43.** cos 760°.   **44.** cos 1000°.

Use Table IV to find the acute angle θ for each of the Exercises 45–50.

45. $\cos \theta = .9483$.   **46.** $\cos \theta = .8732$.

47. $\cos \theta = .3584$.   **48.** $\cos \theta = .5250$.

49. $\cos \theta = .0262$.   **50.** $\cos \theta = .0494$.

306   INTERPOLATION

Interpolation is the process of approximating a number that, although not an entry in the table, lies between two consecutive entries. Several methods of interpolation are used. The method used here is known as **linear interpolation**. In interpolation, fractions of table differences are rounded off to the same number of decimal places as are used in the table. In rounding off a number that is *exactly halfway* between two entries, it is conventional to choose the number that makes the *final result even* rather than odd. This procedure will be followed throughout this book.

Example 1.   Find sin 45°.

Of course sin 45° can be read directly from the table, but we use it here in conjunction with the $y = \sin \theta$ graph to illustrate linear interpolation.

$$\text{difference} = 30° \begin{cases} \sin 30° = .500 \\ \sin 45° = ? \\ \sin 60° = .866 \end{cases} \text{difference} = .366$$

The angle 45° is halfway from 30° to 60°; then, in linear interpolation, sin 45° is taken halfway between sin 30° = .500 and sin 60° = .866. That is,

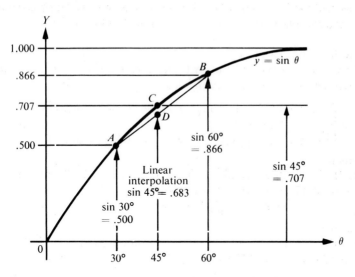

Figure 308

$\sin 45° = .500 + \frac{1}{2}(.366) = .683$. From the tables the actual value of $\sin 45°$ is .707, which shows that linear interpolation is not always accurate.

The error, in linear interpolation, is the vertical distance CD (Fig. 308) between the sine curve and the chord AB, at $\theta = 45°$. For very small differences in values of θ, points A and B become close together, forcing the error distance, CD, to become very small.

Example 2.   Find $\sin 24° 38'$.

Solution.   In linear interpolation we make the assumption (which is not completely accurate) that since $24° 38'$ is eight-tenths of the way from $24° 30'$ to $24° 40'$, then $\sin 24° 38'$ will be eight-tenths of the way from $\sin 24° 30'$ to $\sin 24° 40'$. From Table IV we find the **tabular difference** of $\sin 24° 40' - \sin 24° 30' = .4173 - .4147 = .0026$. Eight-tenths of .0026 = .00208 which we round off to .0021, because the table entries are rounded to four decimal places. Therefore we take

$$\sin 24° 38' = \sin 24° 30' + \tfrac{8}{10}(.0026)$$
$$= .4147 + .0021 = .4168.$$

The following diagrammatic arrangement presents this same operation in tabular form:

$$10'\left\{8'\begin{cases}\sin 24° 30' = .4147\\ \sin 24° 38' = ?\\ \sin 24° 40' = .4173\end{cases}\right.\left.\begin{array}{l}\tfrac{8}{10}(.0026) = .0021\end{array}\right|\begin{array}{l}\text{tabular difference}\\ = .0026\end{array}$$

Therefore, $\sin 24° 38' = .4147 + .0021 = .4168$.

We could have found sin 24° 38′ from Table III directly. Table IV was used to give necessary exercise in interpolation.

Example 3.   Find cos 31° 17′.

*The student is reminded that as θ **increases** from 0° to 90°, the value of* cos θ ***decreases*** *from 1 to 0. (See Fig. 306.)*

$$10'\begin{cases}7'\begin{cases}\cos 31^\circ\ 10' = .8577\\\cos 31^\circ\ 17' = ?\\3'\begin{cases}\cos 31^\circ\ 20' = .8542\end{cases}\end{cases}\end{cases}\begin{array}{l}\tfrac{7}{10}(.0015) = .00105* \approx .0011\\[4pt]\tfrac{3}{10}(.0015) = .00045\dagger \approx .0004\end{array}\Bigg\}.0015$$

FIRST METHOD.

cos 31° 17′ is three-tenths of the way from cos 31° 20′ to cos 31° 10′. Hence, cos 31° 17′ = cos 31° 20′ + $\tfrac{3}{10}$(.0015)

$$= .8542 + .0004 = .8546.$$

SECOND METHOD.

cos 31° 17′ is seven-tenths of the way from cos 31° 10′ to cos 31° 20′. Hence, cos 31° 17′ = cos 31° 10′ − $\tfrac{7}{10}$(.0015)

$$= .8557 - .0011 = .8546.$$

Example 4.   Find the acute angle θ for which

$$\sin \theta = .5217.$$

Solution.

$$10'\begin{cases}\sin 31^\circ\ 20' = .5200\\\sin \theta\qquad = .5217\\\sin 31^\circ\ 30' = .5225\end{cases}\ 17\ \text{(in last two digits)}\Bigg\}25\ \text{(in last two digits)}$$

$$\theta = 31^\circ\ 20' + \tfrac{17}{25}(10')$$
$$= 31^\circ\ 20' + 6.8' = 31^\circ\ 27'.$$

We round off to the nearest minute.

Example 5.   Find the acute angle θ for which

$$\cos \theta = .3700.$$

Solution.

$$10'\begin{cases}\cos 68^\circ\ 10' = .3719\\\cos \theta\qquad = .3700\\\cos 68^\circ\ 20' = .3692\end{cases}\ 19\Bigg\}27$$

$$\theta = 68^\circ\ 10' + \tfrac{19}{27}(10')$$
$$= 68^\circ\ 10' + 7' = 68^\circ\ 17'.$$

* This halfway number is rounded off to an odd number because it is to be subtracted from the odd number .8557. Thus the result is even.

† This halfway number is rounded off to an even number because it is to be added to the even number .8542. Thus the sum is even.

Example 6. Find a second-quadrant angle θ for which

$$\cos \theta = -.9258.$$

Solution. Let α be the related angle, then

$$\cos \alpha = +.9258.$$

$$10' \begin{cases} \cos 22°\ 10' = .9261 \\ \cos \alpha \quad\ \ = .9258 \\ \cos 22°\ 20' = .9250 \end{cases} \Big\}^{3} \Big\} 11$$

$$\alpha = 22°\ 10' + \tfrac{3}{11}(10')$$
$$= 22°\ 10' + 3' = 22°\ 13'.$$

Then $\theta = 180° - 22°\ 13' = 179°\ 60' - 22°\ 13' = 157°\ 47'.$

Example 7. Solve for $x = \sqrt{10.63}$.

Solution. On page *104* of Table VII we find that $\sqrt{10.63}$ lies in the "$\sqrt{10n}$" column between $\sqrt{10.6}$ and $\sqrt{10.7}$.

$$10 \begin{cases} 3 \begin{cases} \sqrt{10.6}\ = 3.25576 \\ \sqrt{10.63} = ? \end{cases} \\ \quad\ \sqrt{10.7}\ = 3.27109 \end{cases} \Bigg\} 1533$$

$$\sqrt{10.63} = 3.25576 + \tfrac{3}{10}(.01533)$$

Therefore $\sqrt{10.63} = 3.26$ (to three significant figures).

EXERCISES

Use Table IV and the given or related angle to find the value of each function in Exercises 1–12.

1. $\cos 22°\ 16'$.
2. $\cos 33°\ 19'$.
3. $\sin 49°\ 22'$.
4. $\sin 54°\ 04'$.
5. $\cos 75°\ 03'$.
6. $\cos 81°\ 57'$.
7. $\sin 186°\ 14'$.
8. $\sin 277°\ 12'$.
9. $\cos(-15°\ 13')$.
10. $\cos 3°\ 07'$.
11. $\sin 4°\ 09'$.
12. $\sin(-17°\ 17')$.

Use Table IV to find the acute angle θ for each of the Exercises 13–18.

13. $\sin \theta = .3293$.
14. $\sin \theta = .5529$.
15. $\cos \theta = .9888$.
16. $\sin \theta = .9880$.
17. $\cos \theta = .9800$.
18. $\cos \theta = .6167$.

Use Table IV to find a second-quadrant angle θ for each of the Exercises 19–24.

19. $\sin \theta = .6950.$ **20.** $\cos \theta = -.3621.$ **21.** $\cos \theta = -.9120.$

22. $\sin \theta = .9472.$ **23.** $\cos \theta = -.9882.$ **24.** $\cos \theta = -.9795.$

Solve each of the following equations for all of the values of θ for which $0° \leqq \theta < 360°$. Do not use tables.

25. $\cos \theta = 1.$ **26.** $\cos \theta = -1.$

27. $\cos \theta = 0.$ **28.** $\cos \theta = \frac{1}{2}\sqrt{2}.$

29. $\cos \theta = \frac{1}{2}\sqrt{3}.$ **30.** $\cos \theta = -\frac{1}{2}\sqrt{2}.$

31. $\cos \theta = -\frac{1}{2}\sqrt{3}.$ **32.** $2 \cos \theta = 1.$

33. $2 \cos \theta = -1.$ **34.** $\cos^2 \theta = 1.$

35. $\cos^2 \theta + \cos \theta = 0.$ **36.** $\cos^2 \theta - \cos \theta = 0.$

37. $4 \cos^2 \theta - 1 = 0.$ **38.** $4 \cos^2 \theta - 3 = 0.$

39. $2 \cos^2 \theta + \cos \theta - 1 = 0.$ **40.** $\cos^2 \theta + 2 \cos \theta + 1 = 0.$

41. $\cos^2 \theta - 2 \cos \theta + 1 = 0.$ **42.** $\cos^3 \theta - \cos \theta = 0.$

43. $2 \cos^3 \theta - \cos^2 \theta - \cos \theta = 0.$ **44.** $\cos^2 \theta + \cos \theta - 2 = 0.$

45. $\cos 2\theta = 0.$ **46.** $\cos 2\theta = 1.$

47. $\cos^2 2\theta + \cos 2\theta = 0.$ **48.** $\cos^2 2\theta - \cos 2\theta = 0.$

49. $2 \cos^2 2\theta + \cos 2\theta = 1.$ **50.** $\cos^2 \frac{1}{2}\theta - 2 \cos \frac{1}{2}\theta + 1 = 0.$

51. $\cos^3 \theta - 1 = 0.$ **52.** $(\cos \theta)(2 \cos \theta + 1)(\cos \theta - 1) = 0.$

53. $(2 \sin \theta - 1)(\cos \theta + 1) = 0.$ **54.** $\sin \theta \cos \theta + \cos \theta = \sin \theta + 1.$

307   ACCURACY IN THE SOLUTION OF TRIANGLES

In general, in any calculation, the computed result should not show any more significant digits than the number warranted by the tables used nor by the least accurate of the measured data. (*See* Section 210.)

A relationship between the accuracy of sides and angles of triangles is given as follows:

ACCURACY IN SIDES	ACCURACY IN ANGLES
2 significant digits	Nearest degree.
3 significant digits	Nearest multiple of 10 minutes.
4 significant digits	Nearest minute.
5 significant digits	Nearest tenth of a minute.

If a problem comes in a form involving *seconds* in angular measure, immediately round off to the nearest tenth of a minute, remembering that $0.1' = 6''$. In using tables, one should carry out accuracy as far as the tables in use permit and *round off figures only after all computation has been completed.* If one rounds off earlier, there is often a serious accumulation of error that goes much beyond that of rounding off a single entry.

308   APPLICATION OF THE SINE AND COSINE FUNCTIONS TO RIGHT TRIANGLES

Example 1.   Solve the right triangle ABC in which $c = 5.000$ and $A = 35°\ 10'$. (Fig. 309.)

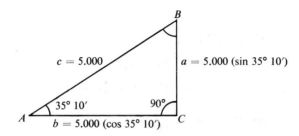

$c = 5.000$

$a = 5.000\ (\sin 35°\ 10')$

$35°\ 10'$

$90°$

$b = 5.000\ (\cos 35°\ 10')$

Figure 309

Solution:

$$\cos 35°\ 10' = \frac{b}{5.000},$$

$$b = 5.000(\cos 35°\ 10') = 5.000(.8175) = 4.088.$$

$$\sin 35°\ 10' = \frac{a}{5.000},$$

$$a = 5.000(\sin 35°\ 10') = 5.000(.5760) = 2.880.$$

$$B = 90° - 35°\ 10' = 54°\ 50'.$$

Example 2.   Solve the right triangle ABC in which $a = 3.621$ and $c = 10.000$. (Fig. 310.)

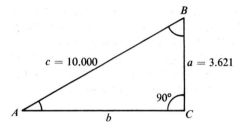

Figure 310

Solution.

$$\sin A = \frac{3.621}{10.000} = .3621;$$

$$\sin 21° \ 10' = .3611$$

$$\sin A = .3621$$

$$\sin 21° \ 20' = .3638,$$

$$A = 21° \ 10' + \tfrac{10}{27}(10')$$

$$= 21° \ 10' + 4' = 21° \ 14'.$$

$$\cos 21° \ 14' = \frac{b}{10.000},$$

$$b = 10.000(\cos 21° \ 14')$$

$$= 10.000(.9321) = 9.321.$$

$$B = 90° - 21° \ 14'$$

$$= 89° \ 60' - 21° \ 14' = 68° \ 46'.$$

309 THE LAW OF COSINES

In any triangle, the square of any side is equal to the sum of the squares of the other two sides, minus twice the product of those sides times the cosine of their included angle. As an equation, the Law of Cosines takes one of the following forms:

$$a^2 = b^2 + c^2 - 2bc \cos \alpha,$$

$$b^2 = c^2 + a^2 - 2ca \cos \beta,$$

$$c^2 = a^2 + b^2 - 2ab \cos \gamma.$$

Proof. In any given triangle let α be any angle. Place α in standard position with respect to a system of coordinates. Then A is at the origin, and one of the other vertices lies on the positive x-axis. As a matter of notation we shall call this vertex B, which then has coordinates $(c, 0)$. (If we denote this vertex by C, we merely interchange b and c in what follows.) It is immaterial whether α is acute, as in Figure 311a, or obtuse, as in Figure 311b, or is a right angle.

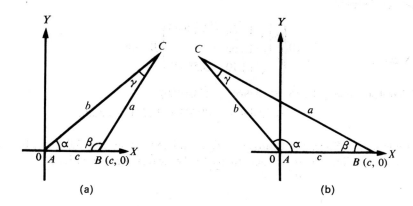

(a)   (b)

Figure 311

The radius vector of the vertex C is b, and by the definition of the trigonometric functions, the coordinates of C are $(b \cos \alpha, b \sin \alpha)$. From the coordinates of B and C, we can find $|BC|^2 = a^2$, by the distance formula (Section 105). We get

$$
\begin{aligned}
a^2 = |BC|^2 &= (b \cos \alpha - c)^2 + (b \sin \alpha - 0)^2 \\
&= b^2 \cos^2 \alpha - 2bc \cos \alpha + c^2 + b^2 \sin^2 \alpha \\
&= b^2(\sin^2 \alpha + \cos^2 \alpha) + c^2 - 2bc \cos \alpha \\
&= b^2 + c^2 - 2bc \cos \alpha,
\end{aligned}
$$

as we wished to show. Since α was *any* angle of the triangle, this proves all three forms of the equation of the Law of Cosines.*

The Law of Cosines is used (1) to solve a triangle when two sides and the included angle are known, and (2) to solve for an angle when all three sides are known.

Example 1.   Given $b = 2.00$, $c = 5.00$, $\alpha = 35°\ 10'$, find a.

* This proof is taken, by permission, from R. W. Brink, *Plane Trigonometry*, Third Edition, Appleton-Century-Crofts, N.Y., 1959, page 167.

Solution.

$$a^2 = (2)^2 + (5)^2 - 2(2)(5)\cos 35° \ 10'$$
$$= 4 + 25 - 20(.8175),$$
$$a = \sqrt{12.65}$$

To find $\sqrt{12.65}$ use the square root table, Table VII, page *104*; interpolate as follows:

$$10\left\{ \begin{matrix} 5\left\{ \begin{matrix} \sqrt{12.60} = 3.54965 \\ \sqrt{12.65} = \qquad ? \end{matrix} \right. \\ \sqrt{12.70} = 3.56371 \end{matrix} \right\}.01406$$

Therefore, $a = \sqrt{12.65} = 3.54965 + .5(.01406)$
$$= 3.54965 + .00703 = 3.55668 \approx 3.56.$$

Example 2. Solve the triangle ABC completely, given $a = 3.0$ $b = 6.0$, and $c = 7.0$, finding the angles accurately to the nearest degree. (Fig. 312.)

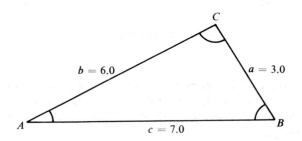

Figure 312

Solution.

We first solve for A. Begin by writing the Law of Cosines for a, the side opposite the desired angle.

$$a^2 = b^2 + c^2 - 2bc \cos A,$$
$$(3)^2 = (6)^2 + (7)^2 - 2(6)(7)\cos A,$$
$$9 = 36 + 49 - 84 \cos A,$$
$$84 \cos A = 76,$$
$$\cos A = \frac{76}{84} = .9048;$$
$$A = 25° \text{ (to the nearest degree).}$$

Having solved for A, we could use the Law of Sines to solve for another angle; however, it is better, when possible, to use the original data in solving for the remaining parts of a problem. Therefore we use the Law of Cosines to find B and then check the results by using the Law of Sines.

$$b^2 = a^2 + c^2 - 2ac \cos B,$$
$$6^2 = 3^2 + 7^2 - 2(3)(7)\cos B,$$
$$36 = 9 + 49 - 42 \cos B,$$
$$42 \cos B = 22,$$
$$\cos B = .5238;$$
$$B = 58° \quad \text{(to the nearest degree)}.$$

The Law of Sines will not distinguish between an acute and an obtuse angle ($\sin \theta = \sin (180° - \theta)$). Therefore, if we use the Law of Sines, we must choose an angle known to be acute. *A triangle can have only one obtuse or right angle, which must lie opposite the longest side*; therefore, B, not being opposite the longest side, is an acute angle.

$$\frac{\sin B}{b} = \frac{\sin A}{a},$$
$$\frac{\sin B}{6} = \frac{\sin 25°}{3},$$
$$\sin B = \frac{6 \sin 25°}{3} = 2(.4226) = .8452;$$
$$B = 58° \quad \text{(to the nearest degree)}.$$

Knowing two angles, we can find the third angle from the relation

$$A + B + C = 180°,$$
$$25° + 58° + C = 180°,$$
$$C = 97°.$$

EXERCISES

$A^2 = b^2 + c^2 - 2bc(\cos A)$

In Exercises 1–4, solve each of the right triangles ABC having the given parts ($C = 90°$).

1. $c = 20.0, A = 24° 10'$.
2. $c = 30.0, A = 41° 20'$.
3. $c = 10.00, A = 67° 43'$.
4. $c = 10.00, A = 74° 17'$.

In Exercises 5–8, solve each of the oblique triangles ABC for the missing side.

5. $A = 27° \; 10', b = 6.00, c = 5.00.$ **6.** $A = 28° \; 40', b = 7.00, c = 6.00.$

7. $B = 110° \; 50', a = 3.00, c = 5.00.$ **8.** $B = 120° \; 10', a = 4.00, c = 5.00.$

In Exercises 9 and 10, solve completely each of the oblique triangles ABC.

9. $a = 5.0, b = 7.0, c = 10.0.$

10. $a = 4.0, b = 8.0, c = 10.0.$

11. Point A is 2000.0 feet from one end of a lake and 3000.0 feet from the other end. The lake subtends (that is, is opposite to) an angle of $114° \; 50'$ at point A. Find the length of the lake.

12. A 50.0-foot flagpole stands beside a road which is inclined $9° \; 10'$ from the horizontal. Point P is 110.0 feet down the road from the foot of the flagpole. Find the distance from the top of the flagpole to point P.

13. Prove that if R is the radius of the earth (regarded as a sphere), the radius of the parallel of latitude that passes through a place P of latitude L is given by $r = R \cos L$. (Fig. 313).

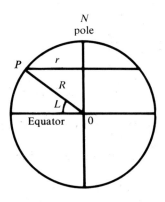

Figure 313

14. Find the speed at which New York City is moving east due solely to the rotation of the earth about its axis. The latitude of New York City is about 40.5°. Assume the radius of the earth to be 4000 miles. *See* Problem 13 and Figures 313 and 314.

15. Find the speed with which Los Angeles, California, is moving east due solely to the rotation of the earth about its axis. The latitude of Los Angeles is about 34°. Assume the radius of the earth to be 4000 miles. *See* Problem 13 and Figures 313 and 314.

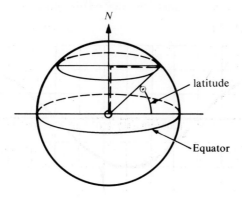

Figure 314

310 GRAPH OF THE COSINE FUNCTION

The unit circle can be used to show the increasing and decreasing values of the cosine function of an angle that changes from 0° to 360°. Let the variable angle θ be in standard position with respect to the coordinate system. Then, as was shown in Figure 304, the value of the cosine of θ is equal to the abscissa (x-value) of the point where the terminal side of θ crosses the unit circle. We will show the relationship of θ and $\cos \theta$ by means of a graph.

Example 1.   Graph $y = \cos \theta$.

We take values of θ along the horizontal axis and the corresponding values of y along the vertical axis. On the θ-axis we choose a convenient length to represent the equal distances between 0°, 30°, 60°, etc. By plotting many points on the curve $y = \cos \theta$, we establish the shape of the curve. The co-ordinates of some of these points are given in the following table:

θ	0°	30°	60°	90°	120°	150°	180°	210°	240°	270°	300°	330°	360°
y	1	$\frac{1}{2}\sqrt{3}$	$\frac{1}{2}$	0	$-\frac{1}{2}$	$-\frac{1}{2}\sqrt{3}$	-1	$-\frac{1}{2}\sqrt{3}$	$-\frac{1}{2}$	0	$\frac{1}{2}$	$\frac{1}{2}\sqrt{3}$	1

In Figure 315 the cosine function is graphed through one *complete cycle.* This curve repeats its cycle in an interval of 360°, and therefore the cosine has a period of 360°. The maximum deviation from the θ-axis is one unit. This deviation is called the *amplitude* of the curve, and thus the amplitude of this curve is 1.

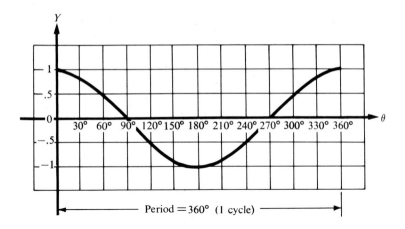

Figure 315

Example 2.   Graph $y = \sin \theta$ and $y = \cos \theta$, on the same coordinate axes, in the interval from $-90°$ to $360°$.

As an aid in determining the points on the respective curves, we keep in mind the unit-circle, the 30°-60° right triangle, the 45° right triangle, and the principle of related angles. By plotting points of each equation, we sketch the curves as shown in Figure 316.

The student will observe that the basic sine curve and cosine curve have the same general shape (Fig. 316). The only difference between them is that the graph of $y = \cos \theta$ is shifted a distance of 90° to the left of the graph of $y = \sin \theta$.

The graph of $y = \cos \theta$ is therefore like a simple sine curve with amplitude 1 and **wave length** 360°, but we say that it **differs in phase** by 90° from the graph of $y = \sin \theta$.

Now that the general shape of the sine and cosine curves is known, the

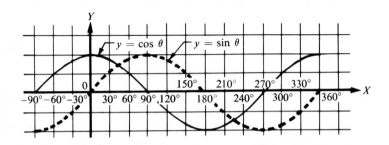

Figure 316

student should be able to sketch either curve quickly. To do this, locate the points where the required curve crosses the θ-axis, its high and low points, and a few general points; then draw a smooth curve through these points.

Example 3. Graph the two curves $y_1 = \cos\theta$ and $y = 2\cos\theta$, on the same axes, in the interval from $-180°$ to $180°$.

Solution. We first graph $y_1 = \cos\theta$ as in Figure 316. For each point of this graph there is a corresponding point of the graph of $y = 2\cos\theta$, having the same abscissa (or value of θ) and an ordinate (or value of y) that is twice the ordinate (or the value of y_1) on the first graph. After plotting a sufficient number of such points, we can draw the graph of $y = 2\cos\theta$ through them. (Fig. 317.)

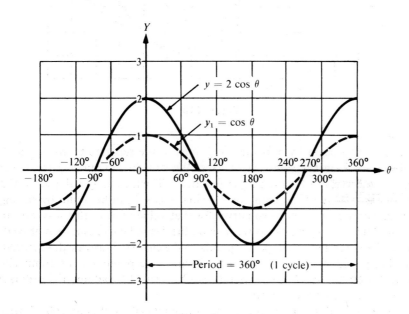

Figure 317

311   GRAPHING BY THE ADDITION OF ORDINATES

If a function is in the form of the sum of two or more simpler functions, its graph is usually best drawn by the method known as the *addition of ordinates* or as the *composition of ordinates*. The method is illustrated in the following example.

Example.   Graph the curve $y = \sin\theta + \cos\theta$ in the interval from 0° to 360°. (*See* Fig. 318.)

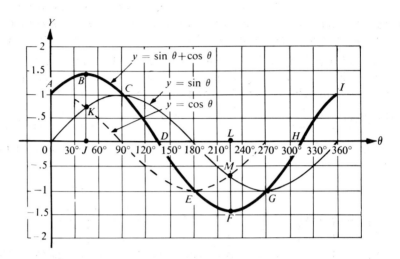

Figure 318

Solution.   We begin by drawing the graphs of the separate terms $y = \sin\theta$ and $y = \cos\theta$ on the same set of axes. We can do this quickly from our general knowledge of these simple functions. Then, for any value of θ, the height or ordinate of the graph of $y = \sin\theta + \cos\theta$ can be found by adding the ordinates of the other two curves for that value of θ. In this way we can quickly find as many points as we wish on the desired curve and then draw it. We shall take a number of specific points on the curve to illustrate this method.

Let points A, B, C, D, E, F, G, H, and I be points on the composite curve $y = \sin\theta + \cos\theta$. These points can be located by adding the ordinate values of the component curves.

The value of y, for $y = \sin\theta$, is zero when $\theta = 0°$, 180°, and 360°. Therefore, points A, E, and I of the composite curve are on the cosine curve for these values of θ.

The value of y, for $y = \cos\theta$, is zero when $\theta = 90°$ and 270°. Therefore, points C and G of the composite curve are on the sine curve for these values of θ.

The ordinates of the component curves are equal and positive when $\theta = 45°$. Therefore, the ordinate value of point B on the composite curve is twice the length of JK.

The ordinates of the component curves are equal and negative when $\theta = 225°$. Therefore, the ordinate value of point F on the composite curve is negative and twice the length of LM.

At points D and H the ordinates of the component curves are numerically equal but opposite in sign, thus the sum of these component ordinates is zero.

For most purposes the composite curve can be determined by locating a sufficient number of points, as we have done, and then drawing a smooth free-hand curve through these points. For more accurate work the student would find a pair of draftsman's dividers helpful.

EXERCISES

In Exercises 1–8, find the maximum and minimum values of each of the functions of θ.

1. $2 + \sin \theta$.   **2.** $1 + \cos \theta$.   **3.** $1 - \cos \theta$.

4. $1 - \sin \theta$.   **5.** $3 + \sin 2\theta$.   **6.** $2 + \cos 2\theta$.

7. $2 - \cos^2 \theta$.   **8.** $2 - \sin^2 \theta$.

In Exercises 9–18, graph each of the following functions through one complete cycle.

9. $y = 3 \cos \theta$.   **10.** $y = \frac{1}{2} \cos \theta$.

11. $y = 1 + \cos \theta$.   **12.** $y = -1 + \cos \theta$.

13. $y = \sin \theta + 2 \cos \theta$.   **14.** $y = 2 \sin \theta + \cos \theta$.

15. $y = \sin \theta - \cos \theta$.   *answers switched*   **16.** $y = \cos \theta - \sin \theta$.

17. $y = \cos \theta + \frac{1}{2} \sin \theta$.   **18.** $y = \sin \theta + \frac{1}{2} \cos \theta$.

Due Friday or Thursday.

4

The tangent function

Let θ be an angle placed in standard position (Fig. 401) and P be a point on the terminal side of θ. Associated with P are three number: the x and y coordinates, and r, the radius vector.

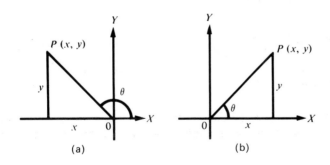

(a)　　　　　(b)

Figure 401

As was explained in Chapters 2 and 3, six ratios called trigonometric functions can be made with the numbers x, y, and r. In this chapter we shall learn about the ratio of the ordinate to the abscissa of P, which is called the **tangent function**. The radius vector, r, does not come into the definition of the tangent function.

401   DEFINITION OF THE TANGENT OF AN ANGLE

1. Place an angle θ in standard position.

2. Let point P, whose coordinates are (x, y) where $x \neq 0$, be a point on the terminal side of angle θ.

Then the tangent of θ (written tan θ) is defined as the ratio of y to x, or,

$$\tan \theta = \frac{y}{x} = \frac{\text{ordinate}}{\text{abscissa}}, \qquad x \neq 0.$$

Special case.   Let θ be an acute angle of a right triangle. Let this triangle be placed in the first quadrant with θ in standard position. Let·the point (x, y) be the end of the hypotenuse in the first quadrant. With the triangle in this position x becomes the side of the triangle adjacent to θ and y the side opposite θ. Then, from the definition of the tangent of an angle,

$$\tan \theta = \frac{y}{x} = \frac{\text{side opposite}}{\text{side adjacent}}.$$

Examples: (*See* Fig. 402.)

Tan $0° = 0$

Tan $90° = (\infty)$

1. $\tan 30° = \dfrac{1}{\sqrt{3}} = \frac{1}{3}\sqrt{3},$ Tan $180° = 0$

Tan $270° = (\infty)$

2. $\tan 60° = \dfrac{\sqrt{3}}{1} = \sqrt{3},$

3. $\tan 45° = \dfrac{1}{1} = 1.$

The student will recall that sin θ and cos θ exist for all values of θ. The tangent function of an odd multiple of $90°$ is not defined; for such an angle,

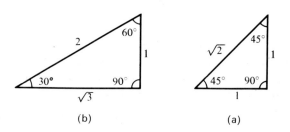

(b)   (a)

Figure 402

the point (x, y) of our definition would lie on the y-axis, and hence x would equal 0 and tan θ would involve $y/0$, which is not defined. The sine and cosine functions vary from minus 1 to plus 1; the tangent function varies from "minus infinity" to "plus infinity". In essence this means that we can make tan θ as large as we please by having $0° < \theta < 90°$ and making θ sufficiently close to 90°. When $\theta = 90°$, tan θ is undefined. These are some of the characteristics of tan θ that distinguish it from sin θ and cos θ.

402   THE SIGN OF THE TANGENT OF AN ANGLE

As defined, tan $\theta = y/x$ and is therefore positive when the values of y and x have the same sign and negative when the values of y and x have opposite signs. Therefore, tan θ is positive when θ is in the first or third quadrant and negative when θ is in the second or fourth quadrant (Fig. 403).

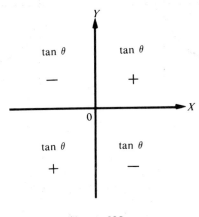

Figure 403

403   THE UNIT CIRCLE AND THE TANGENT FUNCTION

We construct a circle with center at the origin and radius equal to one unit. AB is the horizontal diameter and lines l_1 and l_2 are drawn tangent to the circle at points A and B. Let θ represent any angle placed in standard position and let point $P(x, y)$ be the intersection of the terminal side of θ and one of the lines l_1 or l_2. By definition $\tan \theta = \dfrac{y}{x}$. In these figures (Fig. 404),

$$\tan \theta = \frac{y}{\pm 1} = \pm y.$$

If an angle is in standard position, its tangent is numerically equal to the ordinate of the point of intersection of its terminal side and a line parallel to and at the distance of one unit to the right or to the left of the y-axis; the tangent is positive if the angle terminates in the first or third quadrant and negative if the angle terminates in the second or fourth quadrant.

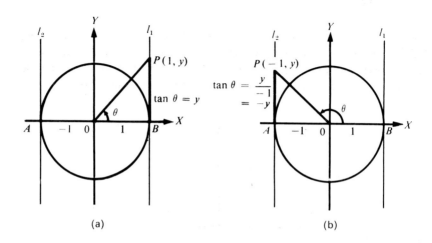

(a)   (b)

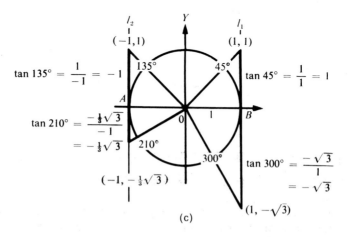

(c)

Figure 404

Examples:

1. tan 0° = 0.

2. tan 30° = $\dfrac{1}{\sqrt{3}}$ ≈ .577.

3. tan 45° = 1.000.

5. tan 80° ≈ 5.671.

7. tan 90° undefined.

9. tan 180° = 0.

4. tan 60° = $\sqrt{3}$ ≈ 1.732.

6. tan 89° 50′ ≈ 343.8.

8. tan 100° ≈ −5.671.

10. tan 270° undefined.

11. tan 300° = $-\sqrt{3}$ ≈ −1.732.

12. tan 330° = $-\dfrac{1}{\sqrt{3}}$ ≈ −.577.

To find the tangent of an angle place the angle in standard position then determine its quadrant and related angle.

Example 13. Find tan 120°.
Tan 120° is negative because 120° is in the second quadrant. The related angle is 60°. Therefore, tan 120° = −tan 60° = $-\sqrt{3}$.

Example 14. Find tan(−120°).
Tan(−120°) is positive because −120° is in the third quadrant. The related angle is 60°. Therefore, tan(−120°) = tan 60° = $\sqrt{3}$.

Example 15. Find tan 339° 10′.
Tan 339° 10′ is negative because 339° 10′ is in the fourth quadrant. The related angle = (360° − 339° 10′) = 20° 50′. Locate 20° 50′ in the degree column on page *94* of Table IV and read the value of tan 20° 50′ = .3805 to the right of 20° 50′ in the tangent column. The tangents of angles in the fourth quadrant are negative; therefore tan 339° 10′ = −.3805.
In Example 16, we see how to find θ when tan θ is known.

Example 16. Find θ where 90° < θ < 180° and tan θ = −.6494.
We first find the related angle of θ from tan θ = +.6494. On page *95* of Table IV, locate .6494 in the tangent column, then read 33° 00′ in the degree column to the left of .6494. The desired angle is (180° − 33° 00′) = 147° 00′.

EXERCISES

In Exercises 1–14, find the value of each term, and then combine the terms.

1. tan 0° + tan² 30° + tan 45° + tan 135° + tan 180°

2. tan 315° + tan 225° + tan 60° + tan(−60°).

3. tan² 210° + tan 420° + tan(−135°).

4. $\tan 750° + \tan(-30°) - \tan 150°$.

5. $\tan 180° - \tan 240° + \tan 225°$.

6. $\sin 30° + \cos 30° + \tan 30°$.

7. $\sin 45° + \tan 45° + \cos 45°$.

8. $\sin 0° + \tan 0° + \cos 0°$.

9. $\sin 315° + \cos 315° + \tan 315°$.

10. $\sin^2(-150°) + \cos^2(-150°) + \tan(-135°)$.

11. $\sin^2 60° + \cos^2 60° + \tan^2 60°$.

12. $\sin^3 45° + \cos^3 45° + \tan^3 60°$.

13. $\sin^2(-135°) - \cos^3(-45°) - \tan^3(-180°)$.

14. $\tan(-150°) + \tan(-210°) - \cos(-120°)$.

Use Table IV to find the value of each of the functions in Exercises 15–30.

15. $\tan 35°$.   **16.** $\sin 35°$.   **17.** $\cos 35°$

18. $\tan 70°$.   **19.** $\tan 85°$.   **20.** $\tan 100°$

21. $\tan 125°$.   **22.** $\tan 200°$.   **23.** $\tan 220°$.

24. $\tan 280°$.   **25.** $\tan 310° \, 20'$.   **26.** $\tan 190° \, 40'$.

27. $\tan(-50°)$.   **28.** $\tan(-80°)$.   **29.** $\tan(-150° \, 50')$.

30. $\tan(-285° \, 20')$.

In each of the Exercises 31–36, determine the quadrant in which θ lies if the functions are positive or negative as indicated.

31. $\sin \theta$ and $\cos \theta$ are both negative.

32. $\sin \theta$ and $\tan \theta$ are both negative.

33. $\cos \theta$ and $\tan \theta$ are both negative.

34. $\sin \theta$ is negative and $\tan \theta$ is positive.

35. $\cos \theta > 0$ and $\tan \theta < 0$.

36. $\sin \theta < 0$ and $\cos \theta < 0$.

In each of the Exercises 37–42, use the given information to find the value of the indicated expressions.

37. Given $\sin \theta = \frac{3}{5}$ and $0° < \theta < 90°$, find the value of $(\cos \theta + \tan \theta)$.

38. Given $\tan \theta = -\frac{3}{4}$ and $90° < \theta < 180°$, find the value of $(5 \sin \theta - 10 \cos \theta)$.

39. Given $\cos \theta = \frac{3}{5}$ and $270° < \theta < 360°$, find the value of $9 \tan^2 \theta - 5 \sin \theta$.

40. Given $\sin \theta = -\frac{5}{13}$ and $180° < \theta < 270°$, find the value of $(13 \cos \theta + 12 \tan \theta)^2$.

41. Given $\tan \theta = \frac{1}{7}$ and θ not in Q I, find the value of $(\sin \theta + \cos \theta)^2$.

42. Given $\tan \theta = -\frac{3}{5}$ and θ not in Q IV, find the value of $(\sin \theta + \cos \theta)^2$.

404   ANGLES OF ELEVATION AND DEPRESSION

Suppose that an observer is at a point O and a certain object is at a point P. The line OP is then called the **line of sight** from O to P. Let OH be drawn horizontally in the vertical plane through OP. Then the acute angle HOP, which the line of sight makes with the horizontal, is called the **angle of elevation** of P from O if P is higher than O, and it is called the **angle of depression** of P from O if P is lower than O. (*See* Fig. 405.)

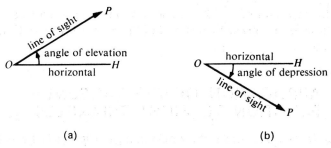

(a)   (b)

Figure 405

Example.   From an airplane the angle of depression of a boat is 40° and at the same time the angle of elevation of the airplane, as observed from the boat, is 40°.

405   THE BEARING OF A LINE

The direction of a line on the earth's surface (or the direction of one point with respect to another) is generally given by means of the angle which the line makes with the true north-south line. This angle is called the **bearing of the line**. The **bearing angle** of a line is the smallest angle which that line makes with the north-south line. It can never be more than 90 degrees. The bearing of a line is written as the bearing angle with the notation of the directional quadrant in which the line lies. Thus, in Figure 406 the bearing of OA is N 40° E, that of OB is N 60° W, that of OC is S 70° W, and that of OD is S 20° E.

In some other systems such as astronomy and navigation, the bearing is measured clockwise from the north up to 360°. Thus the direction N 60° W

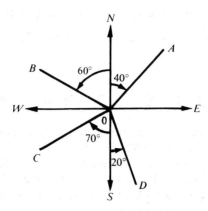

Figure 406

could also be expressed as 300°, a due west direction could be expressed as 270°, and so forth. This method is used by most branches of the United States Armed Forces.

406   APPLICATION OF THE TANGENT FUNCTION TO RIGHT TRIANGLES

Example 1.   Solve for side a of the right triangle ABC in which $b = 10.00$ and $A = 35° 10'$. (Fig. 407.)

Solution.

$$\tan 35° 10' = \frac{a}{10.00},$$

$$a = 10.00(\tan 35° 10')$$

$$= 10.00(.7046) = 7.046 \approx 7.05 \text{ (to two decimal places)}.$$

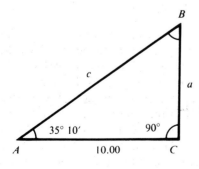

Figure 407

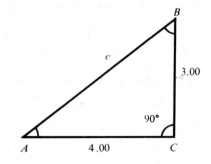

Figure 408

Example 2. Solve for angle A of the right triangle ABC in which $a = 3.00$ and $b = 4.00$. (Fig. 408).

Solution.

$$\tan A = \frac{3.00}{4.00} = .7500.$$

From Table IV, we have $\tan 36° \ 50' = .7490$ and $\tan 37° \ 00' = .7536$. Since $\tan A$ is between these numbers and is nearer to .7490 than to .7536, we take $A = 36° \ 50'$ (accurate to 10′).

The sides were expressed accurately to three significant digits which limits the accuracy of the angle to the nearest 10 minutes. *See* Section 307.

EXERCISES

In Exercises 1–4, solve the right triangles ABC for side a. Angle $C = 90°$.

1. $A = 14° \ 50'$, $b = 10.00$.   2. $A = 28° \ 10'$, $b = 100.0$.

3. $A = 77° \ 10'$, $b = 25.00$.   4. $A = 81' \ 30°$, $b = 20.00$.

In Exercises 5–8, use Table IV to solve the right triangles ABC for angle A. Angle $C = 90°$.

5. $a = 7.465$, $b = 10.00$.   6. $a = 3.814$, $b = 10.00$.

7. $a = 1.043$, $b = 10.00$.   8. $a = 2.079$, $b = 10.00$.

9. A road changes altitude 67.0 feet for every 1000.0 feet of horizontal change. Find the angle of inclination of the road.

10. A tree, standing on level ground, casts a 91-foot shadow when the angle of elevation of the sun is 55°. Find the height of the tree.

11. A flagpole, standing on level ground, casts a 65-foot shadow when the angle of elevation of the sun is 61°. Find the height of the flagpole.

12. A surveyor wishes to find the distance from point A to point B which is across a pond and due south of A. From point C, which is due east of B, he measures the distance BC which is 344 feet. With an instrument at C he finds that angle $BCA = 61° \ 10'$. Find the distance AB.

13. The top of an observation tower is 95 feet above the shore of a lake. Find the distance of a buoy from the foot of the tower if the angle of depression of the buoy, as seen from the top of the tower, is 21° 30′.

14. A ship is 25 miles to the east and 35 miles to the south of a certain port. Find its distance and bearing from the port.

15. A ship is 15 miles to the west and 25 miles to the north of a certain port. Find its distance and bearing from the port.

16. A road of uniform inclination along a mountainside rises 52.3 feet in a distance of 1000.0 feet measured along the road. Find the angle of inclination of the road.

17. From the top of a mountain, the angles of depression of two successive mile-stones in the horizontal plane which is below and in the same vertical plane as the observer are 24° 10′ and 16° 50′. Find the height of the mountain above the horizontal plane.

18. A tower stands at point P which is 845 feet due north of point B. Point C is due west of B. From C the bearing of P is N 32° 20′ E and the angle of elevation of the top of the tower is 20° 20′. Find the height of the tower.

19. A tower stands at point P which is 900.0 feet due north of point B. Point C is due east of B. From C the bearing of P is N 36° 00′ W and the angle of elevation of the top of the tower is 18° 10′. Find the height of the tower.

20. A ship, 15.2 miles west of a shore that runs due north and south, heads for a port on the shore on a course bearing 156° 20′ (using the second system of giving bearings). What is its distance from the port?

Solve the following equations for all of the values of θ for which $0° \leqq \theta < 360°$, using Table IV when necessary.

21. $\tan \theta = 1.$ **22.** $\tan \theta = \sqrt{3}.$

23. $\tan \theta = -1.$ **24.** $\tan \theta = -\sqrt{3}.$

25. $\tan \theta = \dfrac{1}{\sqrt{3}}.$ **26.** $\tan \theta = -\dfrac{1}{\sqrt{3}}.$

27. $\tan \theta = 0.$ **28.** $\tan^2 \theta = 1.$

29. $\tan^2 \theta = 3.$ **30.** $\tan^2 \theta = \frac{1}{3}.$

31. $\tan^2 \theta + \tan \theta = 0.$ **32.** $\tan 2\theta = 1.$

33. $\tan \frac{1}{2}\theta = 1.$ **34.** $\tan 2\theta = 0.$

35. $\tan \frac{1}{2}\theta = 0.$ **36.** $\tan^2 \theta - \tan \theta = 0.$

37. $\tan^2 \theta + 2 \tan \theta + 1 = 0.$ **38.** $\tan^3 \theta - \tan \theta = 0.$

39. $\tan^2 2\theta + \tan 2\theta = 0.$ **40.** $\tan \theta = .3640.$

41. $\tan \theta = .7002.$ **42.** $\tan \theta = 2.7475.$

43. $\tan \theta = 5.6713.$ **44.** $\tan \theta = 57.290.$

45. $\tan \theta = -.3455.$ **46.** $\tan \theta = -.1445.$

407   VECTORS AND VECTOR QUANTITIES

Some physical quantities are completely determined when their magnitudes, in terms of specific units, are given. Such quantities are called **scalars** and are exemplified by volume and length. Other quantities like forces and velocities, in which direction as well as magnitude is important, are called **vector quantities.**

Definition 1. *A vector is a directed line segment; it is characterized by its length and direction, its actual position being immaterial.*

It is customary to represent a vector by an arrow whose direction represents the direction of the vector and whose length (in terms of some chosen unit of length) represents the magnitude.

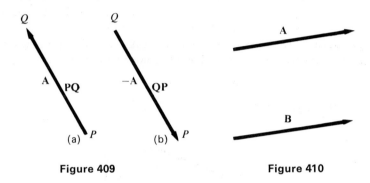

Figure 409　　　　　　　　　　　　　　　Figure 410

A vector is determined by two points, the **initial point** and the **terminal point** of the vector (Fig. 409). Boldface type is usually used to denote vectors. For example, we have labeled the vector in Figure 409a as **A**, and we may also call it **PQ** to emphasize that it is the vector from P to Q. It is customary to denote the magnitude of the vector **A** by the symbol $|A|$. We often refer to this number as the **absolute value** of **A**. If **A** is the vector **PQ**, then the vector **QP** is denoted by $-A$ (*see* Fig. 409b).

Definition 2. *Two vectors are said to be equal if, and only if, they have the same magnitude and the same direction.*

For two vectors **A** and **B**, Figure 410 illustrates the conditions in which **A** = **B**. Notice that we do not require that equal vectors coincide, but they must be parallel, have the same length, and point in the same direction.

We obtain the vector sum **A** + **B** by placing the initial point of **B** on the terminal point of A and joining the initial point of A to the terminal point of **B**, as shown in Figure 411b.

Thus, if P, Q, and R are points such that **A** = **PQ** and **B** = **QR**, then **A** + **B** = **PR**; that is,

$$PQ + QR = PR.$$

In a similar way we get the sum **B** + **A** by placing the initial point of **A** on the terminal point of **B** and joining the initial point of **B** to the terminal

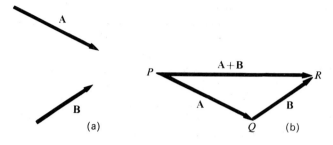

Figure 411

point of *A*. Figure 412 shows how we construct **A** + **B** and **B** + **A**, and from this figure it is apparent that the **commutative law** of addition,

$$\mathbf{A} + \mathbf{B} = \mathbf{B} + \mathbf{A},$$

holds for vector addition.

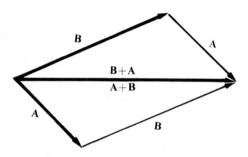

Figure 412

An alternate method for finding the sum of two vectors is known as the *parallelogram law for the composition of forces*. Let vectors **P** and **Q** represent two forces. We move these vectors until the initial point of each falls on some point such as *O* (Fig. 413). We let *A* be the terminal point of **P** and *B* the the terminal point of **Q**. We then complete the parallelogram with the sides adjacent to *OA* and *OB* and draw **OC** = **R**, the diagonal of the parallelogram. It is shown in mechanics that if vectors **P** and **Q** represent two forces they may be replaced in their effect by an equivalent force **R**, called the **resultant** of **P** and **Q**. **P** and **Q** are called **components** of **R**.

It is often required to **resolve** a given vector quantity into components in two given directions. For example, in Figure 414, let it be required to find the components of **R** in the directions *OA′* and *OB′*. We complete the parallelo-

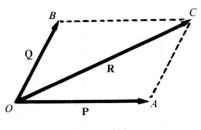

Figure 413

gram by drawing lines through C parallel to $A'O$ and $B'O$. Let B and C be the respective points where these lines intersect $B'O$ and $A'O$. Then the line segments OB and OA represent the required vector components of $\mathbf{R}$.

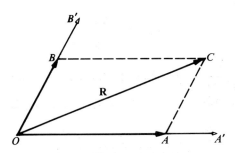

Figure 414

Example 1. An airplane has a heading (bearing) of 37° and an air speed of 550 mph (miles per hour). Find the east and north components of the velocity of the airplane (assuming zero wind velocity).

Solution. In Figure 415, $\mathbf{OC}$ represents the velocity of the airplane. CA and CB are drawn parallel to the N- and E-axes respectively. Then

$\mathbf{OA} = \mathbf{BC} = 550 \times \sin 37° = 331$ mph (the speed that the airplane is travelling east), and
$\mathbf{OB} = 550 \times \cos 37° = 439$ mph (the speed that the airplane is traveling north).

Example 2. Find the tension on each arm of a 150-pound gymnast as he hangs by his hands from a high horizontal bar. His hands are separated such that his arms form 43° angles with the bar.

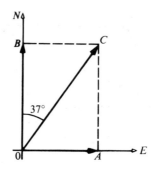

Figure 415

Solution. In Figure 416, **GC** represents the vertical upward force required to support the weight of the gymnast. Since T_1 and T_2 each support half the load, in computing $|T_1|$ we take $|GC| = 75$ pounds. Then

$$\sin 43° = \frac{75}{|T_1|},$$

$|T_1| = \dfrac{75}{\sin 43°} = \dfrac{75}{.68200} = 110$ pounds (to the nearest pound). Thus the pull on each arm is 110 pounds.

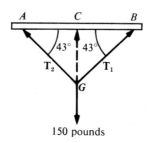

150 pounds

Figure 416

Example 3.   Find the force (independent of friction) required to pull a 1650-pound boat up a ramp that is inclined 17° to the horizontal.

Solution.   In Figure 417, the vertical vector **w** (representing the 1650-pound gravitational force) is drawn, then, from point *B*, lines perpendicular and parallel to the direction of the ramp are drawn. Vectors f_1 and f_2 are components of **w**. Then $|f_3|$, the force required to pull the boat up the ramp, is equal to $|\mathbf{f_1}|$, the force causing the boat to roll down the ramp.

$$\sin 17° = \frac{|\mathbf{f_1}|}{|\mathbf{w}|},$$

$$|\mathbf{f}| = |\mathbf{w}| \cdot \sin 17° = 1650(.2924) = 482 \text{ pounds}$$

(to the nearest pound); this is the force required to pull the boat up the ramp.

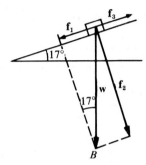

Figure 417

Example 4. Two forces, $f_1 = 250$ pounds and $f_2 = 150$ pounds, act on a point P. Find the magnitude of the equilibrant when the angle between $\mathbf{f_1}$ and $\mathbf{f_2}$ is 64°. Find the angle between the equilibrant and $\mathbf{f_1}$. (The equilibrant is the force that could be applied to prevent the motion that $\mathbf{f_1}$ and $\mathbf{f_2}$ jointly tend to produce. *See* Fig. 418.)

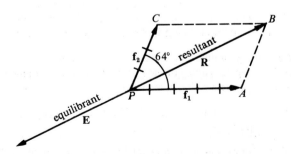

Figure 418

Solution. In Figure 418, we construct $\mathbf{f}_1$ and $\mathbf{f}_2$ with their initial point at P, making an angle of 64° with each other. Then $\mathbf{R}$, the diagonal of the parallelogram constructed on $\mathbf{f}_1$ and $\mathbf{f}_2$, is the resultant of the two given forces. $\mathbf{E}$ is the equilibrant which has the same magnitude as $\mathbf{R}$ but acts in the opposite direction. In triangle PAB, angle $PAB = 180° - 64° = 116°$ and $AB = 150$; therefore, using the Law of Cosines (Sec. 309), we have

$$|\mathbf{R}|^2 = (250)^2 + (150)^2 - 2(250)(150)\cos 116°$$
$$= 62,500 + 22,500 - (75,000)(-.4384)$$
$$= 117,880.$$

Thus $|\mathbf{R}| = \sqrt{117,880} = 343$ pounds (to the nearest pound). Therefore the magnitude of the equilibrant is 343 pounds.

The angle formed by $\mathbf{E}$ and $\mathbf{f}_1$ is the supplement of $\angle APB$. We apply the Law of Sines (Sec. 214) to triangle ABP to find $\angle ABP$.

$$\frac{\sin \angle APB}{150} = \frac{\sin 116°}{343},$$

$$\sin \angle APB = \frac{(150)(\sin 116°)}{343} = .3931.$$

Thus $\angle APB = 23°$ (to the nearest degree). Therefore the angle between $\mathbf{E}$ and $\mathbf{f}_1$ is $(180° - 23°) = 157°$.

EXERCISES

1. Given two forces, 12 and 16 pounds, each acting upon the point P, find their resultant (a) when the angle between them in 180°; (b) when the included angle is 120°; (c) when it is 90°; (d) when it is 51°; (e) when it is zero.

2. A force of 30 pounds acts easterly upon the point P. A second force of 50 pounds acts $S\ 40°\ W$. Represent graphically and show the length and direction (or bearing) of the equilibrant. What is the magnitude of the resultant?

3. Forces of 30 and 40 pounds act on an object at an angle of 70° with respect to each other. Find the magnitude of the resultant force and the angles the resultant force makes with respect to the given forces.

4. A force of 75 pounds makes an angle of 35° with the vertical. Find the horizontal and vertical components of the force.

5. A force of 55 pounds makes an angle of 55° with the horizontal. Find the horizontal and vertical components of the force.

6. Forces of 125 and 150 pounds act on an object at an angle of 140° with respect to each other. Find the magnitude of the resultant force and the angles the resultant force makes with respect to the given forces.   *97°*

7. An inclined plane is 12 feet long and one end is 4 feet higher than the other. A weight of 300 pounds rests on the plane. Find the value of the force perpendicular to the plane, and the value of the force needed to keep the weight from sliding down the plane (disregard the effect of friction).

8. The resultant of two equal forces acting at right angles upon an object is 120 pounds. Find the magnitude of each force.   *85 lb.*

9. A man pushes with a force of 70 pounds against the handle of a lawn mower. If the handle makes an angle of 35° with the level of the ground, find both the horizontal and the vertical components of the 70-pound force.

10. A 30-pound picture is hung from a hook on the wall by means of a wire. Find the tension in the wire if the two divisions of the wire make an angle of 110° at the hook.   *26 lbs.*

11. A person who weighs 180 pounds sits in the center of a hammock. The ropes supporting the hammock make angles of 58° 40′ with the posts to which they are attached. Find the tension on each rope.   *106 lbs.*

12. A child weighing 100 pounds sits on a swing. Find the tension on the ropes when a man pushes against the swing with a horizontal force of 35 pounds.

13. An airplane heads due north at a rate which in still air would carry it 250 miles per hour. A wind, of constant velocity 50 miles per hour, is blowing from the direction of 250°. Find the actual speed of the plane over the ground and its course (direction or bearing of flight).

14. An airplane heads due south at a rate which in still air would carry it 240 miles per hour. A wind, of constant velocity 40 miles per hour, is blowing from the direction 245°. Find the actual speed of the plane over the ground and its course (direction or bearing of flight).   *226 m.p.h.*   *171° course*

15. An airplane is climbing with an air speed of 450 miles per hour at an angle of 12° from the horizontal. Find its rate of vertical rise.

16. Find the force (independent of friction) required to pull a 2500-pound boat up a ramp that is inclined 15° to the horizontal.   *647 ch.*

408   GRAPH OF THE TANGENT FUNCTION

The graph of the tangent function always slopes upward toward the right, except at points where there is a sudden break in the graph.

The student will recall that the tangent function is undefined for all odd numbered multiples of 90° ($\pm 90°$, $\pm 270°$, etc.); therefore there is no point on the curve for any of these values. We use U as a symbol for *undefined* in the following table and draw vertical lines at the places that separate the branches of the tangent curve. (*See* Fig. 419.)

By making use of our previous knowledge of the values of the tangent function of certain angles, we trace out the graph of the curve $y = \tan \theta$.

A brief table of points on the curve $y = \tan \theta$ follows:

θ	−270°	−240°	−225°	−210°	−180°	−150°	−135°	−120°	−90°	−45°	−30°	0°
y	U	$-\sqrt{3}$	-1	$-\frac{1}{3}\sqrt{3}$	0	$\frac{1}{3}\sqrt{3}$	1	$\sqrt{3}$	U	-1	$-\frac{1}{3}\sqrt{3}$	0
$\tan \theta$	U	-1.732	-1	$-.577$	0	$.577$	1	1.732	U	-1	$-.577$	0

θ	30	45	60	90	120	135	150	180	210	225	240	270
y	$\frac{1}{3}\sqrt{3}$	1	$\sqrt{3}$	U	$-\sqrt{3}$	-1	$-\frac{1}{3}\sqrt{3}$	0	$\frac{1}{3}\sqrt{3}$	1	$\sqrt{3}$	U
$\tan \theta$	$.577$	1	1.732	U	-1.732	-1	$-.577$	0	$.577$	1	1.732	U

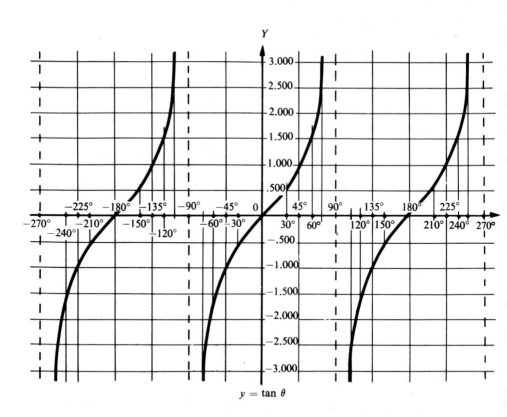

$y = \tan \theta$

Figure 419

EXERCISES

1. Graph $y = \frac{1}{2} \tan \theta$ from $\theta = 90°$ to $270°$.

2. Graph $y = 1 + \tan \theta$ from $\theta = -90°$ to $270°$.

3. Graph $y = \tan \theta - 1$ from $\theta = -90°$ to $270°$.

4. Graph $y = \frac{1}{4} \tan \theta$ from $\theta = 90°$ to $270°$.

5. Graph $y = 3 - \tan \theta$ from $\theta = -90°$ to $270°$.

6. Graph $y = 1 - \tan \theta$ from $\theta = -90°$ to $270°$.

The reciprocal functions

Let θ be an angle placed in standard position (Fig. 501) and P be a point on the terminal side of θ. Associated with P are three numbers: the x and y coordinates, and r, the radius vector.

As was explained previously, six ratios called trigonometric functions can be made with the numbers x, y, and r. In this chapter we will learn about the *reciprocals* of the sine, cosine, and tangent functions. These functions are often called the **reciprocal functions**.

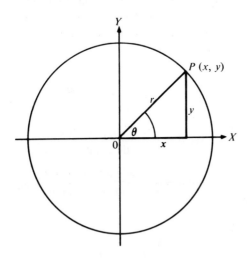

Figure 501

501 DEFINITIONS OF THE COTANGENT, SECANT, AND COSECANT FUNCTIONS

We shall now define the three reciprocal functions of an angle θ, which are the cosecant of θ, the secant of θ, and the cotangent of θ, written csc θ, sec θ, and cot θ, respectively.

1. Place an angle θ in standard position.

2. Let point P, whose coordinates are (x, y), be a point on the terminal side of angle θ, not at the origin.

3. Let r be the distance of P from the origin.

Definition: *For any angle, the sine and the cosecant, the cosine and the secant, and the tangent and the cotangent are, respectively, reciprocals of each other:*

$$\text{csc } \theta = \frac{1}{\sin \theta} = \frac{\text{radius vector}}{\text{ordinate}} = \frac{r}{y}, \qquad (y \neq 0),$$

$$\text{sec } \theta = \frac{1}{\cos \theta} = \frac{\text{radius vector}}{\text{abscissa}} = \frac{r}{x}, \qquad (x \neq 0),$$

$$\text{cot } \theta = \frac{1}{\tan \theta} = \frac{\text{abscissa}}{\text{ordinate}} = \frac{x}{y}, \qquad (y \neq 0).$$

Reciprocal Functions (handwritten annotation)

Special cases. Let θ be an acute angle of a right triangle. Let this triangle be placed in the first quadrant with θ in standard position. Let the point (x, y) be at the end of the hypotenuse in the first quadrant. With the triangle in this position, x becomes the side of the triangle adjacent to θ, and y the side opposite θ. Then, from the definitions of cosecant, secant, and cotangent of an angle,

ah so (handwritten annotation)

$$\sin \theta = \frac{1}{\csc \theta}$$
$$\cos \theta = \frac{1}{\sec \theta}$$
$$\tan \theta = \frac{1}{\cot \theta}$$
(handwritten annotations)

$$\text{csc } \theta = \frac{r}{y} = \frac{\text{hypotenuse}}{\text{side opposite } \theta},$$

$$\text{sec } \theta = \frac{r}{x} = \frac{\text{hypotenuse}}{\text{side adjacent to } \theta},$$

$$\text{cot } \theta = \frac{x}{y} = \frac{\text{side adjacent to } \theta}{\text{side opposite } \theta}.$$

Examples: (*See* Fig. 502.)

1.
$$\text{csc } 30° = \frac{1}{\sin 30°} = \frac{1}{1/2} = \frac{2}{1},$$

2.
$$\sec 30° = \frac{1}{\cos 30°} = \frac{1}{\frac{1}{2}\sqrt{3}} = \frac{2}{\sqrt{3}},$$

3.
$$\cot 30° = \frac{1}{\tan 30°} = \frac{1}{1/\sqrt{3}} = \frac{\sqrt{3}}{1}.$$

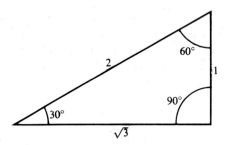

Figure 502

When $P(x, y)$, Figure 501, is not on a coordinate axis, r is always the hypotenuse of a right triangle and thus is greater than $|x|$ or $|y|$. When $P(x, y)$ is on any one of the coordinate axes, then either $r = |x|$ or $r = |y|$. Therefore $r \geq |x|$ and $r \geq |y|$, and as a result,

$$|\csc \theta| = \left|\frac{r}{y}\right| \geq 1,$$

and

$$|\sec \theta| = \left|\frac{r}{x}\right| \geq 1.$$

As $|y| \to 0$, $|\csc \theta| \to \infty$ (read, "as the numerical value of y approaches zero, $|\csc \theta|$ increases without limit").

Sin θ and cos θ exist for all values of θ, but the other four trigonometric functions of θ have limitations.

A fraction has no meaning and is undefined if its denominator is zero. For a quadrantal angle the terminal side lies on one of the coordinate axes and either x or y is zero. When $x = 0$, tan θ and sec θ are therefore undefined, and when $y = 0$, cot θ and csc θ are undefined. Consequently,

tan θ *and* sec θ *are undefined for* 90°, 270°, *and their coterminal angles*;

cot θ *and* csc θ *are undefined for* 0°, 180°, *and their coterminal angles*.

With these exceptions, the domain of definition of the trigonometric functions includes all angles θ.

502   THE SIGNS OF THE COTANGENT, SECANT, AND COSECANT OF AN ANGLE

Inverting a fraction does not change its sign; therefore, a trigonometric function of any angle and the reciprocal function of the same angle have the same sign.

In Figure 503 we show the quadrants in which the respective functions are positive. Thus sin θ *and* csc θ *are positive in the quadrants above the x-axis*; cos θ *and* sec θ *are positive in the quadrants to the right of the y-axis*; and tan θ and cot θ *are positive in the first and third quadrants.*

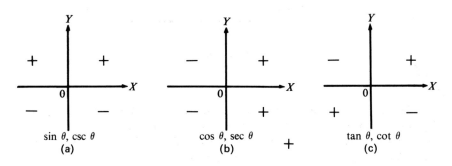

Figure 503

503   TRIGONOMETRIC FUNCTIONS OF COMPLEMENTARY ANGLES

Let *ABC* be a right triangle with *a* and *b* the lengths of the sides opposite the acute angles *A* and *B*, and let *c* be the length of the side opposite the right angle *C*.

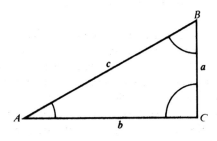

Figure 504

Two angles are said to be **complementary** if their sum is equal to 90°. Thus, in the right triangle ABC (Fig. 504), A and B are complementary angles, since $A + B = 90°$. The sine and cosine are said to be **cofunctions**, each of the other, as are also the tangent and cotangent, and the secant and cosecant.

Theorem. *Any trigonometric function of an acute angle is equal to the corresponding cofunction of the complementary angle.* Thus $\sin 80° = \cos 10°$, $\tan 30° = \cot 60°$, $\sec 20° = \csc 70°$, etc.

Proof. If A and B are any two complementary acute angles, they can be taken as the acute angles of the triangle ABC (Fig. 504). We use the definitions of the trigonometric functions of an acute angle in a right triangle, namely

$$\sin \theta = \frac{\text{side opposite } \theta}{\text{hypotenuse}}, \qquad \cos \theta = \frac{\text{side adjacent to } \theta}{\text{hypotenuse}}, \text{ etc.,}$$

to show that

$$\sin A = \frac{a}{c} = \cos B, \qquad \cos A = \frac{b}{c} = \sin B,$$

$$\tan A = \frac{a}{b} = \cot B, \qquad \cot A = \frac{b}{a} = \tan B,$$

$$\sec A = \frac{c}{b} = \csc B, \qquad \csc A = \frac{c}{a} = \sec B.$$

Example 1. Find the acute angle x for which

$$\sec x = \csc 23° \ 15.4'.$$

Solution. Since secant and cosecant are cofunctions, the statement $\sec x = \csc 23° \ 15.4'$ implies

$$x + 23° \ 15.4' = 90°, \qquad \text{or} \qquad x = 89° \ 60' - 23° \ 15.4'.$$

Therefore

$$x = 66° \ 44.6'.$$

Example 2. Find the acute angle x for which

$$\cos \left(\tfrac{3}{2}x + 20°\right) = \sin(50° - x).$$

Solution. For acute angles, the statement $\cos(\tfrac{3}{2}x + 20°) = \sin(50° - x)$ implies $(\tfrac{3}{2}x + 20°) + (50° - x) = 90°$. Therefore $\tfrac{1}{2}x + 70° = 90°$ or $x = 40°$.

EXERCISES

In Exercises 1–20, find the value of each term and then combine the values.

1. csc 30° + sec 60° + cot 45°.

2. sec 30° + csc 60° + cot 90°

3. csc 90° + sec 0° + cot 270°.

4. cot(−45°) − sec(−60°) − csc(−30°).

5. sec(−180°) − csc(−90°) − cot(−270°).

6. sec² 30° − tan² 30°.

7. sec² 60° − tan² 60°.

8. csc² 45° − cot² 45°.

9. csc² 135° − cot² 135°.

10. sin 150° csc 150° + sec 60° cos 60°.

11. tan 135° cot 135° + csc 210° sin 210°.

12. (sec 150°)(tan 150° + cot 150°).

13. (csc 60°)(cot 60° + tan 60°).

14. sin 720° + cot 225° − sec² 225°.

15. cot(−30°) − sec(−60°) − csc(−30°).

16. csc(−150)° − cot(−135°) − sec(−300°).

17. 2 csc² 60° + 4 sec² 45° + 6 cot 135°.

18. 4 cot² 120° + 2 csc² 150° − 6 sec² 210°.

19. 3 tan 135° + 2 sec² 135° − 2 cot 225°.

20. 2 sec² 60° csc³ 45° − 2 sec³ 45° sin² 30°.

Each of the following points is on the terminal side of an angle θ, in standard position. Find the six trigonometric functions of θ when they exist.

21. (4, 3).   **22.** (12, 5).   **23.** (−12, 5).

24. (−4, 3).   **25.** (4, −3).   **26.** (5, −12).

27. (−12, −5).   **28.** (−4, −3).   **29.** (−3, 2).

30. (−2, 3).   **31.** (−2, 0).   **32.** (0, −2).

Name the quadrant in which θ must be in order to satisfy the condition in each of the following exercises.

33. cos θ is negative and tan θ is positive.

34. csc θ is positive and cos θ is negative.

35. cot θ is negative and cos θ is positive.

36. sec θ is positive and cot θ is negative.

37. cot θ is negative and sin θ is positive.

38. tan θ is positive and csc θ is negative.

Identify each of the following as being possible or impossible:

39. sin $\theta = 2$. **40.** tan $\theta = 0$.

41. cot $\theta = 500$. **42.** sec $\theta = .500$.

43. csc $\theta = .313$. **44.** cos $\theta = 2$.

45. tan $\theta = .001$. **46.** csc $\theta = .614$.

47. cot $\theta < 0$ and sin $\theta < 0$. **48.** tan $\theta < 0$ and csc $\theta < 0$.

In the following exercises $0° \leq \theta \leq 90°$. State whether θ is closer to $0°$ or to $90°$.

49. cos $\theta = .99$. **50.** sin $\theta = .99$. **51.** csc $\theta = 1.01$.

52. sec $\theta = 1.01$. **53.** sin $\theta = .01$. **54.** cos $\theta = .01$.

55. tan $\theta = .01$. **56.** tan $\theta = 45$.

In Exercises 57–62 find the acute angle θ for which each of the following statements is true.

57. sin $\theta = \cos 15° \, 10'$. **58.** cos $\theta = \sin 89°$.

59. cot $\theta = \tan (2\,\theta + 30°)$. **60.** tan $\theta = \cot(50° - \tfrac{1}{2}\theta)$.

61. $\sec(\tfrac{1}{4}\theta + 25°) = \csc(5° + \tfrac{1}{2}\theta)$. **62.** $\csc(\tfrac{1}{2}\theta - 40°) = \sec(\tfrac{1}{4}\theta + 55°)$.

In Exercises 63–66 prove each statement without the use of tables.

63. $\tan 89° = \dfrac{1}{\tan 1°}$. **64.** $\sec 87° = \dfrac{1}{\sin 3°}$.

65. $\cos 87° = \dfrac{1}{\csc 3°}$. **66.** $\sin 88° = \dfrac{1}{\sec 2°}$.

504   GRAPH OF THE COTANGENT FUNCTION

The graph of the cotangent function always slopes downward toward the right except at points where the function is not defined. The cotangent function is undefined for $0°$ and $180°$ and for angles coterminal with those angles.

The student is asked to form a table showing the values, if they exist, of cot θ for $\theta = 0°$, $\pm 30°$, $\pm 45°$, $\pm 60°$, $\pm 90°$, $\pm 120°$, $\pm 135°$, $\pm 150°$, $\pm 180°$, $\pm 210°$, $\pm 225°$, $\pm 240°$, $\pm 270°$, $\pm 300°$, $\pm 315°$, $\pm 330°$, ± 360, and with these values to verify the graph of $y = \cot \theta$ as shown in Figure 505.

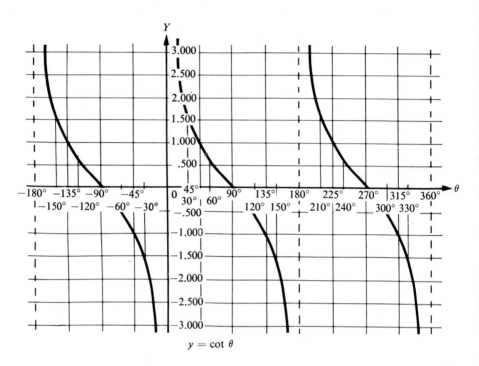

$$y = \cot \theta$$

Figure 505

505   GRAPH OF THE COSECANT FUNCTION

The graphy of $y = \csc \theta$ can be obtained from that of $y = \sin \theta$, of which it is the reciprocal: $y = \csc \theta = 1/\sin \theta$. When $\sin \theta = \pm 1$, $\csc \theta$ has the same value. When $\sin \theta = \pm \frac{1}{2}$, $\csc \theta = \pm 2$. As θ tends toward $0°$, or any whole multiple of $180°$, $|\csc \theta|$ becomes infinite or increases without limit.

Draw the graph of $y = \sin \theta$, and construct vertical lines at the points where it crosses the θ-axis. These vertical lines separate the branches of the graph of $y = \csc \theta$. (*See* Fig. 506.)

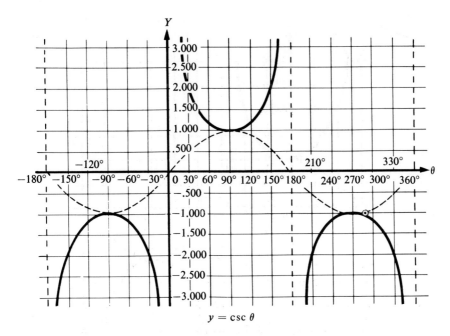

$$y = \csc \theta$$

Figure 506

A brief table of points on the $y = \csc \theta$ curve is given.

θ	$-180°$	$-150°$	$-90°$	$-30°$	$0°$	$30°$	$90°$	$150°$	$180°$	$210°$	$270°$	$330°$	$360°$
y	U	-2	-1	-2	U	2	1	2	U	-2	-1	-2	U

506   GRAPH OF THE SECANT FUNCTION

The graph of $y = \sec \theta$ is obtained from that of its reciprocal, $y = \cos \theta$, in a manner similar to that used to obtain the graph of $y = \csc \theta$: $y = \sec \theta = 1/\cos \theta$.

Draw the graph of $y = \cos \theta$, and construct vertical lines at the points where it crosses the θ-axis. These vertical lines separate the branches of the $y = \sec \theta$ curve. (*See* Fig. 507.)

A brief table of points on the $y = \sec \theta$ curve is given.

θ	$-90°$	$-60°$	$0°$	$60°$	$90°$	$120°$	$180°$	$240°$	$270°$
y	U	2	1	2	U	-2	-1	-2	U

The student should practice drawing the six basic trigonometric curves until he can make a quick rough sketch of each from memory.

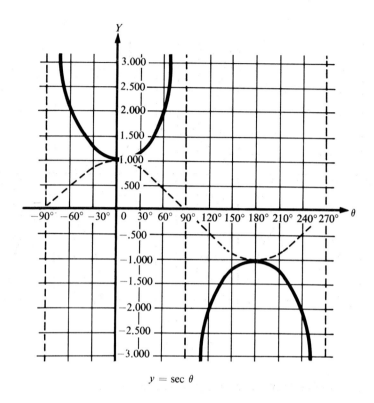

$$y = \sec \theta$$

Figure 507

EXERCISES

Graph each of the following functions in the indicated interval:

1. $y = \frac{1}{3} \cot \theta$ from $\theta = -180°$ to $360°$.

2. $y = 1 + \frac{1}{2} \cot \theta$ from $\theta = -180°$ to $360°$.

3. $y = 1 + \sec \theta$ from $\theta = -90°$ to $270°$.

4. $y = 1 + \csc \theta$ from $\theta = -180°$ to $360°$.

5. $y = -\sec \theta$ from $\theta = -90°$ to $270°$.

6. $y = -\csc \theta$ from $\theta = -180°$ to $360°$.

7. $y = 2 \csc \theta$ from $\theta = -180°$ to $360°$.

8. $y = 2 \sec \theta$ from $\theta = -90°$ to $270°$.

9. $y = \csc 2\theta$ from $0°$ to $180°$.

10. $y = \sec 2\theta$ from $-45°$ to $135°$.

11. $y = \cot 2\theta$ from $0°$ to $90°$.

12. $y = \cot \tfrac{1}{2}\theta$ from $-360°$ to $360°$.

① To change degree measure to radian measure, multiply the degree measure by $\frac{\pi}{180°}$

Example: $60° = (60°)\left(\frac{\pi}{180°}\right) = \frac{\pi}{3}$

② And the converse of that such that $\frac{180°}{\pi} \times \theta = \angle$ in degrees

Example $\frac{\pi}{2} = \left(\frac{\pi}{2}\right)\left(\frac{180°}{\pi}\right) = 90°$

Radian measure of angles

601 THE RADIAN

The student is already familiar with the system of measuring angles in which the unit of measure is the degree. A second system, in which the unit is the *radian*, is the one that is most useful in calculus and in the analytic part of trigonometry.

If an angle is placed with its vertex at the center of a fixed circle, it subtends on the circumference (that is, its sides cut off on the circumference) an arc whose length is proportional to the magnitude of the angle (Fig. 601). This fact affords us the means of defining our units for measuring angles.

Definition. *A **radian** is an angle which, if its vertex is placed at the center of a circle, subtends on the circumference an arc equal to the radius of the circle (Fig. 601a). Thus, the radian measure θ of a central angle of a circle is the ratio of the intercepted arc s to the radius r,*

(1) $$\theta = \frac{s}{r}.$$

In order to arrive at a conversion factor relating the radian-measure system to the degree-measure system, we need to determine the number of radians subtended by a circle at the center of the circle.

As a consequence of the definition of a radian, the number of radians in 360° is equal to the number of times the radius of a circle can be laid off along the circumference of that circle. The length of the circumference $= 2\pi r$. Therefore the central angle (θ) which intercepts 360° of arc contains

$$\frac{2\pi r}{r} = 2\pi \text{ radians} \approx 6.28 \text{ radians.} = \text{circle}$$

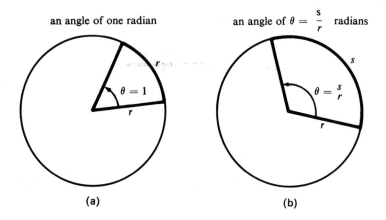

Figure 601

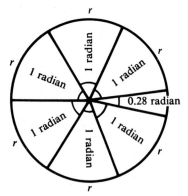

The complete circle subtends
2π (approximately 6.28) radians.

Figure 602

Thus,

$$2\pi \text{ radians} = 360°,$$

and

(2) $$\pi \text{ radians} = 180°.$$

(3) $$1 \text{ radian} = \frac{180°}{\pi} \approx 57.2958° \approx 57° \ 17.75'.$$

Since

$$180° = \pi \text{ radians,}$$

(4) $$1° = \frac{\pi}{180} \approx 0.017453 \text{ radian.}$$

Equation (2) is an important relation to remember, as it can be used for deriving equivalent radian and degree measures.

Examples:
1. $180° = \pi$ radians.
2. $90° = \frac{1}{2}(180°) = \pi/2$ radians.
3. $60° = \frac{1}{3}(180°) = \pi/3$ radians.
4. $45° = \frac{1}{4}(180°) = \pi/4$ radians.
5. $30° = \frac{1}{6}(180°) = \pi/6$ radians.
6. $270° = 3(90°) = 3\pi/2$ radians.

Since they will be used often in this and other related subjects, the student should memorize these radian-degree equivalences; to assist in memorizing these equivalences, it is helpful to place them on a circle, as shown in Figure 603. The radian measures of some of these angles are usually read " π over 6, π over 4, π over 2, 2π over 3," and the like.

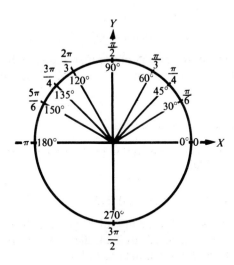

Figure 603

When no other unit of angular measure is indicated, it is assumed that an angle is expressed in radian measure.

Example 7.   Express 37.2° in radians.

First Solution.   By Formula (4), 1° = 0.017453 radian. Therefore

$$37.2° = 37.2 \times 0.017453 \text{ radians}$$
$$= 0.649 \text{ radians (to three decimal places).}$$

Second Solution.   From page *102* of Table VI,

$$37° = 0.64577 \text{ radian,}$$
$$\underline{(.2)° = (.2 \times 60)' = 12' = 0.00349 \text{ radian.}}$$

Therefore $37.2° = 0.64926$ or 0.649 radians (to three
decimal places).

Example 8.   Express 2.45 radians in degree measure accurate to the nearest
tenth of a minute.

First Solution.   By Formula (3), 1 radian = 57.2958°. Therefore,

$$2.45 \text{ radians} = 2.45 \times 57.2958° = 140.3747° = 140° + (.3747 \times 60)'$$
$$= 140° \ 22.5'.$$

Second Solution.   From pages *101* and *102* of Table VI,

$$1.45 \text{ radians} = \quad 83° \ 04.7', \text{ and}$$
$$\underline{1.00 \text{ radian} \ = \quad 57° \ 17.7'.}$$

Therefore $2.45 \text{ radians} = 140° \ 22.4'$.

The slight difference in the results of the first and second solutions is due to
the rounded off conversion numbers used. More accurately, 1 radian =
57.2957795°.

Example 9.   Find the value of $\sin \frac{1}{6}\pi$.

Solution. $\frac{1}{6}\pi$ is assumed to be in radian measure and is equivalent to 30°.
Therefore $\sin \frac{1}{6}\pi = .5000$.

Example 10.   Find the value of cos 1.243.

Solution.   1.243 is assumed to be in radian measure. We use Table VI, page
102, and find that the value of cos 1.243 lies between the table entries for
cos 1.24 and cos 1.25. We interpolate as follows:

$$\left. \begin{array}{l} \cos 1.24 \ = .32480 \\ \cos 1.243 = \\ \cos 1.25 \ = .31532 \end{array} \right\} .00948$$

Therefore $\cos 1.243 = .32480 - .3(.00948) = .32196$.

EXERCISES

The following angles are given in radian measure. Express them in degree measure either exactly or to the nearest tenth of a minute. Draw each angle, indicating its approximate magnitude by a curved arrow. Use Table VI when necessary.

1. $\pi/3$.　　　　　**2.** $\pi/4$.　　　　　**3.** $\pi/2$.　　　　　**4.** $\frac{2}{3}\pi$.

5. $\frac{5}{6}\pi$.　　　　　**6.** π.　　　　　**7.** 2π.　　　　　**8.** $\frac{3}{2}\pi$.

9. $\frac{7}{6}\pi$.　　　　　**10.** $\frac{4}{5}\pi$.　　　　　**11.** $\frac{5}{3}\pi$.　　　　　**12.** $\frac{11}{6}\pi$.

13. $\frac{5}{2}\pi$.　　　　　**14.** $\frac{13}{6}\pi$.　　　　　**15.** 1.15.　　　　　**16.** 1.03.

17. 2.03.　　　　　**18.** 2.14.　　　　　**19.** 0.758.　　　　　**20.** 0.109.

Assume the following angles to be exact and express each in radian measure either exactly in terms of π or to four decimal places. Use Table VI when necessary.

21. 30°.　　　　　**22.** 60°.　　　　　**23.** 45°.　　　　　**24.** 90°.

25. 150°.　　　　　**26.** 135°.　　　　　**27.** 225°.　　　　　**28.** 300°.

29. 270°.　　　　　**30.** 240°.　　　　　**31.** 330°.　　　　　**32.** 315°.

33. 18°.　　　　　**34.** 10°.　　　　　**35.** 75° 15′.　　　　　**36.** 47° 35′.

37. 14° 57′.　　　　　**38.** 21° 53′.　　　　　**39.** 125° 14.7′.　　　　　**40.** 135° 33.6′.

In Exercises 41–50, find the value of each term, and then combine the terms.

41. $\sin \pi/6 + \tan^2 \pi/3 + \cos \pi/2 + \sec \pi/3$.

42. $\cot^2 \pi/6 + \sec \pi/2 + \sin 2\pi + \cos \pi$.

43. $\sec \pi + \sin 2\pi + \cos \pi/2 - \tan \pi/4$.

44. $\cos^2 \frac{5}{6}\pi + \sin^2 \frac{5}{6}\pi + \tan \frac{3}{4}\pi - \cot \frac{5}{4}\pi$.

45. $\sin (-\pi/2) + \cos (-\pi/3) - \sec^2 \frac{5}{6}\pi + \csc \frac{11}{6}\pi$.

46. $\tan (-\pi/4) + \cot (-\frac{3}{4}\pi) + \sin^2 \frac{13}{6}\pi - \cos(-\pi)$.

47. $\sqrt{3} \tan \frac{5}{6}\pi + \sqrt{2} \sin \frac{3}{4}\pi + 4 \sec^2 \frac{5}{4}\pi$.

48. $\sqrt{2} \cos (-\pi/4) - \sqrt{3} \tan \frac{4}{3}\pi - \csc^2(-\frac{5}{4}\pi)$.

49. $\sin^2 \frac{7}{3}\pi + \cos^2 \frac{7}{3}\pi - \tan(-\frac{3}{4}\pi)$.

50. $\sec^2 \pi/4 - \tan^2 \pi/4 - \sin^2 \frac{15}{4}\pi$.

602　ARCS AND ANGLES

Let s be the length of a circular arc that is intercepted by a central angle θ on a circle whose radius is r. Then using Formula (1), Section 601,

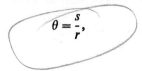

$$\theta = \frac{s}{r},$$

we obtain

(1)   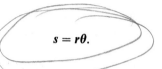 $s = r\theta.$

That is,

arc length = radius × central angle *in radians*.

(Note, s and r may be measured in any unit of length, but they must be expressed in the same unit.)

Example 1.   To the nearest tenth of an inch, find the length of an arc of a circle, of radius 10 inches, which is intercepted by a central angle of 125°.

Solution.   We must convert 125° to radian measure, which can be done by using either Formula (4) Section 601, or Table VI. By Formula (4), 1° = 0.017453 radians. Therefore

$$125° = 125 × 0.017453 = 2.181625 \approx 2.18 \text{ radians.}$$

From Table VI, page *102*, we find

$$90° = 1.57080 \text{ radians,}$$
$$\underline{35° = 0.61087} \text{ radians,}$$

Thus $125° = 2.18167 \approx 2.18$ radians.

Therefore the arc length s is

$s = r\theta = 10 × 2.18$ inches $= 21.8$ inches (to the nearest tenth of an inch).

Example 2.   As a wheel, of 20″ diameter, rotates, a point on the rim travels a linear distance of 200″ each second.
(a) Find the angular velocity in radians and in degrees per second.
(b) How many revolutions does the wheel make each second?

Solution.   *The **angular velocity** is the **angular displacement** of any radius of the wheel, relative to some fixed direction, during a particular unit of time.* It can, for example, be expressed in terms of degrees per second, revolutions per minute, or similar units.

Using Formula (1), Section 601, we have

$$\theta = \frac{s}{r} = \frac{200}{10} = 20 \text{ radians per second.}$$

Now,   1 radian $= 57.2958°$,

therefore, 20 radians $= (20 × 57.2958)° = 1145.9°$ (accurate to the nearest tenth of a degree).
(b) *One revolution is made when the wheel rotates through 2π radians.* Therefore, the number of revolutions (R) can be found by the formula

(2) $$R = \frac{\theta}{2\pi}.$$

(Note, R and θ may be any unit of time, but they must be expressed in the same unit of time. θ must be expressed in radians.)

$$R = \frac{\theta}{2\pi} = \frac{20}{2\pi} = 3.18 \text{ r.p.s. (revolutions per second).}$$

Example 3.   Find the height of a tower that stands 1 mile away, on a horizontal plane, if the angle of elevation of the top of the tower is $1° \, 24'$.

Solution.   When the central angle is relatively small, the length of the intercepted arc may be taken as a close approximation of the length of its chord (Fig. 604).

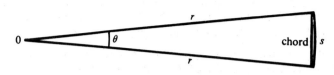

Figure 604

From the tables on page *102*

$$1° \quad = 0.01745 \text{ rad.,}$$
$$\underline{24' = 0.00698 \text{ rad.,}}$$

therefore $1° \, 24' \approx 0.02443 \text{ rad.}$

Height($\approx s$) $\approx r\theta$

$$= (5280)(0.02443)$$
$$= 129.0 \text{ feet (to the nearest tenth of a foot).}$$

EXERCISES

1. Find the number of radians in the central angle that subtends an arc of 8 inches on a circle of diameter 2 feet. Express the angle in radians, accurate to the nearest hundredth of a radian.

2. A pendulum 25 inches long oscillates $3° \, 30'$ on each side of its vertical position. Find the length of arc through which the end of the pendulum swings.

3. The hour hand of a clock is 5 inches long. Through what distance does the tip of the hour hand travel in 55 minutes?

4. The latitude of Los Angeles, California, is 34° 03′. Find the surface (arc) distance from Los Angeles to the equator. Assume the earth to be a sphere of diameter 7,920 statute miles.

5. Find the number of feet in a nautical mile. A nautical mile is the length of arc on a great circle of the earth which is subtended by an angle of 1′ at the center of the earth. Assume the earth to be a sphere of diameter 7,920 statute miles (1 statute mile = 5280 feet).

6 From a distance 1000 feet a building subtends an angle of 5°. Find the approximate height of the building.

7. As viewed from the earth the sun subtends an angle of about 32′. If the sun is 93,000,000 miles from the earth, find the diameter of the sun.

8. An automobile is traveling 60 miles per hour. The effective radius of the wheels is 13 inches. Find (a) the rate at which the wheel turns (angular speed) in radians per second; (b) the rpm (revolutions per minute) of the wheels.

9. A diesel locomotive is traveling 80 miles per hour. The diameter of its drive wheels is 36 inches. Find (a) the rate at which the wheel turns (angular speed) in radians per second; (b) the rpm (revolutions per minute) of the wheels.

10. Determine the speed of a point on the equator of the earth (in mph) due to the rotation of the earth. Assume the diameter of the earth at the equator to be 7926 statute miles.

11. The latitude of Los Angeles, California, is 34° 03′. Find the surface (arc) distance from Los Angeles to the north pole. Assume the earth to be a sphere of diameter 7,920 statute miles.

12. As viewed from the earth the moon subtends an angle of about 31′. If the moon is 240,000 miles from the earth, find the diameter of the moon.

13. The latitude of Los Angeles, California, is 34° 03′ and that of Portland, Oregon, is 45° 30′. Find, in both statute and nautical miles, the surface (arc) distance that Portland is due north of Los Angeles. Assume the earth to be a sphere of diameter 7,920 statute miles. (See Problem 5 for the definition of a nautical mile.)

14. The latitude of Chicago, Illinois, is 41° 50′ and that of Mexico City, Mexico, is 19° 26′. Find, in both statute and nautical miles, the surface (arc) distance that Chicago is due north of Mexico City. Assume the earth to be a sphere of diameter 7,920 statute miles. (See Problem 5 for definition of a nautical mile.)

15. A belt traveling at the rate of 30 feet per second drives a pulley at a speed of 600 revolutions per minute. Find the radius of the pulley.

16. The diameters of the front and rear wheels of a tractor are 30 inches and 50 inches, respectively. Find the revolutions per minute of the wheels when the tractor is traveling 10 miles per hour.

17. If an automobile wheel, with an effective radius of 13 inches, rotates at 900 revolutions per minute, what is the speed of the car in miles per hour?

18. A tower at a distance of 2500 feet subtends an angle of 5°. How high is the tower?

19. Two pulleys of diameters 10 inches and 20 inches, respectively, are connected by an open belt (not crossing). If the centers of the pulleys are 20 inches apart, find the length of the belt.   *(answer)* ✗ *(5)*

20. If the belt, of Problem 19, crossed between the pulleys, what would be its length?

10 points extra

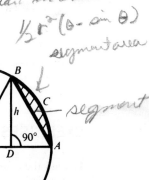

603   AREAS OF SECTORS AND SEGMENTS OF CIRCLES

*A **sector** of a circle is a region bounded by two radii and an arc of the circle.* (*See* Fig. 605.)

   It was proved in plane geometry that the areas of two sectors of the same circle are to each other as their corresponding angles. Let K be the area of a sector in a circle of radius r whose corresponding angle is θ, expressed in radian measure. If we take the entire circle as a sector whose angle is 2π (radians), we have

$$\frac{K}{\pi r^2} = \frac{\theta}{2\pi}.$$

Hence, the area of the sector is

(1)
$$K = \tfrac{1}{2}r^2\theta.$$

sector area — *in radian measure*

$\tfrac{1}{2}r^2(\theta - \sin\theta)$ *segment area*

segment

Figure 605	**Figure 606**

A segment of a circle is a region bounded by an arc and a chord of the circle. (*See* Fig. 606.)

We denote the radius of the circle by r and the central angle of the arc by θ, expressed in radians.

Area of segment ACB = area of sector $OACB$ − area of triangle OAB
$$= \tfrac{1}{2}r^2\theta - \tfrac{1}{2}rh.$$

But h (the altitude of triangle OAB) $= r \sin \theta$.
Therefore,

$$\text{Area of segment } ACB = \tfrac{1}{2}r^2\theta - \tfrac{1}{2}r(r \sin \theta)$$
$$= \tfrac{1}{2}r^2\theta - \tfrac{1}{2}r^2 \sin \theta.$$

Hence,

(2) **Area of segment** $= \tfrac{1}{2}r^2(\theta - \sin \theta)$.

Example.   In a circle of radius 10 inches, find the areas of the sector and the segment whose central angle is 125°.

Solution.   From Example 7, Section 601, 125° = 2.1817 radians.

$$\text{Area of sector} = \tfrac{1}{2}r^2\theta,$$
$$= \tfrac{1}{2}(10^2)2.1817,$$
$$= 109.1 \text{ square inches}$$

(to the nearest tenth of a square inch).

$$\text{Area of segment} = \tfrac{1}{2}r^2(\theta - \sin \theta),$$
$$= \tfrac{1}{2}(10^2)(2.1817 - \sin 125°).$$
$$= \tfrac{1}{2}(10^2)(2.187 - .8192),$$
$$= 68.1 \text{ square inches}$$

(to the nearest tenth of a square inch).

EXERCISES

1. Find the areas of the sector and segment subtended by a central angle of 100° in a circle of radius 10 inches.

2. Find the areas of the sector and segment subtended by a central angle of $\pi/3$ radians in a circle of radius 12 inches.

3. Find the areas of the sector and segment subtended by a central angle of 4 radians in a circle of radius 10 inches.

4. Find the areas of the sector and segment subtended by a central angle of 280° in a circle of radius 10 inches.

5. Find the area of a circular segment whose chord is 5 inches from the center of the circle if the radius is 10 inches.

6. A cylindrical tank placed with its axis horizontal is 10 feet in diameter and 20 feet long. If this tank is partly filled with water to a depth of 4 feet, how many gallons of water are in the tank? (Take $7\frac{1}{2}$ gallons to 1 cubic foot.)

7. Find the number of gallons of water in the tank, of Exercise 6, if the tank is filled to a depth of 7 feet.

8. Find the area common to two intersecting circles of radii 8 feet and 10 feet if their common chord is 10 feet long.

The fundamental relations between the trigonometric functions

701   THE FUNDAMENTAL RELATIONS

We recall, from previous chapters, that if θ is any angle in standard position (Fig. 701a) then,

$$\text{(a) } \sin \theta = \frac{y}{r}, \qquad \text{(b) } \cos \theta = \frac{x}{r}, \qquad \text{(c) } \tan \theta = \frac{y}{x},$$

$$\text{(d) } \cot \theta = \frac{x}{y}, \qquad \text{(e) } \sec \theta = \frac{r}{x}, \qquad \text{(f) } \csc \theta = \frac{r}{y}.$$

$P(x, y)$ is the point where the terminal side of θ meets the circle with radius r and center at $(0, 0)$ (Fig. 204). Exceptions occur for certain quadrantal angles (*see* Sections 401 and 501).

Since x, y, and r are numerically equal to the sides of a right triangle,

(g) $$x^2 + y^2 = r^2,$$

and in the unit circle,

(h) $$x^2 + y^2 = 1.$$

From these eight equations we obtain eight **fundamental relations** or **fundamental identities** that connect the functions of any angle for which both members of the identities are defined. These fundamental relations are

(1) $$\csc \theta = \frac{1}{\sin \theta}, \qquad (\csc \theta \sin \theta = 1)$$

(2) $$\sec \theta = \frac{1}{\cos \theta}, \qquad (\sec \theta \cos \theta = 1)$$

(3) $$\cot\theta = \frac{1}{\tan\theta}, \qquad (\cot\theta\,\tan\theta = 1)$$

(4) $$\tan\theta = \frac{\sin\theta}{\cos\theta},$$

(5) $$\cot\theta = \frac{\cos\theta}{\sin\theta},$$

(6) $$\sin^2\theta + \cos^2\theta = 1,$$

(7) $$1 + \tan^2\theta = \sec^2\theta,$$

(8) $$1 + \cot^2\theta = \csc^2\theta.$$

Proof. Because corresponding sides of similar triangles have equal ratios, we can simplify the task of proving these identities by referring to the unit circle (Fig. 701b).

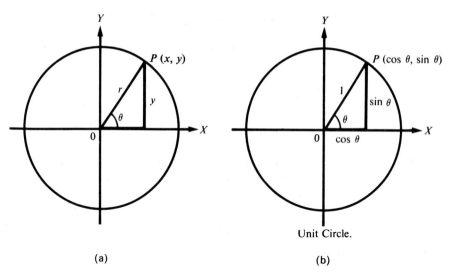

(a)   (b)

Figure 701

Identities (1), (2), (3), the "reciprocal relations," are already familiar to us. To prove these, refer to Figures 701a and 701b.

(1) $$\csc\theta = \frac{r}{y} = \frac{1}{\frac{y}{r}} = \frac{1}{\sin\theta}.$$

(2)
$$\sec \theta = \frac{r}{x} = \frac{1}{\dfrac{x}{r}} = \frac{1}{\cos \theta}.$$

(3)
$$\cot \theta = \frac{x}{y} = \frac{1}{\dfrac{y}{x}} = \frac{1}{\tan \theta} = \frac{\cos \theta}{\sin \theta}.$$

We prove Identities (4) and (5) by using the unit circle (Fig. 701b) and by applying the definitions of $\tan \theta$ and $\cot \theta$:

(4)
$$\tan \theta = \frac{y}{x} = \frac{\text{ordinate}}{\text{abscissa}} = \frac{\sin \theta}{\cos \theta},$$

(5)
$$\cot \theta = \frac{x}{y} = \frac{\text{abscissa}}{\text{ordinate}} = \frac{\cos \theta}{\sin \theta}.$$

The proof for Identity (6) is given in Section 305.
To prove Identity (7), divide each term of Identity (6) by $\cos^2 \theta$. $\quad \rightarrow \mathcal{I}\,(7)$

$$\frac{\sin^2 \theta}{\cos^2 \theta} + \frac{\cos^2 \theta}{\cos^2 \theta} = \frac{1}{\cos^2 \theta},$$

which gives

$$\left(\frac{\sin \theta}{\cos \theta}\right)^2 + \left(\frac{\cos \theta}{\cos \theta}\right)^2 = \left(\frac{1}{\cos \theta}\right)^2.$$

But $\sin \theta / \cos \theta = \tan \theta$ (Identity 5) and $1/\cos \theta = \sec \theta$ (Identity 2)

therefore
$$\tan^2 \theta + 1 = \sec^2 \theta,$$

or
$$1 + \tan^2 \theta = \sec^2 \theta.$$

The proof for Identity (8) can be obtained by dividing each term of $\rightarrow \mathcal{I}\,(8)$
Identity (6) by $\sin^2 \theta$; then simplify as was done in the proof of Identity (7).
These basic fundamental identities should be memorized at once. The student should be able to recognize them in other forms. For example,

$$\cos \theta = \frac{1}{\sec \theta}, \quad \rightarrow \mathcal{I}\,(2)$$

$$\sin \theta = \pm\sqrt{1 - \cos^2 \theta}, \rightarrow \mathcal{I}\,(6)$$

$$\sec^2 \theta - \tan^2 \theta = 1, \quad \rightarrow \mathcal{I}\,(7)$$

$sin^{-2}\theta$

$$\sin^2 \theta = \frac{1}{\csc^2 \theta}. \quad \rightarrow \mathcal{I}$$

Example 1.   Express each of the other functions in terms of $\sin \theta$.

Solution. From Identity (6) $\sin^2 \theta + \cos^2 \theta = 1$; thus

$$\cos \theta = \pm\sqrt{1 - \sin^2 \theta}.$$

From (4)
$$\tan \theta = \frac{\sin \theta}{\cos \theta} = \pm \frac{\sin \theta}{\sqrt{1 - \sin^2 \theta}}.$$

From (3)
$$\cot \theta = \frac{1}{\tan \theta} = \pm \frac{\sqrt{1 - \sin^2 \theta}}{\sin \theta}.$$

From (2)
$$\sec \theta = \frac{1}{\cos \theta} = \pm \frac{1}{\sqrt{1 - \sin^2 \theta}}.$$

From (1)
$$\csc \theta = \frac{1}{\sin \theta}.$$

Example 2. Use the fundamental identities to find the remaining functions when $\tan \theta = 3/4$ and $180° < \theta < 270°$.

Solution. From Identity (7)

$$1 + \tan^2 \theta = \sec^2 \theta,$$

$$1 + (\tfrac{3}{4})^2 = 1 + \tfrac{9}{16} = \tfrac{25}{16} = \sec^2 \theta.$$

Therefore $\sec \theta = \pm\tfrac{5}{4}$. But when $180° < \theta < 270°$ $\sec \theta$ is negative.

Hence
$$\sec \theta = -\frac{5}{4}.$$

From (3)
$$\cot \theta = \frac{1}{\tan \theta} = \frac{4}{3}.$$

From (2)
$$\cos \theta = \frac{1}{\sec \theta} = -\frac{4}{5}.$$

From (6)
$$\sin^2 \theta + \cos^2 \theta = 1,$$

$$\sin^2 \theta + \left(-\frac{4}{5}\right)^2 = 1.$$

Hence, in the third quadrant, $\sin \theta$ is negative and $\sin^2 \theta = 9/25$.

Therefore
$$\sin \theta = -\frac{3}{5}.$$

From (1)
$$\csc \theta = \frac{1}{\sin \theta} = \frac{1}{-3/5} = -\frac{5}{3}.$$

Example 3.   Use the fundamental identities to simplify the following expressions.

(a) $\tan \theta \cot \theta + \dfrac{\sin^2 \theta}{\cos^2 \theta}.$

Solution.   From Identities (3), (4), and (7)

$$\tan \theta \cot \theta + \frac{\sin^2 \theta}{\cos^2 \theta} = 1 + \tan^2 \theta = \sec^2 \theta.$$

(b) $\dfrac{\sin \theta}{1 + \cos \theta} + \dfrac{1 + \cos \theta}{\sin \theta}.$

Solution:

$$\frac{\sin \theta}{1 + \cos \theta} + \frac{1 + \cos \theta}{\sin \theta} = \frac{\sin^2 \theta + (1 + \cos \theta)^2}{\sin \theta \, (1 + \cos \theta)}$$

$$= \frac{\sin^2 \theta + 1 + 2 \cos \theta + \cos^2 \theta}{\sin \theta \, (1 + \cos \theta)}$$

$$= \frac{2 + 2 \cos \theta}{\sin \theta \, (1 + \cos \theta)} = \frac{2(1 + \cos \theta)}{\sin \theta \, (1 + \cos \theta)}$$

$$= \frac{2}{\sin \theta} = 2\left(\frac{1}{\sin \theta}\right) = 2 \csc \theta.$$

EXERCISES

In each of the Exercises 1–6, use the fundamental identities to find the remaining functions of the angle satisfying the given conditions.

1. $\cos \theta = -\frac{3}{5}$, θ in second quadrant.

2. $\sin \theta = -\frac{5}{13}$, θ in fourth quadrant.

3. $\tan \theta = \frac{4}{3}$, θ in first quadrant.

4. $\sec \theta = 2$, θ in fourth quadrant.

5. $\cot \theta = \frac{5}{12}$, $\sin \theta < 0$.

6. $\csc \beta = \frac{5}{4}$, $\cos \beta > 0$.

In each of the Exercises 7–12, reduce the given expression to an equivalent expression involving only $\sin x$ and $\cos x$ and then simplify it.

7. $\dfrac{\tan x \csc x}{\sec x}.$

8. $\dfrac{\cot x \sec x}{\csc x}.$

9. $\tan x + \cot x.$

10. $\dfrac{1}{\csc^2 x} + \dfrac{1}{\sec^2 x}$.

11. $\dfrac{\csc x}{\tan x + \cot x}$.

12. $\dfrac{\sec x \csc x}{\sec^2 x + \csc^2 x}$.

In each of the Exercises 13–24, perform the indicated operations and use the fundamental identities to simplify the expressions.

13. $(1 + \tan \theta)^2 - 2 \sin \theta \sec \theta$.

14. $\sin \theta \csc \theta + \cos^2 \theta + \sec \theta \cos \theta + \sin^2 \theta$.

15. $\csc^2 \theta \, (\sec \theta - \cos \theta)$.

16. $\sin \theta \sec \theta \cot \theta$.

17. $(\sin \theta + \cos \theta)^2 - (\sin \theta - \cos \theta)^2$.

18. $\dfrac{1 + \tan^2 \theta}{\csc^2 \theta}$.

19. $\dfrac{1}{\sin \theta - 1} - \dfrac{1}{\sin \theta + 1}$.

20. $\dfrac{\sin \theta}{\csc \theta} + \dfrac{\cos \theta}{\sec \theta}$.

21. $\cot \theta + \dfrac{\sin \theta}{1 + \cos \theta}$.

22. $\dfrac{3 - \cot \theta}{3 \sec \theta - \csc \theta}$.

23. $\dfrac{\cot^3 x - \tan^3 x}{\cot x - \tan x} - \sec^2 x$.

24. $\dfrac{\sin x + \tan x \sin x + \cos x}{\sec x} - \sin x \cos x$.

702 IDENTITIES AND EQUATIONS

An equation that is true for all numbers for which the two sides are defined is called an *identical equation* or an **identity**. Thus the statement

$$(x + 1)^2 = x^2 + 2x + 1$$

is an identity, since it is true for all values of x. The trigonometric statement

$$\sin^2 x + \cos^2 x = 1$$

is an identity because it is true for every value of x.

An equation that is untrue for at least one number at which both sides are defined is called a *conditional equation,* or just an **equation**. Thus the statement

$$x^2 - x - 6 = 0$$

is a conditional equation, because it is true only for $x = 3$ and for $x = -2$, and we say that both 3 and -2 *satisfy* the equation $x^2 - x - 6 = 0$. To say that a certain value of a variable "satisfies" an equation means that the equation becomes true when that value is substituted for the variable in the equation. If the members of a conditional equation can be made identical by replacing the unknown by a number, that number is a **root** of the equation. The process of finding the roots of an equation is known as **solving** the equation. The trigonometric equation

$$\sin x = \cos x$$

is conditional, because it is true only for the angles $x = 45°$ and $x = 225°$, and for values of x coterminal with those values.

703   PROVING TRIGONOMETRIC IDENTITIES

The process of transforming one member of an identity into the other is called *proving the identity.*

As the student advances through his study of mathematics and science, he will find that the changing of a trigonometric expression, from one form to another, is a necessary step in the solution of many problems. Proving identities is, therefore, an important part of this course.

Basic to proving trigonometric identities is the ability of the student to recognize the fundamental trigonometric identities in all of their various forms. When a student is asked to prove an identity he should strive by one or more changes to transform one member of the identity into the form of the other side *which should be left unchanged. This method should be followed even though other logically correct methods may be available.*

In solving equations we generally follow a definite procedure. In proving trigonometric identities no definite procedure can be given; however the following suggestions should prove helpful:

1. It is usually best to start with the more complicated member and reduce it to the simpler member.

2. The student should keep constantly in mind the expression he is trying to arrive at and look for fundamental relations that will lead him to this final expression.

3. Factoring, adding fractions, multiplying two or more expressions, etc. will often bring obvious simplifications to light.

4. If possible, avoid introducing radicals.

5. Sometimes it is useful to express all of the functions on one side in terms of sines and cosines.

Example 1. Prove the identity

$$\frac{\csc x}{\cot x} = \sec x.$$

Solution. The left-hand side is the more complicated, so we will start with it and, by successive applications of the fundamental identities, reduce it to the form of the right-hand side.

$$\frac{\csc x}{\cot x} = \frac{1/\sin x}{\cos x/\sin x} = \frac{1}{\sin x} \cdot \frac{\sin x}{\cos x} = \frac{1}{\cos x} = \sec x.$$

Example 2. Prove the identity

$$\csc^2 x = \frac{\sin^4 x - \cos^4 x}{\sin^2 x - \cos^2 x} + \cot^2 x.$$

Solution. The right-hand side is the more complicated, so we will start with it and, by factoring and applying the fundamental identities, reduce it to the form of the left-hand side.

$$\csc^2 x = \frac{\sin^4 x - \cos^4 x}{\sin^2 x - \cos^2 x} + \cot^2 x$$

$$= \frac{(\sin^2 x + \cos^2 x)(\sin^2 x - \cos^2 x)}{\sin^2 x - \cos^2 x} + \cot^2 x$$

$$= \sin^2 x + \cos^2 x + \cot^2 x = 1 + \cot^2 x = \csc^2 x.$$

Example 3. Prove the identity

$$\frac{\tan x + \sin x}{\tan x - \sin x} = \frac{\sec x + 1}{\sec x - 1}.$$

Both sides of this identity appear to be as complicated as each other. We show two solutions.

First Solution. Reduce the right-hand side to the form of the left-hand side. As a clue we observe that if the numerator and denominator of (sec x + 1)/ (sec x − 1) are multiplied by sin x, then the last term of both becomes sin x, which is part of the expression we want. We make this multiplication and find that the desired tan x expression appears.

$$\frac{\sin x\,(\sec x + 1)}{\sin x\,(\sec x - 1)} = \frac{\sin x \sec x + \sin x}{\sin x \sec x - \sin x}.$$

But $$\sin x \sec x = \sin x \left(\frac{1}{\cos x}\right) = \frac{\sin x}{\cos x} = \tan x.$$

Thus $$\frac{\sin x \sec x + \sin x}{\sin x \sec x - \sin x} = \frac{\tan x + \sin x}{\tan x - \sin x}.$$

Second Solution. We observe that if the numerator and denominator of $(\tan x + \sin x)/(\tan x - \sin x)$ are divided by $\sin x$ the last term of both becomes 1, which is part of the epxression we want. We make this division and find that the desired expression, $\sec x$, results.

$$\frac{(\tan x/\sin x) + (\sin x/\sin x)}{(\tan x/\sin x) - (\sin x/\sin x)} = \frac{(\tan x/\sin x) + 1}{(\tan x/\sin x) - 1}.$$

But $$\frac{\tan x}{\sin x} = \frac{(\sin x/\cos x)}{\sin x} = \frac{\sin x}{\cos x} \cdot \frac{1}{\sin x} = \frac{1}{\cos x} = \sec x.$$

Thus $$\frac{(\tan x/\sin x) + 1}{(\tan x/\sin x) - 1} = \frac{\sec x + 1}{\sec x - 1}.$$

EXERCISES

Prove the following identities by operating on one member and leaving the other unchanged.

1. $\dfrac{\sin x}{\tan x} = \cos x.$

2. $\dfrac{\cos \alpha}{\cot \alpha} = \sin \alpha.$

3. $\dfrac{\sin^2 \beta + \cos^2 \beta}{1 + \tan^2 \beta} = \cos^2 \beta.$

4. $\dfrac{\sec^2 \theta - \tan^2 \theta}{1 + \cot^2 \theta} = \sin^2 \theta.$

5. $\dfrac{(\sin x + \cos x)^2}{\sin x} = \csc x + 2 \cot x.$

6. $\cos(1 + \tan x)^2 = \sec x + 2 \sin x.$

7. $\dfrac{1 - \cos^2 x}{\sin x \,(\csc x + \cot x)} = 1 - \cos x.$

8. $\dfrac{1 - \sin^2 A}{\cos A \,(\sec A + \tan A)} = 1 - \sin A.$

9. $\sec B \csc B = \tan B + \cot B.$

10. $\dfrac{1}{\tan A} + \tan A = \sec A \csc A.$

11. $\dfrac{\sin^3 x - \cos^3 x}{\sin x - \cos x} = 1 + \sin x \cos x.$

12. $\dfrac{\sec^4 x - \tan^4 x}{\sec^2 x + \tan^2 x} + \tan^2 x = \sec^2 x.$

13. $\dfrac{\csc^4 x - \cot^4 x}{\csc^2 x + \cot^2 x} + \cot^2 x = \csc^2 x.$

14. $\dfrac{1}{1 + \cos A} + \dfrac{1}{1 - \cos A} = 2 \csc^2 A.$

15. $\dfrac{1}{\sin x - 1} - \dfrac{1}{\sin x + 1} = -2 \sec^2 x.$

16. $\dfrac{1 + \cos x}{1 - \cos x} = \dfrac{\sec x + 1}{\sec x - 1}.$

17. $\dfrac{\tan B + 1}{\tan B - 1} = \dfrac{\sec B + \csc B}{\sec B - \csc B}.$

18. $\dfrac{\tan^2 A - 1}{\sec^2 A} = \dfrac{\tan A - \cot A}{\tan A + \cot A}.$

19. $\dfrac{\cos A}{1 + \sin A} + \dfrac{1 + \sin A}{\cos A} = 2 \sec A.$

20. $\dfrac{\sin x}{1 + \cos x} + \dfrac{1 + \cos x}{\sin x} = 2 \csc x.$

21. $\tan \theta + \dfrac{\cos \theta}{1 + \sin \theta} = \sec \theta.$

22. $\csc \alpha - \dfrac{\sin \alpha}{1 + \cos \alpha} = \cot \alpha.$

23. $\dfrac{\tan^2 \theta - 1}{1 + \tan^2 \theta} = 1 - 2 \cos^2 \theta.$

24. $\dfrac{\cot^2 \beta - 1}{1 + \cot^2 \beta} = 1 - 2 \sin^2 \beta.$

25. $\dfrac{1 + \sin \beta}{\cos \beta} = \dfrac{\cos \beta}{1 - \sin \beta}.$

26. $\dfrac{\sec \alpha - 1}{\tan \alpha} = \dfrac{\tan \alpha}{\sec \alpha + 1}.$

27. $\dfrac{\cot \theta - \cos \theta}{\cos^3 \theta} = \dfrac{\csc \theta}{1 + \sin \theta}.$

28. $\dfrac{\tan \theta - \sin \theta}{\sin^3 \theta} = \dfrac{\sec \theta}{1 + \cos \theta}.$

29. $\cot^2 x - \tan^2 x = \dfrac{\cot x - \tan x}{\sin x \cos x}.$

30. $\sin x - \sin x \cos^2 x = \sin^3 x.$

31. $1 - 2 \sin^2 x + \sin^4 x = \cos^4 x.$

32. $\dfrac{\sec^2 x - 1}{\cot x} = \tan^3 x.$

33. $\dfrac{\csc^2 x - 1}{\tan x} = \cot^3 x.$

34. $\dfrac{1 - 2 \cos^2 \theta}{\sin \theta \cos \theta} = \tan \theta - \cot \theta.$

35. $(x \sin \theta + y \cos \theta)^2 + (x \cos \theta - y \sin \theta)^2 = x^2 + y^2.$

36. $(1 + \sin \beta + \cos \beta)^2 = 2(1 + \sin \beta)(1 + \cos \beta).$

37. $\dfrac{\tan x + \sec x}{\sec x - \cos x + \tan x} = \csc x.$

38. $\sec^4 x - \tan^4 x = 1 + 2 \tan^2 x.$

39. $\dfrac{\sin A \cos B + \cos A \sin B}{\cos A \cos B - \sin A \sin B} = \dfrac{\tan A + \tan B}{1 - \tan A \tan B}.$

40. $\dfrac{\cos A \cos B - \sin A \sin B}{\cos A \cos B + \sin A \sin B} = \dfrac{1 - \tan A \tan B}{1 + \tan A \tan B}.$

704 TRIGONOMETRIC EQUATIONS

The student will recall that we solved trigonometric equations in Section 216 and again in Chapters 3, 4, and 5. Now, with our added knowledge of the eight fundamental identities, we can solve more complicated trigonometric equations.

Example 1. Solve the following equation for values of x in the range $0° < x < 360°$.

$$2 \cos^2 x = 1 + \sin x.$$

Solution. In problems of this type it is generally best to express all terms in the same function. Here we replace $\cos^2 x$ by $1 - \sin^2 x$.

$$2 \cos^2 x = 1 + \sin x,$$
$$2(1 - \sin^2 x) = 1 + \sin x,$$
$$2(1 + \sin x)(1 - \sin x) = 1 + \sin x.$$

We transpose all terms to the left-hand member (so as to obtain 0 on the right) and factor. We get

$$(1 + \sin x)[2(1 - \sin x) - 1] = 0,$$
$$(1 + \sin x)(1 - 2 \sin x) = 0.$$

We set each factor equal to 0, for, provided that all of the factors are defined, a product is equal to 0 if and only if at least one of its factors is equal to 0.

$$2 \sin x - 1 = 0, \qquad \text{or} \qquad \sin x + 1 = 0,$$
$$\sin x = \tfrac{1}{2}; \qquad\qquad\qquad\qquad \sin x = -1;$$
$$x = 30° \text{ and } 150°. \qquad\qquad\qquad x = 270°.$$

The complete solution is therefore $x = 30°$, $150°$, and $270°$.

Example 2.   Solve the equation

$$3 \tan^2 \theta = 7 \sec \theta - 5,$$

for $0 < \theta < 2\pi$.

Solution.   We replace $\tan^2 \theta$ with $\sec^2 \theta - 1$ then transpose and factor.

$$3 \tan^2 \theta = 7 \sec \theta - 5,$$
$$3(\sec^2 \theta - 1) = 7 \sec \theta - 5,$$
$$3 \sec^2 \theta - 3 = 7 \sec \theta - 5,$$
$$3 \sec^2 \theta - 7 \sec \theta + 2 = 0,$$
$$(3 \sec \theta - 1)(\sec \theta - 2) = 0.$$

Then, if the equation has a solution, either

$$3 \sec \theta - 1 = 0 \quad \text{or} \quad \sec \theta - 2 = 0.$$

But $3 \sec \theta - 1 \neq 0$, since $\sec \theta$ would then be $\tfrac{1}{3}$, which is impossible; the numerical value of the secant function can not be numerically less than one (*see* Figure 507, Section 506). When the other factor, $\sec \theta - 2$, equals 0,

$$\sec \theta = 2.$$

Hence $$\theta = \frac{\pi}{3} \quad \text{and} \quad \frac{5\pi}{3}.$$

Example 3.   Solve the equation

$$2 \sin x \cos x = \sin x.$$

for $0° \leq x < 360°$.

Solution.   If the student proceeds to simplify this equation by dividing both members by $\sin x$, the resulting equation

$$2 \cos x = 1$$

will not give all of the roots of the original equation. To avoid this loss of roots, solve by factoring.

$$2 \sin x \cos x = \sin x,$$
$$2 \sin x \cos x - \sin x = 0,$$
$$\sin x (2 \cos x - 1) = 0.$$

$$\sin x = 0; \qquad \text{or} \qquad 2 \cos x - 1 = 0;$$
$$x = 0°, 180°. \qquad\qquad \cos x = \tfrac{1}{2}; \quad x = 60°, 300°.$$

The complete solution is therefore $x = 0°, 60°, 180°, 300°$.

Example 4.   Solve the equation $\sin^2 x = 1$ for $0° \leq x < 360°$.

Solution.   If we solve by taking the square root of both sides of the equation then we must use both $\sin x = +1$ and $\sin x = -1$. The solution by factoring will give the same results.

$$\sin^2 x = 1.$$

$$(\sin x + 1)(\sin x - 1) = 0,$$

$$\sin x + 1 = 0, \qquad \text{or} \qquad \sin x - 1 = 0,$$
$$\sin x = -1; \qquad\qquad \sin x = 1;$$
$$x = 270°. \qquad\qquad\quad x = 90°.$$

The complete solution is therefore $x = 90°$ and $270°$.

As in any algebraic solution of an equation, if we divide both members of an equation by a common factor involving the unknown, or if we take the square root of both members, we must be careful not to throw away any roots.

Example 5.   Solve the equation $\tan x = \sec x$ for $0° \leq x < 360°$.

Solution.   Write $\tan x = \sin x / \cos x$ and $\sec x = 1 / \cos x$, and then multiply both members of the resulting equation by $\cos x$. We obtain

$$\frac{\sin x}{\cos x} = \frac{1}{\cos x}; \qquad \text{hence} \qquad \sin x = 1, x = 90°.$$

However, $x = 90°$ does not satisfy the original equation since neither $\tan 90°$ nor $\sec 90°$ is defined. We introduced the extraneous root $x = 90°$ when we multiplied both members of

$$\frac{\sin x}{\cos x} = \frac{1}{\cos x}$$

by $\cos x$, since $\cos 90° = 0$. The equation has no roots.

If both members of an equation are multiplied by an expression involving the unknown (in order, for example, to get rid of fractions), or if we square both members (to free the equation of radicals, perhaps) the resulting equation may have roots that do not satisfy the original equation. The roots that do not satisfy the original equation are called *extraneous roots*. The best test is to substitute each of the supposed roots in the original equation, and then reject any that fail to satisfy the equation.

Example 6.   Solve the equation

$$\sin^2 \theta - 2 \sin \theta - 1 = 0$$

for $0° \leq \theta < 360°$.

Solution.   This trigonometric equation is a quadratic equation that does not have simple factors. We will use the quadratic formula to solve this equation.
     The roots of the quadratic equation

$$ax^2 + bx + c = 0, \qquad \text{where } a \neq 0,$$

are
$$x = \frac{-b \pm \sqrt{b^2 - 4ac}}{2a}.$$

Let $\sin \theta$ represent x in the quadratic formula, then, for

$$(\sin \theta)^2 - 2(\sin \theta) - 1 = 0,$$

$$\sin \theta = \frac{2 \pm \sqrt{4 + 4}}{2} = 1 \pm \sqrt{2}.$$

Hence $\sin \theta = 1 + \sqrt{2}$ or $\sin \theta = 1 - \sqrt{2}$

$\qquad\qquad\qquad = 2.4142.\qquad\qquad\qquad\qquad = -.4142.$

But $\sin \theta$ cannot be numerically greater than 1; therefore we take $\sin \theta = -.4142$. From Table IV we find $\sin 24° 28' = 0.4142$. But $\sin \theta$ is negative; therefore θ is in the third or fourth quadrant. The desired roots are

$$\theta = 180° + 24° 28' = 204° 28',$$

$$\theta = 360° - 24° 28' = 335° 32'.$$

EXERCISES

Solve the following equations for x in the range $0° \leq x < 360°$.

1. $(\tan x - \sqrt{3})(\sin x + 1) = 0$.

2. $(2 \cos x - \sqrt{3})(2 \sin x + \sqrt{2}) = 0$.

3. $(\sec x - 2)(\csc x + 1) = 0$.

4. $\sin x \,(\cot x - 1)(\cos x - 1) = 0$.

5. $2 \sin^2 x = 1 + \cos x$.

6. $2 \cos^2 x = 1 - \sin x$.

7. $4 \sin^2 x = 3$.

8. $2 \cos^2 x = 1$.

9. $2 \tan^2 x = \sec x - 1$.

10. $2 \tan^2 x + 3 \sec x = 0$.

11. $\sin x \cos x = \cos x$.

12. $\sin x = \sin x \cos x$.

13. $\sin x + \csc x + 2 = 0$.

14. $\cos x + \sec x + 2 = 0$.

15. $\sin x - \sqrt{1 - \sin^2 x} = 0$.

16. $\tan x + \sqrt{1 - 2 \tan^2 x} = 0$.

17. $\sin^2 x = 2(\sin x + 1)$.

18. $\cos^2 x = 1 + \cos x$.

19. $\sqrt{3}\,(\sec x + 1) = \tan x$.

20. $\sin x \tan x - 1 = \tan x - \sin x$.

8

Combination of angles

In Chapter 7 we developed eight fundamental trigonometric identities. In this chapter we shall add many other important trigonometric identities to our collection.

801 FUNCTIONS OF $(-\theta)$

The identities that we wish to develop in this section are:

(9) $$\sin(-\theta) = -\sin\theta,$$

(10) $$\cos(-\theta) = \cos\theta,$$

(11) $$\tan(-\theta) = -\tan\theta,$$

(12) $$\cot(-\theta) = -\cot\theta,$$

(13) $$\sec(-\theta) = \sec\theta,$$

(14) $$\csc(-\theta) = -\csc\theta.$$

reciprocals are the plane

In the above identities observe that reversing the sign of the angle reverses the sign of all of the functions except the cosine and secant.

Examples:

1. $\sin(-30°) = -\sin 30°;$
   $-\frac{1}{2} = -(+\frac{1}{2}).$

2. $\cos(-60°) = \cos 60°;$
   $\frac{1}{2} = \frac{1}{2}.$

3. $\tan(-135°) = -\tan 135°;$
   $1 = -(-1).$

4. $\cot(-45°) = -\cot 45°;$
   $-1 = -(+1).$

5. $\sec(-240°) = \sec 240°$;   **6.** $\csc(-90°) = -\csc 90°$;
   $-2 = -2.$ $-1 = -(+1).$

7. $\sin[-(-568°)] = -\sin 28°$ and $\sin(-568°) = \sin 28°$.   Therefore
   $\sin[-(-568°)] = -\sin(-568°).$

8. $\cos[-(-568°)] = -\cos 28°$ and $\cos(-568°) = -\cos 28°$.   Therefore
   $\cos[-(-568°)] = \cos(-568°).$

Proof. Let θ and $(-\theta)$ be angles in standard position. Let the terminal sides of these angles meet the unit circle at the points P and Q, whose coordinates are (x, y) and (x', y') respectively. Then, as defined in Sections 205 and 304,

$$x = \cos\theta, \qquad y = \sin\theta,$$

$$x' = \cos(-\theta), \qquad y' = \sin(-\theta).$$

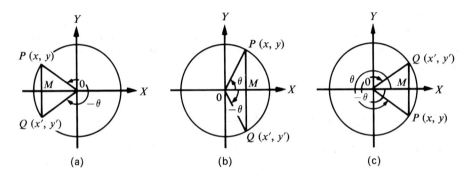

(a)   (b)   (c)

Figure 801

In each figure of Figure 801, the x-axis is the perpendicular bisector of line segment PQ at M. Then

$$x' = x, \qquad \text{and} \qquad y' = -y$$

thus $\cos(-\theta) = \cos\theta, \qquad \sin(-\theta) = -\sin\theta.$

We use these results and Identities (1), (2), (3), and (4) of Section 701 to find the other trigonometric functions of the negative of an angle.

$$\tan(-\theta) = \frac{\sin(-\theta)}{\cos(-\theta)} = \frac{-\sin\theta}{\cos\theta} = -\tan\theta.$$

$$\cot(-\theta) = \frac{1}{\tan(-\theta)} = \frac{1}{-\tan\theta} = -\cot\theta.$$

$$\sec(-\theta) = \frac{1}{\cos(-\theta)} = \frac{1}{\cos \theta} = \sec \theta.$$

$$\csc(-\theta) = \frac{1}{\sin(-\theta)} = \frac{1}{-\sin \theta} = -\csc \theta.$$

802   cos(A + B) AND cos(A − B)

signs reversed

If *A* and *B* are any two angles, then

(15) $\cos(A + B) = \cos A \cos B - \sin A \sin B,$

and

(16) $\cos(A - B) = \cos A \cos B + \sin A \sin B.$

Stated in words:
   The cosine of the sum of two angles is equal to the product of the cosines of the two angles, minus the product of their sines.
   The cosine of the difference of two angles is equal to the product of the cosines of the two angles, plus the product of their sines.

Proof for Identity (16),

$$\cos(A - B) = \cos A \cos B + \sin A \sin B.$$

   Let α and θ be any two angles in standard position in a unit circle. Let P_1, P_2, P_3; and P_4 be points on the unit circle. Point P_1, on the initial side of α and θ, has coordinates $(1, 0)$; point P_2, on the terminal side of α, has coordinates $(\cos \alpha, \sin \alpha)$; point P_3, on the terminal side of θ, has coordinates $(\cos \theta, \sin \theta)$, and point P_4, on the terminal side of $(\alpha + \theta)$, has coordinates $[\cos(\alpha + \theta), \sin(\alpha + \theta)]$. (*See* Fig. 802.)
   Chord $P_1 P_2$ equals chord $P_3 P_4$, and by using the formula for the distance between two points (Sec. 105) we find

$$\sqrt{(\cos \alpha - 1)^2 + (\sin \alpha - 0)^2}$$
$$= \sqrt{[\cos(\alpha + \theta) - \cos \theta]^2 + [\sin(\alpha + \theta) - \sin \theta]^2}.$$

By squaring both members we obtain

$$\cos^2 \alpha - 2 \cos \alpha + 1 + \sin^2 \alpha$$
$$= \cos^2(\alpha + \theta) - 2 \cos(\alpha + \theta) \cos \theta$$
$$+ \cos^2 \theta + \sin^2(\alpha + \theta) - 2 \sin(\alpha + \theta) \sin \theta + \sin^2 \theta,$$

$$2 - 2 \cos \alpha = 2 - 2[\cos(\alpha + \theta) \cos \theta + \sin(\alpha + \theta) \sin \theta],$$

which reduces to

(1) $\cos \alpha = \cos(\alpha + \theta) \cos \theta + \sin(\alpha + \theta) \sin \theta.$

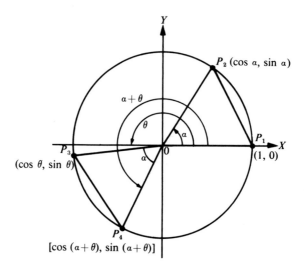

Figure 802

Equation (1) is true for all angles α and θ. Therefore, if A and B are any angles and we set $\alpha = A - B$ and $\theta = B$ in Equation (1) we obtain

$$\cos(A - B) = \cos A \cos B + \sin A \sin B.$$

We use Identity (16) and Identities (9) and (10) of Section 801 to prove

$$\cos(A + B) = \cos A \cos B - \sin A \sin B.$$

Proof. We substitute $(-B)$ for $(+B)$ in Identity (16) and proceed as follows:

$$\cos(A - B) = \cos A \cos B + \sin A \sin B,$$

$$\cos[A - (-B)] = \cos A \cos(-B) + \sin A \sin(-B).$$

But $\cos(-B) = \cos B$ and $(\sin - B) = -\sin B$.

Therefore $\cos(A + B) = \cos A \cos B + \sin A(-\sin B),$

or

$$\cos(A + B) = \cos A \cos B - \sin A \sin B.$$

The student is warned that, in general,

$$\cos(A - B) \neq \cos A - \cos B,$$

and

$$\cos(A + B) \neq \cos A + \cos B.$$

For example,

$$\cos(60° + 30°) \neq \cos 60° + \cos 30°,$$
$$\cos 90° \neq \tfrac{1}{2} + \tfrac{1}{2}\sqrt{3},$$
$$0 \neq \frac{1 + \sqrt{3}}{2}.$$

Example 1.   Show that Identity (15) is true for $A = 30°$ and $B = 60°$.

Solution. $\cos (A + B) = \cos A \cos B - \sin A \sin B,$
$$\cos (30° + 60°) = \cos 30° \cos 60° - \sin 30° \sin 60°,$$
$$\cos 90° = \tfrac{1}{2}\sqrt{3} \cdot \tfrac{1}{2} - \tfrac{1}{2} \cdot \tfrac{1}{2}\sqrt{3},$$
$$0 = 0.$$

Example 2.   Compute the value of $\cos 15°$ from the functions of $60°$ and $45°$.

Solution.   We use Identity (16).
$$\cos(A - B) = \cos A \cos B + \sin A \sin B,$$
$$\cos(60° - 45°) = \cos 60° \cos 45° + \sin 60° \sin 45°,$$
$$\cos 15° = \tfrac{1}{2} \cdot \tfrac{1}{2}\sqrt{2} + \tfrac{1}{2}\sqrt{3} \cdot \tfrac{1}{2}\sqrt{2},$$
$$= \frac{\sqrt{2}(1 + \sqrt{3})}{4}.$$

Example 3.   Given $\sin A = \tfrac{3}{5}$, with A in Q I, and $\cos B = -\tfrac{5}{13}$, with B in Q III. Find (a) $\cos(A + B)$, and (b) $\cos(A - B)$.

Solution.   Cos A and sin B must be determined before we can apply Identities (15) and (16). We find these from the figures shown in Figure 803. Thus

$$\cos A = \frac{4}{5} \quad \text{and} \quad \sin B = -\frac{12}{13}.$$

(a) $\qquad\qquad \cos(A + B) = \cos A \cos B - \sin A \sin B$
$$= \left(\frac{4}{5}\right)\left(-\frac{5}{13}\right) - \left(\frac{3}{5}\right)\left(-\frac{12}{13}\right)$$
$$= -\frac{20}{65} + \frac{36}{65} = \frac{16}{65}.$$

(b) $\qquad\qquad \cos(A - B) = \cos A \cos B + \sin A \sin B$
$$= \left(\frac{4}{5}\right)\left(-\frac{5}{13}\right) + \left(\frac{3}{5}\right)\left(-\frac{12}{13}\right)$$
$$= -\frac{20}{65} - \frac{36}{65} = -\frac{56}{65}.$$

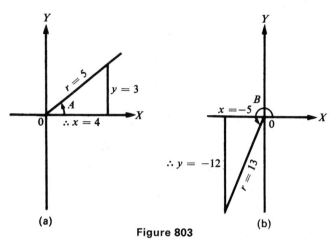

Figure 803

Example 4. Use Identity (16) to prove

$$\cos(90° - \theta) = \sin \theta.$$

Proof. This relation was shown to be true for θ an acute angle (Section 503) but now, by applying Identity (16), we prove the relation to be true for any value of θ. Substitute 90° for A and θ for B, in Identity (16).

$$\cos(A - B) = \cos A \cos B + \sin A \sin B,$$

therefore $\cos(90° - \theta) = \cos 90° \cos \theta + \sin 90° \sin \theta.$

But $\cos 90° = 0$ and $\sin 90° = 1.$

Hence $\cos(90° - \theta) = 0 \cdot \cos \theta - 1 \cdot \sin \theta$
$$= \sin \theta.$$

EXERCISES

1. By substitution, verify that Identity (15) is true for $A = 45°$ and $B = 45°$.

2. By substitution, verify that Identity (16) is true for $A = 60°$ and $B = 30°$.

3. Compute the value of $\cos 105°$ from the functions of 60° and 45°.

4. Compute the value of $\cos 75°$ from the functions of 30° and 45°.

5. Given $\sin A = \frac{4}{5}$, with A in Q II, and $\cos B = \frac{12}{13}$, with B in Q I. Find (a) $\cos(A + B)$, and (b) $\cos (A - B)$.

6. Given $\sin A = -\frac{4}{5}$, with A in Q III, and $\cos B = -\frac{1}{5}$, with B in Q II. Find (a) $\cos(A + B)$, and (b) $\cos(A - B)$.

Prove the identities in Exercises 7–10.

7. $\cos(180° + \theta) = -\cos\theta.$ **8.** $\cos(\tfrac{1}{2}\pi + \theta) = -\sin\theta.$

9. $\cos(\pi - \theta) = -\cos\theta.$ **10.** $\cos(\tfrac{3}{2}\pi - \theta) = -\sin\theta.$

Simplify the following by reducing each to a single term.

11. $\cos\tfrac{1}{3}\pi \cos\tfrac{2}{3}\pi - \sin\tfrac{1}{3}\pi \sin\tfrac{2}{3}\pi.$

12. $\cos\tfrac{4}{3}\pi \cos\tfrac{1}{3}\pi + \sin\tfrac{4}{3}\pi \sin\tfrac{1}{3}\pi.$

13. $\cos 3\theta \cos 2\theta + \sin 3\theta \sin 2\theta.$ **14.** $\cos 100° \cos 40° + \sin 100° \sin 40°.$

15. $\cos 50° \cos 10° - \sin 50° \sin 10°.$ **16.** $\cos\tfrac{3}{2}\theta \cos\tfrac{1}{2}\theta - \sin\tfrac{3}{2}\theta \sin\tfrac{1}{2}\theta.$

803   sin(A + B) AND sin(A − B)

$\sin A = \cos(90° - A)$
$\cos A = \sin(90° - A)$

If A and B are any two angles, then

(17) $\sin(A + B) = \sin A \cos B + \cos A \sin B,$ and

(18) $\sin(A - B) = \sin A \cos B - \cos A \sin B.$

Stated in words:

The sine of the sum of two angles is equal to the sine of the first angle times the cosine of the second, plus the cosine of the first angle times the sine of the second.

The sine of the difference of two angles is equal to the sine of the first angle times the cosine of the second, minus the cosine of the first angle times the sine of the second.

In Example 4, Section 802, it was shown that

(1) $\cos(90° - \theta) = \sin\theta$ for all values of $\theta.$

By substituting $(90° - \theta)$ for θ in (1), we obtain

(2) $\sin(90° - \theta) = \cos\theta$ for all values of $\theta.$

We substitute $(A + B)$ for θ in (1) to prove Identity (17).

$$\sin\theta = \cos(90° - \theta).$$

Then $\sin(A + B) = \cos[90° - (A + B)]$

$$= \cos[(90° - A) - B]$$

$$= \cos(90° - A) \cos B + \sin(90° - A) \sin B.$$

But $\cos(90° - A) = \sin A$ and $\sin(90° - A) = \cos A.$ Therefore

$$\sin(A + B) = \sin A \cos B + \cos A \sin B.$$

In Identity (17) we replace B by $(-B)$ to obtain

$$\sin[A + (-B)] = \sin A \cos(-B) + \cos A \sin(-B).$$

But $\cos(-B) = \cos B$ and $\sin(-B) = -\sin B$; therefore

$$\sin(A - B) = \sin A \cos B - \cos A \sin B.$$

Identities (15), (16), (17), and (18) are basic identities and should be memorized before proceeding any farther.

804 $\tan(A + B)$ AND $\tan(A - B)$

The following identities are true for all values of the angles A and B for which both members are defined:

(19) $$\tan(A + B) = \frac{\tan A + \tan B}{1 - \tan A \tan B},$$

(20) $$\tan(A - B) = \frac{\tan A - \tan B}{1 + \tan A \tan B}.$$

Stated in words:

The tangent of the sum of two angles is equal to the tangent of the first angle plus the tangent of the second, divided by 1 minus the product of the tangents of the two angles (whenever all of these tangents are defined).

The tangent of the difference of two angles is equal to the tangent of the first angle minus the tangent of the second, divided by 1 plus the product of the tangents of the two angles (whenever all of these tangents are defined).

Proof. Our proof depends on the identity $\tan \theta = \sin \theta / \cos \theta$; Identities (15), (16), (17), and (18); and the formulas for $\sin(A \pm B)$ and $\cos(A \pm B)$. Thus, for (19), we set

$$\tan(A + B) = \frac{\sin(A + B)}{\cos(A + B)} = \frac{\sin A \cos B + \cos A \sin B}{\cos A \cos B - \sin A \sin B}.$$

We divide the numerator and denominator of this fraction by $\cos A \cos B$, and obtain

$$\tan(A + B) = \frac{\left(\dfrac{\sin A \cos B}{\cos A \cos B}\right) + \left(\dfrac{\cos A \sin B}{\cos A \cos B}\right)}{\left(\dfrac{\cos A \cos B}{\cos A \cos B}\right) - \left(\dfrac{\sin A \sin B}{\cos A \cos B}\right)}$$

$$= \frac{\left(\dfrac{\sin A}{\cos A}\right) + \left(\dfrac{\sin B}{\cos B}\right)}{1 - \left(\dfrac{\sin A}{\cos A}\right) \cdot \left(\dfrac{\sin B}{\cos B}\right)} = \frac{\tan A + \tan B}{1 - \tan A \tan B}.$$

For Identity (20), we make a similar reduction of

$$\tan(A - B) = \frac{\sin(A - B)}{\cos(A - B)}$$

$$= \frac{\sin A \cos B - \cos A \sin B}{\cos A \cos B + \sin A \sin B} = \frac{\tan A - \tan B}{1 + \tan A \tan B}.$$

Example. Find tan 15° from the known values of the functions of 60° and 45°.

Solution. $\tan 15° = \tan(60° - 45°)$

$$= \frac{\tan 60° - \tan 45°}{1 + \tan 60° \tan 45°}$$

$$= \frac{\sqrt{3} - 1}{1 + \sqrt{3}} = \frac{(\sqrt{3} - 1)(\sqrt{3} - 1)}{(1 + \sqrt{3})(\sqrt{3} - 1)}$$

$$= \frac{3 - 2\sqrt{3} + 1}{3 - 1} = \frac{4 - 2\sqrt{3}}{2}$$

$$= 2 - \sqrt{3} \approx 0.2679.$$

EXERCISES

1. Given $\sin A = -\frac{3}{5}$, with A in Q III, and $\tan B = \frac{4}{3}$, with B in Q I, find the sine, cosine, and tangent of $(A + B)$ and of $(A - B)$, and indicate the quadrant for each of the angles $(A + B)$ and $(A - B)$.

2. Find the sine, cosine, and tangent of $\frac{2}{3}\pi$ from functions of π and $\frac{1}{3}\pi$.

3. Find the sine, cosine, and tangent of $\frac{4}{3}\pi$ from functions of π and $\frac{1}{3}\pi$.

4. Find the sine, cosine, and tangent of 105° from functions of 60° and 45°.

   Prove the following identities.

5. $\tan(\frac{3}{4}\pi - \theta) = \dfrac{\tan \theta + 1}{\tan \theta - 1}.$ 6. $\sin(\pi + \theta) = -\sin \theta.$

7. $\tan(\pi + \theta) = \tan \theta.$ 8. $\sin(\frac{3}{2}\pi - \theta) = -\cos \theta.$

9. $\tan(\frac{1}{4}\pi + \theta) = \dfrac{1 + \tan \theta}{1 - \tan \theta}.$ 10. $\cos(\frac{3}{2}\pi + \theta) = \sin \theta.$

11. $\sin(\theta - \frac{1}{2}\pi) = -\cos \theta.$

12. $\sin(x + y) + \sin(x - y) = 2 \sin x \cos y.$

13. $\sin(x + y) - \sin(x - y) = 2 \cos x \sin y.$

+ identity (20)
  proof

14. $\cos(x+y) - \cos(x-y) = -2 \sin x \sin y$.

15. $\cos(x+y) + \cos(x-y) = 2 \cos x \cos y$.

16. $\sin(x-y) \cos y + \cos(x-y) \sin y = \sin x$.

17. $\sin(x+y) \sin y + \cos(x+y) \cos y = \cos x$.

In each of the Exercises 18–27, simplify the given expression, reducing it to a single term.

18. $\sin 7x \cos 6x - \cos 7x \sin 6x$.

19. $\cos 5x \cos 4x - \sin 5x \sin 4x$.

20. $\dfrac{\tan x + \tan 3x}{1 - \tan x \tan 3x}$.

21. $\dfrac{\tan x - \tan 2x}{1 + \tan x \tan 2x}$.

22. $\cos^2 x - \sin^2 x$. (Write as $\cos x \cos x - \sin x \sin x$.)

23. $2 \sin x \cos x$. (Write as $\sin x \cos x + \cos x \sin x$.)

24. $2 \sin 3x \cos 3x$. (Write as $\sin 3x \cos 3x + \cos 3x \sin 3x$.)

25. $\cos^2 3x - \sin^2 3x$. (Write as $\cos 3x \cos 3x - \sin 3x \sin 3x$.)

26. $\sin x \sin(x+y) + \cos x \cos(x+y)$.

27. $\sin(\theta + 60°) - \cos(\theta + 30°)$.

28. Replace B by A in Identity (17) and thus show $\sin 2A = 2 \sin A \cos A$.

29. Replace B by A in Identity (15) and thus show $\cos 2A = \cos^2 A - \sin^2 A$.

30. Replace B by A in Identity (19) and thus show

$$\tan 2A = \frac{2 \tan A}{1 - \tan^2 A}.$$

805   TRIGONOMETRIC FUNCTIONS OF TWICE AN ANGLE

If we replace B by A in Identities (15), (17), and (19), $(A + B)$ becomes equal to $2A$, and we obtain the following important identities.

(21) $\qquad\qquad\qquad \sin 2A = 2 \sin A \cos A$.

(22a) $\qquad\qquad\qquad \cos 2A = \cos^2 A - \sin^2 A$

(22b) $\qquad\qquad\qquad\qquad\quad = 1 - 2 \sin^2 A$

(22c) $\qquad\qquad\qquad\qquad\quad = 2 \cos^2 A - 1$.

(23)
$$\tan 2A = \frac{2 \tan A}{1 - \tan^2 A}.$$

The second and third forms for cos 2A are obtained by setting

$$\cos 2A = \cos^2 A - \sin^2 A$$
$$= (1 - \sin^2 A) - \sin^2 A$$
$$= 1 - 2 \sin^2 A,$$

and

$$\cos 2A = \cos^2 A - \sin^2 A$$
$$= \cos^2 A - (1 - \cos^2 A)$$
$$= 2 \cos^2 A - 1.$$

Stated in words:

The sine of twice an angle is equal to 2 times the product of the sine and the cosine of the angle.

The cosine of twice an angle is equal to the square of the cosine of the angle minus the square of the sine of the angle.

The tangent of twice an angle is equal to twice the tangent of the angle divided by 1 minus the square of the tangent of the angle.

Example 1.   Show that Identities (21), (22), and (23) are true for $A = 30°$.

Solution.

(21)
$$\sin 2A = 2 \sin A \cos A.$$
$$\sin 60° = 2 \sin 30° \cos 30°,$$
$$\tfrac{1}{2}\sqrt{3} = 2(\tfrac{1}{2})(\tfrac{1}{2}\sqrt{3}),$$
$$\tfrac{1}{2}\sqrt{3} = \tfrac{1}{2}\sqrt{3}.$$

(22)
$$\cos 2A = \cos^2 A - \sin^2 A.$$
$$\cos 60° = \cos^2 30° - \sin^2 30°,$$
$$\tfrac{1}{2} = (\tfrac{1}{2}\sqrt{3})^2 - (\tfrac{1}{2})^2,$$
$$\tfrac{1}{2} = \tfrac{3}{4} - \tfrac{1}{4}.$$

(23)
$$\tan 2A = \frac{2 \tan A}{1 - \tan^2 A}.$$

$$\tan 60° = \frac{2 \tan 30°}{1 - \tan^2 30°},$$

$$\sqrt{3} = \frac{2(\tfrac{1}{3}\sqrt{3})}{1 - (\tfrac{1}{3}\sqrt{3})^2} = \frac{3}{\sqrt{3}} = \sqrt{3}.$$

Example 2. If $\tan \theta = \frac{4}{3}$ and θ is in the first quadrant, find $\sin 2\theta$, $\cos 2\theta$, and $\tan 2\theta$.

Solution. Sketch angle θ in standard position (Fig. 804); then, after finding $r = 5$, we get $\sin \theta = \frac{4}{5}$ and $\cos \theta = \frac{3}{5}$.

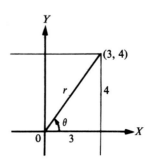

Figure 804

Then

$$\sin 2\theta = 2 \sin \theta \cos \theta = 2(\tfrac{4}{5})(\tfrac{3}{5}) = \tfrac{24}{25},$$
$$\cos 2\theta = \cos^2 \theta - \sin^2 \theta = (\tfrac{3}{5})^2 - (\tfrac{4}{5})^2 = -\tfrac{7}{25}.$$

By dividing the first of these two results by the second, we get

$$\tan 2\theta = \frac{\sin 2\theta}{\cos 2\theta} = -\frac{24}{7}.$$

The $\tan 2\theta$ could also have been obtained by use of Identity (23).

Example 3. Express $\cos^4\theta$ in terms of first degree functions of multiple angles.

Solution. From Identity (22c)

$$2 \cos^2 \theta - 1 = \cos 2\theta,$$

we get

$$\cos^2 \theta = \frac{1 + \cos 2\theta}{2}.$$

Then

$$\cos^4 \theta = (\cos^2 \theta)^2 = \left(\frac{1 + \cos 2\theta}{2}\right)^2$$

$$= \tfrac{1}{4}(1 + 2 \cos 2\theta + \cos^2 2\theta).$$

But

$$\cos^2 2\theta = \frac{1 + \cos 4\theta}{2}.$$

Therefore $\qquad \cos^4 \theta = \frac{1}{4}\left(1 + 2\cos 2\theta + \dfrac{1 + \cos 4\theta}{2}\right)$

$$= \tfrac{1}{8}(3 + 4\cos 2\theta + \cos 4\theta).$$

Example 4. Express $\sin 3\theta$ in terms of $\sin \theta$.

Solution. $\sin 3\theta = \sin(2\theta + \theta) = \sin 2\theta \cos \theta + \cos 2\theta \sin \theta$

$$= (2\sin \theta \cos \theta)\cos \theta + (1 - 2\sin^2 \theta)\sin \theta$$

$$= 2\sin \theta \cos^2 \theta + \sin \theta - 2\sin^3 \theta$$

$$= 2\sin \theta (1 - \sin^2 \theta) + \sin \theta - 2\sin^3 \theta$$

$$= 3\sin \theta - 4\sin^3 \theta.$$

Example 5. Prove the identity $\dfrac{\sin 3x}{\sin x} - \dfrac{\cos 3x}{\cos x} = 2.$

Solution. $\dfrac{\sin 3x}{\sin x} - \dfrac{\cos 3x}{\cos x} = \dfrac{\sin 3x \cos x - \cos 3x \sin x}{\sin x \cos x}$

$$= \dfrac{\sin(3x - x)}{\sin x \cos x} = \dfrac{\sin 2x}{\sin x \cos x}$$

$$= \dfrac{2\sin x \cos x}{\sin x \cos x} = 2.$$

806   TRIGONOMETRIC FUNCTIONS OF HALF AN ANGLE

By using the last two forms of Identity (22) of Section 805 for the cosine of twice an angle, we can prove the following identities:

(24) $\qquad\qquad \sin \tfrac{1}{2}\theta = \pm \sqrt{\dfrac{1 - \cos \theta}{2}},$

(25) $\qquad\qquad \cos \tfrac{1}{2}\theta = \pm \sqrt{\dfrac{1 + \cos \theta}{2}},$

(26) $\qquad\qquad \tan \tfrac{1}{2}\theta = \pm \sqrt{\dfrac{1 - \cos \theta}{1 + \cos \theta}} \qquad$ unless $\cos \theta = -1.$

The double sign in the last member is to be interpreted as either $+$ or $-$; not both signs simultaneously. For example if $90° < \tfrac{1}{2}\theta < 180°$ then $\sin \tfrac{1}{2}\theta$ is positive, $\cos \tfrac{1}{2}\theta$ is negative, and $\tan \tfrac{1}{2}\theta$ is negative.

As an aid in memorizing these half-angle identities it is helpful to remember: (1) that each identity is written in terms of the cosine: (2) that the sign preceding the cosine, under the radical, is minus for the sine of half an angle and plus for the cosine of half an angle: and (3) that the sign before the radical is determined by the quadrant in which the half-angle lies.

Proof.   To obtain Identity (24) we use Identity (22b)

$$\cos 2A = 1 - 2 \sin^2 A,$$

and get $2 \sin^2 A = 1 - \cos 2A.$

Then by substituting $\tfrac{1}{2}\theta$ for A we obtain

$$2 \sin^2 \tfrac{1}{2}\theta = 1 - \cos 2(\tfrac{1}{2}\theta),$$

or $$\sin \tfrac{1}{2}\theta = \pm \sqrt{\frac{1 - \cos \theta}{2}}.$$

The derivations of Identities (25) and (26), from Identities (22c) and (24), are left as exercises for the student.

Example 1.   Given that $90° < \theta < 180°$ and that $\sin \theta = \tfrac{3}{5}$, find $\sin \tfrac{1}{2}\theta$, $\cos \tfrac{1}{2}\theta$, and $\tan \tfrac{1}{2}\theta$.

Solution.   In order to apply the half-angle identities we must find $\cos \theta$ and determine the quadrant in which the half-angle lies. We draw a triangle of reference (Fig. 805), compute $x = -4$, and obtain $\cos \theta = -\tfrac{4}{5}$.

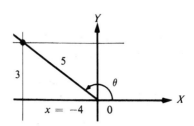

Figure 805

Since $90° < \theta < 180°$, $45° < \tfrac{1}{2}\theta < 90°$, so that the functions of $\tfrac{1}{2}\theta$ are positive, but $\cos \theta < 0$ and by Identity (24)

$$\sin \tfrac{1}{2}\theta = \sqrt{\frac{1 - (-\tfrac{4}{5})}{2}} = \sqrt{\frac{9}{10}} = \frac{3}{10}\sqrt{10}.$$

By making use of Identity (25) we obtain

$$\cos \tfrac{1}{2}\theta = \sqrt{\frac{1 + (-\tfrac{4}{5})}{2}} = \sqrt{\frac{1}{10}} = \frac{1}{10}\sqrt{10}.$$

By making use of Identity (26), or $\tan \tfrac{1}{2}\theta = \dfrac{\sin \tfrac{1}{2}\theta}{\cos \tfrac{1}{2}\theta}$, we find

$$\tan \tfrac{1}{2}\theta = 3.$$

Example 2.   Prove the identity

$$\cos^2 \tfrac{1}{2}x = \frac{\tan x + \sin x}{2 \tan x}.$$

Solution.   We shall reduce the right-hand member of the identity to the form of the left-hand member.

$$\frac{\tan x + \sin x}{2 \tan x} = \frac{\dfrac{\sin x}{\cos x} + \sin x}{2\dfrac{\sin x}{\cos x}} = \frac{\sin x \,(1 + \cos x)}{2 \sin x} = \frac{1 + \cos x}{2}$$

$$= \left(\sqrt{\frac{1 + \cos x}{2}}\right)^2 = \cos^2 \tfrac{1}{2}x, \text{ by Identity (25).}$$

Example 3.   Solve the equation

$$3 \cos x - 1 = 2 \sin^2 \tfrac{1}{2}x$$

for all possible values of x in the interval $0° \le x < 360°$.

Solution.   Express $\sin \tfrac{1}{2}x$ in terms of x and proceed as follows:

$$3 \cos x - 1 = 2 \sin^2 \tfrac{1}{2}x$$

$$= 2\left(\frac{1 - \cos x}{2}\right) = 1 - \cos x.$$

Thus $3 \cos x - 1 = 1 - \cos x,$

and $\cos x = \tfrac{1}{2}.$

Therefore $x = 60°$ and $300°$.

EXERCISES

In Exercises 1–4, use the tables to verify each statement.

1. $\sin 80° \ne 2 \sin 40°$.   **2.** $\tan 86° \ne 2 \tan 43°$.

3. $\cos 70° \neq 2 \cos 35°.$ **4.** $\sin 70° \neq 2 \sin 35°.$

5. By substitution, verify Identities (21), (22), and (23) for $A = 60°$.

6. By substitution, verify Identities (24), (25), and (26) for $\theta = 60°$.

7. Find the values of sine, cosine, and tangent of $105°$ using the value of $\cos 210° = -\frac{1}{2}\sqrt{3}$. Express these values in simplified radical form.

8. Find the values of sine, cosine, and tengent of $157\frac{1}{2}°$ using the value of $\cos 315° = \frac{1}{2}\sqrt{2}$. Express these values in simplified radical form.

In Exercises 9–20, simplify each expression by reducing it to a single term involving only one function of an angle.

9. $2 \sin 40° \cos 40°.$ **10.** $\cos^2 40° - \sin^2 40°.$

11. $\cos^2 50° - \sin^2 50°.$ **12.** $2 \sin 20° \cos 20°.$

13. $\dfrac{2 \tan 10°}{1 - \tan^2 10°}.$ **14.** $\sqrt{\dfrac{1 - \cos 80°}{2}}.$

15. $\sqrt{\dfrac{1 + \cos 70°}{2}}.$ **16.** $\dfrac{2 \tan 3\alpha}{1 - \tan^2 3\alpha}.$

17. $6 \sin 3\alpha \cos 3\alpha.$ **18.** $\sin \frac{1}{2}\beta \cos \frac{1}{2}\beta.$

19. $5 \sqrt{\dfrac{1 - \cos 5\beta}{1 + \cos 5\beta}}.$ **20.** $2 \cos^2 5\beta - 1.$

21. Given that $270° < x < 360°$ and that $\tan x = -\frac{5}{12}$, without using tables find $\sin 2x$, $\cos 2x$, $\tan 2x$, $\sin \frac{1}{2}x$, $\cos \frac{1}{2}x$, and $\tan \frac{1}{2}x$, and name the quadrants in which $2x$ and $\frac{1}{2}x$ lie.

22. Find the value of $z = 8 \cos^2 \theta - 24 \sin \theta \cos \theta + \sin^2 \theta$, when $\tan 2\theta = -\frac{24}{7}$ and $270° < 2\theta < 360°$.

In Exercises 23–40 prove the identities, operating on one member and leaving the other unchanged.

23. $\dfrac{2 \cot x}{\csc^2 x - 2} = \tan 2x.$

24. $\dfrac{1 - \tan^2 x}{1 + \tan^2 x} = \cos 2x.$

25. $\cos 3x = 4 \cos^3 x - 3 \cos x.$

26. $\dfrac{\sin 5x}{\sin x} - \dfrac{\cos 5x}{\cos x} = 4 \cos 2x.$

27. $\dfrac{1 - \tan \frac{1}{2}x}{1 + \tan \frac{1}{2}x} = \dfrac{1 - \sin x}{\cos x}.$

28. $\sin^4 x = \frac{1}{8}(3 - 4 \cos 2x + \cos 4x).$

29. $\dfrac{\cos^3 x + \sin^3 x}{\cos x + \sin x} = 1 - \frac{1}{2}\sin 2x.$

30. $\dfrac{1 - \tan \frac{1}{2}x}{1 + \tan \frac{1}{2}x} = \dfrac{\cos x}{1 + \sin x}.$

31. $\tan(x + 45°) + \tan(x - 45°) = 2\tan 2x.$

32. $\tan(x + 45°) - \tan(x - 45°) = 2\sec 2x.$

33. $2(\cos 3x \cos x + \sin 3x \sin x)^2 = 1 + \cos 4x.$

34. $2(\sin 2x \cos x - \cos 2x \sin x)^2 = 1 - \cos 2x.$

35. $\dfrac{\sin 2x}{2\sin x} = \cos^2 \frac{1}{2}x - \sin^2 \frac{1}{2}x.$

36. $\cos^2 2x - \cos^2 x = \sin^2 x - \sin^2 2x.$

37. $\dfrac{1 + \sin 2x + \cos 2x}{1 + \sin 2x - \cos 2x} = \cot x.$

38. $\tan \frac{1}{2}x = \dfrac{\sec x - 1}{\tan x}.$

39. $\frac{1}{2}[\sin(x + y) + \sin(x - y)] = \sin x \cos y.$

40. $\frac{1}{2}[\cos(x + y) + \cos(x - y)] = \cos x \cos y.$

807   PRODUCTS OF SINES AND COSINES EXPRESSED AS SUMS

The identities that we wish to develop in this section are:

(27) $\qquad\qquad \sin A \cos B = \frac{1}{2}[\sin(A + B) + \sin(A - B)],$

(28) $\qquad\qquad \cos A \sin B = \frac{1}{2}[\sin(A + B) - \sin(A - B)],$

(29) $\qquad\qquad \sin A \sin B = \frac{1}{2}[\cos(A - B) - \cos(A + B)],$

(30) $\qquad\qquad \cos A \cos B = \frac{1}{2}[\cos(A + B) + \cos(A - B)].$

Examples:

1. $\sin 3\theta \cos 2\theta = \frac{1}{2}[\sin(3\theta + 2\theta) + \sin(3\theta - 2\theta)]$

$\qquad\qquad\quad = \frac{1}{2}[\sin 5\theta + \sin \theta].$

2. $\cos 55° \cos 15° = \frac{1}{2}[\cos(55° + 15°) + \cos(55° - 15°)]$

$\qquad\qquad\qquad = \frac{1}{2}[\cos 70° + \cos 40°].$

These identities, and those of Section 808, are important in certain aspects of calculus and applied mathematics; however in this book we shall make little use of them, and it is not necessary to memorize them. The student should, however, be able to derive them. The derivation is not difficult.

To derive Identity (27) we add Identities (17) and (18) of Section 803.

(17) $\sin(A + B) = \sin A \cos B + \cos A \sin B$

(18) $\dfrac{\sin(A - B) = \sin A \cos B - \cos A \sin B}{\sin(A + B) + \sin(A - B) = 2 \sin A \cos B.}$

Dividing by 2, we obtain Identity (27)

$$\sin A \cos B = \tfrac{1}{2}[\sin(A + B) + \sin(A - B)].$$

Subtracting Identity (18) from Identity (17) we get

$$\sin(A + B) - \sin(A - B) = 2 \cos A \sin B.$$

We divide by 2 to obtain Identity (28).

$$\cos A \sin B = \tfrac{1}{2}[\sin(A + B) - \sin(A - B)].$$

To derive Identity (30) we add Identities (15) and (16) of Section 802.

(15) $\cos(A + B) = \cos A \cos B - \sin A \sin B$

(16) $\dfrac{\cos(A - B) = \cos A \cos B + \sin A \sin B}{\cos(A + B) + \cos(A - B) = 2 \cos A \cos B.}$

We divide by 2 to obtain Identity (30).

$$\cos A \cos B = \tfrac{1}{2}[\cos(A + B) + \cos(A - B)].$$

To derive Identity (29) we subtract Identity (16) from Identity (15) and obtain

$$\cos(A + B) - \cos(A - B) = -2 \sin A \sin B.$$

We divide by -2 to obtain identity (29).

$$\sin A \sin B = \tfrac{1}{2}[\cos(A - B) - \cos(A + B)].$$

808   SUMS OF SINES AND COSINES EXPRESSED AS PRODUCTS

In Identities (27), (28), (29), (30) we let $A + B = x$ and $A - B = y$. Then upon adding and subtracting these two equations

$$A + B = x \qquad A + B = x$$
$$\frac{A - B = y}{2A \quad = x + y} \qquad \frac{A - B = y}{2B = x - y}$$

we get $\qquad A = \tfrac{1}{2}(x + y) \qquad$ and $\qquad B = \tfrac{1}{2}(x - y).$

Using these values for $A + B$, $A - B$, A, and B in the same identities, we get

(31) $\qquad\qquad \sin x + \sin y = 2 \sin \tfrac{1}{2}(x + y) \cos \tfrac{1}{2}(x - y),$

(32) $\qquad\qquad \sin x - \sin y = 2 \cos \tfrac{1}{2}(x + y) \sin \tfrac{1}{2}(x - y),$

(33) $\qquad\qquad \cos x + \cos y = 2 \cos \tfrac{1}{2}(x + y) \cos \tfrac{1}{2}(x - y),$

(34) $\qquad\qquad \cos x - \cos y = -2 \sin \tfrac{1}{2}(x + y) \sin \tfrac{1}{2}(x - y).$

Example 1. Prove the identity $\dfrac{\sin 4\theta - \sin 2\theta}{\cos 4\theta + \cos 2\theta} = \tan \theta.$

Solution. Applying Identity (32) to the numerator and Identity (33) to the denominator of the left-hand member, we find

$$\frac{\sin 4\theta - \sin 2\theta}{\cos 4\theta + \cos 2\theta} = \frac{2 \cos \tfrac{1}{2}(4\theta + 2\theta) \sin \tfrac{1}{2}(4\theta - 2\theta)}{2 \cos \tfrac{1}{2}(4\theta + 2\theta) \cos \tfrac{1}{2}(4\theta - 2\theta)}$$

$$= \frac{2 \cos 3\theta \sin \theta}{2 \cos 3\theta \cos \theta} = \tan \theta.$$

Example 2. Solve the equation $\sin 4x + \sin 2x = \cos x$ for all values of x in the interval $0° \le x < 360°$.

Solution. Applying Identity (31) to the left-hand member of the equation, we get

$$\sin 4x + \sin 2x = \cos x,$$
$$2 \sin 3x \cos x = \cos x.$$

Then $\qquad\qquad\qquad 2 \sin 3x \cos x - \cos x = 0,$

or $\qquad\qquad\qquad \cos x(2 \sin 3x - 1) = 0.$

This is equivalent to the two equations

$$\cos x = 0 \qquad \text{and} \qquad \sin 3x = \tfrac{1}{2}.$$

From $\cos x = 0$ we get $x = 90°$ and $270°$.
From $\sin 3x = \tfrac{1}{2}$ we get $3x = 30° + n360°$ and $3x = 150° + n360°$

or $\qquad\qquad x = 10° + n120° \qquad$ and $\qquad x = 50° + n120°.$

Therefore

$$x = 10°, 130°, 250° \qquad \text{and} \qquad x = 50°, 170°, 290°.$$

The values of x are

$$x = 10°, \ 50°, \ 90°, \ 130°, \ 170°, \ 250°, \ 270°, \text{ and } 290°,$$

all of which satisfy the original equation.

EXERCISES

In Exercises 1–6, transform each of the products into a sum or difference of sines or cosines.

1. $2 \sin 35° \cos 25°$.   **2.** $2 \cos 55° \cos 35°$.

3. $6 \cos 5\theta \cos 3\theta$.   **4.** $8 \sin 7\theta \cos \theta$.

5. $\frac{3}{2} \sin 2\beta \sin \beta$.   **6.** $\frac{1}{2} \cos 3\beta \sin 2\beta$.

In Exercises 7–12, transform each of the sums or differences into a product of sines and cosines.

7. $\sin 40° + \sin 30°$.   **8.** $\cos 50° + \cos 20°$.

9. $\cos 3\theta + \cos 5\theta$.   **10.** $\sin 5\theta - \sin 7\theta$.

11. $\sin \frac{7}{2}\alpha - \sin \frac{3}{2}\alpha$.   **12.** $\sin \frac{5}{2}\alpha + \sin \frac{1}{2}\alpha$.

In Exercises 13–30, prove the identities by operating on one member and leaving the other unchanged.

13. $\sin(150° + x) + \sin(150° - x) = \cos x$.

14. $\cos(135° + x) - \cos(135° - x) = -\sqrt{2} \sin x$.

15. $\cos(45° + x) + \cos(45° - x) = \sqrt{2} \cos x$.

16. $\sin(30° + x) + \sin(30° - x) = \cos x$.

17. $\cos(x + 30°) - \cos(x - 30°) = -\sin x$.

18. $\sin 2x + 2 \sin 4x + \sin 6x = 4 \cos^2 x \sin 4x$.

19. $\sin^2 6x - \sin^2 4x = \sin 2x \sin 10x$.

20. $\cos^2 2x - \cos^2 6x = \sin 4x \sin 8x$.

21. $\sin(2x - y) \cos y + \cos(2x - y) \sin y = \sin 2x$.

22. $\cos(2x + y) \cos y + \sin(2x + y) \sin y = \cos 2x$.

23. $\dfrac{\cos 70° + \cos 50°}{\sin 280° + \sin 260°} = -\frac{1}{2}$.

24. $\dfrac{\cos 9x - \cos 5x}{\sin 17x - \sin 3x} = -\sin 2x \sec 10x$.

25. $\dfrac{\sin 5x + \sin 3x}{\cos 5x + \cos 3x} = \tan 4x.$

26. $\dfrac{\sin \alpha - \sin \theta}{\cos \alpha + \cos \theta} = \tan \frac{1}{2}(\alpha - \theta).$

27. $\dfrac{\sin 40° - \sin 20°}{\cos 220° - \cos 200°} = \sqrt{3}.$

28. $\dfrac{\sin x - \sin 3x}{\sin^2 x - \cos^2 x} = 2 \sin x.$

29. $\dfrac{\sin 5x - 2 \sin 3x + \sin x}{\cos 5x - \cos x} =: \tan x.$

30. $\dfrac{(\sin 3x + \sin 9x) + (\sin 5x + \sin 7x)}{(\cos 3x + \cos 9x) + (\cos 5x + \cos 7x)} = \tan 6x.$

In Exercises 31–50, solve the equations for all values of x in the interval $0 \leq x < 2\pi$.

31. $\sin 3x = 1.$ **32.** $\cos 3x = 1.$

33. $\cos \frac{1}{2}x = 0.$ **34.** $\sin \frac{1}{2}x = 1.$

35. $\cos 2x \,(2 \cos x - 1) = 0.$ **36.** $\sin 2x \,(2 \sin x + 1) = 0.$

37. $\sin 2x \,(\tan x - 1) = 0.$ **38.** $\cos x \,(\tan x + 1) = 0.$

39. $\sin 2x - \sin x = 0.$ **40.** $\cos 2x + \cos x = 0.$

41. $\cos 3x + \cos x = 0.$ **42.** $\sin 3x - \sin x = 0.$

43. $\sin 2x - \cos x = 0.$ **44.** $\sin 4x + \sin 2x = \cos x.$

45. $\sin 3x + 4 \sin^2 x = 0.$ **46.** $\cos 7x - \cos x = 0.$

47. $\cos 5x + \cos 3x = 2 \cos 4x.$ **48.** $\sin 4x + \sin 2x = \cos x.$

49. $\sin x + \sin 3x + \sin 5x = 0.$ **50.** $\cos x + \cos 2x + \cos 3x = 0.$

809   REDUCTION OF $a \cos\theta + b \sin\theta$ TO $c \cos(\theta - \alpha)$

It is often desirable to transform a sum of two terms of the form $a \cos\theta + b \sin\theta$ into a single term such as $c \cos(\theta - \alpha)$.

We let the real nonzero numbers a and b, of the expression $a \cos\theta + b \sin\theta$, represent the coordinates of a point (a, b) on the rectangular coordinate plane. Let α be an angle in standard position with its terminal side passing through the point (a, b) (Fig. 806). Then

(1) $\sin \alpha = \dfrac{b}{\sqrt{a^2 + b^2}}, \quad \cos \alpha = \dfrac{a}{\sqrt{a^2 + b^2}}, \quad \tan \alpha = \dfrac{b}{a}.$

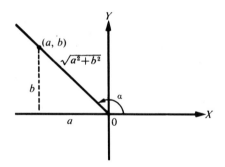

Figure 806

We multiply and divide the expression $a \cos \theta + b \sin \theta$ by $\sqrt{a^2 + b^2}$, substitute the values of $\sin \alpha$ and $\cos \alpha$ of (1) into the resulting expression, and apply Identity (17).

$$a \cos \theta + b \sin \theta = \sqrt{a^2 + b^2} \left[\frac{a}{\sqrt{a^2 + b^2}} \cos \theta + \frac{b}{\sqrt{a^2 + b^2}} \sin \theta \right]$$

$$= \sqrt{a^2 + b^2} \left[\cos \alpha \cos \theta + \sin \alpha \sin \theta \right]$$

$$= \sqrt{a^2 + b^2} \left[\cos \theta \cos \alpha + \sin \theta \sin \alpha \right]$$

$$= \sqrt{a^2 + b^2} \cos (\theta - \alpha).$$

Hence:

   If θ is any angle,

(35) $a \cos \theta + b \sin \theta = c \cos(\theta - \alpha),$

where $c = \sqrt{a^2 + b^2}, \; \sin \alpha = \dfrac{b}{c}, \; \cos \alpha = \dfrac{a}{c}, \; \tan \alpha = \dfrac{b}{a}.$

Example 1.   Transform $4 \cos \theta + 3 \sin \theta$ to the form $c \cos(\theta - \alpha)$, and find c and α.

Solution.   Plot the point $(4, 3)$ and place α in standard position with its terminal side passing through $(4, 3)$. (*See* Fig. 807a.) Then $c = \sqrt{4^2 + 3^2} = 5$, $\sin \alpha = \frac{3}{5}$, $\cos \alpha = \frac{4}{5}$, and $\tan \alpha = \frac{3}{4}$. Therefore $\alpha = 36° \; 52'$ or any angle coterminal with $36° \; 52'$. By applying Formula (35) we obtain

$$4 \cos \theta + 3 \sin \theta = 5 \cos(\theta - 36° \; 52').$$

To further clarify what Formula (35) is doing for us in this problem, we show the following development. We plot the point $(4, 3)$ and compute c; then

$$4 \cos \theta + 3 \sin \theta = 5(\tfrac{4}{5} \cos \theta + \tfrac{3}{5} \sin \theta)$$

$$= 5(\cos \alpha \cos \theta + \sin \alpha \sin \theta)$$

$$= 5 \cos(\theta - \alpha)$$

$$= 5 \cos(\theta - 36° \; 52').$$

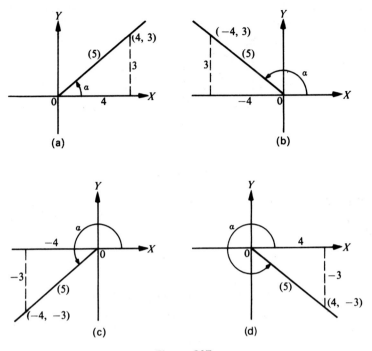

Figure 807

Example 2. Transform −4 cos θ + 3 sin θ to the form *c* cos(θ − α), and find *c* and α.

Solution. Plot the point (−4, 3), and place α in standard position with its terminal side passing through (−4, 3). (*See* Fig. 807b.) Then we have $c = \sqrt{(-4)^2 + 3^2} = 5$, sin α = ⅗, cos α = −⅘, and tan α = −¾. Therefore α = 143° 08′, and by applying Formula (35) we obtain

$$-4 \cos \theta + 3 \sin \theta = 5 \cos(\theta - 143° \, 08').$$

As an alternate approach to this transformation, we plot the point (−4, 3) and apply Formula (1) of Section 809 as follows:

$$
\begin{aligned}
-4 \cos \theta + 3 \sin \theta &= 5(-\tfrac{4}{5} \cos \theta + \tfrac{3}{5} \sin \theta) \\
&= 5(\cos \alpha \cos \theta + \sin \alpha \sin \theta) \\
&= 5 \cos(\theta - \alpha) \\
&= 5 \cos(\theta - 143° \, 08').
\end{aligned}
$$

Example 3. Transform −4 cos θ − 3 sin θ to the form *c* cos(θ − α), and find *c* and α.

Solution. Plot the point $(-4, -3)$, and place α in standard position with its terminal side passing through $(-4, -3)$. (*See* Fig. 807c.) Then $c = \sqrt{(-4)^2 + (-3)^2} = 5$, $\sin \alpha = -\frac{3}{5}$, $\cos \alpha = -\frac{4}{5}$, and $\tan \alpha = \frac{3}{4}$. Therefore $\alpha = 216°\ 52'$, and by applying Formula (35) we obtain

$$-4 \cos \theta - 3 \sin \theta = 5 \cos(\theta - 216°\ 52').$$

An alternate solution.

$$\begin{aligned} -4 \cos \theta - 3 \sin \theta &= 5(-\tfrac{4}{5} \cos \theta - \tfrac{3}{5} \sin \theta) \\ &= -5(\tfrac{4}{5} \cos \theta + \tfrac{3}{5} \sin \theta) \\ &= -5 \cos(\theta - 36°\ 52'). \end{aligned}$$

Example 4. Transform $4 \cos \theta - 3 \sin \theta$ to the form $c \cos(\theta - \alpha)$, and find c and α.

Solution. Plot the point $(4, -3)$, and place α in standard position with its terminal side passing through $(4, -3)$. (*See* Fig. 807d.) Then $c = \sqrt{4^2 + (-3)^2} = 5$, $\sin \alpha = -\frac{3}{5}$, $\cos \alpha = \frac{4}{5}$, and $\tan \alpha = -\frac{3}{4}$. Therefore $\alpha = 323°\ 08'$, and when we apply Formula (35) we obtain

$$4 \cos \theta - 3 \sin \theta = 5 \cos(\theta - 323°\ 08').$$

Example 5. Show that the maximum value of the function $a \cos \theta + b \sin \theta = c \cos(\theta - \alpha)$ is $\sqrt{a^2 + b^2}$ and occurs when $\theta = \alpha$.

Solution. The maximum value of $a \cos \theta + b \sin \theta = c \cos(\theta - \alpha)$ occurs when $\cos(\theta - \alpha)$ equals its greatest value, 1. When

$$\cos(\theta - \alpha) = 1,$$
$$\theta - \alpha = 0,$$

and
$$\theta = \alpha.$$

Hence the maximum value of the function $a \cos \theta + b \sin \theta$ is $\sqrt{a^2 + b^2} = c$, which occurs when $\theta = \alpha$. Of course the function $c \cos(\theta - \alpha)$ is periodic and has an unlimited number of values of θ, all of which yield the maximum value of c for the function. These values of θ can be found as follows:

$$\cos(\theta - \alpha) = 1$$
$$\theta - \alpha = n360°, \qquad (n \text{ is an integer})$$
$$\theta = \alpha + n360°.$$

Example 6. Show that the minimum value of the function $a \cos \theta + b \sin \theta = c \cos(\theta - \alpha)$ is $-\sqrt{a^2 + b^2}$ and occurs when $\theta = \alpha + (2n + 1)180°$ (n is an integer).

Solution. The minimum value of the function

$$a \cos \theta + b \sin \theta = c \cos(\theta - \alpha)$$

occurs when $\cos(\theta - \alpha)$ has its least value, -1.

When $\quad \cos(\theta - \alpha) = -1,$
$$\theta - \alpha = 180° + n360°, \qquad (n \text{ is an integer})$$
$$\theta = \alpha + 180° + n360°,$$
$$\theta = \alpha + (2n + 1)180°.$$

By applying Formula (35) we can quickly transform the function $a \cos \theta + b \sin \theta$ (a composite wave form with two terms) into an equivalent function with wave form and a single term, from which we can easily determine the maximum, minimum, and periodic properties of the function.

EXERCISES

In Exercises 1–10, transform each expression to the form $c \cos(\theta - \alpha)$, and determine c and a value of α in the interval $0 < \alpha < 360°$.

1. $5 \cos \theta + 12 \sin \theta.$ **2.** $12 \cos \theta + 5 \sin \theta.$

3. $5 \cos \theta - 12 \sin \theta.$ **4.** $12 \cos \theta - 5 \sin \theta.$

5. $-12 \cos \theta - 5 \sin \theta.$ **6.** $-5 \cos \theta - 12 \sin \theta.$

7. $-12 \cos \theta + 5 \sin \theta.$ **8.** $-5 \cos \theta + 12 \sin \theta.$

9. $3 \sin \theta - 2 \cos \theta.$ **10.** $2 \sin \theta - 3 \cos \theta.$

11. Find the maximum value of the function $\cos \theta - 2 \sin \theta$ and a value of θ in the interval $0 < \theta < 360°$ at which this maximum occurs.

12. Find the minimum value of the function $-24 \cos \theta + 7 \sin \theta$ and the smallest positive value of θ at which this minimum first occurs.

13. Find the smallest positive value of θ for which $-5 \cos \theta + 12 \sin \theta = 0.$

14. Find the smallest positive value of θ for which $5 \cos \theta - 12 \sin \theta = 0.$

15. Let the real nonzero numbers a and b of the expression $a \cos \theta + b \sin \theta$ represent the coordinates of the point (b, a) on the rectangular coordinate plane. Give the derivation of the formula

$$a \cos \theta + b \sin \theta = k \sin(\theta + \beta)$$

where $k = \sqrt{a^2 + b^2}$, $\sin \beta = a/k$, $\cos \beta = b/k$, and $\tan \beta = a/b$.

16. Express the right-hand member of the expression

$$V = 100 \sin 500t + 200 \cos 500t$$

as a single term, and give the maximum value of V.

9

Graphical methods

901   GENERAL TECHNIQUE

In Sections 217, 309, 406, 504, 505, and 506 we discussed the basic graphs of the sine, cosine, tangent, contagent, secant, and cosecant functions; in Section 310 we discussed graphing by the addition of ordinates, sometimes called, the composition of ordinates. We shall now deal with graphs of the more general forms of these functions.

Part of our previous method of graphing trigonometric functions was to assume arbitrary units of measure for the degree along the horizontal axis. Now that we have a knowledge of radian measure we shall, in general, use identical units of measure on both the horizontal and vertical axis.

In graphing the more general forms of the six basic trigonometric functions, we shall in most cases (1) locate some of the consecutive x-intercepts (the points where the curve crosses the x-axis); (2) find the upper and lower bounds, if these exist, or the vertical lines, if they exist, which separate the branches of the curve; (3) locate a few points on the curve; and (4) then sketch the balance of the curve from our previous knowledge of the general shape of the curve in question.

902   ANALYSIS OF THE GENERAL SINE AND COSINE CURVES

The graph of $y = a \sin bx$, where a and b are constants, is illustrated in the following example.

Example 1.   Graph the function $y = 2 \sin 3x$, where x is expressed in radian measure, and units are the same on the horizontal and vertical axes.

Solution.   We are familiar with the general shape of the graph of $y = \sin \theta$ (Figs. 226 and 227, Section 217), so our concern here is to: (1) locate some of the consecutive x-intercepts; (2) determine the upper and lower bounds; (3) locate a few points on the curve; and then, (4) use this information to sketch the curve. We set $y = 0$ and find some of the consecutive x-intercepts.

$$2 \sin 3x = 0$$

$$\sin 3x = 0,$$

$$3x = n\pi \qquad (n \text{ an integer}),$$

$$x = n\tfrac{1}{3}\pi.$$

Therefore some of the x-intercepts are

$$-\pi, \; -\tfrac{2}{3}\pi, \; -\tfrac{1}{3}\pi, \; 0, \; \tfrac{1}{3}\pi, \; \tfrac{2}{3}\pi, \; \pi, \text{ etc. } (\textit{See } \text{Fig. 901.})$$

Since $\pi \approx 3.1416$, $\tfrac{1}{3}\pi \approx \tfrac{1}{3}(3.1416) = 1.0472 \approx 1$. Therefore we shall make only a scarcely discernible distortion of the curves if, in graphing the trigonometric functions, we use one horizontal unit of distance to respresent $\tfrac{1}{3}\pi$. We agree to make this convenient close approximation.

The amplitude (greatest value of $y = 2 \sin 3x$) is 2, therefore this curve oscillates back and forth between $y = 2$ and $y = -2$. We know now where the curve crosses the x-axis and its upper and lower bounds, but we do not know when it is above or below the x-axis. To find this information we solve

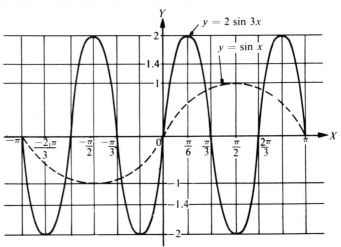

Figure 901

$y = 2 \sin 3x$ for y using some value of x midway between two consecutive x-intercepts. We set $x = \frac{1}{6}\pi$; then

$$y = 2 \sin 3x$$
$$= 2 \sin 3(\tfrac{1}{6}\pi)$$
$$= 2 \sin(\tfrac{1}{2}\pi) = 2.$$

Consequently the curve is above the x-axis when $0 < x < \frac{1}{3}\pi$. Generally one more point, such as the one for $x = \frac{1}{12}\pi$, $y = 2 \sin(3 \cdot \frac{1}{12}\pi) = \sqrt{2} \approx 1.4$, is sufficient for determining the shape of the arc. We draw this arc, and with our previous knowledge of the sine curve we know that similar arcs lie below the x-axis between $x = -\pi$ and $-\frac{2}{3}\pi$, $-\frac{1}{3}\pi$ and 0, etc. The graphs of $y = \sin x$ and $y = \sin 3x$ are shown together in Figure 901.

The function $y = \sin x$ has a period of 2π and an amplitude of 1; the function $y = 2 \sin 3x$ has a period of $\frac{1}{3}(2\pi)$ and an amplitude of 2. For any constant b, the quantity bx may be said to increase "b times as fast" as x. Consequently $\sin bx$ goes through b cycles or periods while $\sin x$ goes through one period. The period of $y = a \sin bx$ is therefore $2\pi/b$; its amplitude, or greatest value, is $|a|$.

The graph of $y = a \cos bx$ can be made in a similar way. It has a period of $2\pi/b$ and amplitude $|a|$. To find its x-intercepts set $y = 0$, then

$$a \cos bx = 0,$$
$$\cos bx = 0,$$

$$bx = \pm\tfrac{1}{2}\pi + n2\pi, \quad \text{or} \quad \tfrac{1}{2}\pi + n\pi = \pi\left(\frac{2n+1}{2}\right),$$

and
$$x = \pm\tfrac{1}{2}\frac{\pi}{b} + n\frac{2\pi}{b}, \quad \text{or} \quad \tfrac{1}{2}\frac{\pi}{b} + n\frac{\pi}{b} = \left(\frac{2n+1}{2b}\right)\pi.$$

The graph of $y = a \sin (bx + c)$, where a, b, and c are constants, is illustrated in the next examples.

Example 2.   Graph the function $y = \sin (x + \frac{1}{2}\pi)$.

Solution.   Our first concern is to find the x-intercepts. To find these x-intercepts we set $y = 0$ then

$$\sin(x + \tfrac{1}{2}\pi) = 0,$$
$$x + \tfrac{1}{2}\pi = n\pi,$$
$$x = n\pi - \tfrac{1}{2}\pi,$$
and
$$x = -\tfrac{1}{2}\pi, \tfrac{1}{2}\pi, \tfrac{3}{2}\pi, \tfrac{5}{2}\pi, \text{ etc.}$$

The amplitude of this curve is 1. The graphs of $y = \sin x$ and $y = \sin (x + \frac{1}{2}\pi)$ are shown together in Figure 902.

It is evident from Example 2 that the graph of $y = \sin(x + \frac{1}{2}\pi)$ may be obtained from that of $y = \sin x$ by shifting the origin a distance of $\frac{1}{2}\pi$ to the left, as shown in Figure 902. It is interesting to note that when we apply Identity (17) to the right-hand member of $y = \sin(x + \frac{1}{2}\pi)$, the function reduces to $y = \cos x$, the graph of which is shifted a distance of $\frac{1}{2}\pi$ to the left of the graph of $y = \sin x$. We say that the graph of $y = \sin(x + \frac{1}{2}\pi)$ *differs in phase* by $-\frac{1}{2}\pi$ from the graph of $y = \sin x$.

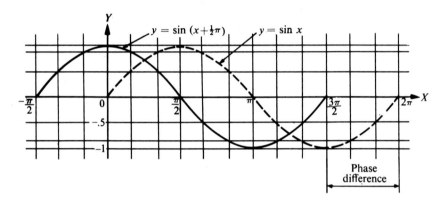

Figure 902

The period of $y = a \sin(bx + c)$ is $2\pi/b$ and its amplitude is $|a|$. The first zero of y comes where $(bx + c) = 0$ or at $x = -c/b$, instead of at $x = 0$ as it does for the graph of $y = a \sin bx$. Hence the graph of $y = a \sin (bx + c)$ is the same as that of $y = a \sin bx$ except that the whole curve is displaced c/b units to the left (Fig. 903). The amount of such a displacement is called the *phase difference* between the two curves.

The graph of $y = a \sin(bx + c)$ *has a simple wave form with amplitude a and period $2\pi/b$ differing in phase from that of $y = \sin x$ by $- c/b$.*

The graph of $y = a \cos(bx + c)$ *has a simple wave form with amplitude a and period $2\pi/b$ differing in phase from that of $y = \cos x$ by $- c/b$.*

Example 3.   Find the period and amplitude of each of the following functions.

(a) $3 \sin \pi x$.
(b) $-2 \sin(\frac{2}{5}x - \frac{1}{3}\pi)$.

Solution.

(a) Period $= 2\pi/b = 2\pi/\pi = 2$. The amplitude $= |a| = |3| = 3$.
(b) Period $= 2\pi/\frac{2}{5} = 5\pi$. The amplitude $= |-2| = 2$.

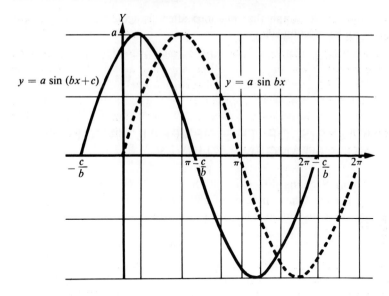

Phase difference $=\frac{c}{b}$

Figure 903

EXERCISES

In Exercises 1–10, find the period and amplitude of each function.

1. $3 \sin 2x$.

2. $2 \cos 3x$.

3. $5 \cos \frac{2}{3}x$.

4. $3 \sin \frac{3}{4}x$.

5. $-2 \sin \frac{1}{6}\pi x$.

6. $-5 \cos \frac{1}{3}\pi x$.

7. $4 \cos(x + \frac{1}{2}\pi)$.

8. $4 \sin(x - \frac{1}{3}\pi)$.

9. $-5 \sin(\frac{1}{3}\pi x - \frac{1}{6}\pi)$.

10. $-3 \sin(\frac{1}{2}\pi x + \frac{1}{3}\pi)$.

In Exercises 11–22, graph each function for one complete period.

11. $y = 3 \sin 2x$.

12. $y = 2 \cos 3x$.

13. $y = \cos \frac{1}{2}x$.

14. $y = \sin \frac{1}{2}x$.

15. $y = -2 \cos 2x$.

16. $y = -2 \sin 2x$.

17. $y = \frac{1}{2} \sin \pi x$.

18. $y = \frac{2}{3} \cos \pi x$.

19. $y = 2 \sin(\frac{2}{3}x + \pi)$.

20. $y = 2 \cos(\frac{1}{3}x - \pi)$.

21. $y = 3 \cos(\pi x - \frac{1}{4}\pi)$.

22. $y = 3 \sin(\frac{1}{2}\pi x + \frac{1}{4}\pi)$.

In Exercises 23–26, graph the equations after changing them to the form $y = c \cos(x - \alpha)$. *See* Section 809.

23. $y = \cos x + 3 \sin x.$ **24.** $y = 3 \cos x - \sin x.$

25. $y = 3 \sin x - 4 \cos x.$ **26.** $y = 2 \sin x + \cos x.$

903 ANALYSIS OF THE GENERAL TANGENT AND COTANGENT CURVES

It was shown in Sections 407 (Fig. 409), and 504 (Fig. 505), that both $y = \tan \theta$ and $y = \cot \theta$ have periods of $180°$. Then (using $180° = \pi$ radians) it follows that the consecutive x-intercepts for either $y = \tan x$ or $y = \cot x$ are separated by an interval of π.

To graph curves of the type $y = a \tan bx$, where a and b are constants, we shall (1) determine some of the consecutive x-intercepts; (2) locate the vertical lines that separate the branches of the curve; (3) locate a few points on the curve; and (4) sketch the balance of the curve from our general knowledge of the $y = \tan \theta$ curve. We illustrate this with an example.

Example 1. Graph the function $y = 2 \tan \frac{1}{2}x$.

Solution. (1) Set $y = 0$ and find some of the consecutive x-intercepts. If $2 \tan \frac{1}{2}x = y$, and if $y = 0$,

$$2 \tan \tfrac{1}{2}x = 0, \ \tan \tfrac{1}{2}x = 0$$
$$\tfrac{1}{2}x = n\pi, \ x = 2n\pi.$$

Thus some of the consecutive x-intercepts are

$$x = -2\pi, 0, 2\pi, \text{ etc.}$$

We plot these points (Fig. 904).

(2) In Section 407 (Fig. 409) we found that $y = \tan \theta$ is undefined for values of θ midway between the consecutive θ-intercepts. This is indeed the case with all curves of the type $y = a \tan bx$. For example, for $y = 2 \tan \frac{1}{2}x$ when $x = \pi$,

$$y = \tan \tfrac{1}{2}x = \tan \tfrac{1}{2}\pi, \text{ which is undefined.}$$

Therefore some of the vertical lines which separate the branches of the curve are: $x = -3\pi$, $x = -\pi$, $x = \pi$, and $x = 3\pi$.

(3) We locate three points on the curve between $x = 0$ and $x = \pi$. It is always desirable to locate the point for which x is midway between the x-intercept and the vertical line which separates the branches of the curve. A brief table of points on this curve is given:

x	$\frac{1}{3}\pi$	$\frac{1}{2}\pi$	$\frac{2}{3}\pi$
y	$\frac{2}{3}\sqrt{3} \approx 1.15$	2	$2\sqrt{3} \approx 3.46$

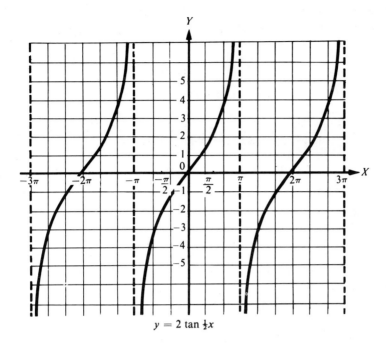

$$y = 2 \tan \tfrac{1}{2}x$$

Figure 904

With these points, the x-intercepts, the vertical lines which separate the branches of the curve, and our knowledge of the $y = \tan \theta$ curve, we are able to sketch the balance of the curve for $y = 2 \tan \tfrac{1}{2}x$.

The function $y = a \cot bx$ can be graphed in a similar way by considering the graph of $y = \cot \theta$.

The functions $y = a \tan(bx + c)$ and $y = a \cot(bx + c)$, where a, b, and c are constants, have periods of π/b.

We have seen that $y = \tan x$ has the period π. That is, for any x_1,

(5) $\tan(x_1 + \pi) = \tan x_1.$

Therefore, the period of $\tan(bx + c)$ is π/b, since, if $x = x_1 + \pi/b$,

$$\tan(bx + c) = \tan(bx_1 + \pi + c) = \tan(bx_1 + c)$$

by Equation (5). Similarly $\cot(bx + c)$ has the period π/b.

Example 2. Graph the function

$$y = 2 \cot(\tfrac{1}{3}x - \tfrac{1}{6}\pi).$$

Solution.

(1) We set $y = 0$ and find some of the consecutive x-intercepts.

$$2 \cot(\tfrac{1}{3}x - \tfrac{1}{6}\pi) = y,$$
$$\cot(\tfrac{1}{3}x - \tfrac{1}{6}\pi) = 0,$$
$$(\tfrac{1}{3}x - \tfrac{1}{6}\pi) = \tfrac{1}{2}\pi + n\pi,$$
$$x = 2\pi + 3n\pi.$$

Therefore some of the consecutive x-intercepts are:

$$x = -4\pi,\ -\pi,\ 2\pi,\ \text{and}\ 5\pi.$$

(*See* Fig. 905.)

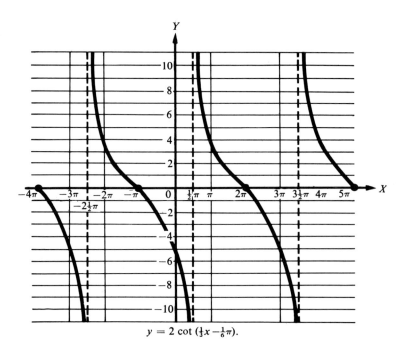

$$y = 2 \cot (\tfrac{1}{3}x - \tfrac{1}{6}\pi).$$

Figure 905

(2) We draw the vertical lines which separate the branches of the curve, mid-way between the x-intercepts. These are

$$x = -2\tfrac{1}{2}\pi,\ x = \tfrac{1}{2}\pi,\ \text{and}\ x = 3\tfrac{1}{2}\pi.$$

(3) We locate a few points as follows:

For $x = \frac{3}{4}\pi$, we have

$$y = 2\cot(\tfrac{1}{3}\cdot\tfrac{3}{4}\pi - \tfrac{1}{6}\pi) = 2\cot(\tfrac{1}{12}\pi) = 2\cot(\tfrac{1}{12}\cdot 180°)$$
$$= 2\cot 150° = 2\times 3.73 = 7.46 \text{ from Table I.}$$

x	$\frac{3}{4}\pi$	π	$1\frac{1}{4}\pi$	2π	$2\frac{3}{4}\pi$	3π
y	7.46	3.46	2	0	-2	-3.46

(4) With the above information and our knowledge of the graph of $y = \cot\theta$, we sketch the curve for $y = 2\cot(\tfrac{1}{3}x - \tfrac{1}{6}\pi)$.

904 ANALYSIS OF THE GENERAL SECANT AND COSECANT CURVES

The graphs for $y = \sec\theta$ and $y = \csc\theta$ were explained in Sections 505 and 506.
The graph of $y = a\csc bx$, where a and b are constants, is illustrated in the following example.

Example. Graph the function $y = 2\csc\tfrac{1}{2}x$.

Solution. The graph of $y = 2\csc\tfrac{1}{2}x$ can be obtained from that of $y = \sin\tfrac{1}{2}x$:
$$y = 2\csc\tfrac{1}{2}x = \frac{2}{\sin\tfrac{1}{2}x}.$$
When $x = \pi$,
$$y = 2\csc\tfrac{1}{2}\pi = \frac{2}{\sin\tfrac{1}{2}\pi} = \frac{2}{1} = 2.$$
When $x = \tfrac{1}{3}\pi$,
$$y = 2\csc\tfrac{1}{6}\pi = \frac{2}{\sin\tfrac{1}{6}\pi} = \frac{2}{\tfrac{1}{2}} = 4.$$

Thus we form the following table, noting that $\csc\tfrac{1}{2}x$ is undefined (U) when $x = 2n\pi$.
We draw the graph of $y = \sin\tfrac{1}{2}x$ and construct vertical lines at the points where it crosses the x-axis. These vertical lines separate the branches of the graph of $y = 2\csc\tfrac{1}{2}x$. (*See* Fig. 906.)
As was done with the tangent function (Section 407) we use U as a symbol for "undefined" in the following, which is a brief table of points on the $y = 2\csc\tfrac{1}{2}x$ curve.

x	-2π	$-1\frac{2}{3}\pi$	$-\pi$	$-\frac{1}{3}\pi$	0	$\frac{1}{3}\pi$	π	$1\frac{2}{3}\pi$	2π	$2\frac{1}{3}\pi$	3π	$3\frac{2}{3}\pi$	4π
y	U	-4	-2	-4	U	4	2	4	U	-4	-2	-4	U

The function $y = a\sec bx$ can be graphed in a similar way by considering the graph of $y = a\cos bx$.

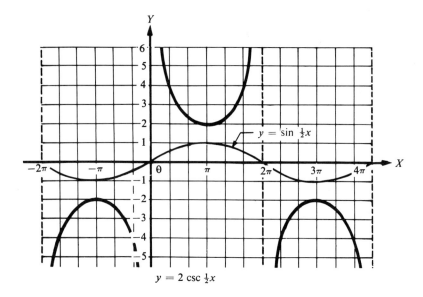

$$y = 2 \csc \tfrac{1}{2}x$$

Figure 906

905 GRAPHING BY THE ADDITION OF ORDINATES

If a function is in the form of the sum of two or more simpler functions, its graph is usually best drawn by the method know as the *addition of ordinates* or as the *composition of ordinates*. This method was discussed in Section 311, but at that time, without the knowledge of radian measure, we could not graph functions of the type $y = ax + b \sin cx$. We demonstrate the graph of this kind of a function, by the method of the addition of ordinates, with an example.

Example.   Graph the function $y = \tfrac{1}{2}x - \sin x$.

Solution.   Rather than work with the function $y = \tfrac{1}{2}x - \sin x$ as a difference between two functions, it may be easier for the student to think of it as a sum such as $y = \tfrac{1}{2}x + (-\sin x)$. We first graph the two functions $y = \tfrac{1}{2}x$ and $y = -\sin x$. The graph of $y = \tfrac{1}{2}x$ is a straight line passing through the points $(0, 0)$, $(1, \tfrac{1}{2})$, $(\pi, \tfrac{1}{2}\pi)$, etc. Some of the x-intercepts of the function $y = -\sin x$ are -2π, $-\pi$, 0, π, 2π, etc. When $x = \tfrac{1}{2}\pi$, $y = -\sin \tfrac{1}{2}\pi = -1$, therefore the arch for the graph of $y = -\sin x$, between $x = 0$ and $x = \pi$, lies below the x-axis. (*See* Fig. 907.)

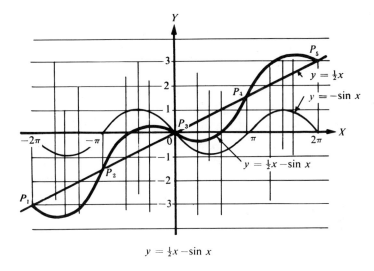

$$y = \tfrac{1}{2}x - \sin x$$

Figure 907

Then, for any value of x, the height or ordinate of the graph of $y = \tfrac{1}{2}x$ $- \sin x$ can be found by adding the ordinates of the other two functions for that value of x. In this way we can quickly find as many points as we wish for the desired curve and then draw it. It is helpful to draw vertical lines from the x-intercepts, of the $y = - \sin x$ curve, to the line $y = \tfrac{1}{2}x$. The points of intersection of these vertical lines and the line $y = \tfrac{1}{2}x$ (P_1, P_2, P_3, P_4, and P_5) are points on the composite curve where it crosses the line $y = \tfrac{1}{2}x$. The composite curve lies below the line $y = \tfrac{1}{2}x$ between P_1 and P_2, and between P_3 and P_4; it lies above the line $y = \tfrac{1}{2}x$ between P_2 and P_3, and between P_4 and P_5.

EXERCISES

In Exercises 1–10, graph each function for one complete period.

1. $y = \tan \tfrac{1}{3}x$. **2.** $y = \cot \tfrac{1}{2}x$. **3.** $y = \sec 2x$.

4. $y = \csc 2x$. **5.** $y = 2 \cot \tfrac{1}{4}x$. **6.** $y = 2 \tan \tfrac{1}{4}x$.

7. $y = 3 \csc x$. **8.** $y = 2 \sec x$. **9.** $y = 2 \tan(x - \tfrac{1}{4}\pi)$.

10. $y = \tfrac{2}{3}\tan(\tfrac{1}{2}x + \tfrac{1}{3}\pi)$.

In Exercises 11–20, use the method of composition of ordinates to graph each of the functions on the specified intervals.

11. $y = x + \cos x$, $x = -\tfrac{1}{2}\pi$ to $x = 2\tfrac{1}{2}\pi$.

12. $y = x + \sin x,$ $x = -\pi$ to $x = 2\pi.$

13. $y = 1 + \sin x,$ $x = -0$ to $x = 2\pi.$

14. $y = 1 - \cos x,$ $x = -\frac{1}{2}x$ to $x = 2\frac{1}{2}\pi.$

15. $y = 2 \sin \frac{1}{2}x - 1,$ $x = 0$ to $x = 4\pi.$

16. $y = 2 \sin \frac{1}{2}x - x,$ $x = 0$ to $x = 4\pi.$

17. $y = \tan x + \sin x,$ $x = -\frac{1}{3}\pi$ to $\boldsymbol{x} = \frac{1}{3}\pi.$

18. $y = \cos x + \cot x,$ $x = \frac{1}{6}\pi$ to $x = \frac{5}{6}\pi.$

19. $y = x + \sin^2 x,$ $x = 0$ to $x = 3\pi.$

20. $y = x + \cos^2 x,$ $x = -\frac{1}{2}\pi$ to $x = \frac{3}{2}\pi.$

10

Logarithms

1001 USE OF LOGARITHMS

In many problems—including the solution of triangles—computations involving multiplication, division, raising to powers, and extraction of roots frequently can be carried out with less labor by the use of logarithms than by direct arithmetical calculation. We shall see that logarithms enable us to replace the processes of multiplication and division by the simpler ones of addition and subtraction, and also to replace the taking of powers and roots by multiplication and division. Certain equations are solved most easily by the use of logarithms.

1002 LAWS OF EXPONENTS

As we shall see, *logarithms are exponents.* In this section we recall certain laws of exponents. These laws of exponents, which are discussed in courses in algebra, hold for all types of real exponents for a positive real base. For the purposes of this course we shall restrict the exponents to rational numbers, and we shall make the following assumptions.

Let a be any positive number and x and y be any two rational numbers. Then

I. $a^x \cdot a^y = a^{x+y}$.   *Example:* $2^3 \cdot 2^4 = 2^7$.

II. $\dfrac{a^x}{a^y} = a^{x-y}$.   *Example:* $\dfrac{2^7}{2^2} = 2^5$.

III. $(a^x)^y = a^{xy}$.   *Example:* $(2^3)^4 = 2^{12}$.

IV. $\sqrt[y]{a^x} = a^{x/y}$.   *Example:* $\sqrt[3]{2^5} = 2^{5/3}$.

V. $a^0 = 1$,   if $a \neq 0$.   *Example:* $2^0 = 1$.

VI. $a^{-x} = \dfrac{1}{a^x}$.   *Example:* $2^{-5} = \dfrac{1}{2^5} = \dfrac{1}{32}$.

VII. If a is any number greater than 1, and if y is positive, it is always possible to find a real exponent x such that

$$a^x = y.$$

But if y is negative or zero, no real value of x exists satisfying this relation.

Definition.   *If a is any real number and n is a positive integer, then*

$$a^n = a \cdot a \cdot \ldots \cdot a, \qquad n \text{ factors of } a;$$

n *is called the* **exponent**, **a** *the* **base**, *and* **a^n** *the* **nth power** *of a.*

1003   DEFINITION OF A LOGARITHM

If b, x, and N are three numbers such that

(1) $N = b^x$

then we say that x is the logarithm of N to the base b, and write

(2) $x = \log_b N.$

Definition.   *The* **logarithm** *of a number to a given base is the exponent of the power to which the base must be raised to give the number.*

We assume in this book that $\log_b N$ is used only for $N > 0$ and $b > 1$, which implies that $\log_b N$ is a real number.

Equations (1) and (2) are different ways of expressing the same relation among the three numbers b (the *base*), N (the *number*), and x (the *exponent* or *logarithm*). Equation (1) states the relation in *exponential form* and Equation (2) says the same thing in *logarithmic form*. If x is the logarithm of N to a given base, then N is called the **antilogarithm** of x to that base.

The following values are very important and should be memorized. Since $b^1 = b$ and $b^0 = 1$, for any base, b,

(3) $\log_b b = 1$ and   (4) $\log_b 1 = 0.$

In each of Examples 1, 2, and 3, we express the same relation between the three respective numbers, first in exponential form and then in logarithmic form.

EXPONENTIAL FORM	LOGARITHMIC FORM

Example 1. $4^3 = 64$.
    4 is the base.
    3 is the exponent of the power.
    4^3 is the third power of 4, and
    64 is the third power of 4.

Log_4 $64 = 3$.
    4 is the base.
    3 is the logarithm of 64 to the
        base 4.

Example 2. $4^{2.5} = 32$.
    4 is the base.
    2.5 is the exponent of the power.
    $4^{2.5}$ is the 2.5 power of 4, and
    32 is the 2.5 power of 4.

Log_4 $32 = 2.5$.
    4 is the base.
    2.5 is the logarithm of 32 to the
        base 4.

Example 3. $e^2 = 7.39$.
    e is the base.
    2 is the exponent of the power.
    e^2 is the second power of e, and
    7.39 is the second power of e.

Log_e $7.39 = 2$.
    e is the base.
    2 is the logarithm of 7.39 to the
        base e.

Example 4.   Find the value of $\log_{32} 8$.

Solution.   Let $$x = \log_{32} 8.$$

Write in exponential form:

$$32^x = 8.$$

Express 32 and 8 as powers of 2:

$$(2^5)^x = 2^3,$$

$$2^{5x} = 2^3.$$

Hence $$5x = 3, x = \tfrac{3}{5}.$$

Therefore $$\log_{32} 8 = \tfrac{3}{5}.$$

Example 5.   Find the number N if $\log_{81} N = \tfrac{3}{4}$.

Solution.

Write in exponential form: $N = (81)^{3/4}$.
Write 81 as a power of 3: $N = (3^4)^{3/4}$.
Apply Law III of Exponents: $N = 3^3 = 27$.

EXERCISES

In Exercises 1–16, express each equation in logarithmic form.

1. $2^3 = 8$.   **2.** $3^2 = 9$.   **3.** $10^2 = 100$.

4. $10^3 = 1000$.   **5.** $25^{1/2} = 5$.   **6.** $16^{1/2} = 4$.

7. $8^{-2/3} = \frac{1}{4}$.   **8.** $4^{-3/2} = \frac{1}{8}$.   **9.** $5^0 = 1$.

10. $7^0 = 1$.   **11.** $3^0 = 1$.   **12.** $6^0 = 1$.

13. $5^1 = 5$.   **14.** $3^1 = 3$.   **15.** $10^1 = 10$.

16. $4^1 = 4$.

In Exercises 17–28, express each equation in exponential form.

17. $\log_4 64 = 3$.   **18.** $\log_8 64 = 2$.   **19.** $\log_2 16 = 4$.

20. $\log_5 125 = 3$.   **21.** $\log_{16} 4 = \frac{1}{2}$.   **22.** $\log_9 3 = \frac{1}{2}$.

23. $\log_{10} 10 = 1$.   **24.** $\log_{10} .01 = -2$.   **25.** $\log_7 7 = 1$.

26. $\log_{10} 100 = 2$.   **27.** $\log_b b = 1$.   **28.** $\log_{10} 0.001 = -3$.

In Exercises 29–37, find the value of each of the logarithms.

29. $\log_3 27$.   **30.** $\log_4 64$.   **31.** $\log_4 8$.

32. $\log_{27} 9$.   **33.** $\log_5 1$.   **34.** $\log_6 1$.

35. $\log_9(\frac{1}{3})$.   **36.** $\log_8(\frac{1}{64})$.   **37.** $\log_4 32$.

In Exercises 39–46, find the unknown b, N, or x.

38. $\log_2 N = 3$.   **39.** $\log_3 N = 2$.   **40.** $\log_{10} 1000 = x$.

41. $\log_{10} 10,000 = x$.   **42.** $\log_b 32 = 2.5$.   **43.** $\log_b 27 = 1.5$.

44. $10^{\log_{10} 3} = x$.   **45.** $3^{\log_3 2} = x$.   **46.** $2^{\log_2 3} = x$.

1004   FUNDAMENTAL LAWS OF LOGARITHMS

Since logarithms are exponents, the laws of exponents enable us to derive certain laws of logarithms that are useful in computation. Any base, b, greater than 1, may be used. We shall assume that M and N are positive numbers.

I. *Multiplication.   The logarithm of the product of two or more numbers is equal to the sum of the logarithms of the numbers.*

$$\log_b (M \cdot N) = \log_b M + \log_b N,$$

$$\log_b(M \cdot N \cdot P \cdot \ldots \cdot R) = \log_b M + \log_b N + \log_b P + \cdots + \log_b R.$$

Proof.   Let $\qquad\qquad\qquad\qquad x = \log_b M \qquad$ and $\qquad y = \log_b N.$

Write in exponential form: $\qquad\quad M = b^x \qquad$ and $\qquad N = b^y.$

Multiply M by N: $\qquad\qquad\qquad MN = b^x \cdot b^y = b^{x+y}.$

Write in logarithmic form: $\qquad\quad \log_b MN = x + y.$

Replace x and y by their values: $\quad \log_b MN = \log_b M + \log_b N.$

Example 1. $\log_{10}(289 \times 356) = \log_{10} 289 + \log_{10} 356.$

II. *Division.*   *The logarithm of the quotient of two numbers is equal to the logarithm of the dividend minus the logarithm of the divisor.*

$$\log_b\left(\frac{M}{N}\right) = \log_b M - \log_b N.$$

Proof.   Let $\qquad\qquad\qquad\qquad x = \log_b M \qquad$ and $\qquad y = \log_b N.$

Write in exponential form: $\qquad\quad M = b^x \qquad$ and $\qquad N = b^y.$

Divide M by N: $\qquad\qquad\qquad \dfrac{M}{N} = \dfrac{b^x}{b^y} = b^{x-y}.$

Write in logarithmic form: $\qquad\quad \log_b\left(\dfrac{M}{N}\right) = x - y.$

Replace x and y by their values: $\quad \log_b\left(\dfrac{M}{N}\right) = \log_b M - \log_b N.$

Example 2. $\log_{10}\left(\dfrac{125}{346}\right) = \log_{10} 125 - \log_{10} 346.$

III. *The logarithm of a power of a number is equal to the logarithm of the number multiplied by the exponent of the power.*

$$\log_b(M^n) = n \log_b M.$$

Proof.   Let $\qquad\qquad\qquad\qquad x = \log_b M.$

Write in exponential form: $\qquad\quad M = b^x.$

Raise to the nth power: $\qquad\qquad M^n = (b^x)^n = b^{nx}.$

Write in logarithmic form: $\qquad\quad \log_b M^n = nx.$

Replace x by its value: $\qquad\quad \log_b M^n = n \log_b M.$

Example 3. $\log_{10}(3.51)^7 = 7 \log_{10} 3.51.$

IV. *The logarithm of a root of a number is equal to the logarithm of the number divided by the index of the root.*

$$\log_b(\sqrt[n]{M}) = \frac{\log_b M}{n}.$$

Proof.   Since, by Law IV of Exponents, $\sqrt[n]{M} = M^{1/n}$, by the above law of logarithms we have

$$\log_b(\sqrt[n]{M}) = \log_b(M^{1/n}) = \frac{1}{n}\log_b M.$$

Example 4. $\log_{10}\sqrt[5]{146} = \log_{10}(146)^{1/5} = \frac{1}{5}\log_{10} 146.$

Example 5.

(a) $\log_{10}\dfrac{345 \times 561}{284} = \log_{10} 345 + \log_{10} 561 - \log_{10} 284;$

(b) $\log_{10}\dfrac{156 \times (4.15)^3}{(1.05)^{12}} = \log_{10} 156 + 3\log_{10} 4.15 - 12\log_{10} 1.05;$

(c) $\log_{10}\dfrac{68.4 \times \sqrt{7.61}}{\sqrt[4]{3.48}} = \log_{10} 68.4 + \frac{1}{2}\log_{10} 7.61 - \frac{1}{4}\log_{10} 3.48.$

Example 6.   Given $\log_{10} 2 = 0.301$, $\log_{10} 3 = 0.477$, and $\log_{10} 7 = 0.845$, find $\log_{10}\frac{54}{7}$.

Solution.

$$\log_{10}\tfrac{54}{7} = \log_{10}\frac{2 \times 3^3}{7}$$

$$= \log_{10} 2 + 3\log_{10} 3 - \log_{10} 7$$

$$= 0.301 + 3(0.477) - 0.845$$

$$= 0.887.$$

Example 7.   Express $3\log_b x + 5\log_b y - \frac{1}{2}\log_b z$ as a single logarithm.

Solution:

$$3\log_b x + 5\log_b y - \tfrac{1}{2}\log_b z = \log_b x^3 + \log_b y^5 - \log_b z^{1/2}$$

$$= \log_b\frac{x^3 y^5}{\sqrt{z}}.$$

EXERCISES

Given $\log_{10} 2 = 0.301$, $\log_{10} 3 = 0.477$, $\log_{10} 7 = 0.845$, find the following logarithms. (Recall that $\log_{10} 10 = 1$.)

1. $\log_{10} 14$.

2. $\log_{10} 21$.

3. $\log_{10} \frac{18}{7}$.

4. $\log_{10} \frac{27}{7}$.

5. $\log_{10} 5$.

6. $\log_{10} 50$.

7. $\log_{10} \sqrt{21}$.

8. $\log_{10} \sqrt{27}$.

9. $\log_{10} 30$.

10. $\log_{10} 60$.

11. $\log_{10} 98$.

12. $\log_{10} 72$.

13. $\log_{10} \sqrt[5]{2}$.

14. $\log_{10} \sqrt[10]{2}$.

15. $\log_{10} 6000$.

Write each expression as a single logarithm. (Assume that all of the logarithms have the same base.)

16. $\log x + \log y$.

17. $\log x - \log y$.

18. $2 \log x - 3 \log y$.

19. $3 \log x + 2 \log y$.

20. $\frac{1}{2} \log x + \log y - \log z$.

21. $\frac{1}{4} \log x - \log y + \log z$.

22. $\log(x^2 - y^2) - \log(x + y)$.

23. $\log(x^2 - y^2) - \log(x - y)$.

In each of the following, solve for y in terms of x:

24. $\log_2 y = 3x$.

25. $\log_3 y = 2x$.

26. $\log_{10} y = 3 \log_{10} x$.

27. $\log_{10} y = -2 \log_{10} x$.

28. $\log_{10} y = \log_{10} x^2 - \log_{10} x$.

29. $\log_{10} y = \log_{10} x^3 - 2 \log_{10} x$.

30. $\frac{1}{3} \log_{10} y = 3 \log_{10} x - \log_{10} x^2$.

31. $\frac{1}{3} \log_{10} y = \log_{10} x^4 - 3 \log_{10} x$.

Solve the following equations for x.

32. $\log_{10}(x + 3) + \log_{10} 4 = \log_{10} 100 + \log_{10} 2$.

33. $\log_{10}(x - 5) + \log_{10} 5 = \log_{10} 10 + \log_{10} 3$.

34. $\log_{10}(x^2 - 4) = 1 + \log_{10}(x + 2)$.

35. $\log_{10}(x^2 - 1) = 2 + \log_{10}(x - 1)$.

1005 SYSTEMS OF LOGARITHMS

There are two important systems of logarithms. The *natural*, or *Napierian*, system uses the base e, where e is approximately 2.71828. This system is used in most advanced applications of mathematics. Many authors use the notation ln x instead of $\log_e x$. The system that is used in the solution of triangles and that is more convenient for computation is the *common*, or *Briggs*, system which uses the base 10. *Hereafter, when we speak of the logarithm of a number without specifying the base, it is to be understood that the base of the logarithm is* 10—and instead of writing $\log_{10} N$, we shall write $\log N$.

1006   LOGARITHMS TO THE BASE 10

The logarithms of integral powers of 10 (10, 10^2, 10^3, 10^{-1}, 10^{-2}, etc.) are integers.

Example 1.

	EXPONENTIAL FORM	LOGARITHMIC FORM
(a)	$10^3 = 1000$	$\log 1000 = 3.$
(b)	$10^2 = 100$	$\log 100\ \ = 2.$
(c)	$10^1 = 10$	$\log 10\ \ \ = 1.$
(d)	$10^0 = 1$	$\log 1\ \ \ = 0.$
(e)	$10^{-1} = 0.10$	$\log 0.10 = -1.$
(f)	$10^{-2} = 0.01$	$\log 0.01 = -2.$

Most of the logarithms we use are irrational numbers that are given with an accuracy specified to a certain number of decimal places. A collection of such numbers is given in Table I.

Example 2.   The following logarithms are given to five decimal accuracy:

	EXPONENTIAL FORM	LOGARITHMIC FORM
(a)	$10^{2.30103} = 200$	$\log 200 = 2.30103.$
(b)	$10^{1.69897} = 50$	$\log 50\ \ = 1.69897.$
(c)	$10^{0.30103} = 2$	$\log 2\ \ = 0.30103.$

The values of logarithms of numbers, such as those in Example 2, are found by methods that are beyond the level of this course.

Any positive number written in ordinary decimal notation can be written in scientific notation, that is, as the product of a number between 1 and 10 and a power of 10. (*See* Section 212 which includes a discussion of how to write a number in scientific notation.)

Example 3.

(a) $2714 = 2.714 \times 10^3.$

(b) $315.6 = 3.156 \times 10^2.$

(c) $21.59 = 2.159 \times 10^1$.
(d) $7.187 = 7.187 \times 10^0$.
(e) $0.8570 = 8.570 \times 10^{-1}$.
(f) $0.01444 = 1.444 \times 10^{-2}$.
(g) $0.00009889 = 9.889 \times 10^{-5}$.

  In scientific notation, two numbers that have the same sequence of digits can differ only in the power of 10 that is used.

Example 4.

(a) $3142 = 3.142 \times 10^3$.
(b) $31.42 = 3.142 \times 10^1$.
(c) $0.3142 = 3.142 \times 10^{-1}$.
(c) $0.00003142 = 3.142 \times 10^{-5}$.

  We recall that $\log 1 = 0$ and $\log 10 = 1$ and that $\log x_1 > \log x_2$, if $x_1 > x_2$; therefore it follows that the value of $\log N$, where $0 < N < 1$, is a number greater than 0 and less than 1.

Example 5.

(a) $\log 1 = 0$.
(b) $\log 1.100 = 0.04139$.
(c) $\log 2.000 = 0.30103$.
(d) $\log 3.142 = 0.49721$.
(e) $\log 9.999 = 0.99996$.
(f) $\log 10.00 = 1.00000$.

  The logarithms of the numbers of Example 4 can be expressed as follows:

Example 6.

(a) $\log 3142 = \log(3.142 \times 10^3)$
                      $= \log 3.142 + \log 10^3$
                      $= \log 3.142 + 3 \log 10$
                      $= \log 3.142 + 3\,(1)$
                      $= 0.49721 + 3 = 3.49721$.

(b) $\log 31.42 = \log(3.142 \times 10^1)$
                      $= \log 3.142 + \log 10^1$
                      $= 0.49721 + 1 = 1.49721$

(c) $$\log 0.3142 = \log(3.142 \times 10^{-1})$$
$$= \log 3.142 + \log 10^{-1}$$
$$= \log 3.142 + (-1) \log 10$$
$$= 0.49721 - 1.$$

By adding 10 and then subtracting 10 from the number $(0.49721 - 1)$, we can write

$$0.49721 - 1 = 0.49721 - 1 + 10 - 10$$
$$= 0.49721 + 9 - 10$$
$$= 9.49721 - 10.$$

(d) $$\log 0.000003142 = \log (3.142 \times 10^{-6})$$
$$= \log 3.142 + \log 10^{-6}$$
$$= 0.49721 - 6$$
$$= 4.49721 - 10.$$

Hence, the common logarithm of any positive number can be written as the sum of two parts: (1) a nonnegative decimal less than 1, and (2) the exponent applied to 10 when the number is written in scientific notation.

The decimal part of the common logarithm of a number is called the **mantissa** of the logarithm. As we saw in Example 6 the *mantissa depends only on the particular sequence of digits in the number*. It is independent of the position of the decimal point.

The exponent of 10, which is the whole part of the logarithm, is called the **characteristic**. *The characteristic depends only on the position of the decimal point in the number*. It is independent of the sequence of digits.

1007   RULE FOR THE CHARACTERISTIC OF A COMMON LOGARITHM

As we stated in the previous section, the characteristic of the logarithm of a given number is equal to the exponent of ten when that number is written in scientific notation.

Some students find the $\wedge$ symbol, called **caret**, an aid in converting a number to its equivalent form in scientific notation. The caret shows the position where the decimal point of the transformed number is to be placed. Some of the examples above are repeated below, showing the use of the caret.

Examples: 2. $2045 = 2_{\wedge}045. = 2.045 \times 10^3.$
3. $204.5 = 2_{\wedge}04.5 = 2.045 \times 10^2.$

Example 1.

	Number	Characteristic of the logarithm of the number	The number in scientific notation
(a)	2045	3	2.045×10^3
(b)	204.5	2	2.045×10^2
(c)	20.45	1	2.045×10^1
(d)	2.045	0	2.045×10^0
(e)	0.2045	-1	2.045×10^{-1}
(f)	0.02045	-2	2.045×10^{-2}
(g)	0.002045	-3	2.045×10^{-3}
(h)	0.0002045	-4	2.045×10^{-4}

Notice that the caret is placed immediately to the right of the first non-zero digit of the number. Then the number is rewritten with the decimal point in the position where the caret was used. This new number is then multiplied by 10^n, where n is the number of places of the decimal point to the right or left of the caret, and is positive when the decimal point is to the right of the caret, as in Examples 2 and 3 above, and negative when the decimal point is to the left of the caret, as in Examples 4 and 5.

Examples: 4. $0.2045 = 0.2_{\wedge}045 = 2.045 \times 10^{-1}$.

5. $0.02045 = 0.02_{\wedge}045 = 2.045 \times 10^{-2}$.

For computational purposes, it is better to write negative characteristics as the difference of two numbers. For example:

the characteristic -1 is written $9 - 10$,
the characteristic -2 is written $8 - 10$,
the characteristic -4 is written $6 - 10$, etc.

EXERCISES

Write the characteristic of the logarithm of each of the following numbers:

1. 542.
2. 125.
3. 93,000,000.
4. 186,000.
5. 0.0745.
6. 0.0815.
7. 0.000904.
8. 0.000516.
9. 8,056.
10. 2,018.
11. 0.00634.
12. 0.808.
13. 3612.
14. 54,000.
15. 0.0684.
16. 0.1006.
17. 0.9008.
18. 3047.
19. 5,880,000,000,000.
20. 0.000000514.

In Exercises 21–30, rewrite each number, and place its decimal point according to the given characteristic of its logarithm.

Number	Characteristic	Number	Characteristic
21. 1278	2	**22** 2354	1
23. 5061	5	**24.** 6043	4
25. 9455	0	**26.** 8544	0
27. 1008	8 — 10	**28.** 2009	7 — 10
29. 7599	5 — 10	**30.** 4988	4 — 10

1008   TABLES OF LOGARITHMS

The mantissa of the logarithm of a number is very often a never-ending decimal. **Tables of common logarithms** contain the *mantissas* of logarithms of numbers, to a specified number of decimal places. In Table I (pages *1–19*) there are listed, to five decimal places, the mantissas of the logarithms of all whole numbers from 1 to 9999. In this table of mantissas, the decimal point which belongs in front of each mantissa is omitted; it should be supplied when the mantissa is taken from the table. This table may be used to find the logarithm of a given number or to find the number corresponding to a given logarithm.

To find the logarithm of a given number: Determine the characteristic, by the methods given in Section 1007, and then find the mantissa from Table I.

To find the mantissa for the logarithm of a number, the first three digits of the number are given in the column at the left under the caption " N." The fourth digit of the number occurs at the top of the table in the same line as N. For example on page *2* we find log 1.500 = 0.17609. The next entry in the same line is log 1.501 = 0.17638. In each case, the mantissa obtained directly from the table is the entire logarithm, since the characteristic is zero. Notice that only the first two digits of the mantissa are printed in the first column of mantissas, and then only when a change occurs. For example, from 1.520 to 1.550 there is a change from 18 to 19 for the first two digits in the mantissa. We shall call these first two digits of the mantissa the *prefix* digits.

In order to read log 1.832, follow down the column under " N " to the number 183. Go across the line to the column under " 2." Then read

$$\log 1.832 = 0.26293.$$

In order to read log 1.863, the same procedure is used. Notice this time that the last three digits of the mantissa are starred; the star means that the first two digits, the prefix digits, are to be read from the following line of the table. Thus,

$$\log 1.863 = 0.27021.$$

To find the number corresponding to a given logarithm: The number that corresponds to a given logarithm is often referred to as the **antilogarithm** of the logarithm. Thus in the statement log 1.501 = 0.17638, the number 1.501 is the antilogarithm of the number 0.17638. The method of finding the anti-logarithm is just the reverse of that explained above for finding the logarithm of a given number.

Since the characteristic of the logarithm depends only on the position of the decimal point in the number, and since the mantissa depends only on the succession of digits in the number, in order to find the antilogarithm of a given logarithm we first disregard the characteristic and look for the given mantissa in the table.

In order to find the number N when log $N = 1.24204$, we first disregard the characteristic 1 then follow down the first column of mantissas (page 3) until the prefix digits, 24, occur. An examination of this section of the table shows 24204 in the line with $N = 174$ and in the column under " 6." Hence if

$$\log N = 1.24204, \qquad \text{then} \qquad N = 17.46.$$

1009   INTERPOLATION

Interpolation is the process of approximating a number that is not an entry in the table but lies between two consecutive entries. Interpolation as applied to numbers involved in the table of trigonometric functions was explained in Section 305. The same type of linear interpolation is used here with our five-place table of logarithms.

The logarithms of numbers with four significant digits are read directly from Table I. By interpolation we can find the logarithm of a number with five significant digits. If we wish to find the logarithm of a number having more than five significant digits we usually round it off to five places before using the table, because the table is not designed to give greater accuracy than this.

Example 1.   Find log 382.47.

Solution.   The characteristic is 2. We know that this logarithm must lie between log 382.40 and log 382.50

$$10\left\{7\begin{cases}\log 382.40 = 2.58252\\[2pt]\log 382.47 = \boxed{2.58260}\\[2pt]\log 382.50 = 2.58263\end{cases}8\right\}11$$

Since 382.47 is "seven-tenths of the way" from 382.40 to 382.50, we assume
that log 382.47 is "seven-tenths of the way" from 2.58252 to 2.58263. Therefore, in the last place we add .7 × 11 = 7.7 (which we call 8) to 2.58252 and
get log 382.47 = 2.58260.

The product .7 × 11 can be found by the tables of proportional parts,
which are at the right of each page of the tables. The numbers above the short
columns are the tabular differences that occur for the page. The nine numbers
in the column are, in order, .1, .2, .3, etc. of the number above the column.
Thus, in Example 1 above, .7 × 11 = 7.7 which we rounded to 8. The student will
find the proportional parts tables more helpful when the tabular difference is a
larger number.

Example 2.   Given log $N = 8.21104 - 10$, find N.

Solution.   We first disregard the characteristic and seek only the succession of
digits in the number N which correspond to the mantissa 0.21104. We find
that this mantissa lies between two table entries (page *3*) as follows:

$$1 \begin{cases} \log 1625 = 21085 \\ \log \quad N = 21104 \\ \log 1626 = 21112 \end{cases} \begin{matrix} 19 \\ \end{matrix} \Big\} 27$$

Therefore the succession of digits for N is

$$1625 + \tfrac{19}{27}(1) = 1625\tfrac{19}{27} = 16257. \quad (\tfrac{19}{27} \approx .7)$$

The characteristic is $(8 - 10)$, that is -2. Therefore

$$N = 0.016257 \qquad \text{(accurate to five significant digits)}.$$

The table of proportional parts (page *3*) can be used to find $\tfrac{19}{27} = .7$ as follows:
Look down the column having 27 at the top for the number nearest 19. It is
18.9, which corresponds to .7.

EXERCISES

In Exercises 1–12, assume the following numbers to be exact numbers and find
the logarithm of each.

1. 125.8.	**2.** 161.6.	**3.** 0.5373.
4. 0.5626.	**5.** 67.356.	**6.** 71.666.
7. 0.0008.	**8.** 0.0040.	**9.** 93,000,000.
10. 186,000.	**11.** 0.050133.	**12.** 0.0676739.

In Exercises 13–24, find the antilogarithms of each of the given numbers.

13. 0.78845.   **14.** 0.82171.   **15.** 3.63002.

16. 2.43008.   **17.** 7.00689 – 10.   **18.** 6.00604 – 10.

19. 1.24315.   **20.** 1.58225.   **21.** 8.68020 – 10.

22. 7.56040 – 10.   **23.** 4.00050.   **24.** 5.00095.

1010   COMPUTATIONS WITH LOGARITHMS

Before any computation is done in a problem, it is important to analyze the problem and make an outline of the procedure to be followed. Without an outline the student is apt to waste time and possibly omit some important detail. This is especially true in solving triangles by the use of logarithms. The characteristics of the logarithms of each given number should be written in the proper location in the outline before going to the tables for the mantissas.

In the five following examples the numbers in the examples are assumed to be exact, and the answer for each example is given correct to five significant figures.

Example 1. Find the value of $x = 21.75 \times 3.146 \times 0.8966$.

Solution. By Law I, Section 1004,

$$\log x = \log 21.75 + \log 3.146 + \log 0.8966.$$

The blank outline, with the characteristics of the numbers written in, is shown at the left. The same outline is also shown at the right after it has been filled in from the tables.

$$
\begin{array}{ll}
\log 21.75 = 1. & \log 21.75 = 1.33746 \\
\log 3.146 = 0. & \log 3.146 = 0.49776 \\
\log 0.8966 = 9. \quad\quad -10(+) & \log 0.8966 = 9.95260 - 10(+) \\
\hline
\quad\quad \log x = & \quad\quad \log x = 11.78782 - 10 \\
\quad\quad\quad x = & \quad\quad\quad x = 61.351.
\end{array}
$$

Example 2. Find the value of $x = \dfrac{9.041}{0.06544}$.

Solution.   Take the logarithms of both sides of the equation.

$$\log x = \log 9.041 - \log 0.06544.$$

$\log 9.041 \quad = 0.$	$\log 9.041 \quad = 10.95622 - 10$
$\log 0.06544 = 8. \qquad - 10(-)$	$\log 0.06544 = \underline{\ 8.81584 - 10(-)}$
$\log x = $	$\log x = 2.14038$
$x = $	$x = 138.19 \approx 138.2.$

Notice that 10 was added to and then subtracted from the logarithm of 9.041 in order to keep the decimal part of $\log x$ positive.

Example 3.   Find the value of $x = \dfrac{(1.500)^4}{(3.461)(0.8960)}.$

Solution.   Take the logarithm of both sides of the equation.

$$\log x = 4 \log 1.500 - [\log 3.461 + \log 0.8960]$$
$$= 4 (0. \qquad) - [(0. \qquad) + (9. \quad -10)].$$

We let the above be our outline and proceed to find the mantissas from the tables.

$$\log x = 4(0.17609) - [0.53920 + (9.95231 - 10)]$$
$$= 0.70436 - [10.49151 - 10]$$
$$= 0.21285.$$

Then $x = 1.632.$

Example 4.   Find the value of $x = \sqrt[3]{0.71654}.$

Solution.   Rewrite the equation in the form

$$x = (0.71654)^{1/3}.$$

Then take the logarithm of both sides of this equation.

$$\log x = \log (0.71654)^{1/3}$$
$$= \tfrac{1}{3} \log 0.71654$$
$$= \tfrac{1}{3}(9.85524 - 10).$$

We are about to divide the logarithm $(9.85524 - 10)$ by 3. In order to make the negative part of the characteristic evenly divisable by 3, we both add and subtract 20 in the characteristic. Thus we rewrite the logarithm $(9.85524 - 10)$ as

$$9.885524 - 10 = 9.885524 + 20 - 20 - 10$$
$$= 29.885524 - 30.$$

Then $\log x = \frac{1}{3}(29.85524 - 30)$

$= 9.95175 - 10.$

Therefore $x = 0.89485.$

Example 5.   Find the value of $x = \dfrac{75 \times (-3.85)}{5.79}$.

Solution.   As stated in Section 1003, we are restricted to the logarithms of positive numbers, whereas the factor (-3.85) is negative. By inspection we observe that

$$x = \frac{75 \times (-3.85)}{5.79} = -\frac{75 \times 3.85}{5.79} = -F,$$

where $F = \dfrac{75 \times 3.85}{5.79}$. Then we proceed as follows:

$$\begin{array}{llll}
\log 75 & = 1. & \qquad & \log 75 & = 1.87506 \\
\log 3.85 & = 0. & (+) & \log 3.85 & = 0.58546(+) \\
\cline{1-2}\cline{4-5}
& & & & 2.46052 \\
\log 5.79 & = 0. & (-) & \log 5.79 & = 0.76268(-) \\
\cline{1-2}\cline{4-5}
\log F & = & & \log F & = 1.69784 \\
F & = & & F & = 49.87.
\end{array}$$

Therefore $x = -F = -49.87.$

EXERCISES

In Exercises 1–26, use logarithms (Table I) to obtain the results correct to four significant figures. Assume the given numbers to be exact numbers.

1. 28.46×3.057.
2. 48.01×7.615.
3. $609.1 \div 25.4$.

4. $711.6 \div 49.1$.
5. $1.916 \div 54.7$.
6. $2.834 \div 65.7$.

7. $(1.166)^5$.
8. $(2.104)^6$
9. $\sqrt[4]{75.6}$.

10. $\sqrt[5]{80.9}$.
11. $\sqrt[3]{0.856}$.
12. $\sqrt[3]{0.199}$.

13. $\sqrt[5]{0.036}$.
14. $\sqrt[6]{0.084}$.
15. $(18.5)^{2/3}$.

16. $(59.4)^{3/4}$.
17. $\sqrt{\dfrac{850}{1.46 \times 7.84}}$.
18. $\sqrt{\dfrac{386}{7.15 \times 25.8}}$.

19. $1.467 + \log 27.36$.
20. $7.86 + \log 14.96$.

21. $38.6 \times \log 187.4$.

22. $58.6 \times \log 784.5$.

23. $\dfrac{85 \times (-6.75)}{38.4}$.

24. $\dfrac{28600}{7.15 \times (-38.7)}$.

25. $(2.8145)^{-2.5}$.

26. $(18.155)^{-3.5}$.

27. $\dfrac{\log 18.5}{1.173}$.

28. $\dfrac{\log 197}{2.864}$.

29. $\dfrac{\log 175}{\log 2845}$.

30. $\dfrac{\log 198}{\log 5747}$.

31. $\dfrac{0.18455 \times 87.141}{35.146 \times 2.1757}$.

32. $\dfrac{0.38777 \times 94.191}{21.123 \times 3.0054}$.

33. $\sqrt{\dfrac{75 \times (3.5)^4}{\sqrt{15.6} \times 2.45}}$.

34. $\sqrt[3]{\dfrac{(1.16)^5 \times 31.7}{\sqrt{18.5} \times 7.75}}$.

35. $(\log 186)(\log 0.065)$.

36. $(\log 175)(\log 0.0187)$.

37. Calculate the volume of a sphere of radius $r = 2.625$ from the formula $V = \frac{4}{3}\pi r^3$. Use $\pi = 3.1416$.

38. Calculate the volume of a right-circular cylinder whose altitude is $h = 7.25$ and the radius of whose base is $r = 1.75$ from the formula $V = \pi r^2 h$. Use $\pi = 3.14$.

1011   THE GRAPH OF $y = \log_b x$

For a given value of the base b we can obtain the graph of the function

$$y = \log_b x$$

by plotting positive values of x as the abscissas and the corresponding values of $\log_b x$ as the ordinates of points on the graph.

We show graphs for the functions $y = \log_{10} x$ and $y = \log_2 x$ (Fig. 1001). Coordinates of points on the graph of $y = \log_{10} x$ may be obtained either from a table of logarithms or from the exponential form of $y = \log_{10} x$, namely, $x = 10^y$. In the exponential form we give y values and solve for x. The coordinates of some of these points are given below

x	.01	.1	1	10	100
y	-2	-1	0	1	2

To find coordinates of points on the graph of the function $y = \log_2 x$, we change the function to exponential form, give values to y and solve for x.

$$y = \log_2 x,$$
$$x = 2^y.$$

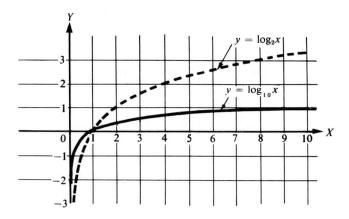

Figure 1001

The coordinates of some of the points on the graph of the function $y = \log_2 x$ are given below

x	$\frac{1}{4}$	$\frac{1}{2}$	1	2	4	8
y	-2	-1	0	1	2	3

Several important properties of the logarithm are illustrated by its graph:

1. The logarithm of a number greater than 1 is positive.
2. The logarithm of 1 is zero.
3. The logarithms of numbers between 0 and 1 are negative.
4. As x approaches zero as a limit from the positive side of 0, log x decreases through negative values, and its numerical value increases beyond all bounds.
5. The logarithm of 0 does not exist.
6. The logarithm of a negative number is not on the graph of real numbers (our definition of the logarithm excludes the logarithm of a negative number).
7. As x increases through positive values, log x increases more and more slowly.

1012 EXPONENTIAL AND LOGARITHMIC EQUATIONS

*An **exponential equation** is an equation in which an unknown quantity appears in one or more exponents.*

For example, $3^x = 50$ is an exponential equation. To solve such an equation we must find the values of the unknown for which the equation is true.

Generally, simple exponential equations can be solved by taking logarithms of both members of the equation, equating them, and then solving the resulting algebraic equation.

Example 1.   Solve the equation $3^x = 50$.

Solution.   Take the logarithm of both members of the equation and equate them.

$$\log 3^x = \log 50,$$
$$x \log 3 = \log 50,$$

$$x = \frac{\log 50}{\log 3} = \frac{1.69897}{0.47712} = 3.5609.$$

We assumed the numbers 3 and 50 to be exact numbers and expressed the value of x to five significant figures (Table I is to five decimal places. The student is warned that to divide $\log 50$ by $\log 3$ we do not subtract the logarithm of 3 from the logarithm of 50, for that would be dividing 50 by 3 and not $\log 50$ by $\log 3$. The solution, by logarithms, of $x = 1.69897/0.47712 = 3.5609$, is left for the student.

Example 2.   Solve the equation $(34.75)^{2x} = (12.55)^{3x-2}$.

Solution.   Take the logarithm of both sides of the equation and equate the two. Then solve the resulting algebraic equation.

$$\log(34.75)^{2x} = \log(12.55)^{3x-2},$$
$$2x \log 34.75 = (3x - 2) \log 12.55,$$
$$2x(1.54095) = (3x - 2)(1.09864),$$
$$3.08190\, x = 3.29592\, x - 2.19728,$$
$$0.21402\, x = 2.19728,$$
$$x = 10.267.$$

Certain equations in which the unknown does not appear in an exponent can be solved by first reducing the equations to exponential form.

Example 3.   Solve for x:

$$2 \log(x + 3) = \log(7x + 1) + \log 2.$$

Solution. Rearrange terms and combine logarithms.

$$2 \log(x + 3) - \log(7x + 1) = \log 2,$$
$$\log(x + 3)^2 - \log(7x + 1) = \log 2,$$

$$\log \frac{(x + 3)^2}{7x + 1} = \log 2.$$

Therefore
$$\frac{(x + 3)^2}{7x + 1} = 2.$$

Solve for x:
$$(x + 3)^2 = 2(7x + 1),$$
$$x^2 + 6x + 9 = 14x + 2,$$
$$x^2 - 8x + 7 = 0,$$
$$(x - 1)(x - 7) = 0.$$

Therefore
$$x = 1, \quad \text{and} \quad 7.$$

Both roots satisfy the original equation.

Example 4. Find $\log_3 35.7$.

Solution. Let $x = \log_3 35.7$, and then write this equation in exponential form.

$$3^x = 35.7.$$

Take the logarithm of both members to the base 10, and then solve the resulting equation for x.

$$\log 3^x = \log 35.7$$
$$x \log 3 = \log 35.7$$

$$x = \frac{\log 35.7}{\log 3} = \frac{1.55267}{0.47712} = 3.2543.$$

Therefore

$$\log_3 35.7 = 3.2543.$$

EXERCISES

In Exercises 1–16, solve each equation, assuming that the given numbers are exact, and carrying out the work as far as possible with Table I.

1. $2^x = 3$.

2. $3^x = 10$.

3. $(7.5)^{x+1} = 100$.

4. $(5.5)^{2x-1} = 100$.

190

5. $5^x = (3.7)^{2x+1}$.

6. $7^{x+1} = (45)^{3x+1}$.

7. $(1.025)^{-x} = 0.725$.

8. $(1.05)^{-x} = 0.850$.

9. $\dfrac{(1.06)^x - 1}{.06} = 20$.

10. $\dfrac{(1.025)^x - 1}{.025} = 10$.

11. $\log x + \log(7 - x) = 1$.

12. $\log x + \log(11 - x) = 1$.

13. $\log(5x - 7) = \log(2x - 3) + \log 3$.

14. $\log(2x + 1) = \log 1 + \log(x + 2)$.

15. $\log(x + 4) - \log 10 = \log 6 - \log x$.

16. $\log(13 - 5x) - \log 3 = \log 2 - \log x$.

In Exercises 17–22, find each of the specified logarithms. Assume the given numbers to be exact and carry out the work as far as possible with Table I.

17. $\log_5 75$.

18. $\log_4 117$.

19. $\log_7 784$.

20. $\log_8 365$.

21. $\log_3 0.157$.

22. $\log_2 0.135$.

In Exercises 23–26, sketch the graph for each function.

23. $y = \log_3 x$.

24. $y = \log_4 x$.

25. $y = \log_5 x$.

26. $y = \log_6 x$.

1013 LOGARITHMS OF TRIGONOMETRIC FUNCTIONS

In Table II (pages *21–66*) there are listed the logarithms of the sine, cosine, tangent, and cotangent functions of angles from 0° to 90° inclusive.

The student will observe that the -10, following the mantissa, is omitted in the tables. Omitting the -10 in tables of trigonometric functions is done to save space.

The quantity -10 is to be appended to all logarithms of the sine and cosine, to logarithms of the tangent from 0–45°, and to logarithms of the cotangent from 45–90°. This results from the fact that for the intervals mentioned these trigonometric functions have values that are less than 1, and their logarithms are therefore negative.

With degrees indicated at the top of the page, use the column headings at the top and the minute column at the left. With degrees stated at the bottom of the page, use the column designations at the bottom and the minute column at the right.

Example 1.   Find log tan 3° 38′.

Solution.   Locate 3° at the top of page *25* (Table II), and look down the left minute column to 38′. Then in the line with 38′ and in the column under caption "Log Tan," read 8.80277. This number is to be followed by −10. Therefore,

$$\log \tan 3° \; 38' = 8.80277 - 10.$$

Example 2.   Find log cos 62° 27.4′.

Solution.   We know that log cos 62° 27.4′ must lie between log cos 62° 27′ and log 62° 28′. In Table II (page *49*) we find

$$1' \begin{cases} \log \cos 62° \; 27' \; = 9.66513 - 10 \\ \log \cos 62° \; 27.4' = \qquad ? \\ \log \cos 62° \; 28' \; = 9.66489 - 10. \end{cases} 24 \qquad \begin{array}{l} \text{(notice that the} \\ \text{``d'' column gives} \\ \text{this 24)} \end{array}$$

Since 62° 27.4′ is .4 of the way from 62° 27′ to 62° 28′, we assume that log 62° 27.4′ is .4 of the way from 9.66513 − 10 to 9.66489 − 10. The tabular difference is 24; therefore .4(24) ≈ 10 (in the last two digits) must be subtracted from the larger entry (9.66513 − 10). We subtract instead of add because for acute angles cos θ decreases, instead of increases, as θ increases. After making this subtraction we get

$$\log \cos 62° \; 27.4' = 9.66503 - 10.$$

Example 3.   Find $x = 37.156 \sin 27° \; 48.9'$.

Solution.   We use Table I (page *7*) to find log 3.7156; Table II (page *49*) for log sin 27° 48.9′; and then Table I to find the antilogarithm of log x.

$$\begin{aligned} \log 37.156 &= 1. \\ \log \sin 27° \; 48.9' &= 9. \qquad -10 \, (+) \\ \hline \log x &= \\ x &= \end{aligned} \qquad\qquad \begin{aligned} \log 37.156 &= 1.57003 \\ \log \sin 27° \; 48.9' &= 9.66897 - 10 \, (+) \\ \hline \log x &= 11.23900 - 10 \\ x &= 17.338. \end{aligned}$$

EXERCISES

In Exercises 1–10, find the logarithms.

1. log sin 1° 25′.

2. log cos 2° 37′.

3. log tan 38° 15′.

4. log sin 75° 55′.

5. log cos 25° 14.7′.

6. log tan 31° 15.8′.

7. log sin 157° 15.3′.

8. log cos 151° 14.1′.

9. log cot 135° 45′.

10. log cot 171° 10′.

In Exercises 11–20, solve the θ.

11. log sin $\theta = 9.64519 - 10$.

12. log tan $\theta = 9.69552 - 10$.

13. log cos $\theta = 9.72320 - 10$.

14. log cos $\theta = 9.95585 - 10$.

15. log tan $\theta = 8.75199 - 10$.

16. log sin $\theta = 8.75130 - 10$.

17. log sin $\theta = 9.21096 - 10$.

18. log cos $\theta = 9.22150 - 10$.

19. log tan $\theta = 1.34946$.

20. log tan $\theta = 1.35802$.

In Exercises 21 through 26, assume the given numbers to be exact and solve for x. Use Tables I and II.

21. $x = 1.355 \sin 45° \ 15'$.

22. $x = 205.4 \cos 15° \ 27'$.

23. $x = \dfrac{25.6 \sin 51° \ 10'}{\sin 35° \ 25'}$.

24. $x = \dfrac{3.84 \sin 78° \ 15'}{\sin 41° \ 55'}$.

25. $x = \dfrac{3.85 \tan 75° \ 10'}{\cos 155°}$.

26. $x = \dfrac{7.84 \tan 81° \ 44'}{\cos 175°}$.

11

Solution of oblique triangles

1101 SOLUTION OF OBLIQUE TRIANGLES

Any triangle which is not a right triangle is an *oblique triangle*.

The solution of oblique triangles is not limited to this chapter. The Law of Sines (Sec. 214) and the Law of Cosines (Sec. 308) were used previously to solve oblique triangles. In this chapter we shall add other formulas, for finding the area of an oblique triangle. For oblique triangles, the necessary three parts may be given in four ways, and solutions for these ways will be discussed under the following four cases:

Case 1. One side and two angles.

Case 2. Two sides and the angle opposite one of them.

Case 3. Two sides and the included angle.

Case 4. Three sides.

As stated in Section 213, we shall denote the vertices and corresponding angles of a triangle in general by capital letters, A, B, and C, and the lengths of the opposite sides by the corresponding small letters, a, b, and c. The formulas developed for solving oblique triangles apply to right triangles but are usually less desirable for right triangles than those given in Chapters 2, 3, and 4.

In solving a problem, a sketch is helpful both as an aid in determining which formula is to be used, and as a check for the detection of gross errors in the solution.

In all cases, before carrying out the numerical calculations, a blank computation form, such as was used in Section 1010, should be made out completely before turning to the tables.

The given parts should be used as much as possible in the calculations, so that possible errors in computed parts may not be carried on into further calculations.

Before going on we recall the Law of Sines

$$\frac{a}{\sin A} = \frac{b}{\sin B} = \frac{c}{\sin C},$$

and the Law of Cosines

$$a^2 = b^2 + c^2 - 2bc \cos A,$$
$$b^2 = a^2 + c^2 - 2ac \cos B,$$
$$c^2 = a^2 + b^2 - 2ab \cos C.$$

1102   THE AREA OF A TRIANGLE

Corresponding to different sets of data (Cases 1, 2, 3, and 4) there are several formulas for the area of a triangle. We shall denote the area of a triangle by K. If a, b, c are the lengths of the sides in inches, for example, K will be expressed in square inches.

I. *Given two sides and the included angle (Case 3). The area of a triangle is equal to one-half the product of any two sides times the sine of the included angle:*

(1)
$$K = \begin{cases} \tfrac{1}{2}ab \sin C, \\ \tfrac{1}{2}ac \sin B, \\ \tfrac{1}{2}bc \sin A. \end{cases}$$

Proof. We construct triangle ABC showing A first as an acute then as an obtuse angle (Fig. 1101). Let h be the altitude drawn from vertex C. Then

$$K = \tfrac{1}{2}ch \qquad \text{and} \qquad h = b \sin A.$$

Therefore
$$K = \tfrac{1}{2}cb \sin A.$$

II. *Given one side and two angles (Case 1)*. The area of a triangle is equal to one-half the square of any side, times the product of the sines of the*

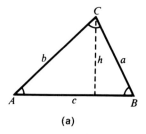
(a)

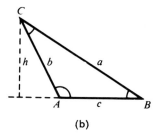
(b)

Figure 1101

* When two angles of a triangle are given, the third angle is also considered as known.

adjacent angles, divided by the sine of the third angle:

(2)
$$K = \begin{cases} \frac{1}{2}a^2 \dfrac{\sin B \sin C}{\sin A}, \\[2mm] \frac{1}{2}b^2 \dfrac{\sin A \sin C}{\sin B}, \\[2mm] \frac{1}{2}c^2 \dfrac{\sin A \sin B}{\sin C}. \end{cases}$$

Proof.   Let b be any side of a triangle. From the Law of Sines we have

$$\frac{b}{\sin B} = \frac{a}{\sin A}, \quad \text{or} \quad b = \frac{a \sin B}{\sin A}.$$

Substituting this value of b in the formula $K = \frac{1}{2}ab \sin C$, we have

$$K = \frac{1}{2}a\left(\frac{a \sin B}{\sin A}\right) \sin C = \frac{1}{2}a^2 \frac{\sin B \sin C}{\sin A}.$$

III. *Given the three sides (Case 4). The area of a triangle is found by the formula*

(3)
$$K = \sqrt{s(s-a)(s-b)(s-c)},$$

where $s = \dfrac{(a+b+c)}{2}$.

Proof.   We square both members of Formula (1), $K = \frac{1}{2}bc \sin A$, and obtain

(a) $K^2 = \frac{1}{4}b^2c^2 \sin^2 A = \frac{1}{4}b^2c^2(1 - \cos^2 A) = \frac{1}{4}b^2c^2(1 + \cos A)(1 - \cos A)$.

From the Law of Cosines,

(b)
$$\cos A = \frac{b^2 + c^2 - a^2}{2bc},$$

We first add then subtract the right member of (b) to and from (1) and obtain

(c)
$$1 + \cos A = 1 + \frac{b^2 + c^2 - a^2}{2bc} = \frac{2bc + b^2 + c^2 - a^2}{2bc}$$

$$= \frac{(b+c)^2 - a^2}{2bc} = \frac{(b+c+a)(b+c-a)}{2bc},$$

(d)
$$1 - \cos A = 1 - \frac{b^2 + c^2 - a^2}{2bc} = \frac{2bc - b^2 - c^2 + a^2}{2bc}$$

$$= \frac{a^2 - (b-c)^2}{2bc} = \frac{(a+b-c)(a-b+c)}{2bc}.$$

We substitute the results of (c) and (d) into (a) and get

(e) $K^2 = \frac{1}{16}(a + b + c)(b + c - a)(a + b - c)(a - b + c).$

When we set $(a + b + c) = 2s$ and subtract $2a$ from both members of the equation, we get $(b + c - a) = 2(s - a)$. In a similar way

$$(a + b - c) = 2(s - c), \quad \text{and} \quad (a - b + c) = 2(s - b).$$

When these expressions are substituted in (e) we get

$$K^2 = s(s - a)(s - b)(s - c).$$

Taking the square root of both members we obtain

$$K = \sqrt{s(s - a)(s - b)(s - c)}$$

This formula is known as **Hero's formula.**

   IV. *Given two sides and the angle opposite one of them (Case 2). Proceed as follows:*

   (i) *Determine whether one or two triangles are possible (See Section 215.)*
   (ii) *Use the Law of Sines to find another side or angle.*
   (iii) *Then use Formula (1): $K = \frac{1}{2}bc \sin A$.*

Example 1. Find the area of the triangle ABC if $a = 75.45$, $b = 39.71$, and $C = 42° \, 37'$. We assume the data to be exact.

Solution. Use $K = \frac{1}{2}ab \sin C$ and Tables I and II.

$\log 75.45 = 1.$	$\log 75.45 = 1.87766$
$\log 39.71 = 1.$	$\log 39.71 = 1.59890$
$\log \sin 42° \, 37' = 9. \qquad -10 \, (+)$	$\log \sin 42° \, 37' = 9.83065 - 10 \, (+)$
	$\overline{3.30721}$
$\log 2 = \underline{\qquad\qquad} \quad (-)$	$\log 2 = \quad .30103 \quad (--)$
$\log K =$	$\log K = 3.\overline{00618}$
$K =$	$K = 1014.3$ square units.

Example 2. Find the area of the triangle ABC if $a = 0.816$, $b = 0.315$, and $c = 0.717$. We assume the given numbers to be exact.

Solution. Use Formula (3)

$$K = \sqrt{s(s - a)(s - b)(s - c)},$$

where $s = \dfrac{a + b + c}{2} = \dfrac{0.816 + 0.315 + 0.717}{2} = 0.924.$

$$s - a = 0.108, \; s - b = 0.609, \; s - c = 0.207.$$

Taking the logarithm of both members of Formula (3) we get

$$\log K = \tfrac{1}{2}[\log s(s-a)(s-b)(s-c)].$$
$$\log s = 9.96567 - 10$$
$$\log(s-a) = 9.03342 - 10$$
$$\log(s-b) = 9.78462 - 10$$
$$\underline{\log(s-c) = 9.31597 - 10 \,(+)}$$
$$\log K = \tfrac{1}{2}(38.09968 - 40)$$
$$= 19.04984 - 20$$
$$K = 0.11216 \text{ square units.}$$

EXERCISES

(Assume the numbers given in the following exercises to be exact and round off the answers to table accuracy. Use Tables I, II, III, or VII.)

In Exercises 1–16, find the areas of the triangles ABC with the given parts.

1. $a = 100, b = 20, C = 41° 35'$.

2. $a = 80, b = 20, C = 38° 15'$.

3. $A = 22° 20', B = 48° 20', c = 10$.

4. $A = 30°, B = 41° 50', c = 10$.

5. $a = 7, b = 8, c = 9$.

6. $a = 9, b = 10, c = 11$.

7. $a = 3.854, b = 5.051, C = 74° 13'$.

8. $a = 1.754, b = 3.191, C = 61° 17'$.

9. $b = 0.760, c = 0.415, A = 157° 45'$.

10. $a = 0.875, c = 0.391, B = 137° 14'$.

11. $b = 1.885, C = 39° 10', A = 105° 50'$.

12. $a = 1.059, B = 41° 10'. C = 110° 45'$.

13. $a = 18.75, b = 9.15, c = 10.48$.

14. $a = 15.87, b = 8.13, c = 11.50$.

15. $c = 15.875, A = 41° 15.7', C = 89° 41.3'$.

16. $a = 17.718, A = 55° 40.8', B = 110° 15.7'$.

17. A triangular corner lot is at the intersection of two streets which meet at an oblique angle. The frontages on the streets are 125 and 65 feet, and the line across the back of the lot measures 160 feet. Find the area of the lot.

18. A triangular lot has one side 125 ft long, another side 185 ft long, and the included angle is 80°. Find the area of the lot.

19. How many acres are there in a triangular field whose sides measure 430, 780, and 550 feet respectively? (One acre equals 43,560 square feet.)

20. How many acres are there in a triangular lot whose sides measure 125, 95, and 160 feet respectively?

21. A farmer wishes to buy a triangular field on the corner formed by two roads which meet at an angle of 115°. The frontage on one road is 44.3 rods and on the other 62.5 rods. What will the field cost him at $350 per acre? (One acre equals 160 square rods.)

22. Find the area of a hexogen inscribed in a circle whose radius is 2.625 inches.

23. A bar of steel with uniform triangular cross section is 12 feet long. The cross-sectional triangle has angles of 50°, 50°, and 80°. The side between the 50° angles measures 6.5 inches. The ends of the bar are perpendicular to its longitudinal axis. (a) Calculate the number of cubic feet in the bar. (b) Steel is about 7.8 times as heavy as water, and one cubic foot of water weighs 62.4 pounds; find the weight of the bar.

24. The area of a triangle is 468 square units, and two of its angles are 58° and 76°. Find the length of the longest side.

25. Prove that the area of a parallelogram is equal to the product of the lengths of a pair of adjacent sides and the sine of their included angle.

26. Two adjacent sides of a parallelogram have lengths of 3.756 and 1.893 and include an angle of 115° 35′. Find the area of the parallelogram.

1103   THE LAW OF TANGENTS

In the solution of an oblique triangle when two sides and the included angle are known (Case 3), either the Law of Cosines or the Law of Tangents may be used. When the lengths of the sides of the triangle are not easily squared and logarithms are to be used, then the Law of Tangents is preferred, as it is well suited to logarithmic computation.

In any triangle the difference between any two sides is to their sum as the tangent of one-half of the difference between the opposite angles is to the tangent of one-half of the sum of those angles:

$$\frac{a-b}{a+b} = \frac{\tan\frac{1}{2}(A-B)}{\tan\frac{1}{2}(A+B)}, \qquad \frac{b-c}{b+c} = \frac{\tan\frac{1}{2}(B-C)}{\tan\frac{1}{2}(B+C)},$$

$$\frac{c-a}{c+a} = \frac{\tan\frac{1}{2}(C-A)}{\tan\frac{1}{2}(C+A)}$$

Proof.   From the Law of Sines, we have

(1)
$$\frac{a}{c} = \frac{\sin A}{\sin C}$$

and

(2)
$$\frac{b}{c} = \frac{\sin B}{\sin C}.$$

Subtracting Equation (2) from Equation (1) and adding Equations (1) and (2) yields

(3)
$$\frac{a-b}{c} = \frac{\sin A - \sin B}{\sin C}$$

and

(4)
$$\frac{a+b}{c} = \frac{\sin A + \sin B}{\sin C}.$$

Dividing each member of (3) by the corresponding member of (4) yields

(5)
$$\frac{a-b}{a+b} = \frac{\sin A - \sin B}{\sin A + \sin B}.$$

Applying Identities (32) and (31) to the right-hand member of (5), we obtain

$$\frac{a-b}{a+b} = \frac{2 \cos \frac{1}{2}(A+B)\sin \frac{1}{2}(A-B)}{2 \sin \frac{1}{2}(A+B)\cos \frac{1}{2}(A-B)}$$

$$= \frac{\sin \frac{1}{2}(A-B)}{\cos \frac{1}{2}(A-B)} \cdot \frac{\cos \frac{1}{2}(A+B)}{\sin \frac{1}{2}(A+B)}.$$

Hence
$$\frac{a-b}{a+b} = \frac{\tan \frac{1}{2}(A-B)}{\tan \frac{1}{2}(A+B)}.$$

In a similar manner we can derive

$$\frac{b-c}{b+c} = \frac{\tan \frac{1}{2}(B-C)}{\tan \frac{1}{2}(B+C)}, \qquad \frac{c-a}{c+a} = \frac{\tan \frac{1}{2}(C-A)}{\tan \frac{1}{2}(C+A)}.$$

1104   APPLICATIONS OF THE LAW OF TANGENTS

When two sides and the included angle of an oblique triangle are known, we use the Law of Tangents to find the other two angles and then apply the Law of Sines to find the third side of the triangle.

Suppose that a and b and the included angle C are given. If $a > b$, we use the Law of Tangents in the form involving $(a - b)$ and $\frac{1}{2}(A - B)$; if $b > a$, we use a form containing $(b - a)$ and $\frac{1}{2}(B - A)$, to avoid negative factors. Let us assume that $a > b$, then the unknown in the Law of Tangents is the expression $\frac{1}{2}(A - B)$. With C known we find $\frac{1}{2}(A + B)$ as follows:

$$A + B + C = 180°.$$

Then
$$A + B = 180° - C,$$

and
$$\tfrac{1}{2}(A + B) = \tfrac{1}{2}(180° - C).$$

We use the law of tangents to find
$$\tfrac{1}{2}(A - B) = \underline{\hspace{2cm}}.$$

The angles A and B can now be found from the relations $A = \frac{1}{2}(A + B) + \frac{1}{2}(A - B)$ and $B = \frac{1}{2}(A + B) - \frac{1}{2}(A - B)$; side c can be found by the use of the Law of Sines.

Example.   Solve the triangle ABC, given
$$a = 591.3, \qquad b = 786.4, \qquad C = 37° \ 17'.$$

Solution.   Since $b > a$ we use the form
$$\frac{b - a}{b + a} = \frac{\tan \frac{1}{2}(B - A)}{\tan \frac{1}{2}(B + A)}$$

$$
\begin{aligned}
b &= 786.4 \\
a &= 591.3 \\
\hline
b + a &= 1377.7 \\
b - a &= 195.1.
\end{aligned}
\qquad
\begin{aligned}
\tfrac{1}{2}(B + A) &= \tfrac{1}{2}(180° - 37° \ 17') \\
&= \tfrac{1}{2}(179° \ 60' - 37° \ 17') \\
&= \tfrac{1}{2}(142° \ 43') = 71° \ 21.5'.
\end{aligned}
$$

$$\tan \tfrac{1}{2}(B - A) = \frac{(b - a)\tan \frac{1}{2}(B + A)}{b + a}$$

$$= \frac{195.1 \tan 71° \ 21.5'}{1377.70}.$$

$$
\begin{aligned}
\log 195.1 &= 2.29026 \\
\log \tan 71° \ 21.5' &= 10.47192 - 10 \ (+) \\
\hline
& 12.76218 - 10
\end{aligned}
$$

$$
\begin{aligned}
\log 1377.7 &= 3.13915 \qquad (-) \\
\hline
\log \tan \tfrac{1}{2}(B - A) &= 9.62303 - 10
\end{aligned}
$$

Therefore,   (1) $\tfrac{1}{2}(B - A) = 22° \ 46.4'$.

But   (2) $\tfrac{1}{2}(B + A) = 71° \ 21.5 \ (+)$

which gives $B = 94° \ 08'$ to the nearest minute.

When we subtract (1) from (2) we find $A = 48°\ 35'$, to the nearest minute. We use the Law of Sines to find c.

$$\frac{c}{\sin C} = \frac{a}{\sin A}$$

$$c = \frac{a \sin C}{\sin A} = \frac{591.3 \sin 37°\ 17'}{\sin 48°\ 35'}.$$

$$
\begin{aligned}
\log 591.3 &= \ 2.77181 \\
\log \sin 37°\ 17' &= \ \underline{9.78230 - 10\ (+)} \\
&\quad\ \ 12.55411 - 10 \\
\log \sin 48°\ 35' &= \ \underline{9.87501 - 10\ (-)} \\
\log c &= \ \ 2.67910 - 10
\end{aligned}
$$

Therefore　　　　　　　　　$c = 477.6$　　　to four significant digits.

EXERCISES

　　　Solve the following triangles. Assume the given numbers to be exact and round off the answers to table accuracy. Use Tables I and II.

1. $b = 23.4,\ c = 28.9,\ A = 35°\ 40'$.

2. $a = 15.8,\ c = 24.9,\ B = 68°\ 10'$.

3. $a = 4.156,\ c = 5.977,\ B = 58°\ 12'$.

4. $b = 2.815,\ c = 3.064,\ A = 41°\ 55'$.

5. $a = 65.10,\ b = 50.99,\ C = 52°\ 48'$.

6. $a = 63.87,\ b = 49.61,\ C = 73°\ 45'$.

7. $b = 1842,\ c = 1721,\ A = 105°\ 14'$.

8. $a = 0.3916,\ c = 0.5155,\ B = 115°\ 34'$.

9. $a = 0.8657,\ b = 0.6316,\ C = 127°\ 21.4'$.

10. $b = 53.177,\ c = 33.654,\ A = 79°\ 52.6'$.

11. Two streets meet at an angle of $96°\ 30'$. A triangular lot at their intersection has a frontage of 158.66 feet on one street and 127.75 feet on the other. Find the angles the third side makes with the two streets and the length of the third side.

12. In the parallelogram $ABCD$, $AB = 375.5$, $AD = 186.9$, and angle $BAD = 75°\ 10'$. Find the length of the diagonal AC.

1105   THE HALF-ANGLE FORMULAS IN TERMS OF THE SIDES OF A TRIANGLE

Let ABC be any triangle with sides equal to a, b, and c. Let AO, BO, and CO be the bisectors of angles A, B, and C, respectively (Fig. 1102). The bisectors of the angles of a triangle meet at a point that is equidistant from the sides of the triangle; O is therefore the center of the circle inscribed in the triangle.

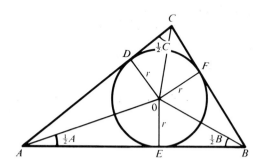

Figure 1102

We draw OE, OF, and OD perpendicular to the respective sides of the triangle. Hence $OE = OF = OD = r$, the radius of the inscribed circle. Then

$$\text{area } \triangle ABC = \text{area } \triangle ABO + \text{area } \triangle BCO + \text{area } \triangle AOC,$$

and it follows that

(1)
$$K = \tfrac{1}{2}cr + \tfrac{1}{2}ar + \tfrac{1}{2}br = r\left(\frac{a + b + c}{2}\right) = rs.$$

By Formula (3) of Section 1101

(2)
$$K = \sqrt{s(s - a)(s - b)(s - c)}.$$

Equating (1) and (2) we get

$$rs = \sqrt{s(s - a)(s - b)(s - c)},$$

and
$$r = \sqrt{\frac{(s - a)(s - b)(s - c)}{s}}.$$

Thus:

The radius of the inscribed circle in any triangle ABC is given by

(3)
$$r = \sqrt{\frac{(s - a)(s - b)(s - c)}{s}},$$

where $s = \tfrac{1}{2}(a + b + c)$.

By elementary geometry, in Figure 1102

$$AE = AD, \qquad BE = BF, \qquad CF = CD.$$

Then

$$AB + BC + AC = 2s,$$
$$AE + BF + CF = s,$$
$$AE = s - (BF + CF)$$
$$= s - a.$$

Thus in the right triangle AEO

$$\tan \tfrac{1}{2}A = \frac{r}{s - a}.$$

When we apply a similar procedure to triangles BOE and CDO we obtain

$$\tan \tfrac{1}{2}B = \frac{r}{s - b} \qquad \text{and} \qquad \tan \tfrac{1}{2}C = \frac{r}{s - c}.$$

Hence:

   In any triangle ABC,

(4) $\tan \tfrac{1}{2}A = \dfrac{r}{s - a}, \qquad \tan \tfrac{1}{2}B = \dfrac{r}{s - b}, \qquad \tan \tfrac{1}{2}C = \dfrac{r}{s - c},$

where $r = \sqrt{\dfrac{(s - a)(s - b)(s - c)}{s}} \qquad and \qquad s = \tfrac{1}{2}(a + b + c).$

   Sines of the half-angles.   If A, B, C are the angles of a triangle, then $\tfrac{1}{2}A$, $\tfrac{1}{2}B$, and $\tfrac{1}{2}C$ are acute angles, and their functions are all positive. Therefore, by Identity (24), Section 806,

(5) $\sin \tfrac{1}{2}A = \sqrt{\dfrac{1 - \cos A}{2}}, \qquad \text{and} \quad \sin^2 \tfrac{1}{2}A = \tfrac{1}{2}(1 - \cos A).$

By the Law of Cosines,

$$\cos A = \frac{b^2 + c^2 - a^2}{2bc}.$$

We substitute this expression for $\cos A$ in (5), and obtain

$$\sin^2 \tfrac{1}{2}A = \frac{1}{2}\left(1 - \frac{b^2 + c^2 - a^2}{2bc}\right) = \frac{a^2 - b^2 + 2bc - c^2}{4bc}$$

$$= \frac{a^2 - (b^2 - 2bc + c^2)}{4bc} = \frac{a^2 - (b - c)^2}{4bc}.$$

We factor the numerator as the difference of two squares, getting

$$\sin^2 \tfrac{1}{2}A = \frac{[a-(b-c)][a+(b-c)]}{4bc} = \frac{a-b+c}{2} \cdot \frac{a+b-c}{2} \cdot \frac{1}{bc}$$

$$= \frac{(s-b)(s-c)}{bc}.$$

In a similar way formulas can be obtained for $\sin^2 \tfrac{1}{2}B$ and $\sin^2 \tfrac{1}{2}C$. Hence:
In any triangle ABC,

(6) $\sin \tfrac{1}{2}A = \sqrt{\dfrac{(s-b)(s-c)}{bc}},$ $\sin \tfrac{1}{2}B = \sqrt{\dfrac{(s-a)(s-c)}{ac}},$

$$\sin \tfrac{1}{2}C = \sqrt{\dfrac{(s-a)(s-b)}{ab}},$$

where $s = \tfrac{1}{2}(a+b+c)$.

If we wish to find all of the angles of a triangle, when the sides are known, it is best to use the half-angle tangent formulas. If only one of the angles is desired, the formula for the sine of the half-angle is satisfactory unless the angle is very large or very small.

Example.   Find the radius of the inscribed circle and the angles of triangle *ABC,* given

$$a = 31.45,$$
$$b = 27.18,$$
$$c = 35.19.$$

Solution.   By addition, $2s = a + b + c = 93.82.$

$$s = 46.91 \qquad r = \sqrt{\frac{(s-a)(s-b)(s-c)}{s}}.$$

$s - a = 15.46$ $\log(s-a) = 1.18921$

$s - b = 19.73$ $\log(s-b) = 1.29513$

$s - c = 11.72$ $\log(s-c) = 1.06893 \ (+)$

$\overline{ 3.55327}$

$\log s = 1.67127 \ (-)$

$\log r^2 = \overline{1.88200} \ (\div 2)$

$\log r = 0.94100$

$r = 8.730.$

$$\tan \tfrac{1}{2}A = \frac{r}{s-a}. \qquad\qquad \tan \tfrac{1}{2}B = \frac{r}{s-b}.$$

$$\log r = 10.94100 - 10 \qquad\qquad \log r = 10.94100 - 10$$
$$\log(s - a) = \underline{\ 1.18921\ } \ (-) \qquad \log(s - b) = \underline{\ 1.29512\ } \ (-)$$
$$\log \tan \tfrac12 A = 9.75179 - 10 \qquad \log \tan \tfrac12 B = 9.64587 - 10$$
$$\tfrac12 A = 29°\,27.1', \qquad\qquad \tfrac12 B = 23°\,52.0',$$
$$A = 58°\,54'. \qquad\qquad\qquad B = 47°\,44'.$$

$$\tan \tfrac12 C = \frac{r}{s - c}.$$

$$\log r = 10.94100 - 10$$
$$\log(s - c) = \underline{\ 1.06893\ } \quad (-)$$
$$\log \tan \tfrac12 C = 9.87207 - 10$$
$$\tfrac12 C = 36°\,40.8'$$
$$C = 73°\,22'.$$

Check.

$$A = 58°\,54'$$
$$B = 47°\,44'$$
$$\underline{C = 73°\,22'}$$
$$A + B + C = 180°\,00'.$$

We solve the above example for angle A using the formula

$$\sin \tfrac12 A = \sqrt{\frac{(s - b)(s - c)}{bc}}.$$

$$\log(s - b) = 1.29513 \qquad\qquad \log b = 1.43425$$
$$\log(s - c) = \underline{1.06893}\ (+) \qquad \log c = \underline{1.54642}\ (+)$$
$$(1) \qquad\qquad 2.36406 \qquad (2) \qquad\qquad 2.98067$$

Subtracting (2) from (1)

$$12.36406 - 10$$
$$\underline{\ 2.98067\ } \quad (-)$$
we get $\qquad \log \sin^2 \tfrac12 A = 9.38339 - 10$
$$= 19.38339 - 20 \quad (\div 2)$$
$$\log \sin \tfrac12 A = 9.69170 - 10,$$
$$\tfrac12 A = 29°\,27',$$
$$A = 58°\,54'.$$

EXERCISES

Solve the following triangles.

1. $a = 116$, $b = 95.5$, $c = 127$.

2. $a = 209$, $b = 184$, $c = 239$.

3. $a = 0.356$, $b = 0.750$, $c = 0.609$.

4. $a = 0.475$, $b = 0.385$, $c = 0.208$.

5. $a = 75.67$, $b = 49.75$, $c = 38.62$.

6. $a = 37.15$, $b = 41.85$, $c = 19.58$.

7. $a = 182.37$, $b = 293.21$, $c = 160.64$.

8. $a = 512.36$, $b = 409.84$, $c = 748.67$.

9. What is the radius of the largest circular track that can be laid out within a triangular field whose sides are 275, 198, and 345 yards long, respectively?

10. A motorboat race follows a triangular course. The first leg of 1760 yards is due north, and the other two legs of 1320 and 2200 yards, respectively, lie to the east of the first one. Find the bearing of the third leg.

12

Inverse trigonometric functions

1201   INVERSE TRIGONOMETRIC RELATIONS

We often need a notation to express an angle in terms of one of its trigonometric functions.

*If $x = \sin \theta$, then θ is an angle (or a number) whose sine is x. We then say
that θ is **an inverse sine of** x, and we express this by the notation*

(1) $\theta = \arcsin x.$

This notation is an abbreviation for "θ is an angle (or a number) whose
sine is x."*

Example 1.   If $\sin \theta = \frac{1}{2}$ then we can say that "θ is a number (or an angle)
whose sine is $\frac{1}{2}$," or

$$\theta = \arcsin \tfrac{1}{2}.$$

Thus, θ may be equal to $\frac{1}{6}\pi$, $\frac{5}{6}\pi$, $\frac{13}{6}\pi$, $-\frac{7}{6}\pi$, and so forth, or, in general,
$\theta = \frac{1}{6}\pi \pm 2n\pi$, or $\frac{5}{6}\pi \pm 2n\pi$, where $n = 0, 1, 2, \ldots$.
From the observation of Example 1, we see that the relation $y = \arcsin x$
is infinitely many-valued. Since y, in the equation $y = \arcsin x$, is not uniquely
determined when x is given, we refrain from calling it a function (if y is a
function of x, at most one value of y corresponds to any one value of x). Such a
pairing in which to each x there may correspond none, one, or several values
of y is called a *relation*. This leads to the need of defining the so-called

---

* Another form in common use is $\theta = \sin^{-1}x$. When this form is used, it must be
understood that the -1 is not an exponent.

principal values of the inverse relations, in terms of which all the related inverse functions may be expressed. This is done in Section 1203.

The remaining inverse trigonometric relations are defined in a corresponding way. The expression arccos x denotes a number whose cosine is x, arctan x denotes a number whose tangent is x, and so forth. The other three inverse trigonometric relations, arccot x, arcsec x, and arccsc x, are of less importance. The arccsc x and arcsec x can be expressed in terms of arcsin x and arccos x, respectively. For example, the expression arccsc x may be considered equivalent to arcsin $1/x$. More specifically,

$$\text{arccsc } 2 = \arcsin \frac{1}{2} = \frac{\pi}{6}.$$

Example 2.   Let θ be an angle placed in standard position with its terminal side passing through the point P(4, 3). (*See* Fig. 1201.)

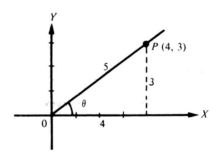

Figure 1201

We describe the relations shown in Figure 1201 in the following different but equivalent ways. Each statement applies to Figure 1201, in which θ is an acute angle.

1. $\sin \theta = 3/5$; θ is the acute angle whose sine is $3/5$;

                             $\theta = \arcsin 3/5.$

2. $\cos \theta = 4/5$; θ is the acute angle whose cosine is $4/5$;

                             $\theta = \arccos 4/5.$

3. $\tan \theta = 3/4$; θ is the acute angle whose tangent is $3/4$;

                             $\theta = \arctan 3/4.$

4. $\cot \theta = 4/3$; θ is the acute angle whose cotangent is $4/3$;

                             $\theta = \text{arccot } 4/3.$

5. $\sec \theta = 5/4$; θ is the acute angle whose secant is 5/4;

$$\theta = \text{arcsec } 5/4.$$

6. $\csc \theta = 5/3$; θ is the acute angle whose cosecant is 5/3;

$$\theta = \text{arccsc } 5/3.$$

Example 3. Solve the equation $y = \sin 2x$ for x in terms of y.

Solution. Since $\sin 2x = y$,

$$2x = \arcsin y.$$

and
$$x = \tfrac{1}{2} \arcsin y.$$

Example 4. Solve the equation $x = \tfrac{1}{3} \arctan 2y$ for y in terms of x.

Solution. Since $x = \tfrac{1}{3} \text{ arc tan } 2y$,

$$3x = \arctan 2y,$$

and
$$2y = \tan 3x,$$

therefore
$$y = \tfrac{1}{2} \tan 3x.$$

Example 5. Restricting arctan $\tfrac{5}{12}$ to the first quadrant, find the value of each of the following expressions:

(a) $\sin(\arctan \tfrac{5}{12})$,

(b) $\cos(\arctan \tfrac{5}{12})$,

(c) $\tan(\arctan \tfrac{5}{12})$.

Solution. (a) This part of the example merely asks for the sine of the acute angle whose tangent is $\tfrac{5}{12}$. Let θ be this angle. Sketch the angle (Fig. 1202); let $P(12, 5)$ be a point on the terminal side of θ; find 13, the radius vector for P; then read the answer for part (a) of the problem, from the figure. Answers for parts (b) and (c) are obtained in the same manner.

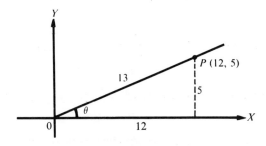

Figure 1202

(a) $\sin(\arctan \frac{5}{12}) = \frac{5}{13}$.

(b) $\cos(\arctan \frac{5}{12}) = \frac{12}{13}$.

(c) $\tan(\arctan \frac{5}{12}) = \frac{5}{12}$.

EXERCISES

In Exercises 1–12, find the value of each expression, in the interval $0 \le \theta < 2\pi$, without the use of tables.

1. arcsin 1.   2. arccos 1.   3. arccos 0.

4. arcsin 0.   5. arctan 1.   6. arccot $\sqrt{3}$.

7. $\arcsin(-\frac{1}{2})$.   8. $\arccos(-\frac{1}{2})$.   9. $\arccos(-1)$.

10. $\arctan(-1)$.   11. arccsc 2.   12. $\text{arcsec}(-2)$.

In Exercises 13–20, use Table IV to find the value of each expression in the interval $0° \le \theta < 360°$.

13. arccos .8746.   14. arcsin .5150.   15. arctan .3906.

16. arctan .9397.   17. $\arcsin(-.0581)$.   18. $\arccos(-.9848)$.

19. $\arctan(-.2126)$.   20. $\arcsin(-.9983)$.

Solve each of the following Exercises for x in terms of y.

21. $y = 2 \sin 3x$.   22. $y = \frac{1}{2} \cos 2x$.

23. $y = \frac{1}{3} \cos 2x$.   24. $y = 3 \sin 4x$.

25. $y = \tan \pi x$.   26. $y = \sin \pi x$.

27. $y = 1 + \cos 3x$.   28. $y = 2 - \sin 2x$.

29. $y = \frac{1}{2}\pi - \arccos 2x$.   30. $y = \pi - \arctan 3x$.

In Exercises 31–40, find the value of each expression. Restrict the arcfunction expressions to acute angles.

31. $\tan(\arcsin \frac{4}{5})$.   32. $\tan(\arcsin \frac{3}{8})$.

33. $\sec(\arccos \frac{2}{3})$.   34. $\csc(\arcsin \frac{1}{6})$.

35. $\sin(\arcsin \frac{3}{7})$.   36. $\cos(\arccos \frac{5}{7})$.

37. $\cos(\arcsin \frac{2}{8})$.   38. $\sin(\arccos \frac{1}{3})$.

39. $\sin(\arcsin \frac{3}{5} + \arccos \frac{3}{5})$. 40. $\cos(\arctan \frac{5}{12} + \arcsin \frac{3}{5})$.

1202 GRAPHS OF THE INVERSE TRIGONOMETRIC RELATIONS

The graphs of the inverse trigonometric relations show their behavior quite clearly.

To obtain the graph of $y = \arcsin x$, write the equivalent direct form of this relation, $x = \sin y$, and proceed as in Section 217. The coordinates of some of the points on the graph are given in the attached table. (*See* Fig. 1203a.)

y	$-\pi$	$-\frac{1}{2}\pi$	0	$\frac{1}{2}\pi$	π	$\frac{3}{2}\pi$	2π
x	0	-1	0	1	0	-1	0

The graphs (Figs. 1203, 1204, 1205) of the remaining inverse trigonometric relations may be obtained in a corresponding way.

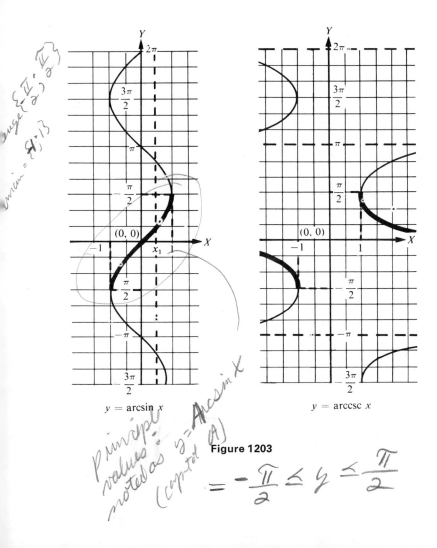

$y = \arcsin x$

$y = \arccsc x$

Figure 1203

Since the sine function only takes on values from -1 to 1, the domain of arcsin x is $-1 \leq x \leq 1$. For example, arcsin 2 is not defined, since there is no angle whose sine is 2. On the other hand, arctan x (Fig. 1205a) is defined for all real values of x, since the tangent may assume any real value. We can say that:

the domain of arcsin x and arccos x is $-1 \leq x \leq 1$, the domain of arctan x and arccot x is $-\infty < x < +\infty$, the domain of arcsec x and arccsc x is $x \geq 1$ and $x \leq -1$.

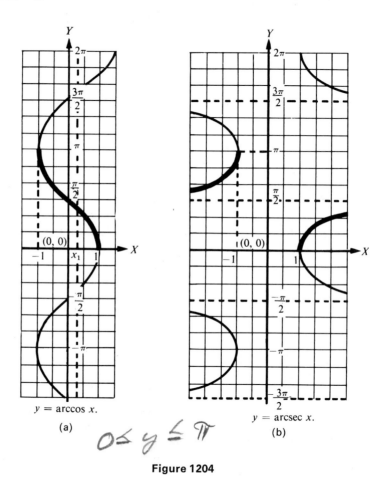

$y = \arccos x.$

(a)

$y = \arcsec x.$

(b)

Figure 1204

A vertical line drawn at any point in the interval $-1 \leq x \leq 1$, on the graphs for $y = \arcsin x$ and $y = \arccos x$ (Figs. 1203a and 1204a), meets the graphs in infinitely many points; the ordinates of these points are the infinitely many values of arcsin x and arccos x for the corresponding value of x.

range = $\{\frac{\pi}{2}; \frac{2\pi}{3}\}$
domain = $\{-I; +J\}$

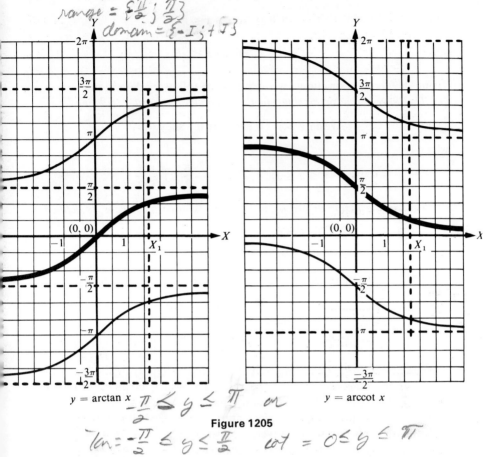

$y = \arctan x$ $-\frac{\pi}{2} \leq y \leq \pi$ or $y = \text{arccot } x$

Figure 1205

$\tan = -\frac{\pi}{2} \leq y \leq \frac{\pi}{2}$ $\cot = 0 \leq y \leq \pi$

1203   PRINCIPAL VALUES OF THE INVERSE TRIGONOMETRIC FUNCTIONS

In our discussion we have emphasized that the trigonometric inverse relations are many-valued. However, under suitable restrictions these inverse relations may be made one-to-one, and in this way we obtain the important **inverse trigonometric** or **arc functions**. As is seen by the graph of $y = \arcsin x$ (Fig. 1203a), if the range of the relation is limited to $-\pi/2 \leq \arcsin x \leq \pi/2$, there is a unique value of y for each x of the domain. In a similar manner, the range for each of the other arc functions is restricted to distinguish between the inverse relations and functions. The capital letter "A" is used in writing the arc functions. Thus *the range of each of the inverse trigonometric functions is defined as follows*:

(i) $-\frac{\pi}{2} \leq \text{Arcsin } x \leq \frac{\pi}{2},$

(ii) $$0 \le \text{Arccos } x \le \pi,$$

(iii) $$-\frac{\pi}{2} < \text{Arctan } x < \frac{\pi}{2}.$$

Using the above restrictions, we arrive at the following definitions:
(1) *The inverse sine or arcsine function, denoted by Arcsin, is defined by*

$$\text{Arcsin } x = y \quad \text{if and only if} \quad \sin y = x,$$

where

$$-1 \le x \le 1 \quad \text{and} \quad -\frac{\pi}{2} \le y \le \frac{\pi}{2}$$

(2) *The inverse cosine or arccosine function, denoted by Arccos, is defined by*

$$\text{Arccos } x = y \quad \text{if and only if} \quad \cos y = x,$$

where $$-1 \le x \le 1 \quad \text{and} \quad 0 \le y \le \pi.$$

(3) *The inverse tangent or arctangent function, denoted by Arctan, is defined by*

$$\text{Arctan } x = y \quad \text{if and only if} \quad \tan y = x,$$

where $$-\infty < x < \infty \quad \text{and} \quad -\frac{\pi}{2} < y < \frac{\pi}{2}.$$

The graphs of these functions are indicated in the figures by the heavy lines.
    Authors sometimes differ in their choice of principal vaiues for the other
three inverse trigonometric functions, Arccot x, Arcsec x, and Arccsc x. As
these are of less importance in elementary work, we shall not be concerned
with them in this course. However, the principal values commonly used for
these are shown by the heavy lines on the graphs of $y = \text{arccot } x$ (Fig. 1205b),
$y = \text{arcsec } x$ (Fig. 1204b), and $y = \text{arccsc } x$ (Fig. 1203b).

Example 1.   In this example we illustrate the difference between the principal
and general values of the expressions. The number n is any integer.

(a)   Arcsin $1 = \pi/2$.
(b)   arc sin $1 = \pi/2 + 2n\pi$.
(c)   Arctan$\sqrt{3} = \pi/3$.
(d)   arctan$\sqrt{3} = \pi/3 + n\pi$.
(e)   Arccos$(-1) = \pi$.
(f)   arccos$(-1) = \pi + 2n\pi$.
(g)   Arcsin$(-1) = -\pi/2$.
(h)   arcsin$(-1) = -\pi/2 + 2n\pi$.

Example 2.   Find the value of sin(2 Arctan $\frac{1}{3}$).

Solution.   Let $\theta = \text{Arctan } \frac{1}{3}$.
Then $\tan \theta = \frac{1}{3}$ and θ is an acute angle. Sketch θ in standard position (Fig. (1206).

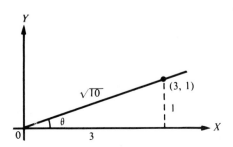

Figure 1206

By Identity (21), Section 805,

$$\sin\left(2 \text{ Arctan } \frac{1}{3}\right) = \sin 2\theta = 2 \sin \theta \cos \theta = 2\left(\frac{1}{\sqrt{10}}\right)\left(\frac{3}{\sqrt{10}}\right) = \frac{3}{5}.$$

Example 3.   Find the value of $\sin(\text{Arcsin } \frac{5}{13} + \text{Arccos } \frac{4}{5})$.

Solution.   Set $\text{Arcsin } \frac{5}{13} = \alpha$ and $\text{Arccos } \frac{4}{5} = \beta$; then α and β are positive acute angles, and $\sin \alpha = \frac{5}{13}$, $\cos \beta = \frac{4}{5}$. Sketch angles α and β in standard position and compute the missing parts of the reference triangles (Fig. 1207). Then, by applying the $\sin(\alpha + \beta)$ identity, we have

$$\sin(\alpha + \beta) = \sin \alpha \cos \beta + \cos \alpha \sin \beta$$

$$= \frac{5}{13} \cdot \frac{4}{5} + \frac{12}{13} \cdot \frac{3}{5} = \frac{56}{65}.$$

Therefore $\sin(\text{Arcsin } \frac{5}{13} + \text{Arccos } \frac{4}{5}) = \dfrac{56}{65}.$

Example 4.   Assume $0 < u < 5$ and show that

$$\tan\left(\text{Arcsin } \frac{u}{5}\right) = \frac{u}{\sqrt{25 - u^2}}.$$

Solution.   Set $\text{Arcsin } u/5 = \theta$,   then $\sin \theta = u/5$.
Sketch θ in standard position and complete the reference triangle as shown in

Figure 1208. Compute the third side of the reference triangle. Then, from the figure,

$$\tan\left(\text{Arcsin}\,\frac{u}{5}\right) = \tan\theta = \frac{u}{\sqrt{25 - u^2}}.$$

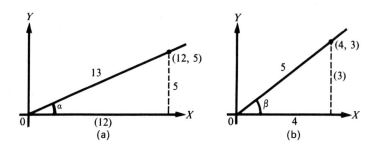

Figure 1207

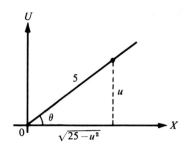

Figure 1208

Example 5. Use inverse trigonometric notation to represent θ where $\frac{1}{2}\pi < \theta < \pi$ and $\tan\theta = -\frac{3}{4}$.

Solution. Sketch the angle in standard position (Fig. 1209). Let α be the related angle of θ, then

$$\alpha = \text{Arctan}\,\tfrac{3}{4}.$$

But $$\theta = \pi - \alpha.$$

Therefore $$\theta = \pi - \text{Arctan}\,\tfrac{3}{4}.$$

Example 6. Solve $\text{Arctan}(3x^2 + 1) = 2\,\text{Arctan}\,\tfrac{1}{2}$.

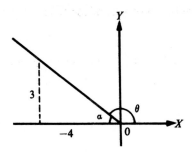

Figure 1209

Solution. Take the tangents of both members of the equation and equate them.

$$\tan[\operatorname{Arctan}(3x^2 + 1)] = \tan(2 \operatorname{Arctan} \tfrac{1}{2}),$$

(1) $$3x^2 + 1 = \tan(2 \operatorname{Arctan} \tfrac{1}{2}).$$

Set $\operatorname{Arctan} \tfrac{1}{2} = \theta$, then $\tan \theta = \tfrac{1}{2}$. Apply the double angle identity to the right member of Equation (1).

$$\tan(2 \operatorname{Arctan} \tfrac{1}{2}) = \tan 2\theta$$

$$= \frac{2 \tan \theta}{1 - \tan^2 \theta} = \frac{2 \cdot \tfrac{1}{2}}{1 - (\tfrac{1}{2})^2} = \frac{4}{3}.$$

Therefore Equation (1) becomes

$$3x^2 + 1 = \tfrac{4}{3}.$$

and $$x = \pm\tfrac{1}{3}.$$

Check.   For $x = \pm\tfrac{1}{3}$, $(3x^2 + 1) = \tfrac{4}{3}$.

Then $$\operatorname{Arctan}(3x^2 + 1) = 2 \operatorname{Arctan} \tfrac{1}{2},$$
$$\operatorname{Arctan} \tfrac{4}{3} = 2 \operatorname{Arctan} \tfrac{1}{2},$$
$$\tan(\operatorname{Arctan} \tfrac{4}{3}) = \tan(2 \operatorname{Arctan} \tfrac{1}{2}),$$
$$\tfrac{4}{3} = \tfrac{4}{3}.$$

EXERCISES

Find the value of each of the following expressions without using tables.

1. Arctan 1.

2. Arccos $\tfrac{1}{2}$.

3. Arcsin $\tfrac{1}{2}\sqrt{2}$.

4. Arcsin $\tfrac{1}{2}$.

5. Arccos$(-\frac{1}{2})$.   **6.** Arcsin$(-\frac{1}{2})$.

7. Arctan$(-\sqrt{3})$.   **8.** Arctan(-1).

9. Arcsin$(-\frac{1}{2})$.   **10.** Arccos$(-\frac{1}{2}\sqrt{2})$.

11. Arccos $\frac{1}{2}\sqrt{3}$.   **12.** Arctan $\frac{1}{3}\sqrt{3}$.

13. $\sin($Arccos $\frac{1}{3})$.   **14.** $\cos($Arctan $\frac{2}{3})$.

15. $\tan($Arcsin $\frac{5}{13})$.   **16.** $\sin($Arcsin $\frac{7}{8})$.

17. $\cos($Arccos $\frac{5}{7})$.   **18.** $\tan($Arccos $\frac{3}{5})$.

19. $\cos(2 $Arctan $\frac{3}{5})$.   **20.** $\sin(2 $Arccos $\frac{1}{3})$.

21. $\sin(2 $Arcsin $\frac{7}{8})$.   **22.** $\tan(2 $Arcsin $\frac{5}{13})$.

23. $\tan(2 $Arccos $\frac{3}{5})$.   **24.** $\cos(2 $Arccos $\frac{5}{7})$.

25. $\sin[$Arcsin$(-\frac{1}{2}) + $Arccos$(-\frac{1}{2})]$.   **26.** $\sin[$Arccos $\frac{1}{3} + $Arcsin $\frac{1}{3}]$.

27. $\cos[$Arcsin $\frac{5}{13} - $Arccos $\frac{4}{5}]$.   **28.** $\sin[$Arcsin $\frac{12}{13} - $Arcsin $\frac{4}{5}]$.

29. $\tan[$Arctan $\frac{9}{40} + $Arctan $\frac{3}{4}]$.   **30.** $\tan[$Arctan $\frac{4}{5} + $Arctan $\frac{1}{4}]$.

31. $\tan[$Arctan $2 - $Arctan $\frac{1}{2}]$.   **32.** $\tan[$Arctan $3 - $Arctan $2]$.

In Exercises 33–41, use inverse-trigonometric notation to represent θ under the specified conditions.

33. $\sin \theta = \frac{3}{5}, \frac{1}{2}\pi < \theta < \pi$.   **34.** $\cos \theta = \frac{3}{5}, \frac{1}{2}\pi < \theta < \pi$.

35. $\cos \theta = -\frac{5}{13}, \pi < \theta < \frac{3}{2}\pi$.   **36.** $\sin \theta = -\frac{5}{13}, \pi < \theta < \frac{3}{2}\pi$.

37. $\tan \theta = -\frac{3}{4}, \frac{3}{2}\pi < \theta < 2\pi$.   **38.** $\tan \theta = -\frac{5}{12}, \frac{3}{2}\pi < \theta < 2\pi$.

39. $\sin \theta = \frac{4}{5}, 2\pi < \theta < \frac{5}{2}\pi$.   **40.** $\sin \theta = \frac{3}{5}, 2\pi < \theta < \frac{5}{2}\pi$.

41. $\tan 2\theta = \frac{24}{7}, \frac{1}{2}\pi < \theta < \pi$.

In Exercises 42–52, show that each of the given statements is true.

42. Arcsin $\frac{4}{5} + $ Arctan $\frac{3}{4} = \frac{1}{2}\pi$.

43. Arccos $\frac{12}{13} + $ Arctan $\frac{1}{4} = $ Arccot $\frac{43}{32}$.

44. Arctan $\frac{3}{4} = 2$ Arctan $\frac{1}{3}$.

45. $\frac{1}{2}$ Arctan $\frac{7}{24} = $ Arccos $\frac{7}{10}\sqrt{2}$.

46. Arctan $\frac{1}{2} - $ Arctan $\frac{1}{3} = $ Arctan $\frac{1}{7}$.

47. Arcsin $x + $ Arccos $x = \frac{1}{2}\pi$, if $-1 \leq x \leq 1$.

48. $\cos($Arcsin $x) = \sqrt{1 - x^2}$, if $-1 \leq x \leq 1$.

49. 2 Arcsin $x = $ Arccos$(1 - 2x^2)$, if $0 \leq x \leq 1$.

50. Arctan $x =$ Arcsin $\dfrac{x}{\sqrt{1+x^2}}$ for all values of x.

51. $\cos(\frac{1}{2} \text{Arccos } x) = \sqrt{\dfrac{1+x}{2}}$, if $-1 \leq x \leq 1$.

52. $\tan 2(\text{Arccos } x) = \dfrac{2x\sqrt{1-x^2}}{2x^2-1}$, if $-1 \leq x \leq 1$.

In Exercises 53–64, solve each equation for x.

53. Arctan $x + 2$ Arctan $1 = \frac{3}{4}\pi$.

54. Arctan $x =$ Arcsin $\frac{12}{13}$.

55. Arccos $2x =$ Arcsin x.

56. Arccos $x + 2$ Arcsin $1 = \pi$.

57. Arcsin $2x = \frac{1}{6}\pi +$ Arccos x.

58. Arcsin $x +$ Arcsin $2x = \frac{1}{3}\pi$.

59. Arcsin $x +$ Arccos $2x = \frac{1}{6}\pi$.

60. Arccos $2x -$ Arccos $x = \frac{1}{3}\pi$.

61. Arcsin $x +$ Arccos$(1 - x) = 0$.

62. Arctan$(x + 1) +$ Arctan$(x - 1) =$ Arctan $\frac{8}{31}$.

63. Arctan $2x +$ Arctan $3x = \frac{3}{4}\pi$.

64. Arctan $x +$ Arctan $(1 - x) =$ Arctan $\frac{4}{3}$.

Complex numbers

1301 PURE IMAGINARY NUMBERS

The student will recall from algebra that a good deal of imagination, in the sense of inventiveness, has been required to construct the real number system. The further invention of a complex number system does not then seem so strange. It is fitting for us to study such a system, since modern engineering has found that it provides a convenient language for expressing vibratory motion, harmonic oscillation, damped vibrations, alternating currents, and other wave phenomena.

The square root of a negative number (for example $\sqrt{-1}, \sqrt{-7}, \sqrt{-4}$) is called a **pure imaginary number**. To operate with such numbers a new symbol $i = \sqrt{-1}$ has been invented, then

$$\sqrt{-7} = \sqrt{7} \cdot \sqrt{-1} = \sqrt{7}\,i,$$

and

$$\sqrt{-4} = \sqrt{4} \cdot \sqrt{-1} = 2i.$$

Accordingly, the symbol $\sqrt{-1}$, called the **imaginary unit** and usually designated by i, represents a new kind of number with the property that

(1)
$$\begin{cases} \sqrt{-1} = i, \\ i^2 = (\sqrt{-1})^2 = -1. \end{cases}$$

From Property (1) it follows that

$i^1 = i$; $\qquad\qquad\qquad i^5 = i^4 \cdot i = i$;

$i^2 = -1$; $\qquad\qquad\quad i^6 = (i^2)^3 = (-1)^3 = -1$;

$i^3 = i^2 \cdot i = -i;$ $i^7 = i^6 \cdot i = -i;$

$i^4 = (i^2)^2 = (-1)^2 = 1;$ $i^8 = (i^2)^4 = (-1)^4 = 1.$

Therefore the integral powers of i have only four different values; i, -1, $-i$, 1.

1302   COMPLEX NUMBERS

If a and b are real numbers, $a + bi$ is called a **complex number**, of which a is the **real part** and bi is the **imaginary part**. Complex numbers may be thought of as including all real numbers and all pure imaginary numbers. For example, if $b = 0$, the complex number $a + bi$ is *real*; if $b \neq 0$, the complex number is said to be *imaginary*; in particular, if $a = 0$ and $b \neq 0$, the complex number is a *pure imaginary* number.

    I. Fundamental principle of operations. *Algebraic operations involving only whole powers of complex numbers may be performed by writing i in place of* $\sqrt{-1}$, *then operating with i as if it were a real number, and, finally, replacing* i^2 *by* -1.

    II. Definition of equality. *Two complex numbers are said to be equal if and only if their real parts are equal and their imaginary parts are equal:*

$$x + yi = a + bi, \quad \text{if and only if} \quad x = a \text{ and } y = b.$$

    III. Conjugate of a complex number. *The conjugate of a complex number* $a + bi$ *is the complex number* $a - bi$, *which is obtained by changing the sign of the coefficient of the imaginary part.*

1303   ALGEBRAIC OPERATIONS ON COMPLEX NUMBERS

    (1) **Addition.**   To add two complex numbers, add the real parts and add the pure imaginary parts.

$$(a + bi) + (c + di) = (a + c) + (b + d)i.$$

Example 1.

$$(3 + 2i) + (4 - 5i) = (3 + 4) + (2 - 5)i = 7 - 3i.$$

    (2) **Subtraction.**   To subtract one complex number from another, subtract its real and pure imaginary parts from the corresponding parts of the other number.

$$(a + bi) - (c + di) = (a - c) + (b - d)i.$$

Example 2.

$$(3 + 2i) - (4 - 5i) = (3 - 4) + [2 - (-5)]i = -1 + 7i.$$

(3) **Multiplication.**   To multiply two complex numbers, multiply as you would in multiplying two binomials, and replace i^2 by -1.

$$(a + bi)(c + di) = ac + (ad + bc)i + bdi^2$$
$$= (ac - bd) + (ad + bc)i.$$

Example 3.

$$(3 + 2i)(4 - 5i) = 12 - 7i - 10i^2$$
$$= 12 - 7i - 10(-1) = 22 - 7i.$$

In multiplying numbers like $\sqrt{-a} \cdot \sqrt{-b}$, where $a > 0$ and $b > 0$, first re-write each factor in the form

$$\sqrt{-a} = \sqrt{(-1)(a)} = \sqrt{-1} \cdot \sqrt{a} = i\sqrt{a},$$
$$\sqrt{-b} = \sqrt{(-1)(b)} = \sqrt{-1} \cdot \sqrt{b} = i\sqrt{b},$$

and then multiply the numbers as follows:

$$(i\sqrt{a})(i\sqrt{b}) = i^2\sqrt{ab} = -\sqrt{ab}.$$

The student is warned that

$$\sqrt{-a} \cdot \sqrt{-b} \neq \sqrt{(-a)(-b)}, \text{ or } \sqrt{ab}.$$

(4) **Division.**   To divide one complex number by another, multiply both dividend and divisor by the conjugate of the divisor.

$$\frac{a + bi}{c + di} = \frac{(a + bi)(c - di)}{(c + di)(c - di)} = \frac{(ac + bd) + (bc - ad)i}{c^2 + d^2}.$$

Example 4. $\dfrac{3 + 2i}{4 - 3i} = \dfrac{(3 + 2i)(4 + 3i)}{(4 - 3i)(4 + 3i)} = \dfrac{6 + 17i}{16 + 9}$

$$= \frac{6}{25} + \frac{23}{25}i.$$

Example 5.   Solve $y + 2i = (2 + xi)(3 - 2i)$ for the real numbers x and y.

Solution.   Express the product $(2 + xi)(3 - 2i)$ in the form $a + bi$.

(1) $y + 2i = (2 + xi)(3 - 2i)$
$$= 6 - 4i + 3xi - 2xi^2$$
$$= (6 + 2x) + (3x - 4)i.$$

Apply the Definition of equality (II of Section 1302) to (1):

$$y + 2i = (6 + 2x) + (3x - 4)i.$$

Therefore $y = 6 + 2x$, and $2 = 3x - 4$ or $x = 2$,

and $y = 6 + 2x = 6 + 4 = 10$.

Example 6. Solve $9x^2 - 6x + 5 = 0$.

Solution. Apply the quadratic formula $x = (-b \pm \sqrt{b^2 - 4ac})/2a$.

$$x = \frac{6 \pm \sqrt{36 - 180}}{18} = \frac{6 \pm \sqrt{-144}}{18} = \frac{6 \pm 12i}{18} = \frac{1}{3} \pm \frac{2}{3}i.$$

Example 7. Expand $(3 + 2i)^3$, simplify, and write the result in the form $a + bi$.

Solution.

$$\begin{aligned}(3 + 2i)^3 &= 3^3 + 3 \cdot 3^2(2i) + 3 \cdot 3(2i)^2 + (2i)^3 \\ &= 27 + 54i + 36i^2 + 8i^3 \\ &= 27 + 54i - 36 - 8i \\ &= -9 + 46i.\end{aligned}$$

EXERCISES

In Exercises 1–16, perform the indicated operations, simplify, and write the result in the form $a + bi$.

1. $(2 + 3i) + (5 - 2i)$.
2. $(1 - i) + (7 + 3i)$.
3. $(4 - 2i) - (7 - i)$.
4. $(5 + i) - (2 - 3i)$.
5. $(\frac{1}{3} + i) + (2 - \frac{1}{3}i)$.
6. $(\frac{2}{5} - 2i) + (\frac{1}{2} + \frac{1}{6}i)$.
7. $(2 + 3i)^2$.
8. $(3 - 2i)^2$.
9. $(3 + i)(3 - i)$.
10. $(7 - 2i)(7 + 2i)$.
11. $(2 + i)^3$.
12. $(2 - \frac{1}{2}i)^3$.
13. $\dfrac{1 - i}{1 + i}$.
14. $\dfrac{1 + i}{1 - i}$.
15. $\dfrac{4 + \sqrt{2}i}{4 - \sqrt{2}i}$.
16. $\dfrac{3 + \sqrt{5}i}{3 - \sqrt{5}i}$.

In Exercises 17–24, solve each equation for the real numbers x and y.

17. $3x + 4yi = 6 + 8i$.

18. $2x - yi = 4 + 3i$.

19. $y + 2i = (4 - 2i)(3 + xi)$

20. $(x + y) + (x - 4y)i = 4 + 19i$.

21. $(2x - 3y) + (x + y)i = 3 + 4i$.

22. $(x + yi)^2 = 5 + 12i$.

23. $(x^2 - 6y) + 6i = -(2x + y)i$.

24. $(2 - xi)(-y + 3i) = 15 + 27i$.

In Exercises 25–28, solve each equation for x.

25. $x^2 - 4x + 5 = 0$.

26. $x^2 + x + 1 = 0$.

27. $9x^2 - 12x + 53 = 0$.

28. $\sqrt{5}\,x^2 + 4x + \sqrt{5} = 0$.

29. Show that the conjugate of the product of two conjugate numbers is equal to the product of their respective conjugates.

30. Show that the conjugate of the quotient of two conjugate numbers is equal to the quotient of their respective conjugates.

1304   GRAPHIC REPRESENTATION OF COMPLEX NUMBERS

In algebra, real numbers are represented by points on a straight line. Since a complex number $x + yi$ depends on two independent real numbers x and y, and since (x, y) may be plotted as a point on the rectangular coordinate system, every complex number $x + yi$ may be associated with some point on the plane, and every point on the plane may be associated with some complex number. Thus, in Figure 1301, the point P, whose rectangular coordinates are $(2, 3)$, represents the complex number $2 + 3i$, and the point P' represents the

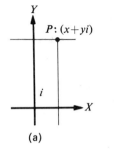
(a)

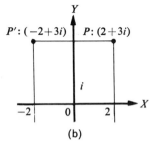
(b)

Figure 1301

number $-2 + 3i$. In such connections, we shall speak of the point which represents the number $x + yi$ as "the point $x + yi$."

As thus used, the plane in Figure 1301 is called the **complex plane**, and the number $(x + yi)$ is said to be the **rectangular** or **algebraic form** for a complex number. Also from this representation, the x-axis is often called the **axis of reals** (or the **real axis**), and the y-axis is called the **axis of imaginaries** (or **imaginary axis**). On this latter axis, the point a unit distance above O represents the imaginary unit i.

Since a real number is regarded as a complex number whose imaginary part is zero, it is represented by a point on the real axis (the x-axis). A pure imaginary number is a complex number whose real part is zero, and it is represented by a point on the imaginary axis (the y-axis). Some representative points (complex numbers) are shown on Figure 1302.

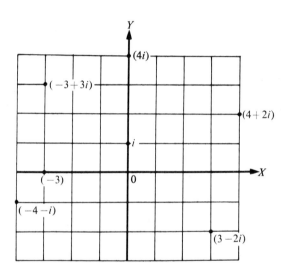

Figure 1302

EXERCISES

Represent graphically each of the following complex numbers and their conjugates:

1. $3 + i$.   **2.** $-2 + i$.   **3.** $-2 + 3i$.   **4.** $3 + 2i$.

5. $2 - 2i$.   **6.** $3 - 3i$.   **7.** $3i$ **8.** 4.

9. -3.   **10.** $2i$.

1305   POLAR COORDINATES

The position of a point in a plane is determined if its rectangular coordinates are known. An alternative method of fixing the position of a point is to give its **polar coordinates**. These are r, the distance of the point from the origin or **pole**, and θ, the angle which the line between the point and the pole makes with the positive x-axis or **polar axis**. The coordinates are written (r, θ); r is called the **radius vector** of the point; θ is its **vectorial angle**.

The radius vector is expressed in any convenient unit of length, and the vectorial angle is commonly measured in radians, though degree measure may be used. In previous chapters, the radius vector of a point has always been taken as positive. In connection with polar coordinates, however, it is often desirable to let r be negative. To plot a point when its polar coordinates (r, θ) are given, we first place the angle θ in standard position, that is, with its vertex at the pole and its initial side on the polar axis. Then, if r is positive, the the point (r, θ) lies on the terminal side of θ and at the distance r from the pole; if r is negative, (r, θ) lies at the distance $|r|$ measured from the pole in the opposite direction, that is, along the line made by extending the terminal side of θ through the pole. Thus, in Figure 1303b, the two points $P_1 : (4, \tfrac{2}{3}\pi)$ and $P_2 : (-4, \tfrac{2}{3}\pi)$ are shown. If $r = 0$, the point (r, θ) is the pole, regardless of the value of θ. Unless there is some contrary indication, we shall assume that $r \neq 0$.

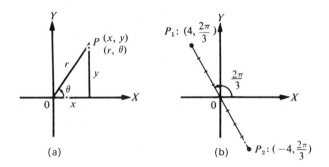

Figure 1303

It is often desirable to use rectangular and polar coordinates together, the polar axis of the polar system coinciding with the positive x-axis, and x, y, and r being measured in the same units. We can always convert from one system to the other by means of the following relations (*see* Fig. 1303a):

$$(1) \quad \begin{cases} x = r\cos\theta, \\ y = r\sin\theta. \end{cases} \qquad (2) \quad \begin{cases} r = \sqrt{x^2 + y^2}, \\ \theta = \arctan\dfrac{y}{x}. \end{cases}$$

In the equations of (2), r is taken as positive, and θ is any one of the infinitely many values of arctan (y/x) whose terminal side lies in the same quadrant (or on the same axis) as the point (x, y). If we wish to do so, we may make $r = -\sqrt{x^2 + y^2}$ and take θ in the opposite quadrant from the one containing (x, y).

Polar coordinates have many uses only one of which is the application to complex numbers.

Example 1. Find the rectangular coordinates of the point whose polar coordinates are $(3, \frac{1}{6}\pi)$. (*See* Fig. 1304.)

Solution. From Equation (1) we have

$$x = r \cos \theta = 3 \cos \tfrac{1}{6}\pi = 3(\tfrac{1}{2}\sqrt{3}) = \tfrac{3}{2}\sqrt{3},$$
$$y = r \sin \theta = 3 \sin \tfrac{1}{6}\pi = 3(\tfrac{1}{2}) = \tfrac{3}{2}.$$

The required coordinates are $(\tfrac{3}{2}\sqrt{3}, \tfrac{3}{2})$.

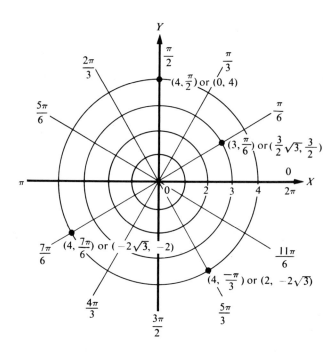

Figure 1304

Example 2. Find the polar coordinates of the point whose rectangular coordinates are $(2, -2\sqrt{3})$. (*See* Fig. 1304.)

Solution. From Equation (2), we can take

$$r = \sqrt{2^2 + (-2\sqrt{3})^2} = \sqrt{4 + 12} = 4,$$

$$\theta = \text{Arctan}(-\sqrt{3}) = -\tfrac{1}{3}\pi.$$

The required coordinates can thus be taken as $(4, -\tfrac{1}{3}\pi)$. We could have taken $r = -4$ and $\theta = \tfrac{2}{3}\pi$. There are, in fact, infinitely many sets of coordinates for every point of the plane. However, it is usually convenient to take r positive and to give θ a numerically small correct value.

EXERCISES

In each of the Exercises 1–12, plot the point whose polar coordinates are given, and find its rectangular coordinates. Tables may be used when necessary.

1. $(4, \tfrac{1}{6}\pi)$.   **2.** $(2, \tfrac{1}{3}\pi)$.   **3.** $(-3, \pi)$.   **4.** $(-2, 2\pi)$.

5. $(8, -\tfrac{1}{2}\pi)$.   **6.** $(6, -\pi)$.   **7.** $(10, 200°)$.   **8.** $(10, 100°)$.

9. $(6, \tfrac{5}{2}\pi)$.   **10.** $(4, \tfrac{3}{4}\pi)$.   **11.** $(-2, \tfrac{5}{4}\pi)$.   **12.** $(6, -\tfrac{13}{6}\pi)$.

In each of the Exercises 13–24, the rectangular coordinates are given. Find one set of polar coordinates and plot the point. Tables may be used when necessary.

13. $(2, 2)$.   **14.** $(-\sqrt{3}, 1)$.   **15.** $(3, 0)$.   **16.** $(0, 2)$.

17. $(-5, 0)$.   **18.** $(0, -2)$.   **19.** $(\sqrt{2}, -\sqrt{2})$.   **20.** $(-1, -2)$.

21. $(0, 2)$.   **22.** $(\tfrac{1}{2}, -\tfrac{1}{2}\sqrt{3})$.   **23.** $(-2, -3)$.   **24.** $(2, 1)$.

1306   POLAR OR TRIGONOMETRIC FORM OF A COMPLEX NUMBER

A trigonometric representation of complex numbers is very useful in studying certain operations with these numbers.

Suppose that (r, θ) are the polar coordinates of the point $x + yi$ on the complex plane as shown in Figure 1305. This number $(x + yi)$ determines both a point on the plane and also the length of a line from the origin to the point. The positive (or zero) length, r, of this line or *radius vector*, is called the **absolute value** or **modulus** of the complex number. The absolute value of the complex number $(x + yi)$ is denoted by

(1) $|x + iy| = r = \sqrt{x^2 + y^2}.$

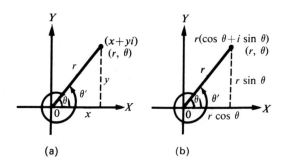

Figure 1305

The **amplitude** or **argument** (θ) of the complex number ($x + yi$) is any angle whose terminal side passes through the point (x, y), where the angle is placed in standard position on the axes. It may be chosen as any one of infinitely many values that differ by multiples of 2π. Thus

(2) $$\theta = \arctan \frac{y}{x}$$

is the **amplitude** or **angle** of the number ($x + yi$).

In Formula (2) the angle θ must be chosen so as to put the point representing the complex number in the proper quadrant. This means that θ must satisfy the two equations $x = r \cos \theta$, $y = r \sin \theta$.

Since $x = r \cos \theta$ and $y = r \sin \theta$, we can write

$$x + iy = r \cos \theta + i(r \sin \theta),$$

or

(3) $$x + iy = r(\cos \theta + i \sin \theta).$$

The left-hand member is the **rectangular form** and the right-hand member is the **polar** or **trigonometric form** of the complex number.

The expression $\cos \theta + i \sin \theta$ is sometimes abbreviated to the more compact symbol cis θ.

Example 1. Write the complex number $\sqrt{3} + i$ in polar form, and find its absolute value and amplitude.

Solution. Plot the point ($\sqrt{3}, 1$) which represents the given number; draw the radius vector and sketch the reference triangle. (Fig. 1306).

Find r and θ as follows:

$$r = \sqrt{x^2 + y^2} = \sqrt{3 + 1} = 2 \qquad \text{(absolute value).}$$

The amplitude is $\theta = \arctan 1/\sqrt{3} = \frac{1}{6}\pi + 2n\pi$.

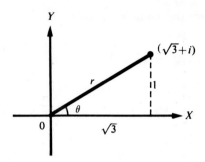

Figure 1306

Applying Formula (3)

$$x + iy = r(\cos \theta + i \sin \theta),$$

we get
$$\sqrt{3} + i = 2(\cos \tfrac{1}{6}\pi + i \sin \tfrac{1}{6}\pi).$$

We check the results as follows:

$$2(\cos \tfrac{1}{6}\pi + i \sin \tfrac{1}{6}\pi) = 2(\tfrac{1}{2}\sqrt{3} + i \cdot \tfrac{1}{2})$$
$$= \sqrt{3} + i.$$

Example 2.   Write the complex number $4i$ in polar form and give several values of its argument.

Solution.   The point that represents the complex number $4i$ is shown on Figure 1302. Here we see that $4i$ is 4 units from the origin, making $r = 4$. The argument

$$\theta = \tfrac{1}{2}\pi \text{ or } \tfrac{1}{2}\pi + 2n\pi.$$

Therefore
$$4i = 4(\cos \tfrac{1}{2}\pi + i \sin \tfrac{1}{2}\pi).$$

A more general polar form for this complex number is

$$4i = [4 \cos(\tfrac{1}{2}\pi + 2n\pi) + i \sin(\tfrac{1}{2}\pi + 2n\pi)].$$

Example 3.   The particular real and pure imaginary numbers 3, -5, and $-2i$ may be written in the polar form as follows:

$$3 = 3(\cos 0 + 1 \sin 0), \qquad -5 = 5(\cos \pi + i \sin \pi),$$
$$-2i = 2(\cos \tfrac{3}{2}\pi + i \sin \tfrac{3}{2}\pi).$$

Example 4.   Express the complex number $8(\cos 120° + i \sin 120°)$ in the rectangular form $x + yi$.

Solution.   Since $\cos 120° = -\frac{1}{2}$ and $\sin 120° = \frac{1}{2}\sqrt{3}$, we have

$$8(\cos 120° + i \sin 120°) = 8(-\tfrac{1}{2} + \tfrac{1}{2}\sqrt{3}\,i)$$
$$= -4 + 4\sqrt{3}\,i.$$

Example 5.   Let $(2, \frac{5}{6}\pi)$ be the polar coordinates of a point on the complex plane. Plot this point, and then write the complex number represented by this point.

Solution.   The polar coordinates of a point are understood to be indicated by (r, θ). Thus for the point $(2, \frac{5}{6}\pi)$ it means that $r = 2$ and $\theta = \frac{5}{6}\pi$. Place $\theta = \frac{5}{6}\pi$ in standard position, and then locate a point on the terminal side of θ two units from the origin. (*See* Fig. 1307.)

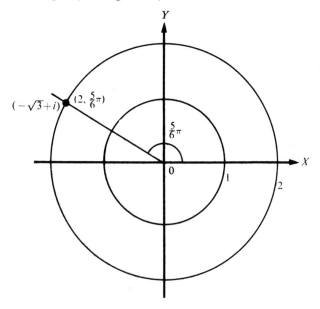

Figure 1307

In polar form this complex number is written

$$2(\cos \tfrac{5}{6}\pi + i \sin \tfrac{5}{6}\pi).$$

We can express this complex number in rectangular form as follows:

$$2(\cos \tfrac{5}{6}\pi + i \sin \tfrac{5}{6}\pi) = 2(-\tfrac{1}{2}\sqrt{3} + i\tfrac{1}{2})$$
$$= -\sqrt{3} + i.$$

EXERCISES

In each of the Exercises 1–16, represent the complex number graphically as a point and draw the radius vector; find the absolute value and one suitable value of the amplitude.

1. $2 + 2i$.

2. $2 - 2i$.

3. $-1 + \sqrt{3}\,i$.

4. $-1 + \sqrt{2}\,i$.

5. $1 - i$.

6. $1 + i$.

7. $-1 - \sqrt{2}\,i$.

8. $-3 - 3i$.

9. $3i$.

10. $-5i$.

11. -5.

12. 7.

13. 6.

14. -2.

15. $2 - 3i$.

16. $1 - 2i$.

In Exercises 17–24, represent graphically each of the complex numbers, and write the number in the rectangular form $x + yi$.

17. $3(\cos 60° + i \sin 60°)$.

18. $2(\cos \pi + i \sin \pi)$.

19. $4(\cos \frac{1}{4}\pi + i \sin \frac{1}{4}\pi)$.

20. $6(\cos \frac{5}{6}\pi + i \sin \frac{5}{6}\pi)$.

21. $\cos \pi + i \sin \pi$.

22. $4(\cos \frac{1}{2}\pi + i \sin \frac{1}{2}\pi)$.

23. $8(\cos 270° + i \sin 270°)$.

24. $2(\cos 4\pi + i \sin 4\pi)$.

In Exercises 25–30, write the complex number, in both polar and rectangular form, corresponding to the given set of polar coordinates (r, θ); draw the figure.

25. $(2, \frac{1}{6}\pi)$.

26. $(4, \frac{1}{4}\pi)$.

27. $(5, \pi)$.

28. $(3, \frac{3}{2}\pi)$.

29. $(4, \frac{3}{4}\pi)$.

30. $(4, 2\pi)$.

31. Prove that $|\cos \theta + i \sin \theta| = 1$.

32. Prove that the conjugate of $\cos \theta + i \sin \theta$ is $\cos(-\theta) + i \sin(-\theta)$.

1307   MULTIPLICATION AND DIVISION OF COMPLEX NUMBERS IN POLAR FORM

Let z_1 and z_2 be two complex numbers, expressed in the polar form by

$$z_1 = r_1(\cos \theta_1 + i \sin \theta_1), \qquad z_2 = r_2(\cos \theta_2 + i \sin \theta_2).$$

*Theorem I.   The product of two complex numbers is a number whose absolute value is the **product** of their absolute values and whose amplitude is the **sum** of the amplitudes of the two factors. Thus*

(1) $$z_1 z_2 = r_1 r_2 [\cos(\theta_1 + \theta_2) + i \sin(\theta_1 + \theta_2)].$$

Proof.   Applying Operation (3) of Section 1303, we have

$$z_1z_2 = [r_1(\cos \theta_1 + i \sin \theta_1)][r_2(\cos \theta_2 + i \sin \theta_2)]$$
$$= (r_1r_2)[(\cos \theta_1 \cos \theta_2 - \sin \theta_1 \sin \theta_2)$$
$$+ i(\sin \theta_1 \cos \theta_2 + \cos \theta_1 \sin \theta_2)].$$

But from Identity (15) of Section 802

$$\cos \theta_1 \cos \theta_2 - \sin \theta_1 \sin \theta_2 = \cos (\theta_1 + \theta_2),$$

and from Identity (17) of Section 803

$$\sin \theta_1 \cos \theta_2 + \cos \theta_1 \sin \theta_2 = \sin (\theta_1 + \theta_2).$$

Making these substitutions we get

$$z_1z_2 = (r_1r_2)[\cos(\theta_1 + \theta_2) + i \sin(\theta_1 + \theta_2)].$$

Example 1.   Multiply $\sqrt{2} - \sqrt{2}i$ and $1 + \sqrt{3}i$.

Solution.   The absolute value of the first number is $\sqrt{2 + 2} = 2$, and its amplitude is Arctan $(-1) = -45°$. The absolute value of the second number is $\sqrt{1 + 3} = 2$, and its amplitude is Arctan$(\sqrt{3}) = 60°$. Therefore, the required product is

$$2[\cos(-45°) + i \sin(-45°)] \cdot 2[\cos 60° + i \sin 60°] = 4(\cos 15° + i \sin 15°).$$

Theorem II.   If one complex number is divided by a second number, the quotient is a number whose absolute value is the quotient of the absolute value of the first divided by the absolute value of the second and whose amplitude is the amplitude of the first minus the amplitude of the second. Thus if

$$z_1 = r_1(\cos \theta_1 + i \sin \theta_1) \text{ and } z_2 = r_2(\cos \theta_2 + i \sin \theta_2)$$

then

(2) $$\frac{z_1}{z_2} = \frac{r_1}{r_2} [\cos (\theta_1 - \theta_2) + i \sin (\theta_1 - \theta_2)].$$

Proof.   Let $z_1 = r_1(\cos \theta_1 + i \sin \theta_1)$, $z_2 = r_2(\cos \theta_2 + i \sin \theta_2) \neq 0$ be any two given complex numbers. Then by multiplying both numerator and denominator of the fraction by the conjugate of the denominator we get

$$\frac{z_1}{z_2} = \frac{r_1(\cos \theta_1 + i \sin \theta_1)}{r_2(\cos \theta_2 + i \sin \theta_2)} = \frac{r_1}{r_2} \frac{(\cos \theta_1 + i \sin \theta_1)(\cos \theta_2 - i \sin \theta_2)}{(\cos \theta_2 + i \sin \theta_2)(\cos \theta_2 - i \sin \theta_2)}$$

$$= \frac{r_1}{r_2} \frac{(\cos \theta_1 + i \sin \theta_1)(\cos \theta_2 - i \sin \theta_2)}{\cos^2 \theta_2 + \sin^2 \theta_2}.$$

Since $\cos(-\theta_2) = \cos \theta_2$ and $\sin(-\theta_2) = -\sin \theta_2$ (Identities (9) and (10) of Section 801), the factor $(\cos \theta_2 - i \sin \theta_2)$ can be written

$$[\cos(-\theta_2) + i \sin (-\theta_2)],$$

and since $\cos^2 \theta_2 + \sin^2 \theta_2 = 1$, the right-hand member of the above equation reduces to

$$\frac{r_1}{r_2} (\cos \theta_1 + i \sin \theta_1)[\cos(-\theta_2) + i \sin(-\theta_2)].$$

Then by applying Theorem I we get

$$\frac{z_1}{z_2} = \frac{r_1}{r_2} [\cos (\theta_1 - \theta_2) + i \sin (\theta_1 - \theta_2)].$$

Example 2.   Divide $6(\cos 170° + i \sin 170°)$ by $3(\cos 110° + i \sin 110°)$.

Solution.   We apply Theorem II and get

$$[6(\cos 170° + i \sin 170°)] \div [3(\cos 110° + i \sin 110°)]$$

$$= \frac{6}{3}[\cos(170° - 110°) + i \sin(170° - 110°)]$$

$$= 2(\cos 60° + i \sin 60°) = 2(\tfrac{1}{2} + i \tfrac{1}{2}\sqrt{3})$$

$$= 1 + \sqrt{3}i.$$

1308   POWERS OF COMPLEX NUMBERS; DE MOIVRE'S THEOREM

As a consequence of Theorem I, Section 1307, when n is a positive integer

$$z_1 z_2 \cdots z_n = (r_1 r_2 \cdots r_n)[\cos(\theta_1 + \theta_2 + \cdots + \theta_n) + i \sin(\theta_1 + \theta_2 + \cdots + \theta_n)].$$

Therefore if $z = r(\cos \theta + i \sin \theta)$, and if n is a positive integer,

(1) $z^n = r^n(\cos n\theta + i \sin n\theta).$

When $r = 1$, this formula reduces to **De Moivre's theorem** for positive integral exponents,

(2) $(\cos \theta + i \sin \theta)^n = \cos n\theta + i \sin n\theta.$

As a special case of Theorem II, Section 1307, it follows that

$$\frac{1}{z} = \frac{\cos 0 + i \sin 0}{r(\cos \theta + i \sin \theta)} = \frac{1}{r} [\cos(0 - \theta) + i \sin (0 - \theta)]$$

$$= \frac{1}{r} (\cos \theta - i \sin \theta),$$

and, in applying Formula (1),

(3) $$z^{-n} = \frac{1}{z^n} = \frac{1}{r^n}\left[\cos(-n\theta) + i\sin(-n\theta)\right].$$

Therefore both Formulas (1) and (2) are true when the exponent is any positive or negative integer.

Example 1. Find $(1 + i)^{10}$.

Solution. We express $1 + i$ in polar form and then apply Formula (1).

$$
\begin{aligned}
(1 + i)^{10} &= \left[\sqrt{2}(\cos 45° + i\sin 45°)\right]^{10}\\
&= (\sqrt{2})^{10}(\cos 450° + i\sin 450°)\\
&= 32(\cos 90° + i\sin 90°)\\
&= 32(0 + i)\\
&= 32\,i.
\end{aligned}
$$

As an alternate solution, one could use the binomial theorem to expand $(1 + i)^{10}$ and then after simplifying and collecting terms we should have only the term $32i$ remaining. This latter solution would be long and certainly is not recommended, which indicates the importance of Formula (1) as applied to problems of this type.

Example 2. Find the value of $z = \dfrac{128}{(1 + i\sqrt{3})^6}$.

Solution. Let $z = \dfrac{128}{(1 + i\sqrt{3})^6} = 128(1 + i\sqrt{3})^{-6}$.

We express $1 + i\sqrt{3}$ in polar form.

$$1 + i\sqrt{3} = 2(\cos 60° + i\sin 60°).$$

Then, by applying Formula (2), we have

$$[2(\cos 60° + i\sin 60°)]^{-6} = 2^{-6}[\cos(-360°) + i\sin(-360°)]$$

$$= \frac{1}{64}(1 + 0).$$

Therefore $$z = \frac{128}{(1 + i\sqrt{3})^6} = 128\left(\frac{1}{64}\right) = 2.$$

Example 3. Derive identities for $\cos 3\theta$ and $\sin 3\theta$ by expanding

$$(\cos\theta + i\sin\theta)^3$$

by De Moivre's theorem and by the binomial theorem, then equating the real parts and the imaginary parts.

Solution.   By De Moivre's theorem

(1) $(\cos \theta + i \sin \theta)^3 = \cos 3\theta + i \sin 3\theta.$

We use the binomial theorem to expand the left-hand member of this equation; replace i^2 by -1, i^3 by $-i$, and arrange in the form $a + bi$.

$\cos^3 \theta + 3(\cos \theta)^2(i \sin \theta) + 3(\cos \theta)(i \sin \theta)^2 + (i \sin \theta)^3$

$$= \cos^3 \theta + (3 \cos^2 \theta \sin \theta)i - 3 \cos \theta \sin^2 \theta - (\sin^3 \theta)i$$

$$= (\cos^3 \theta - 3 \cos \theta \sin^2 \theta) + (3 \cos^2 \theta \sin \theta - \sin^3 \theta)i.$$

By replacing $\sin^2 \theta$ by $(1 - \cos^2 \theta)$ in the first term and $\cos^2 \theta$ by $(1 - \sin^2 \theta)$ in the second term, we reduce the expression as follows:

$[\cos^3 \theta - 3 \cos \theta(1 - \cos^2 \theta)] + [3(1 - \sin^2 \theta) \sin \theta - \sin^3]i$

$$= (4 \cos^3 \theta - 3 \cos \theta) + (3 \sin \theta - 4 \sin^3 \theta)i.$$

Replacing $(\cos \theta + i \sin \theta)^3$ of Equation (1) by this last expression, we have

$$(4 \cos^3 \theta - 3 \cos \theta) + (3 \sin \theta - 4 \sin^3 \theta)i = \cos 3\theta + (\sin 3\theta)i.$$

Applying II (Definition of equality) of Section 1302, we have the identities

$$\cos 3\theta = 4 \cos^3 \theta - 3 \cos \theta,$$
$$\sin 3\theta = 3 \sin \theta - 4 \sin^3 \theta.$$

EXERCISES

In Exercises 1–16, perform the indicated operations, giving the results in both polar and rectangular form; in the latter form, express the roots either exactly or to three decimal places.

1. $2(\cos 50° + i \sin 50°) \cdot 3(\cos 40° + i \sin 40°).$

2. $4(\cos 100° + i \sin 100°) \cdot 5(\cos 80° + i \sin 80°).$

3. $[8(\cos 305° + i \sin 305°)] \div [4(\cos 65° + i \sin 65°)].$

4. $[10(\cos 293° + i \sin 293°)] \div [5(\cos 23° + i \sin 23°)].$

5. $5(\cos 15° + i \sin 15°) \cdot 2(\cos 125° + i \sin 125°).$

6. $5(\cos 50° + i \sin 50°) \cdot 2(\cos 110° + i \sin 110°).$

7. $(\cos 15° + i \sin 15°)^{10}.$

8. $(\cos 75° + i \sin 75°)^{10}.$

9. $[2(\cos 12° + i \sin 12°)]^{-5}.$

10. $2[(1-i)]^{-4}$.

11. $[3(-1+i)]^4$.

12. $\dfrac{16}{(\sqrt{3}+i)^2}$.

13. $\dfrac{8}{(\sqrt{3}-i)^2}$.

14. $\dfrac{50}{(3-4i)^3}$.

15. $\dfrac{169}{(12-5i)^3}$.

16. $(1+\sqrt{3}\,i)(1-i)(-1-i)$.

17. Derive identities for $\cos 2\theta$ and $\sin 2\theta$ by expanding $(\cos\theta + i\sin\theta)^2$ by De Moivre's theorem and by the binomial theorem, then equating real and imaginary parts.

18. Derive identities for $\cos 4\theta$ and $\sin 4\theta$ by a method similar to that in Exercise 17.

1309   ROOTS OF COMPLEX NUMBERS

The statement $\sqrt{9} = x$ implies $x^2 = 9$. The statement $\sqrt[n]{z} = w$ implies $w^n = z$. Thus the problem of extracting the nth roots of a complex number z is one of solving the equation

$$(1) \hspace{4cm} w^n = z$$

for w, when z and the positive integer n are given.

We express z in polar form,

$$z = r(\cos\theta + i\sin\theta),$$

and set

$$w = R(\cos\phi + i\sin\phi),$$

where R and ϕ are presently unknown. Then Equation (1) becomes

$$(2) \hspace{2cm} R^n(\cos n\phi + i\sin n\phi) = r(\cos\theta + i\sin\theta).$$

Since points that represent equal complex numbers coincide (Definition II, Section 1302), it follows that their absolute values are equal and their amplitudes differ only by integral multiples of $360°$. From (2) we then obtain

$$(3) \hspace{3cm} R^n = r \hspace{1cm} \text{and} \hspace{1cm} n\phi = \theta + k \cdot 360°$$

where k is an integer, or

(4) $$R = \sqrt[n]{r} \quad \text{and} \quad \phi = \frac{\theta}{n} + \frac{k\,360°}{n}$$

where $\sqrt[n]{r}$ denotes the principal nth root of the positive number r. Substituting these values of R and ϕ in $w = R(\cos \phi + i \sin \phi)$, we obtain nth roots of z. Write

(5) $$w_k = \sqrt[n]{r}\left[\cos\left(\frac{\theta}{n} + \frac{k\cdot 360°}{n}\right) + i \sin \left(\frac{\theta}{n} + \frac{k \cdot 360°}{n}\right)\right].$$

If we give k the n values 0, 1, 2, ..., $n-1$, we obtain n distinct complex numbers w_k, which are all nth roots of z. If we give k any other integral value, we find that we obtain one of the values w_k previously obtained.

If n is a positive integer and $z \neq 0$, a complex number $z = r(\cos \theta + i \sin \theta)$, real or imaginary, has n and only n distinct nth roots, which are given by the formula

(6) $$w_k = \sqrt[n]{r}\left[\cos\left(\frac{\theta}{n} + \frac{k\cdot 360°}{n}\right) + i \sin \left(\frac{\theta}{n} + \frac{k \cdot 360°}{n}\right)\right],$$

where k takes the values 0, 1, 2, ..., $n-1$.

Example 1.   Find all the fourth roots of $-8 - 8\sqrt{3}\,i$, and plot points which represent these roots on the complex plane.

Solution.   In polar form

$$-8 - 8\sqrt{3}\,i = 16(\cos 240° + i \sin 240°).$$

Therefore $r = 16$ and $\theta = 240°$. Substituting these values for r and θ in Formula (6) we have

$$w_k = \sqrt[4]{16}\left[\cos\left(\frac{240°}{4} + \frac{k \cdot 360°}{4}\right) + i \sin \left(\frac{240°}{4} + \frac{k\cdot 360°}{4}\right)\right],$$

or

$$w_k = 2\cos(60° + k\,90°) + i \sin(60° + k\,90°).$$

The values of k are 0, 1, 2, and 3, giving the following four roots:

$$w_0 = 2(\cos 60° + i \sin 60°) = 2(\tfrac{1}{2} + i\tfrac{1}{2}\sqrt{3}) = 1 + i\sqrt{3},$$

$$w_1 = 2(\cos 150° + i \sin 150°) = 2(-\tfrac{1}{2}\sqrt{3} + i\tfrac{1}{2}) = -\sqrt{3} + i,$$

$$w_2 = 2(\cos 240° + i \sin 240°) = 2[-\tfrac{1}{2} + i(-\tfrac{1}{2}\sqrt{3})] = -1 - i\sqrt{3},$$

$$w_3 = 2(\cos 330° + i \sin 330°) = 2(\tfrac{1}{2}\sqrt{3} - i\tfrac{1}{2}) = \sqrt{3} - i.$$

As the roots have equal amplitudes they can be represented by points on a circle of radius 2 on the complex plane. (*See* Fig. 1308.)

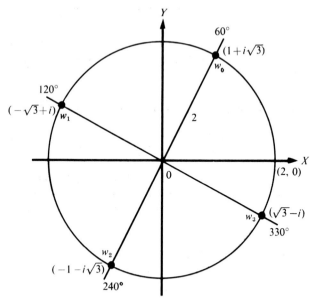

Figure 1308

Example 2.   Find all of the tenth roots of 1.

Solution.   In polar form

$$1 = \cos 0° + i \sin 0°,$$

therefore $r = 1$ and $\theta = 0°$. Substituting these values for r and θ in Formula (6) we have

$$w_k = \sqrt[10]{1}\left[\cos\left(\frac{0°}{10} + \frac{k \cdot 360°}{10}\right) + i \sin\left(\frac{1°}{10} + \frac{k \cdot 360°}{10}\right)\right]$$

or

$$w_k = \cos k36° + i \sin k36°.$$

We assign the values 0, 1, 2, ..., 9 for k, and obtain:

$$w_0 = \cos 0° + i \sin 0° = 1,$$
$$w_1 = \cos 36° + i \sin 36° = 0.8090 + 0.5878\ i,$$
$$w_2 = \cos 72° + i \sin 72° = 0.3090 + 0.9511\ i,$$
$$w_3 = \cos 108° + i \sin 108° = -0.3090 + 0.9511\ i,$$
$$w_4 = \cos 144° + i \sin 144° = -0.8090 + 0.5878\ i,$$
$$w_5 = \cos 180° + i \sin 180° = -1,$$

$$w_6 = \cos 216° + i \sin 216° = -0.8090 - 0.5878\ i,$$
$$w_7 = \cos 252° + i \sin 252° = -0.3090 - 0.9511\ i,$$
$$w_8 = \cos 288° + i \sin 288° = 0.3090 - 0.9511\ i,$$
$$w_9 = \cos 324° + i \sin 324° = 0.8090 - 0.5878\ i.$$

We could have used the compact symbol cis θ (Sec. 1305) to express the above roots as follows:

$$w_0 = \text{cis } 0°,\ w_1 = \text{cis } 36°,\ w_2 = \text{cis } 72°, \text{ etc.}$$

These roots are represented by points, spaced at equal intervals around a unit circle, on the complex plane. (*See* Fig. 1309.)

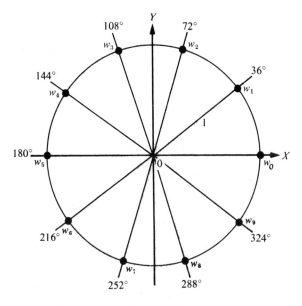

Figure 1309

EXERCISES

In each of the Exercises 1–14, find all of the indicated roots in the rectangular form $a + bi$; in the latter form express the roots either exactly or to three decimal places. Draw a figure, plotting the given number and each of the required roots.

1. Cube roots of i.

2. Cube roots of $-i$.

3. Fourth roots of -1.

4. Sixth roots of 1.

5. Fifth roots of $4 + 4i$.

6. Fifth roots of $4 - 4i$.

7. Cube roots of 1.   **8.** Fourth roots of i.

9. Sixth roots of 64.   **10.** Square roots of $-1 + \sqrt{3}\,i$.

11. Fourth roots of $8 - 8\sqrt{3}\,i$.   **12.** Cube roots of $-4\sqrt{2} + 4\sqrt{2}\,i$.

13. Square roots of $-5 + 12i$.   **14.** Square roots of $4 - 3i$.

In each Exercise 15–20, solve the equation completely.

15. $x^5 + 1 = 0$.   **16.** $x^6 + 64 = 0$.

17. $x^3 + 27 = 0$.   **18.** $x^4 + 81 = 0$.

19. $x^4 - 2x^2 + 4 = 0$.   **20.** $x^4 + 4x^2 + 16 = 0$.

ANSWERS TO ODD-NUMBERED EXERCISES

Section 109   Page 12

1. 2.
3. 3.
5. 4.
7. 2.
9. 1.
11. 2.
13. 80°.
15. 30°.
17. 85°.
19. 85°.
21. 40°.
23. 20°.
25. 30°.
27. 80°.
29. 0°.
31. 90°.
33. 0°.
35. 10°.
37. 10° 15′.
39. 40° 10′.
41. 39° 35′.
43. 75° 10.3′.
45. 59° 49.3′.
47. 28° 44.6′.
49. 5.
51. 13.
53. 5.
55. 5.
57. 11.4018.
59. 8.24621.
61. 13.91119.
63. 4.
65. -12.
67. $-\sqrt{3}$.
69. -1.

Section 207   Page 22

1. $\frac{1}{2}(1 + \sqrt{3})$.
3. -2.
5. $\frac{1}{2}(\sqrt{2} - 1)$.
7. 1.
9. $2\frac{3}{4}$.
11. $-\frac{1}{2}$.
13. $\frac{15}{16}$.
15. $1\frac{1}{2}$.
17. 0.
19. $6\frac{1}{2}$.
21. $\frac{4}{5}$.
23. $\frac{4}{5}$.
25. $-\frac{1}{2}\sqrt{2}$.
29. 0.

Section 209   Page 25

1. 0.1736.
3. 0.3420.
5. -0.6428.
7. -0.9848.
9. -0.1736.
11. -0.9848.
13. 0.3529.
15. 0.9822.
17. -0.9971.
19. 0.6428.
21. 0.7071.
23. 23° 00′.

25. 44° 20′.   **27.** 69° 20′.   **29.** 170° 00′.
31. 123° 50′.   **33.** 213° 00′.   **35.** 240° 10′.
37. 334° 00′.   **39.** 297° 10′.

Section 213   Page 31

1. 4.   **3.** 4.   **5.** 5.
7. 12.2.   **9.** 4.0.   **11.** 2,000.
13. 9.3×10^7.   **15.** 3.147×10^1.   **17.** 1.5×10^{-2}.
19. 4.00×10^3.   **21.** 9.30×10^7.   **23.** 0.00150 or 1.50×10^{-3}.
25. 2.813.   **27.** 1936.   **29.** 10° 50′.
31. 56° 00′.   **33.** 14.34 ft.

Section 214   Page 34

1. $a = 1.02$, $c = 1.17$.   **3.** $b = 1.78$, $c = 3.19$.
5. 85.5 ft.   **7.** 263 ft.

Section 215   Page 39

1. $B = 36° 40′$, $C = 102° 20′$, $c = 16.4$.
3. No triangle.
5. $B_1 = 58° 30′$, $C_1 = 78° 30′$, $c_1 = 11.5$.
 $B_2 = 121° 30′$, $C_2 = 15° 30′$, $c_2 = 3.13$.
7. Right triangle. $A = 90°$, $C = 61° 50′$, $b = 4.72$, $c = 8.82$.
9. $A_1 = 81° 50′$, $B_1 = 35° 10′$, $b_1 = 5.82$.
 $A_2 = 98° 10′$. $B_2 = 18° 50′$, $b_2 = 3.26$.

Section 216   Page 42

1. 90°.   **3.** 0°, 180°.
5. 60°, 120°.   **7.** 225°, 315°.
9. No solution.   **11.** 210°, 330°.
13. 0°, 180°, 270°.   **15.** 30°, 150°, 210°, 330°.
17. 30°, 150°, 270°,   **19.** 270°.
21. 0°, 90°, 180°, 210°, 330°.   **23.** 45°, 225°.
25. 0°, 90°, 135°, 180°, 270°, 315°.   **27.** 15°, 75°, 135°, 195°, 255°, 315°.

Section 217   Page 46

1.

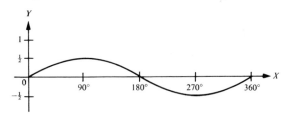

3.

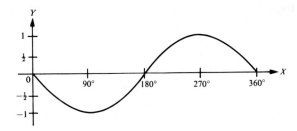

5.

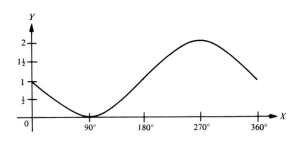

7.

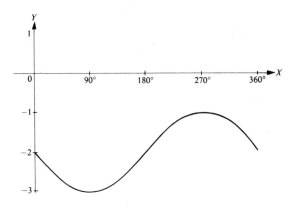

9.

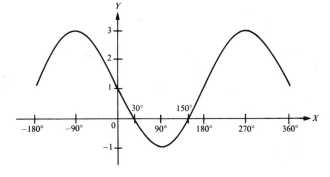

Section 305   Page 53

1. 0.

3. 1.

5. 1.

7. 1.

9. $\sqrt{3}-1$.

11. $\sqrt{3}$.

13. 2.

15. $\frac{1}{2}(4+\sqrt{3})$.

17. $-\frac{1}{2}\sqrt{3}$.

19. $6\frac{1}{2}$.

21. $\sin\theta=\frac{3}{5}$, $\cos\theta=\frac{4}{5}$.

23. $\sin\theta=-\frac{4}{5}$, $\cos\theta=\frac{3}{5}$.

25. $\sin\theta=\frac{1}{2}\sqrt{3}$, $\cos\theta=-\frac{1}{2}$.

27. 0.9848.

29. -0.9397.

31. -0.7660.

33. 0.1736.

35. 0.9848.

37. -0.1736.

39. -0.1880.

41. -0.0843.

43. 0.7660.

45. 18° 30′.

47. 69° 00′.

49. 88° 30′.

Section 306   Page 57

1. 0.9254.

3. 0.7589.

5. 0.2580.

7. -0.1086.

9. 0.9650.

11. 0.0724.

13. 19° 14′.

15. 8° 35′.

17. 11° 28′.

19. 135° 58′.

21. 155° 47′.

23. 171° 12′.

25. 0°.

27. 90°, 270°.

29. 30°, 330°.

31. 150°, 210°.

33. 120°, 240°.

35. 90°, 180°, 270°.

37. 60°, 120°, 240°, 300°.

39. 60°, 180°, 300°.

41. 0°.

43. 0°, 90°, 120°, 240°, 270°.

45. 45°, 135°, 225°, 315°.

47. 45°, 90°, 135°, 225°, 270°, 315°.

49. 30°, 90°, 150°, 210°, 270°, 330°.

51. 0°.

53. 30°, 150°, 180°.

Section 309   Page 63

1. $B=65°\ 50'$, $a=8.19$, $b=18.2$.

3. $B=22°\ 17'$, $a=9.253$, $b=3.792$.

5. 2.76.

7. 6.68.

9. $A=28°$, $B=41°$, $C=111°$.

11. 4,250 ft.

15. 868 m.p.h.

Section 311   Page 69

1. Max. 3; min. 1.

3. Max. 2; min. 0.

5. Max. 4; min. 2.

7. Max. 2; min. 1.

9.

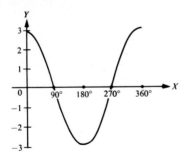

11.

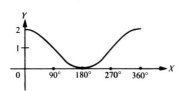

13.

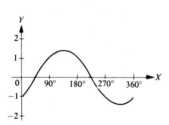

15.

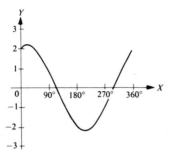

17.

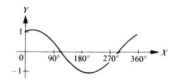

Section 403   Page 75

1. $\frac{1}{3}$

3. $\frac{1}{3}(4 + 3\sqrt{3})$.

5. $1 - \sqrt{3}$.

7. $1 + \sqrt{2}$.

9. -1.

11. 4.

13. $\frac{1}{4}(2 - \sqrt{2})$.

15. 0.7002.

17. 0.8192.

19. 11.430.

21. -1.4281.

23. 0.8391.

25. -1.1778.

27. -1.1918.

29. 0.5581.

31. III.

33. II.

35. IV.

37. $1\frac{11}{20}$.

39. 20.

41. $1\frac{7}{25}$.

Section 406   Page 79

1. 26.5.

3. 110.

5. $36°\,44'$.

7. $5°\,57'$.

9. $3°\,50'$.

11. 117 ft.

13. 241 ft.

15. 29 miles, $N.31°\,W.$

17. 4910 ft.

19. 365 ft.

21. $45°,\,225°$.

23. $135°,\,315°$.

25. $30°,\,210°$.

27. $0°,\,180°$.

29. $60°,\,120°,\,240°,\,300°$.

31. $0°,\,135°,\,180°,\,315°$.

33. $90°$.

35. $0°$.

37. $135°,\,315°$.

39. $0°,\,67\frac{1}{2}°,\,90°,\,157\frac{1}{2}°,\,180°,\,247\frac{1}{2}°,\,270°,$ $337\frac{1}{2}°$.

41. $35°\,00',\,215°\,00'$.

43. $80°\,00',\,260°\,00'$.

45. $160°\,56',\,340°\,56'$.

Section 407   Page 86

1. (a) 4; (b) 14; (c) 20; (d) 25; (e) 28.

3. 58 pounds. $29°$ with 40 pound force and $41°$ with 30 pound force.

5. Horizontal, 32 pounds; vertical, 45 pounds.

7. 283 pounds; 100 pounds.

9. Horizontal, 57 pounds; Vertical, 40 pounds.

11. 173 pounds.

13. Speed, 271 m.p.h.; Course $10°$.

15. 94 m.p.h.

Section 408   Page 89

1.

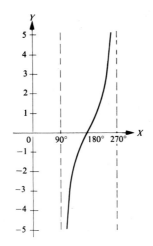

3.

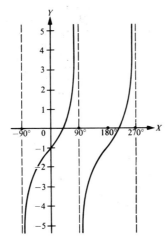

5.

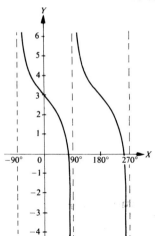

Section 503   Page 96

1. 5. **3.** 2. **5.** 0. **7.** 1.

9. 1. **11.** 2. **13.** $2\frac{2}{3}$. **15.** $-\sqrt{3}$.

17. $4\frac{2}{3}$. **19.** -1.

21. $\sin\theta = \frac{3}{5}$, $\cot\theta = \frac{4}{3}$,
$\cos\theta = \frac{4}{5}$, $\sec\theta = \frac{5}{4}$,
$\tan\theta = \frac{3}{4}$, $\csc\theta = \frac{5}{3}$,

23. $\sin\theta = \frac{5}{13}$, $\cot\theta = -\frac{12}{5}$,
$\cos\theta = -\frac{12}{13}$, $\sec\theta = -\frac{13}{12}$,
$\tan\theta = -\frac{5}{12}$, $\csc\theta = \frac{13}{5}$.

25. $\sin\theta = -\frac{3}{5}$, $\cot\theta = -\frac{4}{3}$.
$\cos\theta = \frac{4}{5}$, $\sec\theta = \frac{5}{4}$.
$\tan\theta = -\frac{3}{4}$, $\csc\theta = -\frac{5}{3}$.

27. $\sin\theta = -\frac{5}{13}$, $\cot\theta = \frac{12}{5}$,
$\cos\theta = -\frac{12}{13}$, $\sec\theta = -\frac{13}{12}$,
$\tan\theta = \frac{5}{12}$, $\csc\theta = -\frac{13}{5}$.

29. $\sin\theta = \frac{2}{13}\sqrt{13}$, $\cot\theta = -\frac{3}{2}$.
$\cos\theta = -\frac{3}{13}\sqrt{13}$, $\sec\theta = -\frac{1}{3}\sqrt{13}$,
$\tan\theta = -\frac{2}{3}$, $\csc\theta = \frac{1}{2}\sqrt{13}$.

31. $\sin\theta = 0$, $\cot\theta$ undefined,
$\cos\theta = -1$, $\sec\theta = -1$,
$\tan\theta = 0$, $\csc\theta$ undefined.

33. III. **35.** IV.

37. II. **39.** Impossible.

41. Possible. **43.** Impossible.

45. Possible. **47.** Possible.

49. Close to 0°. **51.** Close to 90°.

53. Close to 0°. **55.** Close to 0°.

57. 74° 50′. **59.** 20°.

61. 80°.

Section 506   Page 100

1.

3.

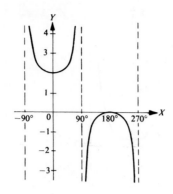

5.

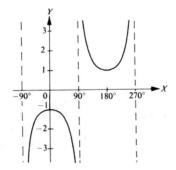

7.

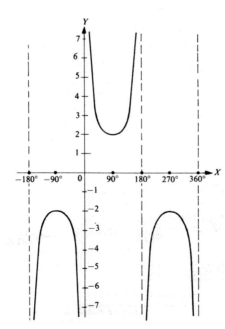

9.

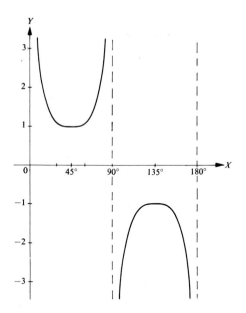

11.

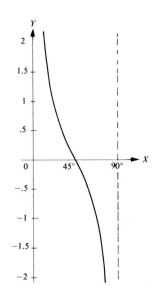

Section 601   Page 107

1. 60°.   3. 90°.   5. 150°.
7. 360°.   9. 210°.   11. 300°.
13. 450°.   15. 65° 53.4′.   17. 116° 18.6′.
19. 43° 25.8′.   21. $\frac{1}{6}\pi$.   23. $\frac{1}{4}\pi$.
25. $\frac{5}{6}\pi$. 27. $\frac{5}{4}\pi$. 29. $\frac{3}{2}\pi$.
31. $\frac{11}{6}\pi$. 33. 0.3142.   35. 1.3134.
37. 0.2609.   39. 2.1859.   41. $5\frac{1}{2}$.
43. -2.   45. $-\frac{9}{2}$.   47. 8.
49. 0.

Section 602   Page 109

1. 0.67.   3. 28.8 inches.
5. 6064 ft when Table VI is used. (More accurate data would give 6080 feet.)
7. 865,800 miles (actually 864,100 miles).
9. (a) 78.2 radians per second; (b) 747 revolutions per minute.
11. 3870 miles.
13. 791 statute miles; 687 nautical miles.
15. 5.73 inches.
17. 69.6 miles per hour.
19. 88.4 inches.

Section 603   Page 112

1. 87 sq. in.; 38 sq. in.   3. 200 sq. in.; 238 sq. in.
5. 61 sq. in.   7. 8,808 gallons.

Section 701   Page 119

1. $\sin \theta = \frac{4}{5}$, $\tan \theta = -\frac{4}{3}$, $\cot \theta = -\frac{3}{4}$, $\sec \theta = -\frac{5}{3}$, $\csc \theta = \frac{5}{4}$.
3. $\sin \theta = \frac{4}{5}$, $\cos \theta = \frac{3}{5}$, $\cot \theta = \frac{3}{4}$, $\sec \theta = \frac{5}{3}$, $\csc \theta = \frac{5}{4}$.
5. $\sin \theta = -\frac{12}{13}$, $\cos \theta = -\frac{5}{13}$, $\tan \theta = \frac{12}{5}$, $\sec \theta = -\frac{13}{5}$, $\csc \theta = -\frac{13}{12}$.
7. 1.   9. $\sec x \csc x$.   11. $\cos x$.
13. $\sec^2 \theta$. 15. $\sec \theta$.   17. $4 \sin \theta \cos \theta$.
19. $-2 \sec^2 \theta$. 21. $\sec \theta$.   23. $\cot^2 x$.

Section 704   Page 129

1. 60°, 240°, 270°.   3. 60°, 270°, 300°.
5. 60°, 180°, 300°.   7. 60°, 120°, 240°, 300°.
9. 0°.   11. 90°, 270°.
13. 270°.   15. 45°, 135°.
17. 227° 04′, 312° 56′.   19. 120°, 180°.

Section 802   Page 136

3. $\frac{1}{4}\sqrt{2}(1-\sqrt{3})$.
11. -1.
15. $\frac{1}{2}$.

5. (a) $-\frac{56}{65}$; (b) $-\frac{16}{65}$.
13. $\cos\theta$.

Section 804   Page 139

1. -1; 0; Undefined; $A+B=270°$. $\frac{7}{25}$; $-\frac{24}{25}$; $-\frac{7}{24}$; $A-B$ in Q II.
3. $-\frac{1}{2}\sqrt{3}$; $-\frac{1}{2}$; $\sqrt{3}$.
21. $-\tan x$.
25. $\cos 6x$.

19. $\cos 9x$.
23. $\sin 2x$.
27. $\sin\theta$.

Section 806   Page 145

7. $\sin 105° = \frac{1}{2}\sqrt{2+\sqrt{3}}$; $\cos 105° = -\frac{1}{2}\sqrt{2-\sqrt{3}}$; $\tan 105° = -2-\sqrt{3}$.
9. $\sin 80°$.
13. $\tan 20°$.
17. $3\sin 6\alpha$.
21. $-\frac{120}{169}$; $\frac{119}{169}$; $-\frac{120}{119}$; $\frac{1}{26}\sqrt{26}$; $-\frac{5}{26}\sqrt{26}$; $-\frac{1}{5}$; IV; II.

11. $\cos 100°$.
15. $\cos 35°$.
19. $5\tan\frac{5}{2}\beta$.

Section 808   Page 150

1. $\sin 60° + \sin 10°$.
5. $\frac{3}{4}(\cos\beta - \cos 3\beta)$.
9. $2\cos 4\theta \cos\theta$.
31. $\frac{1}{6}\pi$, $\frac{5}{6}\pi$, $\frac{3}{2}\pi$.
35. $\frac{1}{4}\pi$, $\frac{1}{3}\pi$, $\frac{5}{4}\pi$, $\frac{5}{3}\pi$.
37. 0, $\frac{1}{4}\pi$, π, $\frac{5}{4}\pi$.
41. $\frac{1}{4}\pi$, $\frac{1}{2}\pi$, $\frac{3}{4}\pi$, $\frac{5}{4}\pi$, $\frac{3}{2}\pi$, $\frac{7}{4}\pi$.
45. 0, π, $\frac{7}{6}\pi$, $\frac{11}{6}\pi$.
47. 0, $\frac{1}{8}\pi$, $\frac{3}{8}\pi$, $\frac{5}{8}\pi$, $\frac{7}{8}\pi$, $\frac{9}{8}\pi$, $\frac{11}{8}\pi$, $\frac{13}{8}\pi$, $\frac{15}{8}\pi$.
49. 0, $\frac{1}{3}\pi$, $\frac{2}{3}\pi$, π, $\frac{4}{3}\pi$, $\frac{5}{3}\pi$.

3. $3(\cos 8\theta + \cos 2\theta)$.
7. $2\sin 35° \cos 5°$.
11. $2\cos\frac{5}{2}\alpha \sin\alpha$.
33. π.
39. 0, $\frac{1}{3}\pi$, π, $\frac{5}{3}\pi$.
43. $\frac{1}{6}\pi$, $\frac{1}{2}\pi$, $\frac{5}{6}\pi$, $\frac{3}{2}\pi$.

Section 809   Page 155

1. $c=13$, $\alpha = 67° 23'$.
5. $c=13$, $\alpha = 202° 37'$.
9. $c=\sqrt{13}$, $\alpha = 123° 41'$.
13. $22° 37'$.

3. $c=13$, $\alpha = 292° 37'$.
7. $c=13$, $\alpha = 157° 23'$.
11. $\sqrt{5}$; $296° 34'$.

Section 902   Page 161

1. π; 3.
7. 2π; 4.

3. 3π; 5.
9. 10; 5.

5. 12; 2.

11.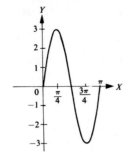

13.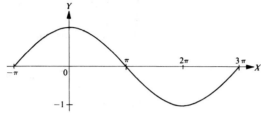

15.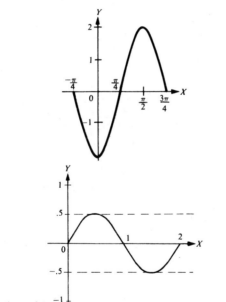

17.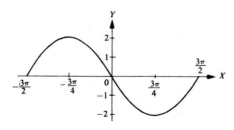

19.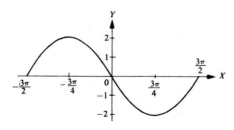

21.

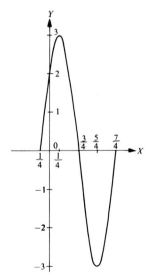

23. $y = \sqrt{10}\, \cos(x - 71°\ 34')$.

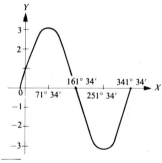

25. $y = 5 \cos(x - 143°\ 08')$.

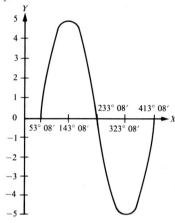

Section 905   Page 167

1.

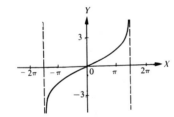

3.

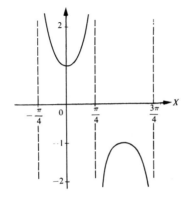

5.

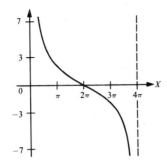

7.

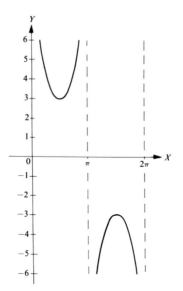

9.

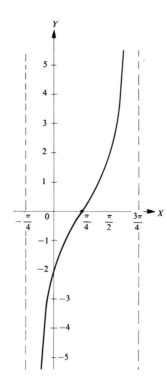

11.

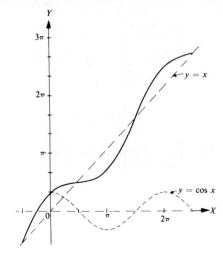

13.

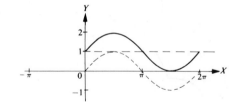

15.

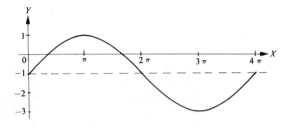

17.

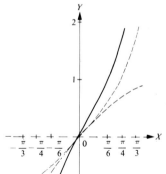

19.

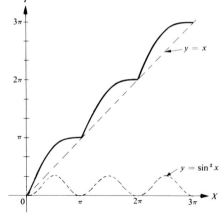

Section 1003   Page 172

1. $\log_2 8 = 3$.

3. $\log_{10} 100 = 2$.

5. $\log_{25} 5 = \frac{1}{2}$.

7. $\log_8 \frac{1}{4} = -\frac{2}{3}$.

9. $\log_5 1 = 0$.

11. $\log_3 1 = 0$.

13. $\log_5 5 = 1$.

15. $\log_{10} 10 = 1$.

17. $4^3 = 64$.

19. $2^4 = 16$.

21. $16^{\frac{1}{2}} = 4$.

23. $10^1 = 10$.

25. $7^1 = 7$.

27. $b^1 = b$.

29. 3.

31. $\frac{3}{2}$.

33. 0.

35. $-\frac{1}{2}$.

37. $\frac{5}{2}$.

39. 9.

41. 4.

43. 9.

45. 2.

Section 1004   Page 175

1. 1.146.

3. 0.410.

5. 0.699.

7. 0.661.

9. 1.477.

11. 1.991.

13. 0.060.

15. 3.778.

17. $\log \dfrac{x}{y}$.

19. $\log x^3 y^2$.

21. $\log \dfrac{\sqrt[4]{xz}}{y}$.

23. $\log(x + y)$.

25. $y = 9^x$.

27. $y = \dfrac{1}{x^2}$.

29. $y = x$.

31. $y = x^3$.

33. 11.

35. 99.

Section 1007   Page 179

1. 2.

3. 7.

5. $-2, 8 - 10$.

7. $-4, 6 - 10$.

9. 3.

11. $-3, 7 - 10$.

13. 3.

15. $-2, 8 - 10$.

17. $-1, 9 - 10$.

19. 12.

21. 127.8.

23. 506, 100.

25. 9.455.

27. 0.01008.

29. 0.00007599.

Section 1009   Page 182

1. 2.09968.

3. $9.73022 - 10$.

5. 1.82838.

7. $6.90309 - 10$.

9. 7.96848.

11. $8.70012 - 10$.

13. 6.1440.

15. 4266.0.

17. 0.0010160.

19. 17.504.

21. 0.047886.

23. 10012.

Section 1010   Page 185

1. 87.00.

3. 23.98.

5. 0.03503.

7. 2.155.

9. 2.949.

11. 0.9495.

13. 0.5144.

15. 6.995.

17. 8.617.

19. 2.904.

21. 87.73.

23. -14.94.

25. 0.07525.

27. 1.080.

29. 0.6494.

31. 0.2103.

33. 10.52.

35. -2.694.

37. 75.76.

Section 1012   Page 189

1. 1.5850.

3. 1.2855.

5. -1.2989.

7. 13.028.

9. 13.529.

11. 5, 2.

13. 2.

15. 6.

17. 2.6826.

19. 3.4248.

21. -1.6853.

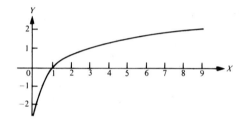

23.

25.

Section 1013   Page 191

1. $8.39310 - 10$.	**3.** $9.89671 - 10$.	**5.** $9.95641 - 10$.
7. $9.58730 - 10$.	**9.** Not a real number.	**11.** $26°\,13'$.
13. $58°\,05'$.	**15.** $3°\,14'$.	**17.** $9°\,21.3'$.
19. $87°\,26.4'$.	**21.** 0.96230.	**23.** 34.411.
25. -16.040		

Section 1102   Page 197

1. 663.71 sq. units.	**3.** 15.042 sq. units.
5. 26.833 sq. units.	**7.** 9.3664 sq. units.
9. 0.059713 sq. units.	**11.** 1.8226 sq. units.
13. 27.173 sq. units.	**15.** 62.770 sq. units.
17. 3,799.6 sq. ft.	**19.** 2.6243 acres.
21. $2,744.60.	**23.** (a) 1.0490 cu. ft.; (b) 510.57 lb.

Section 1104   Page 202

1. $B = 54°\,03.9'$, $C = 90°\,16.1'$, $a = 16.851$.
3. $A = 43°\,00.4'$, $C = 78°\,47.6'$, $b = 5.1784$.
5. $A = 77°\,21.5'$, $B = 49°\,50.5'$, $c = 53.144$.
7. $B = 38°\,52.2'$, $C = 35°\,53.8'$, $a = 2832.1$.
9. $A = 30°\,44.7'$, $B = 21°\,53.9'$, $c = 1.3460$.
11. $36°\,15'$, $47°\,15'$, 214.66 feet.

Section 1105   Page 205

1. $A = 60° 50'$, $B = 46° 00'$, $C = 73° 00'$ (Because of rounding off to the nearest ten minutes the sum $A + B + C \neq 180°$.)

3. $A = 28° 00'$, $B = 98° 40'$, $C = 53° 20'$ (to the nearest ten minutes).

5. $A = 117° 15'$, $B = 35° 46'$, $C = 26° 59'$ (to the nearest minute).

7. $A = 33° 32.4'$, $B = 117° 20.3'$, $C = 29° 7.3'$ (to the nearest tenth of a minute).

9. 66.5 yd.

Section 1201   Page 210

1. $\frac{1}{2}\pi$.

3. $\frac{1}{2}\pi$.

5. $\frac{1}{4}\pi$, $\frac{5}{4}\pi$.

7. $\frac{7}{6}\pi$, $\frac{11}{6}\pi$.

9. π.

11. $\frac{1}{6}\pi$, $\frac{5}{6}\pi$.

13. $29° 00'$, $331° 00'$.

15. $21° 20'$, $201° 20'$.

17. $183° 20'$, $356° 40'$.

19. $168° 00'$, $348° 00'$.

21. $x = \frac{1}{3}$ arc sin $\frac{1}{2}y$.

23. $x = \frac{1}{2}$ arc cos $3y$.

25. $(1/\pi)$ arc tan y.

27. $x = \frac{1}{3}$ arc cos$(y - 1)$.

29. $x = \frac{1}{2}\cos(\frac{1}{2}\pi - y)$.

31. $\frac{4}{3}$.

33. $\frac{3}{2}$.

35. $\frac{3}{7}$.

37. $\frac{1}{5}\sqrt{21}$.

39. 1.

Section 1203   Page 217

1. $\frac{1}{4}\pi$.

3. $\frac{1}{4}\pi$.

5. $\frac{2}{3}\pi$.

7. $-\frac{1}{3}\pi$.

9. $-\frac{1}{6}\pi$.

11. $\frac{1}{6}\pi$.

13. $\frac{2}{3}\sqrt{2}$.

15. $\frac{5}{12}$.

17. $\frac{5}{7}$.

19. $\frac{8}{17}$.

21. $\frac{7}{32}\sqrt{15}$.

23. $-\frac{24}{7}$.

25. 1.

27. $\frac{63}{65}$.

29. $\frac{156}{133}$.

31. $\frac{3}{4}$.

33. $\pi -$ Arcsin $\frac{3}{5}$.

35. $\pi +$ Arccos $\frac{5}{13}$.

37. $2\pi -$ Arctan $\frac{3}{4}$.

39. $2\pi +$ Arcsin $\frac{4}{5}$.

41. Arccos $(-\frac{3}{5})$.

53. 1.

55. $\frac{1}{5}\sqrt{5}$.

57. $\frac{1}{2}$.

59. $\frac{1}{2}$.

61. 0.

63. 1.

Section 1303   Page 224

1. $7 + i$.

3. $-3 - i$.

5. $\frac{11}{15} + \frac{2}{3}i$.

7. $-5 + 12 i$.

9. 10.

11. $2 + 11 i$.

13. $-i$.

15. $\frac{7}{9} + \frac{4}{9}\sqrt{2} i$.

17. $x = 2$, $y = 2$.

19. $x = 2$, $y = 16$.

21. $x = 3$, $y = 1$.

23. $x = -6$, $y = 6$.

25. $2 \pm i$.

27. $\dfrac{2 \pm 7 i}{3}$.

Section 1304   Page 226

1.

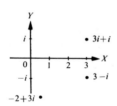

3.

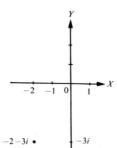

5.

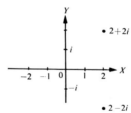

7.

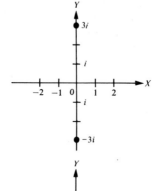

9.

-3 is the conjugate of itself.

Section 1305   Page 229

1. $(2\sqrt{3}, 2)$.   **3.** $(3, 0)$ **5.** $(0, -8)$.
7. $(-9.4, -3.4)$.   **9.** $(0, 6)$.   **11.** $(\sqrt{2}, \sqrt{2})$.
13. $(2\sqrt{2}, \frac{1}{4}\pi)$.   **15.** $(3, 0)$.   **17.** $(5, \pi)$.
19. $(2, \frac{7}{4}\pi)$.   **21.** $(2, \frac{1}{2}\pi)$.   **23.** $(-\sqrt{13}, 56°\ 20')$.

Section 1306   Page 233

1. $2\sqrt{2}, \frac{1}{4}\pi$.   **3.** $2, \frac{2}{3}\pi$,   **5.** $\sqrt{2}, \frac{7}{4}\pi$.
7. $\sqrt{3}, 234°\ 40'$.   **9.** $3, \frac{1}{2}\pi$.   **11.** $5, \pi$.
13. $6, 0$.   **15.** $\sqrt{13}, 303°\ 40'$.   **17.** $\frac{3}{2} + \frac{3}{2}\sqrt{3}\ i$.
19. $2\sqrt{2} + 2\sqrt{2}\ i$.   **21.** -1.   **23.** $-8\ i$.
25. $2(\cos \frac{1}{6}\pi + i \sin \frac{1}{6}\pi), \sqrt{3} + i$.
27. $5(\cos \pi + i \sin \pi), -5$.
29. $4(\cos \frac{3}{4}\pi + i \sin \frac{3}{4}\pi), -2\sqrt{2} + 2\sqrt{2}\ i$.

Section 1308   Page 237

1. $6(\cos 90° + i \sin 90°), 6i$.
3. $2(\cos 240° + i \sin 240°), -1 - \sqrt{3}\ i$.
5. $10(\cos 140° + i \sin 140°), -7.660 + 6.428\ i$
7. $\cos 150° + i \sin 150°, -\frac{1}{2}\sqrt{3} + \frac{1}{2}\ i$.
9. $\frac{1}{32}(\cos 300° + i \sin 300°), \frac{1}{64} - \frac{1}{64}\sqrt{3}\ i$.
11. $324(\cos 180° + i \sin 180°), -324$.
13. $2(\cos 60° + i \sin 60°), 1 + \sqrt{3}\ i$.
15. $\frac{1}{13}(\cos 67°\ 51' + i \sin 67°\ 51'), 0.02900 + 0.07125\ i$.

Section 1309   Page 241

1. $\frac{1}{2}\sqrt{3} + \frac{1}{2}\ i$;
$-\frac{1}{2}\sqrt{3} + \frac{1}{2}\ i$;
$-i$.

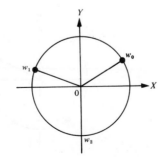

3. $\frac{1}{2}\sqrt{2}+\frac{1}{2}\sqrt{2}\,i;$
$-\frac{1}{2}\sqrt{2}+\frac{1}{2}\sqrt{2}\,i;$
$-\frac{1}{2}\sqrt{2}-\frac{1}{2}\sqrt{2}\,i;$
$\frac{1}{2}\sqrt{2}-\frac{1}{2}\sqrt{2}\,i.$

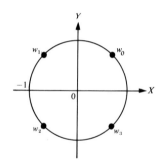

5. $1.398+0.221\,i;$
$0.221+1.398\,i;$
$-1.260+0.642\,i;$
$-1-i;$
$0.642-1.260\,i.$

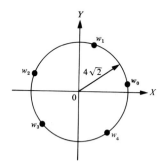

7. $1;$
$-\frac{1}{2}+\frac{1}{2}\sqrt{3}\,i;$
$-\frac{1}{2}-\frac{1}{2}\sqrt{3}\,i.$

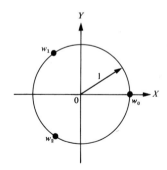

9. $2;$
$1+\sqrt{3}\,i;$
$-1+\sqrt{3}\,i;$
$-2;$
$-1-\sqrt{3}\,i;$
$1-\sqrt{3}\,i.$

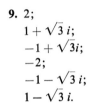
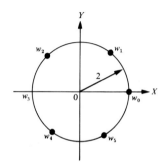

11. $1.932 - 0.518\ i$;
$0.518 + 1.932\ i$;
$-1.932 + 0.518\ i$;
$-0.518 - 1.932\ i$.

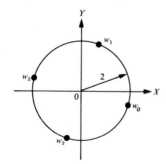

13. $2 + 3i$;
$-2 - 3i$.

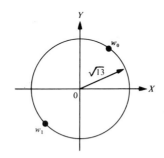

15. $0.809 + 0.588\ i$; $-0.309 + 0.951\ i$; -1; $-0.309 - 0.951\ i$; $+0.809 - 0.588\ i$.

17. $\frac{3}{2} + \frac{3}{2}\sqrt{3}\ i$; -3; $\frac{3}{2} - \frac{3}{2}\sqrt{3}\ i$.

19. $\frac{1}{2}\sqrt{6} + \frac{1}{2}\sqrt{2}\ i$; $-\frac{1}{2}\sqrt{6} - \frac{1}{2}\sqrt{2}\ i$;
$\frac{1}{2}\sqrt{6} - \frac{1}{2}\sqrt{2}\ i$; $-\frac{1}{2}\sqrt{6} + \frac{1}{2}\sqrt{2}\ i$.

INDEX

LOGARITHMIC AND

TRIGONOMETRIC TABLES

CONTENTS

Table I

Common logarithms of numbers

to five decimal places

Table I

N.	Log	N.	Log	N.	Log	N.	Log	N.	Log
1	0.00 000	21	1.32 222	41	1.61 278	61	1.78 533	81	1.90 849
2	0.30 103	22	1.34 242	42	1.62 325	62	1.79 239	82	1.91 381
3	0.47 712	23	1.36 173	43	1.63 347	63	1.79 934	83	1.91 908
4	0.60 206	24	1.38 021	44	1.64 345	64	1.80 618	84	1.92 428
5	0.69 897	25	1.39 794	45	1.65 321	65	1.81 291	85	1.92 942
6	0.77 815	26	1.41 497	46	1.66 276	66	1.81 954	86	1.93 450
7	0.84 510	27	1.43 136	47	1.67 210	67	1.82 607	87	1.93 952
8	0.90 309	28	1.44 716	48	1.68 124	68	1.83 251	88	1.94 448
9	0.95 424	29	1.46 240	49	1.69 020	69	1.83 885	89	1.94 939
10	1.00 000	30	1.47 712	50	1.69 897	70	1.84 510	90	1.95 424
11	1.04 139	31	1.49 136	51	1.70 757	71	1.85 126	91	1.95 904
12	1.07 918	32	1.50 515	52	1.71 600	72	1.85 733	92	1.96 379
13	1.11 394	33	1.51 851	53	1.72 428	73	1.86 332	93	1.96 848
14	1.14 613	34	1.53 148	54	1.73 239	74	1.86 923	94	1.97 313
15	1.17 609	35	1.54 407	55	1.74 036	75	1.87 506	95	1.97.772
16	1.20 412	36	1.55 630	56	1.74 819	76	1.88 081	96	1.98 227
17	1.23 045	37	1.56 820	57	1.75 587	77	1.88 649	97	1.98 677
18	1.25 527	38	1.57 978	58	1.76 343	78	1.89 209	98	1.99 123
19	1.27 875	39	1.59 106	59	1.77 085	79	1.89 763	99	1.99 564
20	1.30 103	40	1.60 206	60	1.77 815	80	1.90 309	100	2.00 000
N.	Log	N.	Log	N.	Log	N.	Log	N.	Log

100 — 150

N.	0	1	2	3	4	5	6	7	8	9
100	00 000	043	087	130	173	217	260	303	346	389
01	432	475	518	561	604	647	689	732	775	817
02	860	903	945	988	*030	*072	*115	*157	*199	*242
03	01 284	326	368	410	452	494	536	578	620	662
04	703	745	787	828	870	912	953	995	*036	*078
05	02 119	160	202	243	284	325	366	407	449	490
06	531	572	612	653	694	735	776	816	857	898
07	938	979	*019	*060	*100	*141	*181	*222	*262	*302
08	03 342	383	423	463	503	543	583	623	663	703
09	743	782	822	862	902	941	981	*021	*060	*100
110	04 139	179	218	258	297	336	376	415	454	493
11	532	571	610	650	689	727	766	805	844	883
12	922	961	999	*038	*077	*115	*154	*192	*231	*269
13	05 308	346	385	423	461	500	538	576	614	652
14	690	729	767	805	843	881	918	956	994	*032
15	06 070	108	145	183	221	258	296	333	371	408
16	446	483	521	558	595	633	670	707	744	781
17	819	856	893	930	967	*004	*041	*078	*115	*151
18	07 188	225	262	298	335	372	408	445	482	518
19	555	591	628	664	700	737	773	809	846	882
120	918	954	990	*027	*063	*099	*135	*171	*207	*243
21	08 279	314	350	386	422	458	493	529	565	600
22	636	672	707	743	778	814	849	884	920	955
23	991	*026	*061	*096	*132	*167	*202	*237	*272	*307
24	09 342	377	412	447	482	517	552	587	621	656
25	691	726	760	795	830	864	899	934	968	*003
26	10 037	072	106	140	175	209	243	278	312	346
27	380	415	449	483	517	551	585	619	653	687
28	721	755	789	823	857	890	924	958	992	*025
29	11 059	093	126	160	193	227	261	294	327	361
130	394	428	461	494	528	561	594	628	661	694
31	727	760	793	826	860	893	926	959	992	*024
32	12 057	090	123	156	189	222	254	287	320	352
33	385	418	450	483	516	548	581	613	646	678
34	710	743	775	808	840	872	905	937	969	*001
35	13 033	066	098	130	162	194	226	258	290	322
36	354	386	418	450	481	513	545	577	609	640
37	672	704	735	767	799	830	862	893	925	956
38	988	*019	*051	*082	*114	*145	*176	*208	*239	*270
39	14 301	333	364	395	426	457	489	520	551	582
140	613	644	675	706	737	768	799	829	860	891
41	922	953	983	*014	*045	*076	*106	*137	*168	*198
42	15 229	259	290	320	351	381	412	442	473	503
43	534	564	594	625	655	685	715	746	776	806
44	836	866	897	927	957	987	*017	*047	*077	*107
45	16 137	167	197	227	256	286	316	346	376	406
46	435	465	495	524	554	584	613	643	673	702
47	732	761	791	820	850	879	909	938	967	997
48	17 026	056	085	114	143	173	202	231	260	289
49	319	348	377	406	435	464	493	522	551	580
150	609	638	667	696	725	754	782	811	840	869
N.	0	1	2	3	4	5	6	7	8	9

Prop. Parts

	44	43	42
1	4.4	4.3	4.2
2	8.8	8.6	8.4
3	13.2	12.9	12.6
4	17.6	17.2	16.8
5	22.0	21.5	21.0
6	26.4	25.8	25.2
7	30.8	30.1	29.4
8	35.2	34.4	33.6
9	39.6	38.7	37.8

	41	40	39
1	4.1	4.0	3.9
2	8.2	8.0	7.8
3	12.3	12.0	11.7
4	16.4	16.0	15.6
5	20.5	20.0	19.5
6	24.6	24.0	23.4
7	28.7	28.0	27.3
8	32.8	32.0	31.2
9	36.9	36.0	35.1

	38	37	36
1	3.8	3.7	3.6
2	7.6	7.4	7.2
3	11.4	11.1	10.8
4	15.2	14.8	14.4
5	19.0	18.5	18.0
6	22.8	22.2	21.6
7	26.6	25.9	25.2
8	30.4	29.6	28.8
9	34.2	33.3	32.4

	35	34	33
1	3.5	3.4	3.3
2	7.0	6.8	6.6
3	10.5	10.2	9.9
4	14.0	13.6	13.2
5	17.5	17.0	16.5
6	21.0	20.4	19.8
7	24.5	23.8	23.1
8	28.0	27.2	26.4
9	31.5	30.6	29.7

	32	31	30
1	3.2	3.1	3.0
2	6.4	6.2	6.0
3	9.6	9.3	9.0
4	12.8	12.4	12.0
5	16.0	15.5	15.0
6	19.2	18.6	18.0
7	22.4	21.7	21.0
8	25.6	24.8	24.0
9	28.8	27.9	27.0

$\log \sqrt{2} = .150515$

150 — 200

N.	0	1	2	3	4	5	6	7	8	9
150	17 609	638	667	696	725	754	782	811	840	869
51	898	926	955	984	*013	*041	*070	*099	*127	*156
52	18 184	213	241	270	298	327	355	384	412	441
53	469	498	526	554	583	611	639	667	696	724
54	752	780	808	837	865	893	921	949	977	*005
55	19 033	061	089	117	145	173	201	229	257	285
56	312	340	368	396	424	451	479	507	535	562
57	590	618	645	673	700	728	756	783	811	838
58	866	893	921	948	976	*003	*030	*058	*085	*112
59	20 140	167	194	222	249	276	303	330	358	385
160	412	439	466	493	520	548	575	602	629	656
61	683	710	737	763	790	817	844	871	898	925
62	952	978	*005	*032	*059	*085	*112	*139	*165	*192
63	21 219	245	272	299	325	352	378	405	431	458
64	484	511	537	564	590	617	643	669	696	722
65	748	775	801	827	854	880	906	932	958	985
66	22 011	037	063	089	115	141	167	194	220	246
67	272	298	324	350	376	401	427	453	479	505
68	531	557	583	608	634	660	686	712	737	763
69	789	814	840	866	891	917	943	968	994	*019
170	23 045	070	096	121	147	172	198	223	249	274
71	300	325	350	376	401	426	452	477	502	528
72	553	578	603	629	654	679	704	729	754	779
73	805	830	855	880	905	930	955	980	*005	*030
74	24 055	080	105	130	155	180	204	229	254	279
75	304	329	353	378	403	428	452	477	502	527
76	551	576	601	625	650	674	699	724	748	773
77	797	822	846	871	895	920	944	969	993	*018
78	25 042	066	091	115	139	164	188	212	237	261
79	285	310	334	358	382	406	431	455	479	503
180	527	551	575	600	624	648	672	696	720	744
81	768	792	816	840	864	888	912	935	959	983
82	26 007	031	055	079	102	126	150	174	198	221
83	245	269	293	316	340	364	387	411	435	458
84	482	505	529	553	576	600	623	647	670	694
85	717	741	764	788	811	834	858	881	905	928
86	951	975	998	*021	*045	*068	*091	*114	*138	*161
87	27 184	207	231	254	277	300	323	346	370	393
88	416	439	462	485	508	531	554	577	600	623
89	646	669	692	715	738	761	784	807	830	852
190	875	898	921	944	967	989	*012	*035	*058	*081
91	28 103	126	149	171	194	217	240	262	285	307
92	330	353	375	398	421	443	466	488	511	533
93	556	578	601	623	646	668	691	713	735	758
94	780	803	825	847	870	892	914	937	959	981
95	29 003	026	048	070	092	115	137	159	181	203
96	226	248	270	292	314	336	358	380	403	425
97	447	469	491	513	535	557	579	601	623	645
98	667	688	710	732	754	776	798	820	842	863
99	885	907	929	951	973	994	*016	*038	*060	*081
200	30 103	125	146	168	190	211	233	255	276	298
N.	0	1	2	3	4	5	6	7	8	9

Prop. Parts

	29	28
1	2.9	2.8
2	5.8	5.6
3	8.7	8.4
4	11.6	11.2
5	14.5	14.0
6	17.4	16.8
7	20.3	19.6
8	23.2	22.4
9	26.1	25.2

	27	26
1	2.7	2.6
2	5.4	5.2
3	8.1	7.8
4	10.8	10.4
5	13.5	13.0
6	16.2	15.6
7	18.9	18.2
8	21.6	20.8
9	24.3	23.4

$\log\sqrt{3} = .238561$

	25	24
1	2.5	2.4
2	5.0	4.8
3	7.5	7.2
4	10.0	9.6
5	12.5	12.0
6	15.0	14.4
7	17.5	16.8
8	20.0	19.2
9	22.5	21.6

	23	22
1	2.3	2.2
2	4.6	4.4
3	6.9	6.6
4	9.2	8.8
5	11.5	11.0
6	13.8	13.2
7	16.1	15.4
8	18.4	17.6
9	20.7	19.8

	21
1	2.1
2	4.2
3	6.3
4	8.4
5	10.5
6	12.6
7	14.7
8	16.8
9	18.9

Prop. Parts

200 — 250

N.	0	1	2	3	4	5	6	7	8	9	Prop. Parts
200	30 103	125	146	168	190	211	233	255	276	298	
01	320	341	363	384	406	428	449	471	492	514	
02	535	557	578	600	621	643	664	685	707	728	
03	750	771	792	814	835	856	878	899	920	942	
04	963	984	*006	*027	*048	*069	*091	*112	*133	*154	
05	31 175	197	218	239	260	281	302	323	345	366	

		22	**21**
06	387 408 429 450 471 492 513 534 555 576		

N.	0	1	2	3	4	5	6	7	8	9		22	21
06	387	408	429	450	471	492	513	534	555	576			
07	597	618	639	660	681	702	723	744	765	785	1	2.2	2.1
08	806	827	848	869	890	911	931	952	973	994	2	4.4	4.2
09	32 015	035	056	077	098	118	139	160	181	201	3	6.6	6.3
210	222	243	263	284	305	325	346	366	387	408	4	8.8	8.4
											5	11.0	10.5
11	428	449	469	490	510	531	552	572	593	613	6	13.2	12.6
12	634	654	675	695	715	736	756	777	797	818	7	15.4	14.7
13	838	858	879	899	919	940	960	980	*001	*021	8	17.6	16.8
14	33 041	062	082	102	122	143	163	183	203	224	9	19.8	18.9
15	244	264	284	304	325	345	365	385	405	425			
16	445	465	486	506	526	546	566	586	606	626			
17	646	666	686	706	726	746	766	786	806	826			
18	846	866	885	905	925	945	965	985	*005	*025			
19	34 044	064	084	104	124	143	163	183	203	223			
220	242	262	282	301	321	341	361	380	400	420		**20**	**19**
21	439	459	479	498	518	537	557	577	596	616	1	2.0	1.9
22	635	655	674	694	713	733	753	772	792	811	2	4.0	3.8
23	830	850	869	889	908	928	947	967	986	*005	3	6.0	5.7
24	35 025	044	064	083	102	122	141	160	180	199	4	8.0	7.6
25	218	238	257	276	295	315	334	353	372	392	5	10.0	9.5
26	411	430	449	468	488	507	526	545	564	583	6	12.0	11.4
27	603	622	641	660	679	698	717	736	755	774	7	14.0	13.3
28	793	813	832	851	870	889	908	927	946	965	8	16.0	15.2
29	984	*003	*021	*040	*059	*078	*097	*116	*135	*154	9	18.0	17.1
230	36 173	192	211	229	248	267	286	305	324	342			
31	361	380	399	418	436	455	474	493	511	530			
32	549	568	586	605	624	642	661	680	698	717			
33	736	754	773	791	810	829	847	866	884	903			
34	922	940	959	977	996	*014	*033	*051	*070	*088			
35	37 107	125	144	162	181	199	218	236	254	273		**18**	**17**
36	291	310	328	346	365	383	401	420	438	457			
37	475	493	511	530	548	566	585	603	621	639	1	1.8	1.7
38	658	676	694	712	731	749	767	785	803	822	2	3.6	3.4
39	840	858	876	894	912	931	949	967	985	*003	3	5.4	5.1
240	38 021	039	057	075	093	112	130	148	166	184	4	7.2	6.8
											5	9.0	8.5
41	202	220	238	256	274	292	310	328	346	364	6	10.8	10.2
42	382	399	417	435	453	471	489	507	525	543	7	12.6	11.9
43	561	578	596	614	632	650	668	686	703	721	8	14.4	13.6
44	739	757	775	792	810	828	846	863	881	899	9	16.2	15.3
45	917	934	952	970	987	*005	*023	*041	*058	*076			
46	39 094	111	129	146	164	182	199	217	235	252			
47	270	287	305	322	340	358	375	393	410	428			
48	445	463	480	498	515	533	550	568	585	602			
49	620	637	655	672	690	707	724	742	759	777			
250	794	811	829	846	863	881	898	915	933	950			
N.	0	1	2	3	4	5	6	7	8	9	Prop. Parts		

250 — 300

N.	0	1	2	3	4	5	6	7	8	9
250	39 794	811	829	846	863	881	898	915	933	950
51	967	985	*002	*019	*037	*054	*071	*088	*106	*123
52	40 140	157	175	192	209	226	243	261	278	295
53	312	329	346	364	381	398	415	432	449	466
54	483	500	518	535	552	569	586	603	620	637
55	654	671	688	705	722	739	756	773	790	807
56	824	841	858	875	892	909	926	943	960	976
57	993	*010	*027	*044	*061	*078	*095	*111	*128	*145
58	41 162	179	196	212	229	246	263	280	296	313
59	330	347	363	380	397	414	430	447	464	481
260	497	514	531	547	564	581	597	614	631	647
61	664	681	697	714	731	747	764	780	797	814
62	830	847	863	880	896	913	929	946	963	979
63	996	*012	*029	*045	*062	*078	*095	*111	*127	*144
64	42 160	177	193	210	226	243	259	275	292	308
65	325	341	357	374	390	406	423	439	455	472
66	488	504	521	537	553	570	586	602	619	635
67	651	667	684	700	716	732	749	765	781	797
68	813	830	846	862	878	894	911	927	943	959
69	975	991	*008	*024	*040	*056	*072	*088	*104	*120
270	43 136	152	169	185	201	217	233	249	265	281
71	297	313	329	345	361	377	393	409	425	441
72	457	473	489	505	521	537	553	569	584	600
73	616	632	648	664	680	696	712	727	743	759
74	775	791	807	823	838	854	870	886	902	917
75	933	949	965	981	996	*012	*028	*044	*059	*075
76	44 091	107	122	138	154	170	185	201	217	232
77	248	264	279	295	311	326	342	358	373	389
78	404	420	436	451	467	483	498	514	529	545
79	560	576	592	607	623	638	654	669	685	700
280	716	731	747	762	778	793	809	824	840	855
81	871	886	902	917	932	948	963	979	994	*010
82	45 025	040	056	071	086	102	117	133	148	163
83	179	194	209	225	240	255	271	286	301	317
84	332	347	362	378	393	408	423	439	454	469
85	484	500	515	530	545	561	576	591	606	621
86	637	652	667	682	697	712	728	743	758	773
87	788	803	818	834	849	864	879	894	909	924
88	939	954	969	984	*000	*015	*030	*045	*060	*075
89	46 090	105	120	135	150	165	180	195	210	225
290	240	255	270	285	300	315	330	345	359	374
91	389	404	419	434	449	464	479	494	509	523
92	538	553	568	583	598	613	627	642	657	672
93	687	702	716	731	746	761	776	790	805	820
94	835	850	864	879	894	909	923	938	953	967
95	982	997	*012	*026	*041	*056	*070	*085	*100	*114
96	47 129	144	159	173	188	202	217	232	246	261
97	276	290	305	319	334	349	363	378	392	407
98	422	436	451	465	480	494	509	524	538	553
99	567	582	596	611	625	640	654	669	683	698
300	712	727	741	756	770	784	799	813	828	842
N.	0	1	2	3	4	5	6	7	8	9

Prop. Parts

	18	17
1	1.8	1.7
2	3.6	3.4
3	5.4	5.1
4	7.2	6.8
5	9.0	8.5
6	10.8	10.2
7	12.6	11.9
8	14.4	13.6
9	16.2	15.3

$M = \log_{10} e = \log_{10} 2.71828... = .43429$

	16	15
1	1.6	1.5
2	3.2	3.0
3	4.8	4.5
4	6.4	6.0
5	8.0	7.5
6	9.6	9.0
7	11.2	10.5
8	12.8	12.0
9	14.4	13.5

	14
1	1.4
2	2.8
3	4.2
4	5.6
5	7.0
6	8.4
7	9.8
8	11.2
9	12.6

300 — 350

N.	0	1	2	3	4	5	6	7	8	9
300	47 712	727	741	756	770	784	799	813	828	842
01	857	871	885	900	914	929	943	958	972	986
02	48 001	015	029	044	058	073	087	101	116	130
03	144	159	173	187	202	216	230	244	259	273
04	287	302	316	330	344	359	373	387	401	416
05	430	444	458	473	487	501	515	530	544	558
06	572	586	601	615	629	643	657	671	686	700
07	714	728	742	756	770	785	799	813	827	841
08	855	869	883	897	911	926	940	954	968	982
09	996	*010	*024	*038	*052	*066	*080	*094	*108	*122
310	49 136	150	164	178	192	206	220	234	248	262
11	276	290	304	318	332	346	360	374	388	402
12	415	429	443	457	471	485	499	513	527	541
13	554	568	582	596	610	624	638	651	665	679
14	693	707	721	734	748	762	776	790	803	817
15	831	845	859	872	886	900	914	927	941	955
16	969	982	996	*010	*024	*037	*051	*065	*079	*092
17	50 106	120	133	147	161	174	188	202	215	229
18	243	256	270	284	297	311	325	338	352	365
19	379	393	406	420	433	447	461	474	488	501
320	515	529	542	556	569	583	596	610	623	637
21	651	664	678	691	705	718	732	745	759	772
22	786	799	813	826	840	853	866	880	893	907
23	920	934	947	961	974	987	*001	*014	*028	*041
24	51 055	068	081	095	108	121	135	148	162	175
25	188	202	215	228	242	255	268	282	295	308
26	322	335	348	362	375	388	402	415	428	441
27	455	468	481	495	508	521	534	548	561	574
28	587	601	614	627	640	654	667	680	693	706
29	720	733	746	759	772	786	799	812	825	838
330	851	865	878	891	904	917	930	943	957	970
31	983	996	*009	*022	*035	*048	*061	*075	*088	*101
32	52 114	127	140	153	166	179	192	205	218	231
33	244	257	270	284	297	310	323	336	349	362
34	375	388	401	414	427	440	453	466	479	492
35	504	517	530	543	556	569	582	595	608	621
36	634	647	660	673	686	699	711	724	737	750
37	763	776	789	802	815	827	840	853	866	879
38	892	905	917	930	943	956	969	982	994	*007
39	53 020	033	046	058	071	084	097	110	122	135
340	148	161	173	186	199	212	224	237	250	263
41	275	288	301	314	326	339	352	364	377	390
42	403	415	428	441	453	466	479	491	504	517
43	529	542	555	567	580	593	605	618	631	643
44	656	668	681	694	706	719	732	744	757	769
45	782	794	807	820	832	845	857	870	882	895
46	908	920	933	945	958	970	983	995	*008	*020
47	54 033	045	058	070	083	095	108	120	133	145
48	158	170	183	195	208	220	233	245	258	270
49	283	295	307	320	332	345	357	370	382	394
350	407	419	432	444	456	469	481	494	506	518
N.	0	1	2	3	4	5	6	7	8	9

Prop. Parts

	15	14
1	1.5	1.4
2	3.0	2.8
3	4.5	4.2
4	6.0	5.6
5	7.5	7.0
6	9.0	8.4
7	10.5	9.8
8	12.0	11.2
9	13.5	12.6

$\log \pi = .49715$

	13	12
1	1.3	1.2
2	2.6	2.4
3	3.9	3.6
4	5.2	4.8
5	6.5	6.0
6	7.8	7.2
7	9.1	8.4
8	10.4	9.6
9	11.7	10.8

350 — 400

N.	0	1	2	3	4	5	6	7	8	9
350	54 407	419	432	444	456	469	481	494	506	518
51	531	543	555	568	580	593	605	617	630	642
52	654	667	679	691	704	716	728	741	753	765
53	777	790	802	814	827	839	851	864	876	888
54	900	913	925	937	949	962	974	986	998	*011
55	55 023	035	047	060	072	084	096	108	121	133
56	145	157	169	182	194	206	218	230	242	255
57	267	279	291	303	315	328	340	352	364	376
58	388	400	413	425	437	449	461	473	485	497
59	509	522	534	546	558	570	582	594	606	618
360	630	642	654	666	678	691	703	715	727	739
61	751	763	775	787	799	811	823	835	847	859
62	871	883	895	907	919	931	943	955	967	979
63	991	*003	*015	*027	*038	*050	*062	*074	*086	*098
64	56 110	122	134	146	158	170	182	194	205	217
65	229	241	253	265	277	289	301	312	324	336
66	348	360	372	384	396	407	419	431	443	455
67	467	478	490	502	514	526	538	549	561	573
68	585	597	608	620	632	644	656	667	679	691
69	703	714	726	738	750	761	773	785	797	808
370	820	832	844	855	867	879	891	902	914	926
71	937	949	961	972	984	996	*008	*019	*031	*043
72	57 054	066	078	089	101	113	124	136	148	159
73	171	183	194	206	217	229	241	252	264	276
74	287	299	310	322	334	345	357	368	380	392
75	403	415	426	438	449	461	473	484	496	507
76	519	530	542	553	565	576	588	600	611	623
77	634	646	657	669	680	692	703	715	726	738
78	749	761	772	784	795	807	818	830	841	852
79	864	875	887	898	910	921	933	944	955	967
380	978	990	*001	*013	*024	*035	*047	*058	*070	*081
81	58 092	104	115	127	138	149	161	172	184	195
82	206	218	229	240	252	263	274	286	297	309
83	320	331	343	354	365	377	388	399	410	422
84	433	444	456	467	478	490	501	512	524	535
85	546	557	569	580	591	602	614	625	636	647
86	659	670	681	692	704	715	726	737	749	760
87	771	782	794	805	816	827	838	850	861	872
88	883	894	906	917	928	939	950	961	973	984
89	995	*006	*017	*028	*040	*051	*062	*073	*084	*095
390	59 106	118	129	140	151	162	173	184	195	207
91	218	229	240	251	262	273	284	295	306	318
92	329	340	351	362	373	384	395	406	417	428
93	439	450	461	472	483	494	506	517	528	539
94	550	561	572	583	594	605	616	627	638	649
95	660	671	682	693	704	715	726	737	748	759
96	770	780	791	802	813	824	835	846	857	868
97	879	890	901	912	923	934	945	956	966	977
98	988	999	*010	*021	*032	*043	*054	*065	*076	*086
99	60 097	108	119	130	141	152	163	173	184	195
400	206	217	228	239	249	260	271	282	293	304
N.	0	1	2	3	4	5	6	7	8	9

Prop. Parts

	13	12
1	1.3	1.2
2	2.6	2.4
3	3.9	3.6
4	5.2	4.8
5	6.5	6.0
6	7.8	7.2
7	9.1	8.4
8	10.4	9.6
9	11.7	10.8

	11	10
1	1.1	1.0
2	2.2	2.0
3	3.3	3.0
4	4.4	4.0
5	5.5	5.0
6	6.6	6.0
7	7.7	7.0
8	8.8	8.0
9	9.9	9.0

Table I

400 — 450

N.	0	1	2	3	4	5	6	7	8	9	Prop. Parts
400	60 206	217	228	239	249	260	271	282	293	304	
01	314	325	336	347	358	369	379	390	401	412	
02	423	433	444	455	466	477	487	498	509	520	
03	531	541	552	563	574	584	595	606	617	627	
04	638	649	660	670	681	692	703	713	724	735	
05	746	756	767	778	788	799	810	821	831	842	
06	853	863	874	885	895	906	917	927	938	949	
07	959	970	981	991	*002	*013	*023	*034	*045	*055	
08	61 066	077	087	098	109	119	130	140	151	162	
09	172	183	194	204	215	225	236	247	257	268	
410	278	289	300	310	321	331	342	352	363	374	

	11	10
1	1.1	1.0
2	2.2	2.0
3	3.3	3.0
4	4.4	4.0
5	5.5	5.0
6	6.6	6.0
7	7.7	7.0
8	8.8	8.0
9	9.9	9.0

N.	0	1	2	3	4	5	6	7	8	9
11	384	395	405	416	426	437	448	458	469	479
12	490	500	511	521	532	542	553	563	574	584
13	595	606	616	627	637	648	658	669	679	690
14	700	711	721	731	742	752	763	773	784	794
15	805	815	826	836	847	857	868	878	888	899
16	909	920	930	941	951	962	972	982	993	*003
17	62 014	024	034	045	055	066	076	086	097	107
18	118	128	138	149	159	170	180	190	201	211
19	221	232	242	252	263	273	284	294	304	315
420	325	335	346	356	366	377	387	397	408	418
21	428	439	449	459	469	480	490	500	511	521
22	531	542	552	562	572	583	593	603	613	624
23	634	644	655	665	675	685	696	706	716	726
24	737	747	757	767	778	788	798	808	818	829
25	839	849	859	870	880	890	900	910	921	931
26	941	951	961	972	982	992	*002	*012	*022	*033
27	63 043	053	063	073	083	094	104	114	124	134
28	144	155	165	175	185	195	205	215	225	236
29	246	256	266	276	286	296	306	317	327	337
430	347	357	367	377	387	397	407	417	428	438
31	448	458	468	478	488	498	508	518	528	538
32	548	558	568	579	589	599	609	619	629	639
33	649	659	669	679	689	699	709	719	729	739
34	749	759	769	779	789	799	809	819	829	839
35	849	859	869	879	889	899	909	919	929	939
36	949	959	969	979	988	998	*008	*018	*028	*038
37	64 048	058	068	078	088	098	108	118	128	137
38	147	157	167	177	187	197	207	217	227	237
39	246	256	266	276	286	296	306	316	326	335
440	345	355	365	375	385	395	404	414	424	434

	9
1	0.9
2	1.8
3	2.7
4	3.6
5	4.5
6	5.4
7	6.3
8	7.2
9	8.1

N.	0	1	2	3	4	5	6	7	8	9
41	444	454	464	473	483	493	503	513	523	532
42	542	552	562	572	582	591	601	611	621	631
43	640	650	660	670	680	689	699	709	719	729
44	738	748	758	768	777	787	797	807	816	826
45	836	846	856	865	875	885	895	904	914	924
46	933	943	953	963	972	982	992	*002	*011	*021
47	65 031	040	050	060	070	079	089	099	108	118
48	128	137	147	157	167	176	186	196	205	215
49	225	234	244	254	263	273	283	292	302	312
450	321	331	341	350	360	369	379	389	398	408
N.	**0**	**1**	**2**	**3**	**4**	**5**	**6**	**7**	**8**	**9**	Prop. Parts

450—500

N.	0	1	2	3	4	5	6	7	8	9	Prop. Parts
450	65 321	331	341	350	360	369	379	389	398	408	
51	418	427	437	447	456	466	475	485	495	504	
52	514	523	533	543	552	562	571	581	591	600	
53	610	619	629	639	648	658	667	677	686	696	
54	706	715	725	734	744	753	763	772	782	792	
55	801	811	820	830	839	849	858	868	877	887	
56	896	906	916	925	935	944	954	963	973	982	
57	992	*001	*011	*020	*030	*039	*049	*058	*068	*077	
58	66 087	096	106	115	124	134	143	153	162	172	
59	181	191	200	210	219	229	238	247	257	266	
460	276	285	295	304	314	323	332	342	351	361	
61	370	380	389	398	408	417	427	436	445	455	**10** / **9**
62	464	474	483	492	502	511	521	530	539	549	
63	558	567	577	586	596	605	614	624	633	642	1 1.0 0.9
64	652	661	671	680	689	699	708	717	727	736	2 2.0 1.8
65	745	755	764	773	783	792	801	811	820	829	3 3.0 2.7 4 4.0 3.6
66	839	848	857	867	876	885	894	904	913	922	5 5.0 4.5
67	932	941	950	960	969	978	987	997	*006	*015	6 6.0 5.4
68	67 025	034	043	052	062	071	080	089	099	108	7 7.0 6.3
69	117	127	136	145	154	164	173	182	191	201	8 8.0 7.2 9 9.0 8.1
470	210	219	228	237	247	256	265	274	284	293	
71	302	311	321	330	339	348	357	367	376	385	
72	394	403	413	422	431	440	449	459	468	477	
73	486	495	504	514	523	532	541	550	560	569	
74	578	587	596	605	614	624	633	642	651	660	
75	669	679	688	697	706	715	724	733	742	752	
76	761	770	779	788	797	806	815	825	834	843	
77	852	861	870	879	888	897	906	916	925	934	
78	943	952	961	970	979	988	997	*006	*015	*024	
79	68 034	043	052	061	070	079	088	097	106	115	
480	124	133	142	151	160	169	178	187	196	205	
81	215	224	233	242	251	260	269	278	287	296	**8**
82	305	314	323	332	341	350	359	368	377	386	1 0.8
83	395	404	413	422	431	440	449	458	467	476	2 1.6
84	485	494	502	511	520	529	538	547	556	565	3 2.4
85	574	583	592	601	610	619	628	637	646	655	4 3.2 5 4.0
86	664	673	681	690	699	708	717	726	735	744	6 4.8
87	753	762	771	780	789	797	806	815	824	833	7 5.6
88	842	851	860	869	878	886	895	904	913	922	8 6.4
89	931	940	949	958	966	975	984	993	*002	*011	9 7.2
490	69 020	028	037	046	055	064	073	082	090	099	
91	108	117	126	135	144	152	161	170	179	188	
92	197	205	214	223	232	241	249	258	267	276	
93	285	294	302	311	320	329	338	346	355	364	
94	373	381	390	399	408	417	425	434	443	452	
95	461	469	478	487	496	504	513	522	531	539	
96	548	557	566	574	583	592	601	609	618	627	
97	636	644	653	662	671	679	688	697	705	714	
98	723	732	740	749	758	767	775	784	793	801	
99	810	819	827	836	845	854	862	871	880	888	
500	897	906	914	923	932	940	949	958	966	975	
N.	0	1	2	3	4	5	6	7	8	9	Prop. Parts

Table I

500 — 550

N.	0	1	2	3	4	5	6	7	8	9
500	69 897	906	914	923	932	940	949	958	966	975
01	984	992	*001	*010	*018	*027	*036	*044	*053	*062
02	70 070	079	088	096	105	114	122	131	140	148
03	157	165	174	183	191	200	209	217	226	234
04	243	252	260	269	278	286	295	303	312	321
05	329	338	346	355	364	372	381	389	398	406
06	415	424	432	441	449	458	467	475	484	492
07	501	509	518	526	535	544	552	561	569	578
08	586	595	603	612	621	629	638	646	655	663
09	672	680	689	697	706	714	723	731	740	749
510	757	766	774	783	791	800	808	817	825	834
11	842	851	859	868	876	885	893	902	910	919
12	927	935	944	952	961	969	978	986	995	*003
13	71 012	020	029	037	046	054	063	071	079	088
14	096	105	113	122	130	139	147	155	164	172
15	181	189	198	206	214	223	231	240	248	257
16	265	273	282	290	299	307	315	324	332	341
17	349	357	366	374	383	391	399	408	416	425
18	433	441	450	458	466	475	483	492	500	508
19	517	525	533	542	550	559	567	575	584	592
520	600	609	617	625	634	642	650	659	667	675
21	684	692	700	709	717	725	734	742	750	759
22	767	775	784	792	800	809	817	825	834	842
23	850	858	867	875	883	892	900	908	917	925
24	933	941	950	958	966	975	983	991	999	*008
25	72 016	024	032	041	049	057	066	074	082	090
26	099	107	115	123	132	140	148	156	165	173
27	181	189	198	206	214	222	230	239	247	255
28	263	272	280	288	296	304	313	321	329	337
29	346	354	362	370	378	387	395	403	411	419
530	428	436	444	452	460	469	477	485	493	501
31	509	518	526	534	542	550	558	567	575	583
32	591	599	607	616	624	632	640	648	656	665
33	673	681	689	697	705	713	722	730	738	746
34	754	762	770	779	787	795	803	811	819	827
35	835	843	852	860	868	876	884	892	900	908
36	916	925	933	941	949	957	965	973	981	989
37	997	*006	*014	*022	*030	*038	*046	*054	*062	*070
38	73 078	086	094	102	111	119	127	135	143	151
39	159	167	175	183	191	199	207	215	223	231
540	239	247	255	263	272	280	288	296	304	312
41	320	328	336	344	352	360	368	376	384	392
42	400	408	416	424	432	440	448	456	464	472
43	480	488	496	504	512	520	528	536	544	552
44	560	568	576	584	592	600	608	616	624	632
45	640	648	656	664	672	679	687	695	703	711
46	719	727	735	743	751	759	767	775	783	791
47	799	807	815	823	830	838	846	854	862	870
48	878	886	894	902	910	918	926	933	941	949
49	957	965	973	981	989	997	*005	*013	*020	*028
550	74 036	044	052	060	068	076	084	092	099	107
N.	0	1	2	3	4	5	6	7	8	9

Prop. Parts

	9	8
1	0.9	0.8
2	1.8	1.6
3	2.7	2.4
4	3.6	3.2
5	4.5	4.0
6	5.4	4.8
7	6.3	5.6
8	7.2	6.4
9	8.1	7.2

	7
1	0.7
2	1.4
3	2.1
4	2.8
5	3.5
6	4.2
7	4.9
8	5.6
9	6.3

550 — 600

N.	0	1	2	3	4	5	6	7	8	9	Prop. Parts
550	74 036	044	052	060	068	076	084	092	099	107	
51	115	123	131	139	147	155	162	170	178	186	
52	194	202	210	218	225	233	241	249	257	265	
53	273	280	288	296	304	312	320	327	335	343	
54	351	359	367	374	382	390	398	406	414	421	
55	429	437	445	453	461	468	476	484	492	500	
56	507	515	523	531	539	547	554	562	570	578	
57	586	593	601	609	617	624	632	640	648	656	
58	663	671	679	687	695	702	710	718	726	733	
59	741	749	757	764	772	780	788	796	803	811	
560	819	827	834	842	850	858	865	873	881	889	
61	896	904	912	920	927	935	943	950	958	966	
62	974	981	989	997	*005	*012	*020	*028	*035	*043	
63	75 051	059	066	074	082	089	097	105	113	120	
64	128	136	143	151	159	166	174	182	189	197	
65	205	213	220	228	236	243	251	259	266	274	
66	282	289	297	305	312	320	328	335	343	351	
67	358	366	374	381	389	397	404	412	420	427	
68	435	442	450	458	465	473	481	488	496	504	
69	511	519	526	534	542	549	557	565	572	580	
570	587	595	603	610	618	626	633	641	648	656	
71	664	671	679	686	694	702	709	717	724	732	
72	740	747	755	762	770	778	785	793	800	808	
73	815	823	831	838	846	853	861	868	876	884	
74	891	899	906	914	921	929	937	944	952	959	
75	967	974	982	989	997	*005	*012	*020	*027	*035	
76	76 042	050	057	065	072	080	087	095	103	110	
77	118	125	133	140	148	155	163	170	178	185	
78	193	200	208	215	223	230	238	245	253	260	
79	268	275	283	290	298	305	313	320	328	335	
580	343	350	358	365	373	380	388	395	403	410	
81	418	425	433	440	448	455	462	470	477	485	
82	492	500	507	515	522	530	537	545	552	559	
83	567	574	582	589	597	604	612	619	626	634	
84	641	649	656	664	671	678	686	693	701	708	
85	716	723	730	738	745	753	760	768	775	782	
86	790	797	805	812	819	827	834	842	849	856	
87	864	871	879	886	893	901	908	916	923	930	
88	938	945	953	960	967	975	982	989	997	*004	
89	77 012	019	026	034	041	048	056	063	070	078	
590	085	093	100	107	115	122	129	137	144	151	
91	159	166	173	181	188	195	203	210	217	225	
92	232	240	247	254	262	269	276	283	291	298	
93	305	313	320	327	335	342	349	357	364	371	
94	379	386	393	401	408	415	422	430	437	444	
95	452	459	466	474	481	488	495	503	510	517	
96	525	532	539	546	554	561	568	576	583	590	
97	597	605	612	619	627	634	641	648	656	663	
98	670	677	685	692	699	706	714	721	728	735	
99	743	750	757	764	772	779	786	793	801	808	
600	815	822	830	837	844	851	859	866	873	880	
N.	0	1	2	3	4	5	6	7	8	9	Prop. Parts

Prop. Parts:

	8	7
1	0.8	0.7
2	1.6	1.4
3	2.4	2.1
4	3.2	2.8
5	4.0	3.5
6	4.8	4.2
7	5.6	4.9
8	6.4	5.6
9	7.2	6.3

600 — 650

N.	0	1	2	3	4	5	6	7	8	9
600	77 815	822	830	837	844	851	859	866	873	880
01	887	895	902	909	916	924	931	938	945	952
02	960	967	974	981	988	996	*003	*010	*017	*025
03	78 032	039	046	053	061	068	075	082	089	097
04	104	111	118	125	132	140	147	154	161	168
05	176	183	190	197	204	211	219	226	233	240
06	247	254	262	269	276	283	290	297	305	312
07	319	326	333	340	347	355	362	369	376	383
08	390	398	405	412	419	426	433	440	447	455
09	462	469	476	483	490	497	504	512	519	526
610	533	540	547	554	561	569	576	583	590	597
11	604	611	618	625	633	640	647	654	661	668
12	675	682	689	696	704	711	718	725	732	739
13	746	753	760	767	774	781	789	796	803	810
14	817	824	831	838	845	852	859	866	873	880
15	888	895	902	909	916	923	930	937	944	951
16	958	965	972	979	986	993	*000	*007	*014	*021
17	79 029	036	043	050	057	064	071	078	085	092
18	099	106	113	120	127	134	141	148	155	162
19	169	176	183	190	197	204	211	218	225	232
620	239	246	253	260	267	274	281	288	295	302
21	309	316	323	330	337	344	351	358	365	372
22	379	386	393	400	407	414	421	428	435	442
23	449	456	463	470	477	484	491	498	505	511
24	518	525	532	539	546	553	560	567	574	581
25	588	595	602	609	616	623	630	637	644	650
26	657	664	671	678	685	692	699	706	713	720
27	727	734	741	748	754	761	768	775	782	789
28	796	803	810	817	824	831	837	844	851	858
29	865	872	879	886	893	900	906	913	920	927
630	934	941	948	955	962	969	975	982	989	996
31	80 003	010	017	024	030	037	044	051	058	065
32	072	079	085	092	099	106	113	120	127	134
33	140	147	154	161	168	175	182	188	195	202
34	209	216	223	229	236	243	250	257	264	271
35	277	284	291	298	305	312	318	325	332	339
36	346	353	359	366	373	380	387	393	400	407
37	414	421	428	434	441	448	455	462	468	475
38	482	489	496	502	509	516	523	530	536	543
39	550	557	564	570	577	584	591	598	604	611
640	618	625	632	638	645	652	659	665	672	679
41	686	693	699	706	713	720	726	733	740	747
42	754	760	767	774	781	787	794	801	808	814
43	821	828	835	841	848	855	862	868	875	882
44	889	895	902	909	916	922	929	936	943	949
45	956	963	969	976	983	990	996	*003	*010	*017
46	81 023	030	037	043	050	057	064	070	077	084
47	090	097	104	111	117	124	131	137	144	151
48	158	164	171	178	184	191	198	204	211	218
49	224	231	238	245	251	258	265	271	278	285
650	291	298	305	311	318	325	331	338	345	351
N.	0	1	2	3	4	5	6	7	8	9

Prop. Parts

	8	7
1	0.8	0.7
2	1.6	1.4
3	2.4	2.1
4	3.2	2.8
5	4.0	3.5
6	4.8	4.2
7	5.6	4.9
8	6.4	5.6
9	7.2	6.3

	6
1	0.6
2	1.2
3	1.8
4	2.4
5	3.0
6	3.6
7	4.2
8	4.8
9	5.4

650 — 700

N.	0	1	2	3	4	5	6	7	8	9
650	81 291	298	305	311	318	325	331	338	345	351
51	358	365	371	378	385	391	398	405	411	418
52	425	431	438	445	451	458	465	471	478	485
53	491	498	505	511	518	525	531	538	544	551
54	558	564	571	578	584	591	598	604	611	617
55	624	631	637	644	651	657	664	671	677	684
56	690	697	704	710	717	723	730	737	743	750
57	757	763	770	776	783	790	796	803	809	816
58	823	829	836	842	849	856	862	869	875	882
59	889	895	902	908	915	921	928	935	941	948
660	954	961	968	974	981	987	994	*000	*007	*014
61	82 020	027	033	040	046	053	060	066	073	079
62	086	092	099	105	112	119	125	132	138	145
63	151	158	164	171	178	184	191	197	204	210
64	217	223	230	236	243	249	256	263	269	276
65	282	289	295	302	308	315	321	328	334	341
66	347	354	360	367	373	380	387	393	400	406
67	413	419	426	432	439	445	452	458	465	471
68	478	484	491	497	504	510	517	523	530	536
69	543	549	556	562	569	575	582	588	595	601
670	607	614	620	627	633	640	646	653	659	666
71	672	679	685	692	698	705	711	718	724	730
72	737	743	750	756	763	769	776	782	789	795
73	802	808	814	821	827	834	840	847	853	860
74	866	872	879	885	892	898	905	911	918	924
75	930	937	943	950	956	963	969	975	982	988
76	995	*001	*008	*014	*020	*027	*033	*040	*046	*052
77	83 059	065	072	078	085	091	097	104	110	117
78	123	129	136'	142	149	155	161	168	174	181
79	187	193	200	206	213	219	225	232	238	245
680	251	257	264	270	276	283	289	296	302	308
81	315	321	327	334	340	347	353	359	366	372
82	378	385	391	398	404	410	417	423	429	436
83	442	448	455	461	467	474	480	487	493	499
84	506	512	518	525	531	537	544	550	556	563
85	569	575	582	588	594	601	607	613	620	626
86	632	639	645	651	658	664	670	677	683	689
87	696	702	708	715	721	727	734	740	746	753
88	759	765	771	778	784	790	797	803	809	816
89	822	828	835	841	847	853	860	866	872	879
690	885	891	897	904	910	916	923	929	935	942
91	948	954	960	967	973	979	985	992	998	*004
92	84 011	017	023	029	036	042	048	055	061	067
93	073	080	086	092	098	105	111	117	123	130
94	136	142	148	155	161	167	173	180	186	192
95	198	205	211	217	223	230	236	242	248	255
96	261	267	273	280	286	292	298	305	311	317
97	323	330	336	342	348	354	361	367	373	379
98	386	392	398	404	410	417	423	429	435	442
99	448	454	460	466	473	479	485	491	497	504
700	510	516	522	528	535	541	547	553	559	566
N.	0	1	2	3	4	5	6	7	8	9

Prop. Parts

	7	6
1	0.7	0.6
2	1.4	1.2
3	2.1	1.8
4	2.8	2.4
5	3.5	3.0
6	4.2	3.6
7	4.9	4.2
8	5.6	4.8
9	6.3	5.4

700 — 750

N.	0	1	2	3	4	5	6	7	8	9	Prop. Parts
700	84 510	516	522	528	535	541	547	553	559	566	
01	572	578	584	590	597	603	609	615	621	628	
02	634	640	646	652	658	665	671	677	683	689	
03	696	702	708	714	720	726	733	739	745	751	
04	757	763	770	776	782	788	794	800	807	813	
05	819	825	831	837	844	850	856	862	868	874	
06	880	887	893	899	905	911	917	924	930	936	
07	942	948	954	960	967	973	979	985	991	997	
08	85 003	009	016	022	028	034	040	046	052	058	
09	065	071	077	083	089	095	101	107	114	120	
710	126	132	138	144	150	156	163	169	175	181	
11	187	193	199	205	211	217	224	230	236	242	
12	248	254	260	266	272	278	285	291	297	303	
13	309	315	321	327	333	339	345	352	358	364	
14	370	376	382	388	394	400	406	412	418	425	
15	431	437	443	449	455	461	467	473	479	485	
16	491	497	503	509	516	522	528	534	540	546	
17	552	558	564	570	576	582	588	594	600	606	
18	612	618	625	631	637	643	649	655	661	667	
19	673	679	685	691	697	703	709	715	721	727	
720	733	739	745	751	757	763	769	775	781	788	
21	794	800	806	812	818	824	830	836	842	848	
22	854	860	866	872	878	884	890	896	902	908	
23	914	920	926	932	938	944	950	956	962	968	
24	974	980	986	992	998	*004	*010	*016	*022	*028	
25	86 034	040	046	052	058	064	070	076	082	088	
26	094	100	106	112	118	124	130	136	141	147	
27	153	159	165	171	177	183	189	195	201	207	
28	213	219	225	231	237	243	249	255	261	267	
29	273	279	285	291	297	303	308	314	320	326	
730	332	338	344	350	356	362	368	374	380	386	
31	392	398	404	410	415	421	427	433	439	445	
32	451	457	463	469	475	481	487	493	499	504	
33	510	516	522	528	534	540	546	552	558	564	
34	570	576	581	587	593	599	605	611	617	623	
35	629	635	641	646	652	658	664	670	676	682	
36	688	694	700	705	711	717	723	729	735	741	
37	747	753	759	764	770	776	782	788	794	800	
38	806	812	817	823	829	835	841	847	853	859	
39	864	870	876	882	888	894	900	906	911	917	
740	923	929	935	941	947	953	958	964	970	976	
41	982	988	994	999	*005	*011	*017	*023	*029	*035	
42	87 040	046	052	058	064	070	075	081	087	·093	
43	099	105	111	116	122	128	134	140	146	151	
44	157	163	169	175	181	186	192	198	204	210	
45	216	221	227	233	239	245	251	256	262	268	
46	274	280	286	291	297	303	309	315	320	326	
47	332	338	344	349	355	361	367	373	379	384	
48	390	396	402	408	413	419	425	431	437	442	
49	448	454	460	466	471	477	483	489	495	500	
750	506	512	518	523	529	535	541	547	552	558	
N.	0	1	2	3	4	5	6	7	8	9	Prop. Parts

Prop. Parts

	7	6
1	0.7	0.6
2	1.4	1.2
3	2.1	1.8
4	2.8	2.4
5	3.5	3.0
6	4.2	3.6
7	4.9	4.2
8	5.6	4.8
9	6.3	5.4

	5
1	0.5
2	1.0
3	1.5
4	2.0
5	2.5
6	3.0
7	3.5
8	4.0
9	4.5

750 — 800

N.	0	1	2	3	4	5	6	7	8	9	Prop. Parts
750	87 506	512	518	523	529	535	541	547	552	558	
51	564	570	576	581	587	593	599	604	610	616	
52	622	628	633	639	645	651	656	662	668	674	
53	679	685	691	697	703	708	714	720	726	731	
54	737	743	749	754	760	766	772	777	783	789	
55	795	800	806	812	818	823	829	835	841	846	
56	852	858	864	869	875	881	887	892	898	904	
57	910	915	921	927	933	938	944	950	955	961	
58	967	973	978	984	990	996	*001	*007	*013	*018	
59	88 024	030	036	041	047	053	058	064	070	076	
760	081	087	093	098	104	110	116	121	127	133	
61	138	144	150	156	161	167	173	178	184	190	
62	195	201	207	213	218	224	230	235	241	247	
63	252	258	264	270	275	281	287	292	298	304	
64	309	315	321	326	332	338	343	349	355	360	
65	366	372	377	383	389	395	400	406	412	417	
66	423	429	434	440	446	451	457	463	468	474	
67	480	485	491	497	502	508	513	519	525	530	
68	536	542	547	553	559	564	570	576	581	587	
69	593	598	604	610	615	621	627	632	638	643	
770	649	655	660	666	672	677	683	689	694	700	
71	705	711	717	722	728	734	739	745	750	756	
72	762	767	773	779	784	790	795	801	807	812	
73	818	824	829	835	840	846	852	857	863	868	
74	874	880	885	891	897	902	908	913	919	925	
75	930	936	941	947	953	958	964	969	975	981	
76	986	992	997	*003	*009	*014	*020	*025	*031	*037	
77	89 042	048	053	059	064	070	076	081	087	092	
78	098	104	109	115	120	126	131	137	143	148	
79	154	159	165	170	176	182	187	193	198	204.	
780	209	215	221	226	232	237	243	248	254	260	
81	265	271	276	282	287	293	298	304	310	315	
82	321	326	332	337	343	348	354	360	365	371	
83	376	382	387	393	398	404	409	415	421	426	
84	432	437	443	448	454	459	465	470	476	481	
85	487	492	498	504	509	515	520	526	531	537	
86	542	548	553	559	564	570	575	581	586	592	
87	597	603	609	614	620	625	631	636	642	647	
88	653	658	664	669	675	680	686	691	697	702	
89	708	713	719	724	730	735	741	746	752	757	
790	763	768	774	779	785	790	796	801	807	812	
91	818	823	829	834	840	845	851	856	862	867	
92	873	878	883	889	894	900	905	911	916	922	
93	927	933	938	944	949	955	960	966	971	977	
94	982	988	993	998	*004	*009	*015	*020	*026	*031	
95	90 037	042	048	053	059	064	069	075	080	086	
96	091	097	102	108	113	119	124	129	135	140	
97	146	151	157	162	168	173	179	184	189	195	
98	200	206	211	217	222	227	233	238	244	249	
99	255	260	266	271	276	282	287	293	298	304	
800	309	314	320	325	331	336	342	347	352	358	
N.	0	1	2	3	4	5	6	7	8	9	Prop. Parts

Prop. Parts:

	6	5
1	0.6	0.5
2	1.2	1.0
3	1.8	1.5
4	2.4	2.0
5	3.0	2.5
6	3.6	3.0
7	4.2	3.5
8	4.8	4.0
9	5.4	4.5

800 — 850

N.	0	1	2	3	4	5	6	7	8	9	Prop. Parts
800	90 309	314	320	325	331	336	342	347	352	358	
01	363	369	374	380	385	390	396	401	407	412	
02	417	423	428	434	439	445	450	455	461	466	
03	472	477	482	488	493	499	504	509	515	520	
04	526	531	536	542	547	553	558	563	569	574	
05	580	585	590	596	601	607	612	617	623	628	
06	634	639	644	650	655	660	666	671	677	682	
07	687	693	698	703	709	714	720	725	730	736	
08	741	747	752	757	763	768	773	779	784	789	
09	795	800	806	811	816	822	827	832	838	843	
810	849	854	859	865	870	875	881	886	891	897	
11	902	907	913	918	924	929	934	940	945	950	
12	956	961	966	972	977	982	988	993	998	*004	
13	91 009	014	020	025	030	036	041	046	052	057	
14	062	068	073	078	084	089	094	100	105	110	
15	116	121	126	132	137	142	148	153	158	164	
16	169	174	180	185	190	196	201	206	212	217	
17	222	228	233	238	243	249	254	259	265	270	
18	275	281	286	291	297	302	307	312	318	323	
19	328	334	339	344	350	355	360	365	371	376	
820	381	387	392	397	403	408	413	418	424	429	
21	434	440	445	450	455	461	466	471	477	482	
22	487	492	498	503	508	514	519	524	529	535	
23	540	545	551	556	561	566	572	577	582	587	
24	593	598	603	609	614	619	624	630	635	640	
25	645	651	656	661	666	672	677	682	687	693	
26	698	703	709	714	719	724	730	735	740	745	
27	751	756	761	766	772	777	782	787	793	798	
28	803	808	814	819	824	829	834	840	845	850	
29	855	861	866	871	876	882	887	892	897	903	
830	908	913	918	924	929	934	939	944	950	955	
31	960	965	971	976	981	986	991	997	*002	*007	
32	92 012	018	023	028	033	038	044	049	054	059	
33	065	070	075	080	085	091	096	101	106	111	
34	117	122	127	132	137	143	148	153	158	163	
35	169	174	179	184	189	195	200	205	210	215	
36	221	226	231	236	241	247	252	257	262	267	
37	273	278	283	288	293	298	304	309	314	319	
38	324	330	335	340	345	350	355	361	366	371	
39	376	381	387	392	397	402	407	412	418	423	
840	428	433	438	443	449	454	459	464	469	474	
41	480	485	490	495	500	505	511	516	521	526	
42	531	536	542	547	552	557	562	567	572	578	
43	583	588	593	598	603	609	614	619	624	629	
44	634	639	645	650	655	660	665	670	675	681	
45	686	691	696	701	706	711	716	722	727	732	
46	737	742	747	752	758	763	768	773	778	783	
47	788	793	799	804	809	814	819	824	829	834	
48	840	845	850	855	860	865	870	875	881	886	
49	891	896	901	906	911	916	921	927	932	937	
850	942	947	952	957	962	967	973	978	983	988	
N.	0	1	2	3	4	5	6	7	8	9	Prop. Parts

Prop. Parts

	6	5
1	0.6	0.5
2	1.2	1.0
3	1.8	1.5
4	2.4	2.0
5	3.0	2.5
6	3.6	3.0
7	4.2	3.5
8	4.8	4.0
9	5.4	4.5

850 — 900

N.	0	1	2	3	4	5	6	7	8	9
850	92 942	947	952	957	962	967	973	978	983	988
51	993	998	*003	*008	*013	*018	*024	*029	*034	*039
52	93 044	049	054	059	064	069	075	080	085	090
53	095	100	105	110	115	120	125	131	136	141
54	146	151	156	161	166	171	176	181	186	192
55	197	202	207	212	217	222	227	232	237	242
56	247	252	258	263	268	273	278	283	288	293
57	298	303	308	313	318	323	328	334	339	344
58	349	354	359	364	369	374	379	384	389	394
59	399	404	409	414	420	425	430	435	440	445
860	450	455	460	465	470	475	480	485	490	495
61	500	505	510	515	520	526	531	536	541	546
62	551	556	561	566	571	576	581	586	591	596
63	601	606	611	616	621	626	631	636	641	646
64	651	656	661	666	671	676	682	687	692	697
65	702	707	712	717	722	727	732	737	742	747
66	752	757	762	767	772	777	782	787	792	797
67	802	807	812	817	822	827	832	837	842	847
68	852	857	862	867	872	877	882	887	892	897
69	902	907	912	917	922	927	932	937	942	947
870	952	957	962	967	972	977	982	987	992	997
71	94 002	007	012	017	022	027	032	037	042	047
72	052	057	062	067	072	077	082	086	091	096
73	101	106	111	116	121	126	131	136	141	146
74	151	156	161	166	171	176	181	186	191	196
75	201	206	211	216	221	226	231	236	240	245
76	250	255	260	265	270	275	280	285	290	295
77	300	305	310	315	320	325	330	335	340	345
78	349	354	359	364	369	374	379	384	389	394
79	399	404	409	414	419	424	429	433	438	443
880	448	453	458	463	468	473	478	483	488	493
81	498	503	507	512	517	522	527	532	537	542
82	547	552	557	562	567	571	576	581	586	591
83	596	601	606	611	616	621	626	630	635	640
84	645	650	655	660	665	670	675	680	685	689
85	694	699	704	709	714	719	724	729	734	738
86	743	748	753	758	763	768	773	778	783	787
87	792	797	802	807	812	817	822	827	832	836
88	841	846	851	856	861	866	871	876	880	885
89	890	895	900	905	910	915	919	924	929	934
890	939	944	949	954	959	963	968	973	978	983
91	988	993	998	*002	*007	*012	*017	*022	*027	*032
92	95 036	041	046	051	056	061	066	071	075	080
93	085	090	095	100	105	109	114	119	124	129
94	134	139	143	148	153	158	163	168	173	177
95	182	187	192	197	202	207	211	216	221	226
96	231	236	240	245	250	255	260	265	270	274
97	279	284	289	294	299	303	308	313	318	323
98	328	332	337	342	347	352	357	361	366	371
99	376	381	386	390	395	400	405	410	415	419
900	424	429	434	439	444	448	453	458	463	468
N.	0	1	2	3	4	5	6	7	8	9

Prop. Parts

	6	5
1	0.6	0.5
2	1.2	1.0
3	1.8	1.5
4	2.4	2.0
5	3.0	2.5
6	3.6	3.0
7	4.2	3.5
8	4.8	4.0
9	5.4	4.5

	4
1	0.4
2	0.8
3	1.2
4	1.6
5	2.0
6	2.4
7	2.8
8	3.2
9	3.6

Table I

900 — 950

N.	0	1	2	3	4	5	6	7	8	9	Prop. Parts
900	95 424	429	434	439	444	448	453	458	463	468	
01	472	477	482	487	492	497	501	506	511	516	
02	521	525	530	535	540	545	550	554	559	564	
03	569	574	578	583	588	593	598	602	607	612	
04	617	622	626	631	636	641	646	650	655	660	
05	665	670	674	679	684	689	694	698	703	708	
06	713	718	722	727	732	737	742	746	751	756	
07	761	766	770	775	780	785	789	794	799	804	
08	809	813	818	823	828	832	837	842	847	852	
09	856	861	866	871	875	880	885	890	895	899	
910	904	909	914	918	923	928	933	938	942	947	
11	952	957	961	966	971	976	980	985	990	995	
12	999	*004	*009	*014	*019	*023	*028	*033	*038	*042	
13	96 047	052	057	061	066	071	076	080	085	090	
14	095	099	104	109	114	118	123	128	133	137	
15	142	147	152	156	161	166	171	175	180	185	
16	190	194	199	204	209	213	218	223	227	232	
17	237	242	246	251	256	261	265	270	275	280	
18	284	289	294	298	303	308	313	317	322	327	
19	332	336	341	346	350	355	360	365	369	374	
920	379	384	388	393	398	402	407	412	417	421	
21	426	431	435	440	445	450	454	459	464	468	
22	473	478	483	487	492	497	501	506	511	515	
23	520	525	530	534	539	544	548	553	558	562	
24	567	572	577	581	586	591	595	600	605	609	
25	614	619	624	628	633	638	642	647	652	656	
26	661	666	670	675	680	685	689	694	699	703	
27	708	713	717	722	727	731	736	741	745	750	
28	755	759	764	769	774	778	783	788	792	797	
29	802	806	811	816	820	825	830	834	839	844	
930	848	853	858	862	867	872	876	881	886	890	
31	895	900	904	909	914	918	923	928	932	937	
32	942	946	951	956	960	965	970	974	979	984	
33	988	993	997	*002	*007	*011	*016	*021	*025	*030	
34	97 035	039	044	049	053	058	063	067	072	077	
35	081	086	090	095	100	104	109	114	118	123	
36	128	132	137	142	146	151	155	160	165	169	
37	174	179	183	188	192	197	202	206	211	216	
38	220	225	230	234	239	243	248	253	257	262	
39	267	271	276	280	285	290	294	299	304	308	
940	313	317	322	327	331	336	340	345	350	354	
41	359	364	368	373	377	382	387	391	396	400	
42	405	410	414	419	424	428	433	437	442	447	
43	451	456	460	465	470	474	479	483	488	493	
44	497	502	506	511	516	520	525	529	534	539	
45	543	548	552	557	562	566	571	575	580	585	
46	589	594	598	603	607	612	617	621	626	630	
47	635	640	644	649	653	658	663	667	672	676	
48	681	685	690	695	699	704	708	713	717	722	
49	727	731	736	740	745	749	754	759	763	768	
950	772	777	782	786	791	795	800	804	809	813	
N.	0	1	2	3	4	5	6	7	8	9	Prop. Parts

Prop. Parts:

	5	4
1	0.5	0.4
2	1.0	0.8
3	1.5	1.2
4	2.0	1.6
5	2.5	2.0
6	3.0	2.4
7	3.5	2.8
8	4.0	3.2
9	4.5	3.6

950 — 1000

N.	0	1	2	3	4	5	6	7	8	9
950	97 772	777	782	786	791	795	800	804	809	813
51	818	823	827	832	836	841	845	850	855	859
52	864	868	873	877	882	886	891	896	900	905
53	909	914	918	923	928	932	937	941	946	950
54	955	959	964	968	973	978	982	987	991	996
55	98 000	005	009	014	019	023	028	032	037	041
56	046	050	055	059	064	068	073	078	082	087
57	091	096	100	105	109	114	118	123	127	132
58	137	141	146	150	155	159	164	168	173	177
59	182	186	191	195	200	204	209	214	218	223
960	227	232	236	241	245	250	254	259	263	268
61	272	277	281	286	290	295	299	304	308	313
62	318	322	327	331	336	340	345	349	354	358
63	363	367	372	376	381	385	390	394	399	403
64	408	412	417	421	426	430	435	439	444	448
65	453	457	462	466	471	475	480	484	489	493
66	498	502	507	511	516	520	525	529	534	538
67	543	547	552	556	561	565	570	574	579	583
68	588	592	597	601	605	610	614	619	623	628
69	632	637	641	646	650	655	659	664	668	673
970	677	682	686	691	695	700	704	709	713	717
71	722	726	731	735	740	744	749	753	758	762
72	767	771	776	780	784	789	793	798	802	807
73	811	816	820	825	829	834	838	843	847	851
74	856	860	865	869	874	878	883	887	892	896
75	900	905	909	914	918	923	927	932	936	941
76	945	949	954	958	963	967	972	976	981	985
77	989	994	998	*003	*007	*012	*016	*021	*025	*029
78	99 034	038	043	047	052	056	061	065	069	074
79	078	083	087	092	096	100	105	109	114	118
980	123	127	131	136	140	145	149	154	158	162
81	167	171	176	180	185	189	193	198	202	207
82	211	216	220	224	229	233	238	242	247	251
83	255	260	264	269	273	277	282	286	291	295
84	300	304	308	313	317	322	326	330	335	339
85	344	348	352	357	361	366	370	374	379	383
86	388	392	396	401	405	410	414	419	423	427
87	432	436	441	445	449	454	458	463	467	471
88	476	480	484	489	493	498	502	506	511	515
89	520	524	528	533	537	542	546	550	555	559
990	564	568	572	577	581	585	590	594	599	603
91	607	612	616	621	625	629	634	638	642	647
92	651	656	660	664	669	673	677	682	686	691
93	695	699	704	708	712	717	721	726	730	734
94	739	743	747	752	756	760	765	769	774	778
95	782	787	791	795	800	804	808	813	817	822
96	826	830	835	839	843	848	852	856	861	865
97	870	874	878	883	887	891	896	900	904	909
98	913	917	922	926	930	935	939	944	948	952
99	957	961	965	970	974	978	983	987	991	996
1000	00 000	004	009	013	017	022	026	030	035	039
N.	0	1	2	3	4	5	6	7	8	9

Prop. Parts

	5	4
1	0.5	0.4
2	1.0	0.8
3	1.5	1.2
4	2.0	1.6
5	2.5	2.0
6	3.0	2.4
7	3.5	2.8
8	4.0	3.2
9	4.5	3.6

Natural (Napierian) Logarithms of Prime Numbers

N.	Log$_e$	N.	Log$_e$	N.	Log$_e$	N.	Log$_e$	N.	Log$_e$
2	0.69 315	13	2.56 495	31	3.43 399	53	3.97 029	73	4.29 046
3	1.09 861	17	2.83 321	37	3.61 092	59	4.07 754	79	4.36 945
5	1.60 944	19	2.94 444	41	3.71 357	61	4.11 087	83	4.41 884
7	1.94 591	23	3.13 549	43	3.76 120	67	4.20 469	89	4.48 864
11	2.39 790	29	3.36 730	47	3.85 015	71	4.26 268	97	4.57 471

For other numbers combine these, or use the formula

$$\log_e N = (\log_e 10)(\log_{10} N) = \frac{1}{M}\log_{10} N = 2.30258509 \log_{10} N.$$

$$\log_{10}\left(\frac{1}{M}\right) = \log_{10}(2.30258509) = 0.36221569.$$

$$\log_e \pi = 1.14472989.$$

Useful Constants and Their Logarithms

N	Log$_{10}$ N
$e = \lim\limits_{x=\infty}\left(1+\frac{1}{x}\right)^x = 2.718\ 2818$	0.434 2945
$M = \log_{10} e = 0.434\ 2945$	9.637 7843 − 10
$\frac{1}{M} = \log_e 10 = 2.302\ 5851$	0.362 2157
$\pi = 3.141\ 5927$	0.497 1499
$\frac{1}{\pi} = 0.318\ 3099$	9.502 8501 − 10
$\frac{180}{\pi}$ = degrees in 1 radian = 57.295 7795	1.758 1226
$\frac{\pi}{180}$ = radians in 1° = 0.017 4533	8.241 8774 − 10

Table II

Logarithms of the trigonometric functions

to five decimal places

0°

′	Log Sin	d	Log Tan	c d	Log Cot	Log Cos	
0	—		—		—	0.00 000	60
1	6.46 373	30103	6.46 373	30103	3.53 627	0.00 000	59
2	6.76 476	17609	6.76 476	17609	3.23 524	0.00 000	58
3	6.94 085	12494	6.94 085	12494	3.05 915	0.00 000	57
4	7.06 579	9691	7.06 579	9691	2.93 421	0.00 000	56
5	7.16 270	7918	7.16 270	7918	2.83 730	0.00 000	55
6	7.24 188	6694	7.24 188	6694	2.75 812	0.00 000	54
7	7.30 882	5800	7.30 882	5800	2.69 118	0.00 000	53
8	7.36 682	5115	7.36 682	5115	2.63 318	0.00 000	52
9	7.41 797	4576	7.41 797	4576	2.58 203	0.00 000	51
10	7.46 373	4139	7.46 373	4139	2.53 627	0.00 000	50
11	7.50 512	3779	7.50 512	3779	2.49 488	0.00 000	49
12	7.54 291	3476	7.54 291	3476	2.45 709	0.00 000	48
13	7.57 767	3218	7.57 767	3219	2.42 233	0.00 000	47
14	7.60 985	2997	7.60 986	2996	2.39 014	0.00 000	46
15	7.63 982	2802	7.63 982	2803	2.36 018	0.00 000	45
16	7.66 784	2633	7.66 785	2633	2.33 215	0.00 000	44
17	7.69 417	2483	7.69 418	2482	2.30 582	9.99 999	43
18	7.71 900	2348	7.71 900	2348	2.28 100	9.99 999	42
19	7.74 248	2227	7.74 248	2228	2.25 752	9.99 999	41
20	7.76 475	2119	7.76 476	2119	2.23 524	9.99 999	40
21	7.78 594	2021	7.78 595	2020	2.21 405	9.99 999	39
22	7.80 615	1930	7.80 615	1931	2.19 385	9.99 999	38
23	7.82 545	1848	7.82 546	1848	2.17 454	9.99 999	37
24	7.84 393	1773	7.84 394	1773	2.15 606	9.99 999	36
25	7.86 166	1704	7.86 167	1704	2.13 833	9.99 999	35
26	7.87 870	1639	7.87 871	1639	2.12 129	9.99 999	34
27	7.89 509	1579	7.89 510	1579	2.10 490	9.99 999	33
28	7.91 088	1524	7.91 089	1524	2.08 911	9.99 999	32
29	7.92 612	1472	7.92 613	1473	2.07 387	9.99 998	31
30	7.94 084	1424	7.94 086	1424	2.05 914	9.99 998	30
31	7.95 508	1379	7.95 510	1379	2.04 490	9.99 998	29
32	7.96 887	1336	7.96 889	1336	2.03 111	9.99 998	28
33	7.98 223	1297	7.98 225	1297	2.01 775	9.99 998	27
34	7.99 520	1259	7.99 522	1259	2.00 478	9.99 998	26
35	8.00 779	1223	8.00 781	1223	1.99 219	9.99 998	25
36	8.02 002	1190	8.02 004	1190	1.97 996	9.99 998	24
37	8.03 192	1158	8.03 194	1159	1.96 806	9.99 997	23
38	8.04 350	1128	8.04 353	1128	1.95 647	9.99 997	22
39	8.05 478	1100	8.05 481	1100	1.94 519	9.99 997	21
40	8.06 578	1072	8.06 581	1072	1.93 419	9.99 997	20
41	8.07 650	1046	8.07 653	1047	1.92 347	9.99 997	19
42	8.08 696	1022	8.08 700	1022	1.91 300	9.99 997	18
43	8.09 718	999	8.09 722	998	1.90 278	9.99 997	17
44	8.10 717	976	8.10 720	976	1.89 280	9.99 996	16
45	8.11 693	954	8.11 696	955	1.88 304	9.99 996	15
46	8.12 647	934	8.12 651	934	1.87 349	9.99 996	14
47	8.13 581	914	8.13 585	915	1.86 415	9.99 996	13
48	8.14 495	896	8.14 500	895	1.85 500	9.99 996	12
49	8.15 391	877	8.15 395	878	1.84 605	9.99 996	11
50	8.16 268	860	8.16 273	860	1.83 727	9.99 995	10
51	8.17 128	843	8.17 133	843	1.82 867	9.99 995	9
52	8.17 971	827	8.17 976	828	1.82 024	9.99 995	8
53	8.18 798	812	8.18 804	812	1.81 196	9.99 995	7
54	8.19 610	797	8.19 616	797	1.80 384	9.99 995	6
55	8.20 407	782	8.20 413	782	1.79 587	9.99 994	5
56	8.21 189	769	8.21 195	769	1.78 805	9.99 994	4
57	8.21 958	755	8.21 964	756	1.78 036	9.99 994	3
58	8.22 713	743	8.22 720	742	1.77 280	9.99 994	2
59	8.23 456	730	8.23 462	730	1.76 538	9.99 994	1
60	8.24 186		8.24 192		1.75 808	9.99 993	0
	Log Cos	d	Log Cot	c d	Log Tan	Log Sin	′

Interpolation for Log Sin and Log Tan of Small Angles and for Log Cos and Log Cot of Large Angles.

When ordinary interpolation is not sufficiently accurate,

For Small Angles,

Let N = No. of Minutes in θ.

Then

$$\log \sin \theta = \log N + S - 10,$$
$$\log \tan \theta = \log N + T - 10.$$

For Large Angles,

Let N = No. of Minutes in (90° − θ).

Then

$$\log \cos \theta = \log N + S - 10,$$
$$\log \cot \theta = \log N + T - 10.$$

Find log N from Table I, and find S and T below.

N	S	N	S
0′— 13′	6.46373	129′—134′	6.46362
14′— 42′	72	135′—140′	61
43′— 58′	71	141′—146′	60
59′— 71′	6.46370	147′—151′	6.46359
72′— 81′	69	152′—157′	58
82′— 91′	68	158′—162′	57
92′— 99′	6.46367	163′—167′	6.46356
100′—107′	66	168′—171′	55
108′—115′	65	172′—176′	54
116′—121′	6.46364	177′—180′	6.46353
122′—128′	63		

N	T	N	T
0′— 26′	6.46373	128′—130′	6.46393
27′— 39′	74	131′—133′	94
40′— 48′	75	134′—136′	95
49′— 56′	6.46376	137′—139′	6.46396
57′— 63′	77	140′—142′	97
64′— 69′	78	143′—145′	98
70′— 74′	6.46379	146′—148′	6.46399
75′— 80′	80	149′—150′	6.46400
81′— 85′	81	151′—153′	01
86′— 89′	6.46382	154′—156′	6.46402
90′— 94′	83	157′—158′	03
95′— 98′	84	159′—161′	04
99′—102′	6.46385	162′—163′	6.46405
103′—106′	86	164′—165′	06
107′—110′	87	166′—168′	07
111′—113′	6.46388	169′—171′	6.46408
114′—117′	89	172′—173′	09
118′—120′	90	174′—175′	10
121′—124′	6.46391	176′—178′	6.46411
125′—127′	92	179′—180′	6.46412

89°

1°

′	Log Sin	d	Log Tan	c d	Log Cot	Log Cos	
0	8.24 186		8.24 192		1.75 808	9.99 993	60
1	8.24 903	717	8.24 910	718	1.75 090	9.99 993	59
2	8.25 609	706	8.25 616	706	1.74 384	9.99 993	58
3	8.26 304	695	8.26 312	696	1.73 688	9.99 993	57
4	8.26 988	684	8.26 996	684	1.73 004	9.99 992	56
		673		673			
5	8.27 661		8.27 669		1.72 331	9.99 992	55
6	8.28 324	663	8.28 332	663	1.71 668	9.99 992	54
7	8.28 977	653	8.28 986	654	1.71 014	9.99 992	53
8	8.29 621	644	8.29 629	643	1.70 371	9.99 992	52
9	8.30 255	634	8.30 263	634	1.69 737	9.99 991	51
		624		625			
10	8.30 879		8.30 888		1.69 112	9.99 991	50
11	8.31 495	616	8.31 505	617	1.68 495	9.99 991	49
12	8.32 103	608	8.32 112	607	1.67 888	9.99 990	48
13	8.32 702	599	8.32 711	599	1.67 289	9.99 990	47
14	8.33 292	590	8.33 302	591	1.66 698	9.99 990	46
		583		584			
15	8.33 875		8.33 886		1.66 114	9.99 990	45
16	8.34 450	575	8.34 461	575	1.65 539	9.99 989	44
17	8.35 018	568	8.35 029	568	1.64 971	9.99 989	43
18	8.35 578	560	8.35 590	561	1.64 410	9.99 989	42
19	8.36 131	553	8.36 143	553	1.63 857	9.99 989	41
		547		546			
20	8.36 678		8.36 689		1.63 311	9.99 988	40
21	8.37 217	539	8.37 229	540	1.62 771	9.99 988	39
22	8.37 750	533	8.37 762	533	1.62 238	9.99 988	38
23	8.38 276	526	8.38 289	527	1.61 711	9.99 987	37
24	8.38 796	520	8.38 809	520	1.61 191	9.99 987	36
		514		514			
25	8.39 310		8.39 323		1.60 677	9.99 987	35
26	8.39 818	508	8.39 832	509	1.60 168	9.99 986	34
27	8.40 320	502	8.40 334	502	1.59 666	9.99 986	33
28	8.40 816	496	8.40 830	496	1.59 170	9.99 986	32
29	8.41 307	491	8.41 321	491	1.58 679	9.99 985	31
		485		486			
30	8.41 792		8.41 807		1.58 193	9.99 985	30
31	8.42 272	480	8.42 287	480	1.57 713	9.99 985	29
32	8.42 746	474	8.42 762	475	1.57 238	9.99 984	28
33	8.43 216	470	8.43 232	470	1.56 768	9.99 984	27
34	8.43 680	464	8.43 696	464	1.56 304	9.99 984	26
		459		460			
35	8.44 139		8.44 156		1.55 844	9.99 983	25
36	8.44 594	455	8.44 611	455	1.55 389	9.99 983	24
37	8.45 044	450	8.45 061	450	1.54 939	9.99 983	23
38	8.45 489	445	8.45 507	446	1.54 493	9.99 982	22
39	8.45 930	441	8.45 948	441	1.54 052	9.99 982	21
		436		437			
40	8.46 366		8.46 385		1.53 615	9.99 982	20
41	8.46 799	433	8.46 817	432	1.53 183	9.99 981	19
42	8.47 226	427	8.47 245	428	1.52 755	9.99 981	18
43	8.47 650	424	8.47 669	424	1.52 331	9.99 981	17
44	8.48 069	419	8.48 089	420	1.51 911	9.99 980	16
		416		416			
45	8.48 485		8.48 505		1.51 495	9.99 980	15
46	8.48 896	411	8.48 917	412	1.51 083	9.99 979	14
47	8.49 304	408	8.49 325	408	1.50 675	9.99 979	13
48	8.49 708	404	8.49 729	404	1.50 271	9.99 979	12
49	8.50 108	400	8.50 130	401	1.49 870	9.99 978	11
		396		397			
50	8.50 504		8.50 527		1.49 473	9.99 978	10
51	8.50 897	393	8.50 920	393	1.49 080	9.99 977	9
52	8.51 287	390	8.51 310	390	1.48 690	9.99 977	8
53	8.51 673	386	8.51 696	386	1.48 304	9.99 977	7
54	8.52 055	382	8.52 079	383	1.47 921	9.99 976	6
		379		380			
55	8.52 434		8.52 459		1.47 541	9.99 976	5
56	8.52 810	376	8.52 835	376	1.47 165	9.99 975	4
57	8.53 183	373	8.53 208	373	1.46 792	9.99 975	3
58	8.53 552	369	8.53 578	370	1.46 422	9.99 974	2
59	8.53 919	367	8.53 945	367	1.46 055	9.99 974	1
		363		363			
60	8.54 282		8.54 308		1.45 692	9.99 974	0
	Log Cos	d	Log Cot	c d	Log Tan	Log Sin	′

Prop. Parts

	720	710	690	680	670
2	144	142	138	136	134
3	216	213	207	204	201
4	288	284	276	272	268
5	360	355	345	340	335
6	432	426	414	408	402
7	504	497	483	476	469
8	576	568	552	544	536
9	648	639	621	612	603

	660	650	640	630	620
2	132	130	128	126	124
3	198	195	192	189	186
4	264	260	256	252	248
5	330	325	320	315	310
6	396	390	384	378	372
7	462	455	448	441	434
8	528	520	512	504	496
9	594	585	576	567	558

	610	600	590	580	570
2	122	120	118	116	114
3	183	180	177	174	171
4	244	240	236	232	228
5	305	300	295	290	285
6	366	360	354	348	342
7	427	420	413	406	399
8	488	480	472	464	456
9	549	540	531	522	513

	560	550	540	530	520
2	112	110	108	106	104
3	168	165	162	159	156
4	224	220	216	212	208
5	280	275	270	265	260
6	336	330	324	318	312
7	392	385	378	371	364
8	448	440	432	424	416
9	504	495	486	477	468

	510	500	490	480	470
2	102	100	98	96	94
3	153	150	147	144	141
4	204	200	196	192	188
5	255	250	245	240	235
6	306	300	294	288	282
7	357	350	343	336	329
8	408	400	392	384	376
9	459	450	441	432	423

	460	450	440	430	420
2	92	90	88	86	84
3	138	135	132	129	126
4	184	180	176	172	168
5	230	225	220	215	210
6	276	270	264	258	252
7	322	315	308	301	294
8	368	360	352	344	336
9	414	405	396	387	378

	410	400	395	390	385
2	82	80	79.0	78	77.0
3	123	120	118.5	117	115.5
4	164	160	158.0	156	154.0
5	205	200	197.5	195	192.5
6	246	240	237.0	234	231.0
7	287	280	276.5	273	269.5
8	328	320	316.0	312	308.0
9	369	360	355.5	351	346.5

	380	375	370	365	360
2	76	75.0	74	73.0	72
3	114	112.5	111	109.5	108
4	152	150.0	148	146.0	144
5	190	187.5	185	182.5	180
6	228	225.0	222	219.0	216
7	266	262.5	259	255.5	252
8	304	300.0	296	292.0	288
9	342	337.5	333	328.5	324

Prop. Parts

88°

2°

′	Log Sin	d	Log Tan	c d	Log Cot	Log Cos	
0	8.54 282	360	8.54 308	361	1.45 692	9.99 974	60
1	8.54 642	357	8.54 669	358	1.45 331	9.99 973	59
2	8.54 999	355	8.55 027	355	1.44 973	9.99 973	58
3	8.55 354	351	8.55 382	352	1.44 618	9.99 972	57
4	8.55 705	349	8.55 734	349	1.44 266	9.99 972	56
5	8.56 054	346	8.56 083	346	1.43 917	9.99 971	55
6	8.56 400	343	8.56 429	344	1.43 571	9.99 971	54
7	8.56 743	341	8.56 773	341	1.43 227	9.99 970	53
8	8.57 084	337	8.57 114	338	1.42 886	9.99 970	52
9	8.57 421	336	8.57 452	336	1.42 548	9.99 969	51
10	8.57 757	332	8.57 788	333	1.42 212	9.99 969	50
11	8.58 089	330	8.58 121	330	1.41 879	9.99 968	49
12	8.58 419	328	8.58 451	328	1.41 549	9.99 968	48
13	8.58 747	325	8.58 779	326	1.41 221	9.99 967	47
14	8.59 072	323	8.59 105	323	1.40 895	9.99 967	46
15	8.59 395	320	8.59 428	321	1.40 572	9.99 967	45
16	8.59 715	318	8.59 749	319	1.40 251	9.99 966	44
17	8.60 033	316	8.60 068	316	1.39 932	9.99 966	43
18	8.60 349	313	8.60 384	314	1.39 616	9.99 965	42
19	8.60 662	311	8.60 698	311	1.39 302	9.99 964	41
20	8.60 973	309	8.61 009	310	1.38 991	9.99 964	40
21	8.61 282	307	8.61 319	307	1.38 681	9.99 963	39
22	8.61 589	305	8.61 626	305	1.38 374	9.99 963	38
23	8.61 894	302	8.61 931	303	1.38 069	9.99 962	37
24	8.62 196	301	8.62 234	301	1.37 766	9.99 962	36
25	8.62 497	298	8.62 535	299	1.37 465	9.99 961	35
26	8.62 795	296	8.62 834	297	1.37 166	9.99 961	34
27	8.63 091	294	8.63 131	295	1.36 869	9.99 960	33
28	8.63 385	293	8.63 426	292	1.36 574	9.99 960	32
29	8.63 678	290	8.63 718	291	1.36 282	9.99 959	31
30	8.63 968	288	8.64 009	289	1.35 991	9.99 959	30
31	8.64 256	287	8.64 298	287	1.35 702	9.99 958	29
32	8.64 543	284	8.64 585	285	1.35 415	9.99 958	28
33	8.64 827	283	8.64 870	284	1.35 130	9.99 957	27
34	8.65 110	281	8.65 154	281	1.34 846	9.99 956	26
35	8.65 391	279	8.65 435	280	1.34 565	9.99 956	25
36	8.65 670	277	8.65 715	278	1.34 285	9.99 955	24
37	8.65 947	276	8.65 993	276	1.34 007	9.99 955	23
38	8.66 223	274	8.66 269	274	1.33 731	9.99 954	22
39	8.66 497	272	8.66 543	273	1.33 457	9.99 954	21
40	8.66 769	270	8.66 816	271	1.33 184	9.99 953	20
41	8.67 039	269	8.67 087	269	1.32 913	9.99 952	19
42	8.67 308	267	8.67 356	268	1.32 644	9.99 952	18
43	8.67 575	266	8.67 624	266	1.32 376	9.99 951	17
44	8.67 841	263	8.67 890	264	1.32 110	9.99 951	16
45	8.68 104	263	8.68 154	263	1.31 846	9.99 950	15
46	8.68 367	260	8.68 417	261	1.31 583	9.99 949	14
47	8.68 627	259	8.68 678	260	1.31 322	9.99 949	13
48	8.68 886	258	8.68 938	258	1.31 062	9.99 948	12
49	8.69 144	256	8.69 196	257	1.30 804	9.99 948	11
50	8.69 400	254	8.69 453	255	1.30 547	9.99 947	10
51	8.69 654	253	8.69 708	254	1.30 292	9.99 946	9
52	8.69 907	252	8.69 962	252	1.30 038	9.99 946	8
53	8.70 159	250	8.70 214	251	1.29 786	9.99 945	7
54	8.70 409	249	8.70 465	249	1.29 535	9.99 944	6
55	8.70 658	247	8.70 714	248	1.29 286	9.99 944	5
56	8.70 905	246	8.70 962	246	1.29 038	9.99 943	4
57	8.71 151	244	8.71 208	245	1.28 792	9.99 942	3
58	8.71 395	243	8.71 453	244	1.28 547	9.99 942	2
59	8.71 638	242	8.71 697	243	1.28 303	9.99 941	1
60	8.71 880		8.71 940		1.28 060	9.99 940	0
	Log Cos	d	Log Cot	c d	Log Tan	Log Sin	′

Prop. Parts

	360	355	350	345
2	72	71.0	70	69.0
3	108	106.5	105	103.5
4	144	142.0	140	138.0
5	180	177.5	175	172.5
6	216	213.0	210	207.0
7	252	248.5	245	241.5
8	288	284.0	280	276.0
9	324	319.5	315	310.5

	340	335	330	325
2	68	67.0	66	65.0
3	102	100.5	99	97.5
4	136	134.0	132	130.0
5	170	167.5	165	162.5
6	204	201.0	198	195.0
7	238	234.5	231	227.5
8	272	268.0	264	260.0
9	306	301.5	297	292.5

	320	315	310	305
2	64	63.0	62	61.0
3	96	94.5	93	91.5
4	128	126.0	124	122.0
5	160	157.5	155	152.5
6	192	189.0	186	183.0
7	224	220.5	217	213.5
8	256	252.0	248	244.0
9	288	283.5	279	274.5

	300	295	290	285
2	60	59.0	58	57.0
3	90	88.5	87	85.5
4	120	118.0	116	114.0
5	150	147.5	145	142.5
6	180	177.0	174	171.0
7	210	206.5	203	199.5
8	240	236.0	232	228.0
9	270	265.5	261	256.5

	280	275	270	265
2	56	55.0	54	53.0
3	84	82.5	81	79.5
4	112	110.0	108	106.0
5	140	137.5	135	132.5
6	168	165.0	162	159.0
7	196	192.5	189	185.5
8	224	220.0	216	212.0
9	252	247.5	243	238.5

	260	255	250	245
2	52	51.0	50	49.0
3	78	76.5	75	73.5
4	104	102.0	100	98.0
5	130	127.5	125	122.5
6	156	153.0	150	147.0
7	182	178.5	175	171.5
8	208	204.0	200	196.0
9	234	229.5	225	220.5

3°

′	Log Sin	d	Log Tan	c d	Log Cot	Log Cos	
0	8.71 880	240	8.71 940	241	1.28 060	9.99 940	60
1	8.72 120	239	8.72 181	239	1.27 819	9.99 940	59
2	8.72 359	238	8.72 420	239	1.27 580	9.99 939	58
3	8.72 597	237	8.72 659	237	1.27 341	9.99 938	57
4	8.72 834	235	8.72 896	236	1.27 104	9.99 938	56
5	8.73 069	234	8.73 132	234	1.26 868	9.99 937	55
6	8.73 303	232	8.73 366	234	1.26 634	9.99 936	54
7	8.73 535	232	8.73 600	232	1.26 400	9.99 936	53
8	8.73 767	230	8.73 832	231	1.26 168	9.99 935	52
9	8.73 997	229	8.74 063	229	1.25 937	9.99 934	51
10	8.74 226	228	8.74 292	229	1.25 708	9.99 934	50
11	8.74 454	226	8.74 521	227	1.25 479	9.99 933	49
12	8.74 680	226	8.74 748	226	1.25 252	9.99 932	48
13	8.74 906	224	8.74 974	225	1.25 026	9.99 932	47
14	8.75 130	223	8.75 199	224	1.24 801	9.99 931	46
15	8.75 353	222	8.75 423	222	1.24 577	9.99 930	45
16	8.75 575	220	8.75 645	222	1.24 355	9.99 929	44
17	8.75 795	220	8.75 867	220	1.24 133	9.99 929	43
18	8.76 015	219	8.76 087	219	1.23 913	9.99 928	42
19	8.76 234	217	8.76 306	219	1.23 694	9.99 927	41
20	8.76 451	216	8.76 525	217	1.23 475	9.99 926	40
21	8.76 667	216	8.76 742	216	1.23 258	9.99 926	39
22	8.76 883	214	8.76 958	215	1.23 042	9.99 925	38
23	8.77 097	213	8.77 173	214	1.22 827	9.99 924	37
24	8.77 310	212	8.77 387	213	1.22 613	9.99 923	36
25	8.77 522	211	8.77 600	211	1.22 400	9.99 923	35
26	8.77 733	210	8.77 811	211	1.22 189	9.99 922	34
27	8.77 943	209	8.78 022	210	1.21 978	9.99 921	33
28	8.78 152	208	8.78 232	209	1.21 768	9.99 920	32
29	8.78 360	208	8.78 441	208	1.21 559	9.99 920	31
30	8.78 568	206	8.78 649	206	1.21 351	9.99 919	30
31	8.78 774	205	8.78 855	206	1.21 145	9.99 918	29
32	8.78 979	204	8.79 061	205	1.20 939	9.99 917	28
33	8.79 183	203	8.79 266	204	1.20 734	9.99 917	27
34	8.79 386	202	8.79 470	203	1.20 530	9.99 916	26
35	8.79 588	201	8.79 673	202	1.20 327	9.99 915	25
36	8.79 789	201	8.79 875	201	1.20 125	9.99 914	24
37	8.79 990	199	8.80 076	201	1.19 924	9.99 913	23
38	8.80 189	199	8.80 277	199	1.19 723	9.99 913	22
39	8.80 388	197	8.80 476	198	1.19 524	9.99 912	21
40	8.80 585	197	8.80 674	198	1.19 326	9.99 911	20
41	8.80 782	196	8.80 872	196	1.19 128	9.99 910	19
42	8.80 978	195	8.81 068	196	1.18 932	9.99 909	18
43	8.81 173	194	8.81 264	195	1.18 736	9.99 909	17
44	8.81 367	193	8.81 459	194	1.18 541	9.99 908	16
45	8.81 560	192	8.81 653	193	1.18 347	9.99 907	15
46	8.81 752	192	8.81 846	192	1.18 154	9.99 906	14
47	8.81 944	190	8.82 038	192	1.17 962	9.99 905	13
48	8.82 134	190	8.82 230	190	1.17 770	9.99 904	12
49	8.82 324	189	8.82 420	190	1.17 580	9.99 904	11
50	8.82 513	188	8.82 610	189	1.17 390	9.99 903	10
51	8.82 701	187	8.82 799	188	1.17 201	9.99 902	9
52	8.82 888	187	8.82 987	188	1.17 013	9.99 901	8
53	8.83 075	186	8.83 175	186	1.16 825	9.99 900	7
54	8.83 261	185	8.83 361	186	1.16 639	9.99 899	6
55	8.83 446	184	8.83 547	185	1.16 453	9.99 898	5
56	8.83 630	183	8.83 732	184	1.16 268	9.99 898	4
57	8.83 813	183	8.83 916	184	1.16 084	9.99 897	3
58	8.83 996	181	8.84 100	182	1.15 900	9.99 896	2
59	8.84 177	181	8.84 282	182	1.15 718	9.99 895	1
60	8.84 358		8.84 464		1.15 536	9.99 894	0
	Log Cos	d	Log Cot	c d	Log Tan	Log Sin	′

Prop. Parts

	241	239	237	235
2	48.2	47.8	47.4	47.0
3	72.3	71.7	71.1	70.5
4	96.4	95.6	94.8	94.0
5	120.5	119.5	118.5	117.5
6	144.6	143.4	142.2	141.0
7	168.7	167.3	165.9	164.5
8	192.8	191.2	189.6	188.0
9	216.9	215.1	213.3	211.5

	234	232	229	227
2	46.8	46.4	45.8	45.4
3	70.2	69.6	68.7	68.1
4	93.6	92.8	91.6	90.8
5	117.0	116.0	114.5	113.5
6	140.4	139.2	137.4	136.2
7	163.8	162.4	160.3	158.9
8	187.2	185.6	183.2	181.6
9	210.6	208.8	206.1	204.3

	226	224	222	220
2	45.2	44.8	44.4	44.0
3	67.8	67.2	66.6	66.0
4	90.4	89.6	88.8	88.0
5	113.0	112.0	111.0	110.0
6	135.6	134.4	133.2	132.0
7	158.2	156.8	155.4	154.0
8	180.8	179.2	177.6	176.0
9	203.4	201.6	199.8	198.0

	219	217	215	213
2	43.8	43.4	43.0	42.6
3	65.7	65.1	64.5	63.9
4	87.6	86.8	86.0	85.2
5	109.5	108.5	107.5	106.5
6	131.4	130.2	129.0	127.8
7	153.3	151.9	150.5	149.1
8	175.2	173.6	172.0	170.4
9	197.1	195.3	193.5	191.7

	211	208	206	203
2	42.2	41.6	41.2	40.6
3	63.3	62.4	61.8	60.9
4	84.4	83.2	82.4	81.2
5	105.5	104.0	103.0	101.5
6	126.6	124.8	123.6	121.8
7	147.7	145.6	144.2	142.1
8	168.8	166.4	164.8	162.4
9	189.9	187.2	185.4	182.7

	201	199	197	195
2	40.2	39.8	39.4	39.0
3	60.3	59.7	59.1	58.5
4	80.4	79.6	78.8	78.0
5	100.5	99.5	98.5	97.5
6	120.6	119.4	118.2	117.0
7	140.7	139.3	137.9	136.5
8	160.8	159.2	157.6	156.0
9	180.9	179.1	177.3	175.5

	193	192	190	188
2	38.6	38.4	38.0	37.6
3	57.9	57.6	57.0	56.4
4	77.2	76.8	76.0	75.2
5	96.5	96.0	95.0	94.0
6	115.8	115.2	114.0	112.8
7	135.1	134.4	133.0	131.6
8	154.4	153.6	152.0	150.4
9	173.7	172.8	171.0	169.2

	186	184	182	181
2	37.2	36.8	36.4	36.2
3	55.8	55.2	54.6	54.3
4	74.4	73.6	72.8	72.4
5	93.0	92.0	91.0	90.5
6	111.6	110.4	109.2	108.6
7	130.2	128.8	127.4	126.7
8	148.8	147.2	145.6	144.8
9	167.4	165.6	163.8	162.9

86°

Table II

4°

′	Log Sin	d	Log Tan	c d	Log Cot	Log Cos		Prop. Parts			
0	8.84 358	181	8.84 464	182	1.15 536	9.99 894	60	**182**	**181**	**180**	**179**
1	8.84 539	179	8.84 646	180	1.15 354	9.99 893	59	2 36.4	36.2	36.0	35.8
2	8.84 718	179	8.84 826	180	1.15 174	9.99 892	58	3 54.6	54.3	54.0	53.7
3	8.84 897	178	8.85 006	179	1.14 994	9.99 891	57	4 72.8	72.4	72.0	71.6
4	8.85 075	177	8.85 185	178	1.14 815	9.99 891	56	5 91.0	90.5	90.0	89.5
								6 109.2	108.6	108.0	107.4
5	8.85 252	177	8.85 363	177	1.14 637	9.99 890	55	7 127.4	126.7	126.0	125.3
6	8.85 429	176	8.85 540	177	1.14 460	9.99 889	54	8 145.6	144.8	144.0	143.2
7	8.85 605	175	8.85 717	176	1.14 283	9.99 888	53	9 163.8	162.9	162.0	161.1
8	8.85 780	175	8.85 893	176	1.14 107	9.99 887	52	**178**	**177**	**176**	**175**
9	8.85 955	173	8.86 069	174	1.13 931	9.99 886	51	2 35.6	35.4	35.2	35.0
								3 53.4	53.1	52.8	52.5
10	8.86 128	173	8.86 243	174	1.13 757	9.99 885	50	4 71.2	70.8	70.4	70.0
11	8.86 301	173	8.86 417	174	1.13 583	9.99 884	49	5 89.0	88.5	88.0	87.5
12	8.86 474	171	8.86 591	172	1.13 409	9.99 883	48	6 106.8	106.2	105.6	105.0
13	8.86 645	171	8.86 763	172	1.13 237	9.99 882	47	7 124.6	123.9	123.2	122.5
14	8.86 816	171	8.86 935	171	1.13 065	9.99 881	46	8 142.4	141.6	140.8	140.0
								9 160.2	159.3	158.4	157.5
15	8.86 987	169	8.87 106	171	1.12 894	9.99 880	45	**174**	**173**	**172**	**171**
16	8.87 156	169	8.87 277	170	1.12 723	9.99 879	44	2 34.8	34.6	34.4	34.2
17	8.87 325	169	8.87 447	169	1.12 553	9.99 879	43	3 52.2	51.9	51.6	51.3
18	8.87 494	167	8.87 616	169	1.12 384	9.99 878	42	4 69.6	69.2	68.8	68.4
19	8.87 661	168	8.87 785	168	1.12 215	9.99 877	41	5 87.0	86.5	86.0	85.5
								6 104.4	103.8	103.2	102.6
20	8.87 829	166	8.87 953	167	1.12 047	9.99 876	40	7 121.8	121.1	120.4	119.7
21	8.87 995	166	8.88 120	167	1.11 880	9.99 875	39	8 139.2	138.4	137.6	136.8
22	8.88 161	165	8.88 287	166	1.11 713	9.99 874	38	9 156.6	155.7	154.8	153.9
23	8.88 326	164	8.88 453	165	1.11 547	9.99 873	37	**170**	**169**	**168**	**167**
24	8.88 490	164	8.88 618	165	1.11 382	9.99 872	36	2 34.0	33.8	33.6	33.4
								3 51.0	50.7	50.4	50.1
25	8.88 654	163	8.88 783	165	1.11 217	9.99 871	35	4 68.0	67.6	67.2	66.8
26	8.88 817	163	8.88 948	163	1.11 052	9.99 870	34	5 85.0	84.5	84.0	83.5
27	8.88 980	162	8.89 111	163	1.10 889	9.99 869	33	6 102.0	101.4	100.8	100.2
28	8.89 142	162	8.89 274	163	1.10 726	9.99 868	32	7 119.0	118.3	117.6	116.9
29	8.89 304	160	8.89 437	161	1.10 563	9.99 867	31	8 136.0	135.2	134.4	133.6
								9 153.0	152.1	151.2	150.3
30	8.89 464	161	8.89 598	162	1.10 402	9.99 866	30	**166**	**165**	**164**	**163**
31	8.89 625	159	8.89 760	160	1.10 240	9.99 865	29	2 33.2	33.0	32.8	32.6
32	8.89 784	159	8.89 920	160	1.10 080	9.99 864	28	3 49.8	49.5	49.2	48.9
33	8.89 943	159	8.90 080	160	1.09 920	9.99 863	27	4 66.4	66.0	65.6	65.2
34	8.90 102	158	8.90 240	159	1.09 760	9.99 862	26	5 83.0	82.5	82.0	81.5
								6 99.6	99.0	98.4	97.8
35	8.90 260	157	8.90 399	158	1.09 601	9.99 861	25	7 116.2	115.5	114.8	114.1
36	8.90 417	157	8.90 557	158	1.09 443	9.99 860	24	8 132.8	132.0	131.2	130.4
37	8.90 574	156	8.90 715	157	1.09 285	9.99 859	23	9 149.4	148.5	147.6	146.7
38	8.90 730	155	8.90 872	157	1.09 128	9.99 858	22	**162**	**161**	**160**	**159**
39	8.90 885	155	8.91 029	156	1.08 971	9.99 857	21	2 32.4	32.2	32.0	31.8
								3 48.6	48.3	48.0	47.7
40	8.91 040	155	8.91 185	155	1.08 815	9.99 856	20	4 64.8	64.4	64.0	63.6
41	8.91 195	154	8.91 340	155	1.08 660	9.99 855	19	5 81.0	80.5	80.0	79.5
42	8.91 349	153	8.91 495	155	1.08 505	9.99 854	18	6 97.2	96.6	96.0	95.4
43	8.91 502	153	8.91 650	153	1.08 350	9.99 853	17	7 113.4	112.7	112.0	111.3
44	8.91 655	152	8.91 803	154	1.08 197	9.99 852	16	8 129.6	128.8	128.0	127.2
								9 145.8	144.9	144.0	143.1
45	8.91 807	152	8.91 957	153	1.08 043	9.99 851	15	**158**	**157**	**156**	**155**
46	8.91 959	151	8.92 110	152	1.07 890	9.99 850	14	2 31.6	31.4	31.2	31.0
47	8.92 110	151	8.92 262	152	1.07 738	9.99 848	13	3 47.4	47.1	46.8	46.5
48	8.92 261	150	8.92 414	151	1.07 586	9.99 847	12	4 63.2	62.8	62.4	62.0
49	8.92 411	150	8.92 565	151	1.07 435	9.99 846	11	5 79.0	78.5	78.0	77.5
								6 94.8	94.2	93.6	93.0
50	8.92 561	149	8.92 716	150	1.07 284	9.99 845	10	7 110.6	109.9	109.2	108.5
51	8.92 710	149	8.92 866	150	1.07 134	9.99 844	9	8 126.4	125.6	124.8	124.0
52	8.92 859	148	8.93 016	149	1.06 984	9.99 843	8	9 142.2	141.3	140.4	139.5
53	8.93 007	147	8.93 165	148	1.06 835	9.99 842	7	**154**	**153**	**152**	**151**
54	8.93 154	147	8.93 313	149	1.06 687	9.99 841	6	2 30.8	30.6	30.4	30.2
								3 46.2	45.9	45.6	45.3
55	8.93 301	147	8.93 462	147	1.06 538	9.99 840	5	4 61.6	61.2	60.8	60.4
56	8.93 448	146	8.93 609	147	1.06 391	9.99 839	4	5 77.0	76.5	76.0	75.5
57	8.93 594	146	8.93 756	147	1.06 244	9.99 838	3	6 92.4	91.8	91.2	90.6
58	8.93 740	145	8.93 903	146	1.06 097	9.99 837	2	7 107.8	107.1	106.4	105.7
59	8.93 885	145	8.94 049	146	1.05 951	9.99 836	1	8 123.2	122.4	121.6	120.8
60	8.94 030		8.94 195		1.05 805	9.99 834	0	9 138.6	137.7	136.8	135.9
	Log Cos	d	Log Cot	c d	Log Tan	Log Sin	′	Prop. Parts			

85°

5°

'	Log Sin	d	Log Tan	c d	Log Cot	Log Cos	
0	8.94 030	144	8.94 195	145	1.05 805	9.99 834	60
1	8.94 174	143	8.94 340	145	1.05 660	9.99 833	59
2	8.94 317	144	8.94 485	145	1.05 515	9.99 832	58
3	8.94 461	142	8.94 630	143	1.05 370	9.99 831	57
4	8.94 603	143	8.94 773	144	1.05 227	9.99 830	56
5	8.94 746	141	8.94 917	143	1.05 083	9.99 829	55
6	8.94 887	142	8.95 060	142	1.04 940	9.99 828	54
7	8.95 029	141	8.95 202	142	1.04 798	9.99 827	53
8	8.95 170	140	8.95 344	142	1.04 656	9.99 825	52
9	8.95 310	140	8.95 486	141	1.04 514	9.99 824	51
10	8.95 450	139	8.95 627	140	1.04 373	9.99 823	50
11	8.95 589	139	8.95 767	141	1.04 233	9.99 822	49
12	8.95 728	139	8.95 908	139	1.04 092	9.99 821	48
13	8.95 867	138	8.96 047	140	1.03 953	9.99 820	47
14	8.96 005	138	8.96 187	138	1.03 813	9.99 819	46
15	8.96 143	137	8.96 325	139	1.03 675	9.99 817	45
16	8.96 280	137	8.96 464	138	1.03 536	9.99 816	44
17	8.96 417	136	8.96 602	137	1.03 398	9.99 815	43
18	8.96 553	136	8.96 739	138	1.03 261	9.99 814	42
19	8.96 689	136	8.96 877	136	1.03 123	9.99 813	41
20	8.96 825	135	8.97 013	137	1.02 987	9.99 812	40
21	8.96 960	135	8.97 150	135	1.02 850	9.99 810	39
22	8.97 095	134	8.97 285	136	1.02 715	9.99 809	38
23	8.97 229	134	8.97 421	135	1.02 579	9.99 808	37
24	8.97 363	133	8.97 556	135	1.02 444	9.99 807	36
25	8.97 496	133	8.97 691	134	1.02 309	9.99 806	35
26	8.97 629	133	8.97 825	134	1.02 175	9.99 804	34
27	8.97 762	132	8.97 959	133	1.02 041	9.99 803	33
28	8.97 894	132	8.98 092	133	1.01 908	9.99 802	32
29	8.98 026	131	8.98 225	133	1.01 775	9.99 801	31
30	8.98 157	131	8.98 358	132	1.01 642	9.99 800	30
31	8.98 288	131	8.98 490	132	1.01 510	9.99 798	29
32	8.98 419	130	8.98 622	131	1.01 378	9.99 797	28
33	8.98 549	130	8.98 753	131	1.01 247	9.99 796	27
34	8.98 679	129	8.98 884	131	1.01 116	9.99 795	26
35	8.98 808	129	8.99 015	130	1.00 985	9.99 793	25
36	8.98 937	129	8.99 145	130	1.00 855	9.99 792	24
37	8.99 066	128	8.99 275	130	1.00 725	9.99 791	23
38	8.99 194	128	8.99 405	129	1.00 595	9.99 790	22
39	8.99 322	128	8.99 534	128	1.00 466	9.99 788	21
40	8.99 450	127	8.99 662	129	1.00 338	9.99 787	20
41	8.99 577	127	8.99 791	128	1.00 209	9.99 786	19
42	8.99 704	126	8.99 919	127	1.00 081	9.99 785	18
43	8.99 830	126	9.00 046	128	0.99 954	9.99 783	17
44	8.99 956	126	9.00 174	127	0.99 826	9.99 782	16
45	9.00 082	125	9.00 301	126	0.99 699	9.99 781	15
46	9.00 207	125	9.00 427	126	0.99 573	9.99 780	14
47	9.00 332	124	9.00 553	126	0.99 447	9.99 778	13
48	9.00 456	125	9.00 679	126	0.99 321	9.99 777	12
49	9.00 581	123	9.00 805	125	0.99 195	9.99 776	11
50	9.00 704	124	9.00 930	125	0.99 070	9.99 775	10
51	9.00 828	123	9.01 055	124	0.98 945	9.99 773	9
52	9.00 951	123	9.01 179	124	0.98 821	9.99 772	8
53	9.01 074	122	9.01 303	124	0.98 697	9.99 771	7
54	9.01 196	122	9.01 427	123	0.98 573	9.99 769	6
55	9.01 318	122	9.01 550	123	0.98 450	9.99 768	5
56	9.01 440	121	9.01 673	123	0.98 327	9.99 767	4
57	9.01 561	121	9.01 796	122	0.98 204	9.99 765	3
58	9.01 682	121	9.01 918	122	0.98 082	9.99 764	2
59	9.01 803	120	9.02 040	122	0.97 960	9.99 763	1
60	9.01 923		9.02 162		0.97 838	9.99 761	0
	Log Cos	d	Log Cot	c d	Log Tan	Log Sin	'

Prop. Parts

	150	149	148	147
2	30.0	29.8	29.6	29.4
3	45.0	44.7	44.4	44.1
4	60.0	59.6	59.2	58.8
5	75.0	74.5	74.0	73.5
6	90.0	89.4	88.8	88.2
7	105.0	104.3	103.6	102.9
8	120.0	119.2	118.4	117.6
9	135.0	134.1	133.2	132.3

	146	145	144	143
2	29.2	29.0	28.8	28.6
3	43.8	43.5	43.2	42.9
4	58.4	58.0	57.6	57.2
5	73.0	72.5	72.0	71.5
6	87.6	87.0	86.4	85.8
7	102.2	101.5	100.8	100.1
8	116.8	116.0	115.2	114.4
9	131.4	130.5	129.6	128.7

	142	141	140	139
2	28.4	28.2	28.0	27.8
3	42.6	42.3	42.0	41.7
4	56.8	56.4	56.0	55.6
5	71.0	70.5	70.0	69.5
6	85.2	84.6	84.0	83.4
7	99.4	98.7	98.0	97.8
8	113.6	112.8	112.0	111.2
9	127.8	126.9	126.0	125.1

	138	137	136	135
2	27.6	27.4	27.2	27.0
3	41.4	41.1	40.8	40.5
4	55.2	54.8	54.4	54.0
5	69.0	68.5	68.0	67.5
6	82.8	82.2	81.6	81.0
7	96.6	95.9	95.2	94.5
8	110.4	109.6	108.8	108.0
9	124.2	123.3	122.4	121.5

	134	133	132	131
2	26.8	26.6	26.4	26.2
3	40.2	39.9	39.6	39.3
4	53.6	53.2	52.8	52.4
5	67.0	66.5	66.0	65.5
6	80.4	79.8	79.2	78.6
7	93.8	93.1	92.4	91.7
8	107.2	106.4	105.6	104.8
9	120.6	119.7	118.8	117.9

	130	129	128	127
2	26.0	25.8	25.6	25.4
3	39.0	38.7	38.4	38.1
4	52.0	51.6	51.2	50.8
5	65.0	64.5	64.0	63.5
6	78.0	77.4	76.8	76.2
7	91.0	90.3	89.6	88.9
8	104.0	103.2	102.4	101.6
9	117.0	116.1	115.2	114.3

	126	125	124	123
2	25.2	25.0	24.8	24.6
3	37.8	37.5	37.2	36.9
4	50.4	50.0	49.6	49.2
5	63.0	62.5	62.0	61.5
6	75.6	75.0	74.4	73.8
7	88.2	87.5	86.8	86.1
8	100.8	100.0	99.2	98.4
9	113.4	112.5	111.6	110.7

	122	121	120
2	24.4	24.2	24.0
3	36.6	36.3	36.0
4	48.8	48.4	48.0
5	61.0	60.5	60.0
6	73.2	72.6	72.0
7	85.4	84.7	84.0
8	97.6	96.8	96.0
9	109.8	108.9	108.0

84°

Table II

6°

′	Log Sin	d	Log Tan	c d	Log Cot	Log Cos	′
0	9.01 923	120	9.02 162	121	0.97 838	9.99 761	60
1	9.02 043	120	9.02 283	121	0.97 717	9.99 760	59
2	9.02 163	120	9.02 404	121	0.97 596	9.99 759	58
3	9.02 283	119	9.02 525	120	0.97 475	9.99 757	57
4	9.02 402	118	9.02 645	121	0.97 355	9.99 756	56
5	9.02 520	119	9.02 766	119	0.97 234	9.99 755	55
6	9.02 639	118	9.02 885	120	0.97 115	9.99 753	54
7	9.02 757	117	9.03 005	119	0.96 995	9.99 752	53
8	9.02 874	118	9.03 124	118	0.96 876	9.99 751	52
9	9.02 992	117	9.03 242	119	0.96 758	9.99 749	51
10	9.03 109	117	9.03 361	118	0.96 639	9.99 748	50
11	9.03 226	116	9.03 479	118	0.96 521	9.99 747	49
12	9.03 342	116	9.03 597	117	0.96 403	9.99 745	48
13	9.03 458	116	9.03 714	118	0.96 286	9.99 744	47
14	9.03 574	116	9.03 832	116	0.96 168	9.99 742	46
15	9.03 690	115	9.03 948	117	0.96 052	9.99 741	45
16	9.03 805	115	9.04 065	116	0.95 935	9.99 740	44
17	9.03 920	114	9.04 181	116	0.95 819	9.99 738	43
18	9.04 034	115	9.04 297	116	0.95 703	9.99 737	42
19	9.04 149	113	9.04 413	115	0.95 587	9.99 736	41
20	9.04 262	114	9.04 528	115	0.95 472	9.99 734	40
21	9.04 376	114	9.04 643	115	0.95 357	9.99 733	39
22	9.04 490	113	9.04 758	115	0.95 242	9.99 731	38
23	9.04 603	112	9.04 873	114	0.95 127	9.99 730	37
24	9.04 715	113	9.04 987	114	0.95 013	9.99 728	36
25	9.04 828	112	9.05 101	113	0.94 899	9.99 727	35
26	9.04 940	112	9.05 214	114	0.94 786	9.99 726	34
27	9.05 052	112	9.05 328	113	0.94 672	9.99 724	33
28	9.05 164	111	9.05 441	112	0.94 559	9.99 723	32
29	9.05 275	111	9.05 553	113	0.94 447	9.99 721	31
30	9.05 386	111	9.05 666	112	0.94 334	9.99 720	30
31	9.05 497	110	9.05 778	112	0.94 222	9.99 718	29
32	9.05 607	110	9.05 890	112	0.94 110	9.99 717	28
33	9.05 717	110	9.06 002	111	0.93 998	9.99 716	27
34	9.05 827	110	9.06 113	111	0.93 887	9.99 714	26
35	9.05 937	109	9.06 224	111	0.93 776	9.99 713	25
36	9.06 046	109	9.06 335	110	0.93 665	9.99 711	24
37	9.06 155	109	9.06 445	111	0.93 555	9.99 710	23
38	9.06 264	108	9.06 556	110	0.93 444	9.99 708	22
39	9.06 372	109	9.06 666	109	0.93 334	9.99 707	21
40	9.06 481	108	9.06 775	110	0.93 225	9.99 705	20
41	9.06 589	107	9.06 885	109	0.93 115	9.99 704	19
42	9.06 696	108	9.06 994	109	0.93 006	9.99 702	18
43	9.06 804	107	9.07 103	108	0.92 897	9.99 701	17
44	9.06 911	107	9.07 211	109	0.92 789	9.99 699	16
45	9.07 018	106	9.07 320	108	0.92 680	9.99 698	15
46	9.07 124	107	9.07 428	108	0.92 572	9.99 696	14
47	9.07 231	106	9.07 536	107	0.92 464	9.99 695	13
48	9.07 337	105	9.07 643	108	0.92 357	9.99 693	12
49	9.07 442	106	9.07 751	107	0.92 249	9.99 692	11
50	9.07 548	105	9.07 858	106	0.92 142	9.99 690	10
51	9.07 653	105	9.07 964	107	0.92 036	9.99 689	9
52	9.07 758	105	9.08 071	106	0.91 929	9.99 687	8
53	9.07 863	105	9.08 177	106	0.91 823	9.99 686	7
54	9.07 968	104	9.08 283	106	0.91 717	9.99 684	6
55	9.08 072	104	9.08 389	106	0.91 611	9.99 683	5
56	9.08 176	104	9.08 495	105	0.91 505	9.99 681	4
57	9.08 280	103	9.08 600	105	0.91 400	9.99 680	3
58	9.08 383	103	9.08 705	105	0.91 295	9.99 678	2
59	9.08 486	103	9.08 810	104	0.91 190	9.99 677	1
60	9.08 589		9.08 914		0.91 086	9.99 675	0
	Log Cos	d	Log Cot	c d	Log Tan	Log Sin	′

Prop. Parts

	121	120	119
1	12.1	12.0	11.9
2	24.2	24.0	23.8
3	36.3	36.0	35.7
4	48.4	48.0	47.6
5	60.5	60.0	59.5
6	72.6	72.0	71.4
7	84.7	84.0	83.3
8	96.8	96.0	95.2
9	108.9	108.0	107.1

	118	117	116
1	11.8	11.7	11.6
2	23.6	23.4	23.2
3	35.4	35.1	34.8
4	47.2	46.8	46.4
5	59.0	58.5	58.0
6	70.8	70.2	69.6
7	82.6	81.9	81.2
8	94.4	93.6	92.8
9	106.2	105.3	104.4

	115	114	113
1	11.5	11.4	11.3
2	23.0	22.8	22.6
3	34.5	34.2	33.9
4	46.0	45.6	45.2
5	57.5	57.0	56.5
6	69.0	68.4	67.8
7	80.5	79.8	79.1
8	92.0	91.2	90.4
9	103.5	102.6	101.7

	112	111	110
1	11.2	11.1	11.0
2	22.4	22.2	22.0
3	33.6	33.3	33.0
4	44.8	44.4	44.0
5	56.0	55.5	55.0
6	67.2	66.6	66.0
7	78.4	77.7	77.0
8	89.6	88.8	88.0
9	100.8	99.9	99.0

	109	108	107	106
1	10.9	10.8	10.7	10.6
2	21.8	21.6	21.4	21.2
3	32.7	32.4	32.1	31.8
4	43.6	43.2	42.8	42.4
5	54.5	54.0	53.5	53.0
6	65.4	64.8	64.2	63.6
7	76.3	75.6	74.9	74.2
8	87.2	86.4	85.6	84.8
9	98.1	97.2	96.3	95.4

7°

′	Log Sin	d	Log Tan	c d	Log Cot	Log Cos	
0	9.08 589	103	9.08 914	105	0.91 086	9.99 675	60
1	9.08 692	103	9.09 019	104	0.90 981	9.99 674	59
2	9.08 795	102	9.09 123	104	0.90 877	9.99 672	58
3	9.08 897	102	9.09 227	103	0.90 773	9.99 670	57
4	9.08 999	102	9.09 330	104	0.90 670	9.99 669	56
5	9.09 101	101	9.09 434	103	0.90 566	9.99 667	55
6	9.09 202	102	9.09 537	103	0.90 463	9.99 666	54
7	9.09 304	101	9.09 640	102	0.90 360	9.99 664	53
8	9.09 405	101	9.09 742	103	0.90 258	9.99 663	52
9	9.09 506	100	9.09 845	102	0.90 155	9.99 661	51
10	9.09 606	101	9.09 947	102	0.90 053	9.99 659	50
11	9.09 707	100	9.10 049	101	0.89 951	9.99 658	49
12	9.09 807	100	9.10 150	102	0.89 850	9.99 656	48
13	9.09 907	99	9.10 252	101	0.89 748	9.99 655	47
14	9.10 006	100	9.10 353	101	0.89 647	9.99 653	46
15	9.10 106	99	9.10 454	101	0.89 546	9.99 651	45
16	9.10 205	99	9.10 555	101	0.89 445	9.99 650	44
17	9.10 304	98	9.10 656	100	0.89 344	9.99 648	43
18	9.10 402	99	9.10 756	100	0.89 244	9.99 647	42
19	9.10 501	98	9.10 856	100	0.89 144	9.99 645	41
20	9.10 599	98	9.10 956	100	0.89 044	9.99 643	40
21	9.10 697	98	9.11 056	99	0.88 944	9.99 642	39
22	9.10 795	98	9.11 155	99	0.88 845	9.99 640	38
23	9.10 893	97	9.11 254	99	0.88 746	9.99 638	37
24	9.10 990	97	9.11 353	99	0.88 647	9.99 637	36
25	9.11 087	97	9.11 452	99	0.88 548	9.99 635	35
26	9.11 184	97	9.11 551	98	0.88 449	9.99 633	34
27	9.11 281	96	9.11 649	98	0.88 351	9.99 632	33
28	9.11 377	97	9.11 747	98	0.88 253	9.99 630	32
29	9.11 474	96	9.11 845	98	0.88 155	9.99 629	31
30	9.11 570	96	9.11 943	97	0.88 057	9.99 627	30
31	9.11 666	95	9.12 040	98	0.87 960	9.99 625	29
32	9.11 761	96	9.12 138	97	0.87 862	9.99 624	28
33	9.11 857	95	9.12 235	97	0.87 765	9.99 622	27
34	9.11 952	95	9.12 332	96	0.87 668	9.99 620	26
35	9.12 047	95	9.12 428	97	0.87 572	9.99 618	25
36	9.12 142	94	9.12 525	96	0.87 475	9.99 617	24
37	9.12 236	95	9.12 621	96	0.87 379	9.99 615	23
38	9.12 331	94	9.12 717	96	0.87 283	9.99 613	22
39	9.12 425	94	9.12 813	96	0.87 187	9.99 612	21
40	9.12 519	93	9.12 909	95	0.87 091	9.99 610	20
41	9.12 612	94	9.13 004	95	0.86 996	9.99 608	19
42	9.12 706	93	9.13 099	95	0.86 901	9.99 607	18
43	9.12 799	93	9.13 194	95	0.86 806	9.99 605	17
44	9.12 892	93	9.13 289	95	0.86 711	9.99 603	16
45	9.12 985	93	9.13 384	94	0.86 616	9.99 601	15
46	9.13 078	93	9.13 478	95	0.86 522	9.99 600	14
47	9.13 171	92	9.13 573	94	0.86 427	9.99 598	13
48	9.13 263	92	9.13 667	94	0.86 333	9.99 596	12
49	9.13 355	92	9.13 761	93	0.86 239	9.99 595	11
50	9.13 447	92	9.13 854	94	0.86 146	9.99 593	10
51	9.13 539	91	9.13 948	93	0.86 052	9.99 591	9
52	9.13 630	92	9.14 041	93	0.85 959	9.99 589	8
53	9.13 722	91	9.14 134	93	0.85 866	9.99 588	7
54	9.13 813	91	9.14 227	93	0.85 773	9.99 586	6
55	9.13 904	90	9.14 320	92	0.85 680	9.99 584	5
56	9.13 994	91	9.14 412	92	0.85 588	9.99 582	4
57	9.14 085	90	9.14 504	93	0.85 496	9.99 581	3
58	9.14 175	91	9.14 597	91	0.85 403	9.99 579	2
59	9.14 266	90	9.14 688	92	0.85 312	9.99 577	1
60	9.14 356		9.14 780		0.85 220	9.99 575	0
	Log Cos	d	Log Cot	c d	Log Tan	Log Sin	′

82°

Prop. Parts

	105	104	103	102
1	10.5	10.4	10.3	10.2
2	21.0	20.8	20.6	20.4
3	31.5	31.2	30.9	30.6
4	42.0	41.6	41.2	40.8
5	52.5	52.0	51.5	51.0
6	63.0	62.4	61.8	61.2
7	73.5	72.8	72.1	71.4
8	84.0	83.2	82.4	81.6
9	94.5	93.6	92.7	91.8

	101	99	98	97
1	10.1	9.9	9.8	9.7
2	20.2	19.8	19.6	19.4
3	30.3	29.7	29.4	29.1
4	40.4	39.6	39.2	38.8
5	50.5	49.5	49.0	48.5
6	60.6	59.4	58.8	58.2
7	70.7	69.3	68.6	67.9
8	80.8	79.2	78.4	77.6
9	90.9	89.1	88.2	87.3

	96	95	94	93
1	9.6	9.5	9.4	9.3
2	19.2	19.0	18.8	18.6
3	28.8	28.5	28.2	27.9
4	38.4	38.0	37.6	37.2
5	48.0	47.5	47.0	46.5
6	57.6	57.0	56.4	55.8
7	67.2	66.5	65.8	65.1
8	76.8	76.0	75.2	74.4
9	86.4	85.5	84.6	83.7

	92	91	90
1	9.2	9.1	9.0
2	18.4	18.2	18.0
3	27.6	27.3	27.0
4	36.8	36.4	36.0
5	46.0	45.5	45.0
6	55.2	54.6	54.0
7	64.4	63.7	63.0
8	73.6	72.8	72.0
9	82.8	81.9	81.0

8°

'	Log Sin	d	Log Tan	c d	Log Cot	Log Cos	
0	9.14 356	89	9.14 780	92	0.85 220	9.99 575	60
1	9.14 445	90	9.14 872	91	0.85 128	9.99 574	59
2	9.14 535	89	9.14 963	91	0.85 037	9.99 572	58
3	9.14 624	90	9.15 054	91	0.84 946	9.99 570	57
4	9.14 714	89	9.15 145	91	0.84 855	9.99 568	56
5	9.14 803	88	9.15 236	91	0.84 764	9.99 566	55
6	9.14 891	89	9.15 327	90	0.84 673	9.99 565	54
7	9.14 980	89	9.15 417	91	0.84 583	9.99 563	53
8	9.15 069	88	9.15 508	90	0.84 492	9.99 561	52
9	9.15 157	88	9.15 598	90	0.84 402	9.99 559	51
10	9.15 245	88	9.15 688	89	0.84 312	9.99 557	50
11	9.15 333	88	9.15 777	90	0.84 223	9.99 556	49
12	9.15 421	87	9.15 867	89	0.84 133	9.99 554	48
13	9.15 508	88	9.15 956	90	0.84 044	9.99 552	47
14	9.15 596	87	9.16 046	89	0.83 954	9.99 550	46
15	9.15 683	87	9.16 135	89	0.83 865	9.99 548	45
16	9.15 770	87	9.16 224	88	0.83 776	9.99 546	44
17	9.15 857	87	9.16 312	89	0.83 688	9.99 545	43
18	9.15 944	86	9.16 401	88	0.83 599	9.99 543	42
19	9.16 030	86	9.16 489	88	0.83 511	9.99 541	41
20	9.16 116	87	9.16 577	88	0.83 423	9.99 539	40
21	9.16 203	86	9.16 665	88	0.83 335	9.99 537	39
22	9.16 289	85	9.16 753	88	0.83 247	9.99 535	38
23	9.16 374	86	9.16 841	87	0.83 159	9.99 533	37
24	9.16 460	85	9.16 928	88	0.83 072	9.99 532	36
25	9.16 545	86	9.17 016	87	0.82 984	9.99 530	35
26	9.16 631	85	9.17 103	87	0.82 897	9.99 528	34
27	9.16 716	85	9.17 190	87	0.82 810	9.99 526	33
28	9.16 801	85	9.17 277	86	0.82 723	9.99 524	32
29	9.16 886	84	9.17 363	87	0.82 637	9.99 522	31
30	9.16 970	85	9.17 450	86	0.82 550	9.99 520	30
31	9.17 055	84	9.17 536	86	0.82 464	9.99 518	29
32	9.17 139	84	9.17 622	86	0.82 378	9.99 517	28
33	9.17 223	84	9.17 708	86	0.82 292	9.99 515	27
34	9.17 307	84	9.17 794	86	0.82 206	9.99 513	26
35	9.17 391	83	9.17 880	85	0.82 120	9.99 511	25
36	9.17 474	84	9.17 965	86	0.82 035	9.99 509	24
37	9.17 558	83	9.18 051	85	0.81 949	9.99 507	23
38	9.17 641	83	9.18 136	85	0.81 864	9.99 505	22
39	9.17 724	83	9.18 221	85	0.81 779	9.99 503	21
40	9.17 807	83	9.18 306	85	0.81 694	9.99 501	20
41	9.17 890	83	9.18 391	84	0.81 609	9.99 499	19
42	9.17 973	82	9.18 475	85	0.81 525	9.99 497	18
43	9.18 055	82	9.18 560	84	0.81 440	9.99 495	17
44	9.18 137	83	9.18 644	84	0.81 356	9.99 494	16
45	9.18 220	82	9.18 728	84	0.81 272	9.99 492	15
46	9.18 302	81	9.18 812	84	0.81 188	9.99 490	14
47	9.18 383	82	9.18 896	83	0.81 104	9.99 488	13
48	9.18 465	82	9.18 979	84	0.81 021	9.99 486	12
49	9.18 547	81	9.19 063	83	0.80 937	9.99 484	11
50	9.18 628	81	9.19 146	83	0.80 854	9.99 482	10
51	9.18 709	81	9.19 229	83	0.80 771	9.99 480	9
52	9.18 790	81	9.19 312	83	0.80 688	9.99 478	8
53	9.18 871	81	9.19 395	83	0.80 605	9.99 476	7
54	9.18 952	81	9.19 478	83	0.80 522	9.99 474	6
55	9.19 033	80	9.19 561	82	0.80 439	9.99 472	5
56	9.19 113	80	9.19 643	82	0.80 357	9.99 470	4
57	9.19 193	80	9.19 725	82	0.80 275	9.99 468	3
58	9.19 273	80	9.19 807	82	0.80 193	9.99 466	2
59	9.19 353	80	9.19 889	82	0.80 111	9.99 464	1
60	9.19 433		9.19 971		0.80 029	9.99 462	0
	Log Cos	d	Log Cot	c d	Log Tan	Log Sin	'

Prop. Parts

	92	91	90	89
1	9.2	9.1	9.0	8.9
2	18.4	18.2	18.0	17.8
3	27.6	27.3	27.0	26.7
4	36.8	36.4	36.0	35.6
5	46.0	45.5	45.0	44.5
6	55.2	54.6	54.0	53.4
7	64.4	63.7	63.0	62.3
8	73.6	72.8	72.0	71.2
9	82.8	81.9	81.0	80.1

	88	87	86
1	8.8	8.7	8.6
2	17.6	17.4	17.2
3	26.4	26.1	25.8
4	35.2	34.8	34.4
5	44.0	43.5	43.0
6	52.8	52.2	51.6
7	61.6	60.9	60.2
8	70.4	69.6	68.8
9	79.2	78.3	77.4

	85	84	83
1	8.5	8.4	8.3
2	17.0	16.8	16.6
3	25.5	25.2	24.9
4	34.0	33.6	33.2
5	42.5	42.0	41.5
6	51.0	50.4	49.8
7	59.5	58.8	58.1
8	68.0	67.2	66.4
9	76.5	75.6	74.7

	82	81	80
1	8.2	8.1	8.0
2	16.4	16.2	16.0
3	24.6	24.3	24.0
4	32.8	32.4	32.0
5	41.0	40.5	40.0
6	49.2	48.6	48.0
7	57.4	56.7	56.0
8	65.6	64.8	64.0
9	73.8	72.9	72.0

81°

9°

′	Log Sin	d	Log Tan	c d	Log Cot	Log Cos		Prop. Parts		
0	9.19 433	80	9.19 971	82	0.80 029	9.99 462	60			
1	9.19 513	79	9.20 053	81	0.79 947	9.99 460	59			
2	9.19 592	80	9.20 134	82	0.79 866	9.99 458	58			
3	9.19 672	79	9.20 216	81	0.79 784	9.99 456	57			
4	9.19 751	79	9.20 297	81	0.79 703	9.99 454	56			
5	9.19 830	79	9.20 378	81	0.79 622	9.99 452	55		**82**	**81**	**80**
6	9.19 909	79	9.20 459	81	0.79 541	9.99 450	54	1	8.2	8.1	8.0
7	9.19 988	79	9.20 540	81	0.79 460	9.99 448	53	2	16.4	16.2	16.0
8	9.20 067	78	9.20 621	80	0.79 379	9.99 446	52	3	24.6	24.3	24.0
9	9.20 145	78	9.20 701	81	0.79 299	9.99 444	51	4	32.8	32.4	32.0
10	9.20 223	79	9.20 782	80	0.79 218	9.99 442	50	5	41.0	40.5	40.0
11	9.20 302	78	9.20 862	80	0.79 138	9.99 440	49	6	49.2	48.6	48.0
12	9.20 380	78	9.20 942	80	0.79 058	9.99 438	48	7	57.4	56.7	56.0
13	9.20 458	77	9.21 022	80	0.78 978	9.99 436	47	8	65.6	64.8	64.0
14	9.20 535	78	9.21 102	80	0.78 898	9.99 434	46	9	73.8	72.9	72.0
15	9.20 613	78	9.21 182	79	0.78 818	9.99 432	45			
16	9.20 691	77	9.21 261	80	0.78 739	9.99 429	44			
17	9.20 768	77	9.21 341	79	0.78 659	9.99 427	43			
18	9.20 845	77	9.21 420	79	0.78 580	9.99 425	42			
19	9.20 922	77	9.21 499	79	0.78 501	9.99 423	41		**79**	**78**	**77**
20	9.20 999	77	9.21 578	79	0.78 422	9.99 421	40	1	7.9	7.8	7.7
21	9.21 076	77	9.21 657	79	0.78 343	9.99 419	39	2	15.8	15.6	15.4
22	9.21 153	76	9.21 736	78	0.78 264	9.99 417	38	3	23.7	23.4	23.1
23	9.21 229	77	9.21 814	79	0.78 186	9.99 415	37	4	31.6	31.2	30.8
24	9.21 306	76	9.21 893	78	0.78 107	9.99 413	36	5	39.5	39.0	38.5
25	9.21 382	76	9.21 971	78	0.78 029	9.99 411	35	6	47.4	46.8	46.2
26	9.21 458	76	9.22 049	78	0.77 951	9.99 409	34	7	55.3	54.6	53.9
27	9.21 534	76	9.22 127	78	0.77 873	9.99 407	33	8	63.2	62.4	61.6
28	9.21 610	75	9.22 205	78	0.77 795	9.99 404	32	9	71.1	70.2	69.3
29	9.21 685	76	9.22 283	78	0.77 717	9.99 402	31			
30	9.21 761	75	9.22 361	77	0.77 639	9.99 400	30			
31	9.21 836	76	9.22 438	78	0.77 562	9.99 398	29			
32	9.21 912	75	9.22 516	77	0.77 484	9.99 396	28		**76**	**75**	**74**
33	9.21 987	75	9.22 593	77	0.77 407	9.99 394	27	1	7.6	7.5	7.4
34	9.22 062	75	9.22 670	77	0.77 330	9.99 392	26	2	15.2	15.0	14.8
35	9.22 137	74	9.22 747	77	0.77 253	9.99 390	25	3	22.8	22.5	22.2
36	9.22 211	75	9.22 824	77	0.77 176	9.99 388	24	4	30.4	30.0	29.6
37	9.22 286	75	9.22 901	76	0.77 099	9.99 385	23	5	38.0	37.5	37.0
38	9.22 361	74	9.22 977	77	0.77 023	9.99 383	22	6	45.6	45.0	44.4
39	9.22 435	74	9.23 054	76	0.76 946	9.99 381	21	7	53.2	52.5	51.8
40	9.22 509	74	9.23 130	76	0.76 870	9.99 379	20	8	60.8	60.0	59.2
41	9.22 583	74	9.23 206	77	0.76 794	9.99 377	19	9	68.4	67.5	66.6
42	9.22 657	74	9.23 283	76	0.76 717	9.99 375	18			
43	9.22 731	74	9.23 359	76	0.76 641	9.99 372	17			
44	9.22 805	73	9.23 435	75	0.76 565	9.99 370	16			
45	9.22 878	74	9.23 510	76	0.76 490	9.99 368	15		**73**	**72**	**71**
46	9.22 952	73	9.23 586	75	0.76 414	9.99 366	14	1	7.3	7.2	7.1
47	9.23 025	73	9.23 661	76	0.76 339	9.99 364	13	2	14.6	14.4	14.2
48	9.23 098	73	9.23 737	75	0.76 263	9.99 362	12	3	21.9	21.6	21.3
49	9.23 171	73	9.23 812	75	0.76 188	9.99 359	11	4	29.2	28.8	28.4
50	9.23 244	73	9.23 887	75	0.76 113	9.99 357	10	5	36.5	36.0	35.5
51	9.23 317	73	9.23 962	75	0.76 038	9.99 355	9	6	43.8	43.2	42.6
52	9.23 390	72	9.24 037	75	0.75 963	9.99 353	8	7	51.1	50.4	49.7
53	9.23 462	73	9.24 112	74	0.75 888	9.99 351	7	8	58.4	57.6	56.8
54	9.23 535	72	9.24 186	75	0.75 814	9.99 348	6	9	65.7	64.8	63.9
55	9.23 607	72	9.24 261	74	0.75 739	9.99 346	5			
56	9.23 679	73	9.24 335	75	0.75 665	9.99 344	4			
57	9.23 752	71	9.24 410	74	0.75 590	9.99 342	3			
58	9.23 823	72	9.24 484	74	0.75 516	9.99 340	2			
59	9.23 895	72	9.24 558	74	0.75 442	9.99 337	1			
60	9.23 967		9.24 632		0.75 368	9.99 335	0			
	Log Cos	d	Log Cot	c d	Log Tan	Log Sin	′		Prop. Parts	

80°

Table II

10°

′	Log Sin	d	Log Tan	c d	Log Cot	Log Cos	d	
0	9.23 967	72	9.24 632	74	0.75 368	9.99 335	2	60
1	9.24 039	71	9.24 706	73	0.75 294	9.99 333	2	59
2	9.24 110	71	9.24 779	74	0.75 221	9.99 331	3	58
3	9.24 181	72	9.24 853	73	0.75 147	9.99 328	2	57
4	9.24 253	71	9.24 926	74	0.75 074	9.99 326	2	56
5	9.24 324	71	9.25 000	73	0.75 000	9.99 324	2	55
6	9.24 395	71	9.25 073	73	0.74 927	9.99 322	3	54
7	9.24 466	70	9.25 146	73	0.74 854	9.99 319	2	53
8	9.24 536	71	9.25 219	73	0.74 781	9.99 317	2	52
9	9.24 607	70	9.25 292	73	0.74 708	9.99 315	2	51
10	9.24 677	71	9.25 365	72	0.74 635	9.99 313	3	50
11	9.24 748	70	9.25 437	73	0.74 563	9.99 310	2	49
12	9.24 818	70	9.25 510	72	0.74 490	9.99 308	2	48
13	9.24 888	70	9.25 582	73	0.74 418	9.99 306	2	47
14	9.24 958	70	9.25 655	72	0.74 345	9.99 304	3	46
15	9.25 028	70	9.25 727	72	0.74 273	9.99 301	2	45
16	9.25 098	70	9.25 799	72	0.74 201	9.99 299	2	44
17	9.25 168	69	9.25 871	72	0.74 129	9.99 297	3	43
18	9.25 237	70	9.25 943	72	0.74 057	9.99 294	2	42
19	9.25 307	69	9.26 015	71	0.73 985	9.99 292	2	41
20	9.25 376	69	9.26 086	72	0.73 914	9.99 290	2	40
21	9.25 445	69	9.26 158	71	0.73 842	9.99 288	3	39
22	9.25 514	69	9.26 229	72	0.73 771	9.99 285	2	38
23	9.25 583	69	9.26 301	71	0.73 699	9.99 283	2	37
24	9.25 652	69	9.26 372	71	0.73 628	9.99 281	3	36
25	9.25 721	69	9.26 443	71	0.73 557	9.99 278	2	35
26	9.25 790	68	9.26 514	71	0.73 486	9.99 276	2	34
27	9.25 858	69	9.26 585	70	0.73 415	9.99 274	2	33
28	9.25 927	68	9.26 655	71	0.73 345	9.99 271	2	32
29	9.25 995	68	9.26 726	71	0.73 274	9.99 269	2	31
30	9.26 063	68	9.26 797	70	0.73 203	9.99 267	3	30
31	9.26 131	68	9.26 867	70	0.73 133	9.99 264	2	29
32	9.26 199	68	9.26 937	71	0.73 063	9.99 262	2	28
33	9.26 267	68	9.27 008	70	0.72 992	9.99 260	2	27
34	9.26 335	68	9.27 078	70	0.72 922	9.99 257	2	26
35	9.26 403	67	9.27 148	70	0.72 852	9.99 255	3	25
36	9.26 470	68	9.27 218	70	0.72 782	9.99 252	2	24
37	9.26 538	67	9.27 288	69	0.72 712	9.99 250	2	23
38	9.26 605	67	9.27 357	70	0.72 643	9.99 248	3	22
39	9.26 672	67	9.27 427	69	0.72 573	9.99 245	2	21
40	9.26 739	67	9.27 496	70	0.72 504	9.99 243	2	20
41	9.26 806	67	9.27 566	69	0.72 434	9.99 241	3	19
42	9.26 873	67	9.27 635	69	0.72 365	9.99 238	2	18
43	9.26 940	67	9.27 704	69	0.72 296	9.99 236	3	17
44	9.27 007	66	9.27 773	69	0.72 227	9.99 233	2	16
45	9.27 073	67	9.27 842	69	0.72 158	9.99 231	2	15
46	9.27 140	66	9.27 911	69	0.72 089	9.99 229	3	14
47	9.27 206	67	9.27 980	69	0.72 020	9.99 226	2	13
48	9.27 273	66	9.28 049	68	0.71 951	9.99 224	3	12
49	9.27 339	66	9.28 117	69	0.71 883	9.99 221	2	11
50	9.27 405	66	9.28 186	68	0.71 814	9.99 219	2	10
51	9.27 471	66	9.28 254	69	0.71 746	9.99 217	3	9
52	9.27 537	65	9.28 323	68	0.71 677	9.99 214	2	8
53	9.27 602	66	9.28 391	68	0.71 609	9.99 212	3	7
54	9.27 668	66	9.28 459	68	0.71 541	9.99 209	2	6
55	9.27 734	65	9.28 527	68	0.71 473	9.99 207	3	5
56	9.27 799	65	9.28 595	67	0.71 405	9.99 204	2	4
57	9.27 864	66	9.28 662	68	0.71 338	9.99 202	2	3
58	9.27 930	65	9.28 730	68	0.71 270	9.99 200	3	2
59	9.27 995	65	9.28 798	67	0.71 202	9.99 197	2	1
60	9.28 060		9.28 865		0.71 135	9.99 195		0
	Log Cos	d	Log Cot	c d	Log Tan	Log Sin	d	′

Prop. Parts

	74	73	72
1	7.4	7.3	7.2
2	14.8	14.6	14.4
3	22.2	21.9	21.6
4	29.6	29.2	28.8
5	37.0	36.5	36.0
6	44.4	43.8	43.2
7	51.8	51.1	50.4
8	59.2	58.4	57.6
9	66.6	65.7	64.8

	71	70	69
1	7.1	7.0	6.9
2	14.2	14.0	13.8
3	21.3	21.0	20.7
4	28.4	28.0	27.6
5	35.5	35.0	34.5
6	42.6	42.0	41.4
7	49.7	49.0	48.3
8	56.8	56.0	55.2
9	63.9	63.0	62.1

	68	67	66
1	6.8	6.7	6.6
2	13.6	13.4	13.2
3	20.4	20.1	19.8
4	27.2	26.8	26.4
5	34.0	33.5	33.0
6	40.8	40.2	39.6
7	47.6	46.9	46.2
8	54.4	53.6	52.8
9	61.2	60.3	59.4

	65	3
1	6.5	0.3
2	13.0	0.6
3	19.5	0.9
4	26.0	1.2
5	32.5	1.5
6	39.0	1.8
7	45.5	2.1
8	52.0	2.4
9	58.5	2.7

Prop. Parts

79°

11°

′	Log Sin	d	Log Tan	c d	Log Cot	Log Cos	d	
0	9.28 060	65	9.28 865	68	0.71 135	9.99 195	3	60
1	9.28 125	65	9.28 933	67	0.71 067	9.99 192	2	59
2	9.28 190	64	9.29 000	67	0.71 000	9.99 190	3	58
3	9.28 254	65	9.29 067	67	0.70 933	9.99 187	2	57
4	9.28 319	65	9.29 134	67	0.70 866	9.99 185	3	56
5	9.28 384	64	9.29 201	67	0.70 799	9.99 182	2	55
6	9.28 448	64	9.29 268	67	0.70 732	9.99 180	3	54
7	9.28 512	65	9.29 335	67	0.70 665	9.99 177	2	53
8	9.28 577	64	9.29 402	66	0.70 598	9.99 175	3	52
9	9.28 641	64	9.29 468	67	0.70 532	9.99 172	2	51
10	9.28 705	64	9.29 535	66	0.70 465	9.99 170	3	50
11	9.28 769	64	9.29 601	67	0.70 399	9.99 167	2	49
12	9.28 833	63	9.29 668	66	0.70 332	9.99 165	3	48
13	9.28 896	64	9.29 734	66	0.70 266	9.99 162	2	47
14	9.28 960	64	9.29 800	66	0.70 200	9.99 160	3	46
15	9.29 024	63	9.29 866	66	0.70 134	9.99 157	2	45
16	9.29 087	63	9.29 932	66	0.70 068	9.99 155	3	44
17	9.29 150	64	9.29 998	66	0.70 002	9.99 152	2	43
18	9.29 214	63	9.30 064	66	0.69 936	9.99 150	3	42
19	9.29 277	63	9.30 130	65	0.69 870	9.99 147	2	41
20	9.29 340	63	9.30 195	66	0.69 805	9.99 145	3	40
21	9.29 403	63	9.30 261	65	0.69 739	9.99 142	2	39
22	9.29 466	63	9.30 326	65	0.69 674	9.99 140	3	38
23	9.29 529	62	9.30 391	66	0.69 609	9.99 137	2	37
24	9.29 591	63	9.30 457	65	0.69 543	9.99 135	3	36
25	9.29 654	62	9.30 522	65	0.69 478	9.99 132	2	35
26	9.29 716	63	9.30 587	65	0.69 413	9.99 130	3	34
27	9.29 779	62	9.30 652	65	0.69 348	9.99 127	3	33
28	9.29 841	62	9.30 717	65	0.69 283	9.99 124	2	32
29	9.29 903	63	9.30 782	64	0.69 218	9.99 122	3	31
30	9.29 966	62	9.30 846	65	0.69 154	9.99 119	2	30
31	9.30 028	62	9.30 911	64	0.69 089	9.99 117	3	29
32	9.30 090	61	9.30 975	65	0.69 025	9.99 114	2	28
33	9.30 151	62	9.31 040	64	0.68 960	9.99 112	3	27
34	9.30 213	62	9.31 104	64	0.68 896	9.99 109	3	26
35	9.30 275	61	9.31 168	65	0.68 832	9.99 106	2	25
36	9.30 336	62	9.31 233	64	0.68 767	9.99 104	3	24
37	9.30 398	61	9.31 297	64	0.68 703	9.99 101	2	23
38	9.30 459	62	9.31 361	64	0.68 639	9.99 099	3	22
39	9.30 521	61	9.31 425	64	0.68 575	9.99 096	3	21
40	9.30 582	61	9.31 489	63	0.68 511	9.99 093	2	20
41	9.30 643	61	9.31 552	64	0.68 448	9.99 091	3	19
42	9.30 704	61	9.31 616	63	0.68 384	9.99 088	2	18
43	9.30 765	61	9.31 679	64	0.68 321	9.99 086	3	17
44	9.30 826	61	9.31 743	63	0.68 257	9.99 083	3	16
45	9.30 887	60	9.31 806	64	0.68 194	9.99 080	2	15
46	9.30 947	61	9.31 870	63	0.68 130	9.99 078	3	14
47	9.31 008	60	9.31 933	63	0.68 067	9.99 075	3	13
48	9.31 068	61	9.31 996	63	0.68 004	9.99 072	2	12
49	9.31 129	60	9.32 059	63	0.67 941	9.99 070	3	11
50	9.31 189	61	9.32 122	63	0.67 878	9.99 067	3	10
51	9.31 250	60	9.32 185	63	0.67 815	9.99 064	2	9
52	9.31 310	60	9.32 248	63	0.67 752	9.99 062	3	8
53	9.31 370	60	9.32 311	62	0.67 689	9.99 059	3	7
54	9.31 430	60	9.32 373	63	0.67 627	9.99 056	2	6
55	9.31 490	59	9.32 436	62	0.67 564	9.99 054	3	5
56	9.31 549	60	9.32 498	63	0.67 502	9.99 051	3	4
57	9.31 609	60	9.32 561	62	0.67 439	9.99 048	3	3
58	9.31 669	59	9.32 623	62	0.67 377	9.99 046	2	2
59	9.31 728	60	9.32 685	62	0.67 315	9.99 043	3	1
60	9.31 788		9.32 747		0.67 253	9.99 040		0
	Log Cos	d	Log Cot	c d	Log Tan	Log Sin	d	′

Prop. Parts

	68	67	66
1	6.8	6.7	6.6
2	13.6	13.4	13.2
3	20.4	20.1	19.8
4	27.2	26.8	26.4
5	34.0	33.5	33.0
6	40.8	40.2	39.6
7	47.6	46.9	46.2
8	54.4	53.6	52.8
9	61.2	60.3	59.4

	65	64	63
1	6.5	6.4	6.3
2	13.0	12.8	12.6
3	19.5	19.2	18.9
4	26.0	25.6	25.2
5	32.5	32.0	31.5
6	39.0	38.4	37.8
7	45.5	44.8	44.1
8	52.0	51.2	50.4
9	58.5	57.6	56.7

	62	61	60
1	6.2	6.1	6.0
2	12.4	12.2	12.0
3	18.6	18.3	18.0
4	24.8	24.4	24.0
5	31.0	30.5	30.0
6	37.2	36.6	36.0
7	43.4	42.7	42.0
8	49.6	48.8	48.0
9	55.8	54.9	54.0

	59	3	2
1	5.9	0.3	0.2
2	11.8	0.6	0.4
3	17.7	0.9	0.6
4	23.6	1.2	0.8
5	29.5	1.5	1.0
6	35.4	1.8	1.2
7	41.3	2.1	1.4
8	47.2	2.4	1.6
9	53.1	2.7	1.8

78°

12°

'	Log Sin	d	Log Tan	c d	Log Cot	Log Cos	d	
0	9.31 788	59	9.32 747	63	0.67 253	9.99 040	2	60
1	9.31 847	60	9.32 810	62	0.67 190	9.99 038	3	59
2	9.31 907	59	9.32 872	61	0.67 128	9.99 035	3	58
3	9.31 966	59	9.32 933	62	0.67 067	9.99 032	2	57
4	9.32 025	59	9.32 995	62	0.67 005	9.99 030		56
5	9.32 084	59	9.33 057	62	0.66 943	9.99 027	3	55
6	9.32 143	59	9.33 119	61	0.66 881	9.99 024	2	54
7	9.32 202	59	9.33 180	62	0.66 820	9.99 022	3	53
8	9.32 261	58	9.33 242	61	0.66 758	9.99 019	3	52
9	9.32 319	59	9.33 303	62	0.66 697	9.99 016	3	51
10	9.32 378	59	9.33 365	61	0.66 635	9.99 013	2	50
11	9.32 437	58	9.33 426	61	0.66 574	9.99 011	3	49
12	9.32 495	58	9.33 487	61	0.66 513	9.99 008	3	48
13	9.32 553	59	9.33 548	61	0.66 452	9.99 005	3	47
14	9.32 612	58	9.33 609	61	0.66 391	9.99 002	2	46
15	9.32 670	58	9.33 670	61	0.66 330	9.99 000	3	45
16	9.32 728	58	9.33 731	61	0.66 269	9.98 997	3	44
17	9.32 786	58	9.33 792	61	0.66 208	9.98 994	3	43
18	9.32 844	58	9.33 853	60	0.66 147	9.98 991	2	42
19	9.32 902	58	9.33 913	61	0.66 087	9.98 989	3	41
20	9.32 960	58	9.33 974	60	0.66 026	9.98 986	3	40
21	9.33 018	57	9.34 034	61	0.65 966	9.98 983	3	39
22	9.23 075	58	9.34 095	60	0.65 905	9.98 980	2	38
23	9.33 133	57	9.34 155	60	0.65 845	9.98 978	3	37
24	9.33 190	58	9.34 215	61	0.65 785	9.98 975	3	36
25	9.33 248	57	9.34 276	60	0.65 724	9.98 972	3	35
26	9.33 305	57	9.34 336	60	0.65 664	9.98 969	2	34
27	9.33 362	57	9.34 396	60	0.65 604	9.98 967	3	33
28	9.33 420	57	9.34 456	60	0.65 544	9.98 964	3	32
29	9.33 477	57	9.34 516	60	0.65 484	9.98 961	3	31
30	9.33 534	57	9.34 576	59	0.65 424	9.98 958	3	30
31	9.33 591	56	9.34 635	60	0.65 365	9.98 955	2	29
32	9.33 647	57	9.34 695	60	0.65 305	9.98 953	3	28
33	9.33 704	57	9.34 755	59	0.65 245	9.98 950	3	27
34	9.33 761	57	9.34 814	60	0.65 186	9.98 947	3	26
35	9.33 818	56	9.34 874	59	0.65 126	9.98 944	3	25
36	9.33 874	57	9.34 933	59	0.65 067	9.98 941	3	24
37	9.33 931	56	9.34 992	59	0.65 008	9.98 938	2	23
38	9.33 987	56	9.35 051	60	0.64 949	9.98 936	3	22
39	9.34 043	57	9.35 111	59	0.64 889	9.98 933	3	21
40	9.34 100	56	9.35 170	59	0.64 830	9.98 930	3	20
41	9.34 156	56	9.35 229	59	0.64 771	9.98 927	3	19
42	9.34 212	56	9.35 288	59	0.64 712	9.98 924	3	18
43	9.34 268	56	9.35 347	58	0.64 653	9.98 921	2	17
44	9.34 324	56	9.35 405	59	0.64 595	9.98 919	3	16
45	9.34 380	56	9.35 464	59	0.64 536	9.98 916	3	15
46	9.34 436	55	9.35 523	58	0.64 477	9.98 913	3	14
47	9.34 491	56	9.35 581	59	0.64 419	9.98 910	3	13
48	9.34 547	55	9.35 640	58	0.64 360	9.98 907	3	12
49	9.34 602	55	9.35 698	59	0.64 302	9.98 904	3	11
50	9.34 658	55	9.35 757	58	0.64 243	9.98 901	3	10
51	9.34 713	56	9.35 815	58	0.64 185	9.98 898	2	9
52	9.34 769	55	9.35 873	58	0.64 127	9.98 896	3	8
53	9.34 824	55	9.35 931	58	0.64 069	9.98 893	3	7
54	9.34 879	55	9.35 989	58	0.64 011	9.98 890	3	6
55	9.34 934	55	9.36 047	58	0.63 953	9.98 887	3	5
56	9.34 989	55	9.36 105	58	0.63 895	9.98 884	3	4
57	9.35 044	55	9.36 163	58	0.63 837	9.98 881	3	3
58	9.35 099	55	9.36 221	58	0.63 779	9.98 878	3	2
59	9.35 154	55	9.36 279	57	0.63 721	9.98 875	3	1
60	9.35 209		9.36 336		0.63 664	9.98 872		0
	Log Cos	d	Log Cot	c d	Log Tan	Log Sin	d	'

Prop. Parts

	63	62	61
1	6.3	6.2	6.1
2	12.6	12.4	12.2
3	18.9	18.6	18.3
4	25.2	24.8	24.4
5	31.5	31.0	30.5
6	37.8	37.2	36.6
7	44.1	43.4	42.7
8	50.4	49.6	48.8
9	56.7	55.8	54.9

	60	59	58
1	6.0	5.9	5.8
2	12.0	11.8	11.6
3	18.0	17.7	17.4
4	24.0	23.6	23.2
5	30.0	29.5	29.0
6	36.0	35.4	34.8
7	42.0	41.3	40.6
8	48.0	47.2	46.4
9	54.0	53.1	52.2

	57	56	55
1	5.7	5.6	5.5
2	11.4	11.2	11.0
3	17.1	16.8	16.5
4	22.8	22.4	22.0
5	28.5	28.0	27.5
6	34.2	33.6	33.0
7	39.9	39.2	38.5
8	45.6	44.8	44.0
9	51.3	50.4	49.5

	3	2
1	0.3	0.2
2	0.6	0.4
3	0.9	0.6
4	1.2	0.8
5	1.5	1.0
6	1.8	1.2
7	2.1	1.4
8	2.4	1.6
9	2.7	1.8

Prop. Parts

13°

'	Log Sin	d	Log Tan	c d	Log Cot	Log Cos	d	
0	9.35 209	54	9.36 336	58	0.63 664	9.98 872	3	60
1	9.35 263	55	9.36 394	58	0.63 606	9.98 869	2	59
2	9.35 318	55	9.36 452	57	0.63 548	9.98 867	3	58
3	9.35 373	54	9.36 509	57	0.63 491	9.98 864	3	57
4	9.35 427	54	9.36 566	58	0.63 434	9.98 861	3	56
5	9.35 481	55	9.36 624	57	0.63 376	9.98 858	3	55
6	9.35 536	54	9.36 681	57	0.63 319	9.98 855	3	54
7	9.35 590	54	9.36 738	57	0.63 262	9.98 852	3	53
8	9.35 644	54	9.36 795	57	0.63 205	9.98 849	3	52
9	9.35 698	54	9.36 852	57	0.63 148	9.98 846	3	51
10	9.35 752	54	9.36 909	57	0.63 091	9.98 843	3	50
11	9.35 806	54	9.36 966	57	0.63 034	9.98 840	3	49
12	9.35 860	54	9.37 023	57	0.62 977	9.98 837	3	48
13	9.35 914	54	9.37 080	57	0.62 920	9.98 834	3	47
14	9.35 968	54	9.37 137	56	0.62 863	9.98 831	3	46
15	9.36 022	53	9.37 193	57	0.62 807	9.98 828	3	45
16	9.36 075	54	9.37 250	56	0.62 750	9.98 825	3	44
17	9.36 129	53	9.37 306	57	0.62 694	9.98 822	3	43
18	9.36 182	54	9.37 363	56	0.62 637	9.98 819	3	42
19	9.36 236	53	9.37 419	57	0.62 581	9.98 816	3	41
20	9.36 289	53	9.37 476	56	0.62 524	9.98 813	3	40
21	9.36 342	53	9.37 532	56	0.62 468	9.98 810	3	39
22	9.36 395	54	9.37 588	56	0.62 412	9.98 807	3	38
23	9.36 449	53	9.37 644	56	0.62 356	9.98 804	3	37
24	9.36 502	53	9.37 700	56	0.62 300	9.98 801	3	36
25	9.36 555	53	9.37 756	56	0.62 244	9.98 798	3	35
26	9.36 608	52	9.37 812	56	0.62 188	9.98 795	3	34
27	9.36 660	53	9.37 868	56	0.62 132	9.98 792	3	33
28	9.36 713	53	9.37 924	56	0.62 076	9.98 789	3	32
29	9.36 766	53	9.37 980	55	0.62 020	9.98 786	3	31
30	9.36 819	52	9.38 035	56	0.61 965	9.98 783	3	30
31	9.36 871	53	9.38 091	56	0.61 909	9.98 780	3	29
32	9.36 924	52	9.38 147	55	0.61 853	9.98 777	3	28
33	9.36 976	52	9.38 202	55	0.61 798	9.98 774	3	27
34	9.37 028	53	9.38 257	56	0.61 743	9.98 771	3	26
35	9.37 081	52	9.38 313	55	0.61 687	9.98 768	3	25
36	9.37 133	52	9.38 368	55	0.61 632	9.98 765	3	24
37	9.37 185	52	9.38 423	56	0.61 577	9.98 762	3	23
38	9.37 237	52	9.38 479	55	0.61 521	9.98 759	3	22
39	9.37 289	52	9.38 534	55	0.61 466	9.98 756	3	21
40	9.37 341	52	9.38 589	55	0.61 411	9.98 753	3	20
41	9.37 393	52	9.38 644	55	0.61 356	9.98 750	4	19
42	9.37 445	52	9.38 699	55	0.61 301	9.98 746	3	18
43	9.37 497	52	9.38 754	54	0.61 246	9.98 743	3	17
44	9.37 549	51	9.38 808	55	0.61 192	9.98 740	3	16
45	9.37 600	52	9.38 863	55	0.61 137	9.98 737	3	15
46	9.37 652	51	9.38 918	54	0.61 082	9.98 734	3	14
47	9.37 703	52	9.38 972	55	0.61 028	9.98 731	3	13
48	9.37 755	51	9.39 027	55	0.60 973	9.98 728	3	12
49	9.37 806	52	9.39 082	54	0.60 918	9.98 725	3	11
50	9.37 858	51	9.39 136	54	0.60 864	9.98 722	3	10
51	9.37 909	51	9.39 190	55	0.60 810	9.98 719	4	9
52	9.37 960	51	9.39 245	54	0.60 755	9.98 715	3	8
53	9.38 011	51	9.39 299	54	0.60 701	9.98 712	3	7
54	9.38 062	51	9.39 353	54	0.60 647	9.98 709	3	6
55	9.38 113	51	9.39 407	54	0.60 593	9.98 706	3	5
56	9.38 164	51	9.39 461	54	0.60 539	9.98 703	3	4
57	9.38 215	51	9.39 515	54	0.60 485	9.98 700	3	3
58	9.38 266	51	9.39 569	54	0.60 431	9.98 697	3	2
59	9.38 317	51	9.39 623	54	0.60 377	9.98 694	4	1
60	9.38 368		9.39 677		0.60 323	9.98 690		0
	Log Cos	d	Log Cot	c d	Log Tan	Log Sin	d	'

Prop. Parts

	58	57	56
1	5.8	5.7	5.6
2	11.6	11.4	11.2
3	17.4	17.1	16.8
4	23.2	22.8	22.4
5	29.0	28.5	28.0
6	34.8	34.2	33.6
7	40.6	39.9	39.2
8	46.4	45.6	44.8
9	52.2	51.3	50.4

	55	54	53
1	5.5	5.4	5.3
2	11.0	10.8	10.6
3	16.5	16.2	15.9
4	22.0	21.6	21.2
5	27.5	27.0	26.5
6	33.0	32.4	31.8
7	38.5	37.8	37.1
8	44.0	43.2	42.4
9	49.5	48.6	47.7

	52	51
1	5.2	5.1
2	10.4	10.2
3	15.6	15.3
4	20.8	20.4
5	26.0	25.5
6	31.2	30.6
7	36.4	35.7
8	41.6	40.8
9	46.8	45.9

	4	3	2
1	0.4	0.3	0.2
2	0.8	0.6	0.4
3	1.2	0.9	0.6
4	1.6	1.2	0.8
5	2.0	1.5	1.0
6	2.4	1.8	1.2
7	2.8	2.1	1.4
8	3.2	2.4	1.6
9	3.6	2.7	1.8

76°

14°

′	Log Sin	d	Log Tan	c d	Log Cot	Log Cos	d	
0	9.38 368	50	9.39 677	54	0.60 323	9.98 690	3	60
1	9.38 418	51	9.39 731	54	0.60 269	9.98 687	3	59
2	9.38 469	50	9.39 785	53	0.60 215	9.98 684	3	58
3	9.38 519	51	9.39 838	54	0.60 162	9.98 681	3	57
4	9.38 570	50	9.39 892	53	0.60 108	9.98 678	3	56
5	9.38 620	50	9.39 945	54	0.60 055	9.98 675	4	55
6	9.38 670	51	9.39 999	53	0.60 001	9.98 671	3	54
7	9.38 721	50	9.40 052	54	0.59 948	9.98 668	3	53
8	9.38 771	50	9.40 106	53	0.59 894	9.98 665	3	52
9	9.38 821	50	9.40 159	53	0.59 841	9.98 662	3	51
10	9.38 871	50	9.40 212	54	0.59 788	9.98 659	3	50
11	9.38 921	50	9.40 266	53	0.59 734	9.98 656	4	49
12	9.38 971	50	9.40 319	53	0.59 681	9.98 652	3	48
13	9.39 021	50	9.40 372	53	0.59 628	9.98 649	3	47
14	9.39 071	50	9.40 425	53	0.59 575	9.98 646	3	46
15	9.39 121	49	9.40 478	53	0.59 522	9.98 643	3	45
16	9.39 170	50	9.40 531	53	0.59 469	9.98 640	3	44
17	9.39 220	50	9.40 584	52	0.59 416	9.98 636	3	43
18	9.39 270	49	9.40 636	53	0.59 364	9.98 633	3	42
19	9.39 319	50	9.40 689	53	0.59 311	9.98 630	3	41
20	9.39 369	49	9.40 742	53	0.59 258	9.98 627	4	40
21	9.39 418	49	9.40 795	52	0.59 205	9.98 623	3	39
22	9.39 467	50	9.40 847	53	0.59 153	9.98 620	3	38
23	9.39 517	49	9.40 900	52	0.59 100	9.98 617	3	37
24	9.39 566	49	9.40 952	53	0.59 048	9.98 614	4	36
25	9.39 615	49	9.41 005	52	0.58 995	9.98 610	3	35
26	9.39 664	49	9.41 057	52	0.58 943	9.98 607	3	34
27	9.39 713	49	9.41 109	52	0.58 891	9.98 604	3	33
28	9.39 762	49	9.41 161	53	0.58 839	9.98 601	3	32
29	9.39 811	49	9.41 214	52	0.58 786	9.98 597	3	31
30	9.39 860	49	9.41 266	52	0.58 734	9.98 594	3	30
31	9.39 909	49	9.41 318	52	0.58 682	9.98 591	3	29
32	9.39 958	48	9.41 370	52	0.58 630	9.98 588	4	28
33	9.40 006	49	9.41 422	52	0.58 578	9.98 584	3	27
34	9.40 055	48	9.41 474	52	0.58 526	9.98 581	3	26
35	9.40 103	49	9.41 526	52	0.58 474	9.98 578	4	25
36	9.40 152	48	9.41 578	51	0.58 422	9.98 574	3	24
37	9.40 200	49	9.41 629	52	0.58 371	9.98 571	3	23
38	9.40 249	48	9.41 681	52	0.58 319	9.98 568	3	22
39	9.40 297	49	9.41 733	51	0.58 267	9.98 565	4	21
40	9.40 346	48	9.41 784	52	0.58 216	9.98 561	3	20
41	9.40 394	48	9.41 836	51	0.58 164	9.98 558	3	19
42	9.40 442	48	9.41 887	52	0.58 113	9.98 555	4	18
43	9.40 490	48	9.41 939	51	0.58 061	9.98 551	3	17
44	9.40 538	48	9.41 990	51	0.58 010	9.98 548	3	16
45	9.40 586	48	9.42 041	52	0.57 959	9.98 545	4	15
46	9.40 634	48	9.42 093	51	0.57 907	9.98 541	3	14
47	9.40 682	48	9.42 144	51	0.57 856	9.98 538	3	13
48	9.40 730	48	9.42 195	51	0.57 805	9.98 535	4	12
49	9.40 778	47	9.42 246	51	0.57 754	9.98 531	3	11
50	9.40 825	48	9.42 297	51	0.57 703	9.98 528	3	10
51	9.40 873	48	9.42 348	51	0.57 652	9.98 525	3	9
52	9.40 921	47	9.42 399	51	0.57 601	9.98 521	3	8
53	9.40 968	48	9.42 450	51	0.57 550	9.98 518	3	7
54	9.41 016	47	9.42 501	51	0.57 499	9.98 515	4	6
55	9.41 063	48	9.42 552	51	0.57 448	9.98 511	3	5
56	9.41 111	47	9.42 603	50	0.57 397	9.98 508	3	4
57	9.41 158	47	9.42 653	51	0.57 347	9.98 505	4	3
58	9.41 205	47	9.42 704	51	0.57 296	9.98 501	3	2
59	9.41 252	48	9.42 755	50	0.57 245	9.98 498	4	1
60	9.41 300		9.42 805		0.57 195	9.98 494		0
	Log Cos	d	Log Cot	c d	Log Tan	Log Sín	d	′

Prop. Parts

	54	53	52
1	5.4	5.3	5.2
2	10.8	10.6	10.4
3	16.2	15.9	15.6
4	21.6	21.2	20.8
5	27.0	26.5	26.0
6	32.4	31.8	31.2
7	37.8	37.1	36.4
8	43.2	42.4	41.6
9	48.6	47.7	46.8

	51	50	49
1	5.1	5.0	4.9
2	10.2	10.0	9.8
3	15.3	15.0	14.7
4	20.4	20.0	19.6
5	25.5	25.0	24.5
6	30.6	30.0	29.4
7	35.7	35.0	34.3
8	40.8	40.0	39.2
9	45.9	45.0	44.1

	48	47
1	4.8	4.7
2	9.6	9.4
3	14.4	14.1
4	19.2	18.8
5	24.0	23.5
6	28.8	28.2
7	33.6	32.9
8	38.4	37.6
9	43.2	42.3

	4	3
1	0.4	0.3
2	0.8	0.6
3	1.2	0.9
4	1.6	1.2
5	2.0	1.5
6	2.4	1.8
7	2.8	2.1
8	3.2	2.4
9	3.6	2.7

75°

15°

′	Log Sin	d	Log Tan	c d	Log Cot	Log Cos	d		Prop. Parts		
0	9.41 300	47	9.42 805	51	0.57 195	9.98 494	3	**60**			
1	9.41 347	47	9.42 856	50	0.57 144	9.98 491	3	59			
2	9.41 394	47	9.42 906	51	0.57 094	9.98 488	4	58			
3	9.41 441	47	9.42 957	50	0.57 043	9.98 484	3	57			
4	9.41 488	47	9.43 007	50	0.56 993	9.98 481	4	56			
5	9.41 535	47	9.43 057	51	0.56 943	9.98 477	3	**55**		**51**	**50**	**49**
6	9.41 582	46	9.43 108	50	0.56 892	9.98 474	3	54	1	5.1	5.0	4.9
7	9.41 628	47	9.43 158	50	0.56 842	9.98 471	4	53	2	10.2	10.0	9.8
8	9.41 675	47	9.43 208	50	0.56 792	9.98 467	3	52	3	15.3	15.0	14.7
9	9.41 722	46	9.43 258	50	0.56 742	9.98 464	4	51	4	20.4	20.0	19.6
10	9.41 768	47	9.43 308	50	0.56 692	9.98 460	3	**50**	5	25.5	25.0	24.5
11	9.41 815	46	9.43 358	50	0.56 642	9.98 457	4	49	6	30.6	30.0	29.4
12	9.41 861	47	9.43 408	50	0.56 592	9.98 453	3	48	7	35.7	35.0	34.3
13	9.41 908	46	9.43 458	50	0.56 542	9.98 450	3	47	8	40.8	40.0	39.2
14	9.41 954	47	9.43 508	50	0.56 492	9.98 447	4	46	9	45.9	45.0	44.1
15	9.42 001	46	9.43 558	49	0.56 442	9.98 443	3	**45**				
16	9.42 047	46	9.43 607	50	0.56 393	9.98 440	4	44				
17	9.42 093	47	9.43 657	50	0.56 343	9.98 436	3	43				
18	9.42 140	46	9.43 707	49	0.56 293	9.98 433	4	42				
19	9.42 186	46	9.43 756	50	0.56 244	9.98 429	3	41		**48**	**47**	**46**
20	9.42 232	46	9.43 806	49	0.56 194	9.98 426	4	**40**	1	4.8	4.7	4.6
21	9.42 278	46	9.43 855	50	0.56 145	9.98 422	3	39	2	9.6	9.4	9.2
22	9.42 324	46	9.43 905	49	0.56 095	9.98 419	4	38	3	14.4	14.1	13.8
23	9.42 370	46	9.43 954	50	0.56 046	9.98 415	3	37	4	19.2	18.8	18.4
24	9.42 416	45	9.44 004	49	0.55 996	9.98 412	3	36	5	24.0	23.5	23.0
25	9.42 461	46	9.44 053	49	0.55 947	9.98 409	4	**35**	6	28.8	28.2	27.6
26	9.42 507	46	9.44 102	49	0.55 898	9.98 405	3	34	7	33.6	32.9	32.2
27	9.42 553	46	9.44 151	50	0.55 849	9.98 402	4	33	8	38.4	37.6	36.8
28	9.42 599	45	9.44 201	49	0.55 799	9.98 398	3	32	9	43.2	42.3	41.4
29	9.42 644	46	9.44 250	49	0.55 750	9.98 395	4	31				
30	9.42 690	45	9.44 299	49	0.55 701	9.98 391	3	**30**				
31	9.42 735	46	9.44 348	49	0.55 652	9.98 388	4	29				
32	9.42 781	45	9.44 397	49	0.55 603	9.98 384	3	28		**45**	**44**	
33	9.42 826	46	9.44 446	49	0.55 554	9.98 381	4	27	1	4.5	4.4	
34	9.42 872	45	9.44 495	49	0.55 505	9.98 377	4	26	2	9.0	8.8	
35	9.42 917	45	9.44 544	48	0.55 456	9.98 373	3	**25**	3	13.5	13.2	
36	9.42 962	46	9.44 592	49	0.55 408	9.98 370	4	24	4	18.0	17.6	
37	9.43 008	45	9.44 641	49	0.55 359	9.98 366	3	23	5	22.5	22.0	
38	9.43 053	45	9.44 690	48	0.55 310	9.98 363	4	22	6	27.0	26.4	
39	9.43 098	45	9.44 738	49	0.55 262	9.98 359	3	21	7	31.5	30.8	
40	9.43 143	45	9.44 787	49	0.55 213	9.98 356	4	**20**	8	36.0	35.2	
41	9.43 188	45	9.44 836	48	0.55 164	9.98 352	3	19	9	40.5	39.6	
42	9.43 233	45	9.44 884	49	0.55 116	9.98 349	4	18				
43	9.43 278	45	9.44 933	48	0.55 067	9.98 345	3	17				
44	9.43 323	44	9.44 981	48	0.55 019	9.98 342	4	16				
45	9.43 367	45	9.45 029	49	0.54 971	9.98 338	4	**15**				
46	9.43 412	45	9.45 078	48	0.54 922	9.98 334	3	14		**4**	**3**	
47	9.43 457	45	9.45 126	48	0.54 874	9.98 331	4	13	1	0.4	0.3	
48	9.43 502	44	9.45 174	48	0.54 826	9.98 327	3	12	2	0.8	0.6	
49	9.43 546	45	9.45 222	49	0.54 778	9.98 324	4	11	3	1.2	0.9	
50	9.43 591	44	9.45 271	48	0.54 729	9.98 320	3	**10**	4	1.6	1.2	
51	9.43 635	45	9.45 319	48	0.54 681	9.98 317	4	9	5	2.0	1.5	
52	9.43 680	44	9.45 367	48	0.54 633	9.98 313	4	8	6	2.4	1.8	
53	9.43 724	45	9.45 415	48	0.54 585	9.98 309	3	7	7	2.8	2.1	
54	9.43 769	44	9.45 463	48	0.54 537	9.98 306	4	6	8	3.2	2.4	
55	9.43 813	44	9.45 511	48	0.54 489	9.98 302	3	**5**	9	3.6	2.7	
56	9.43 857	44	9.45 559	47	0.54 441	9.98 299	4	4				
57	9.43 901	45	9.45 606	48	0.54 394	9.98 295	4	3				
58	9.43 946	44	9.45 654	48	0.54 346	9.98 291	3	2				
59	9.43 990	44	9.45 702	48	0.54 298	9.98 288	4	1				
60	9.44 034		9.45 750		0.54 250	9.98 284		**0**				
	Log Cos	d	Log Cot	c d	Log Tan	Log Sin	d	′		Prop. Parts		

Table II

16°

′	Log Sin	d	Log Tan	c d	Log Cot	Log Cos	d	′
0	9.44 034	44	9.45 750	47	0.54 250	9.98 284	3	60
1	9.44 078	44	9.45 797	48	0.54 203	9.98 281	4	59
2	9.44 122	44	9.45 845	47	0.54 155	9.98 277	4	58
3	9.44 166	44	9.45 892	48	0.54 108	9.98 273	3	57
4	9.44 210	43	9.45 940	47	0.54 060	9.98 270	4	56
5	9.44 253	44	9.45 987	48	0.54 013	9.98 266	4	55
6	9.44 297	44	9.46 035	47	0.53 965	9.98 262	3	54
7	9.44 341	44	9.46 082	48	0.53 918	9.98 259	4	53
8	9.44 385	43	9.46 130	47	0.53 870	9.98 255	4	52
9	9.44 428	44	9.46 177	47	0.53 823	9.98 251	3	51
10	9.44 472	44	9.46 224	47	0.53 776	9.98 248	4	50
11	9.44 516	43	9.46 271	48	0.53 729	9.98 244	4	49
12	9.44 559	43	9.46 319	47	0.53 681	9.98 240	4	48
13	9.44 602	44	9.46 366	47	0.53 634	9.98 237	4	47
14	9.44 646	43	9.46 413	47	0.53 587	9.98 233	4	46
15	9.44 689	44	9.46 460	47	0.53 540	9.98 229	3	45
16	9.44 733	43	9.46 507	47	0.53 493	9.98 226	4	44
17	9.44 776	43	9.46 554	47	0.53 446	9.98 222	4	43
18	9.44 819	43	9.46 601	47	0.53 399	9.98 218	3	42
19	9.44 862	43	9.46 648	46	0.53 352	9.98 215	4	41
20	9.44 905	43	9.46 694	47	0.53 306	9.98 211	4	40
21	9.44 948	44	9.46 741	47	0.53 259	9.98 207	3	39
22	9.44 992	43	9.46 788	47	0.53 212	9.98 204	4	38
23	9.45 035	42	9.46 835	46	0.53 165	9.98 200	4	37
24	9.45 077	43	9.46 881	47	0.53 119	9.98 196	4	36
25	9.45 120	43	9.46 928	47	0.53 072	9.98 192	3	35
26	9.45 163	43	9.46 975	46	0.53 025	9.98 189	4	34
27	9.45 206	43	9.47 021	47	0.52 979	9.98 185	4	33
28	9.45 249	43	9.47 068	46	0.52 932	9.98 181	4	32
29	9.45 292	42	9.47 114	46	0.52 886	9.98 177	3	31
30	9.45 334	43	9.47 160	47	0.52 840	9.98 174	4	30
31	9.45 377	42	9.47 207	46	0.52 793	9.98 170	4	29
32	9.45 419	43	9.47 253	46	0.52 747	9.98 166	4	28
33	9.45 462	42	9.47 299	47	0.52 701	9.98 162	4	27
34	9.45 504	43	9.47 346	46	0.52 654	9.98 159	4	26
35	9.45 547	42	9.47 392	46	0.52 608	9.98 155	4	25
36	9.45 589	43	9.47 438	46	0.52 562	9.98 151	4	24
37	9.45 632	42	9.47 484	46	0.52 516	9.98 147	3	23
38	9.45 674	42	9.47 530	46	0.52 470	9.98 144	4	22
39	9.45 716	42	9.47 576	46	0.52 424	9.98 140	4	21
40	9.45 758	43	9.47 622	46	0.52 378	9.98 136	4	20
41	9.45 801	42	9.47 668	46	0.52 332	9.98 132	3	19
42	9.45 843	42	9.47 714	46	0.52 286	9.98 129	4	18
43	9.45 885	42	9.47 760	46	0.52 240	9.98 125	4	17
44	9.45 927	42	9.47 806	46	0.52 194	9.98 121	4	16
45	9.45 969	42	9.47 852	45	0.52 148	9.98 117	4	15
46	9.46 011	42	9.47 897	46	0.52 103	9.98 113	3	14
47	9.46 053	42	9.47 943	46	0.52 057	9.98 110	4	13
48	9.46 095	41	9.47 989	46	0.52 011	9.98 106	4	12
49	9.46 136	42	9.48 035	45	0.51 965	9.98 102	4	11
50	9.46 178	42	9.48 080	46	0.51 920	9.98 098	4	10
51	9.46 220	42	9.48 126	45	0.51 874	9.98 094	4	9
52	9.46 262	41	9.48 171	46	0.51 829	9.98 090	3	8
53	9.46 303	42	9.48 217	45	0.51 783	9.98 087	4	7
54	9.46 345	41	9.48 262	45	0.51 738	9.98 083	4	6
55	9.46 386	42	9.48 307	46	0.51 693	9.98 079	4	5
56	9.46 428	41	9.48 353	45	0.51 647	9.98 075	4	4
57	9.46 469	42	9.48 398	45	0.51 602	9.98 071	4	3
58	9.46 511	41	9.48 443	46	0.51 557	9.98 067	4	2
59	9.46 552	42	9.48 489	45	0.51 511	9.98 063	3	1
60	9.46 594		9.48 534		0.51 466	9.98 060		0
	Log Cos	d	Log Cot	c d	Log Tan	Log Sin	d	′

Prop. Parts

	48	47	46
1	4.8	4.7	4.6
2	9.6	9.4	9.2
3	14.4	14.1	13.8
4	19.2	18.8	18.4
5	24.0	23.5	23.0
6	28.8	28.2	27.6
7	33.6	32.9	32.2
8	38.4	37.6	36.8
9	43.2	42.3	41.4

	45	44	43
1	4.5	4.4	4.3
2	9.0	8.8	8.6
3	13.5	13.2	12.9
4	18.0	17.6	17.2
5	22.5	22.0	21.5
6	27.0	26.4	25.8
7	31.5	30.8	30.1
8	36.0	35.2	34.4
9	40.5	39.6	38.7

	42	41
1	4.2	4.1
2	8.4	8.2
3	12.6	12.3
4	16.8	16.4
5	21.0	20.5
6	25.2	24.6
7	29.4	28.7
8	33.6	32.8
9	37.8	36.9

	4	3
1	0.4	0.3
2	0.8	0.6
3	1.2	0.9
4	1.6	1.2
5	2.0	1.5
6	2.4	1.8
7	2.8	2.1
8	3.2	2.4
9	3.6	2.7

73°

17°

′	Log Sin	d	Log Tan	c d	Log Cot	Log Cos	d	
0	9.46 594	41	9.48 534	45	0.51 466	9.98 060	4	60
1	9.46 635	41	9.48 579	45	0.51 421	9.98 056	4	59
2	9.46 676	41	9.48 624	45	0.51 376	9.98 052	4	58
3	9.46 717	41	9.48 669	45	0.51 331	9.98 048	4	57
4	9.46 758	42	9.48 714	45	0.51 286	9.98 044	4	56
5	9.46 800	41	9.48 759	45	0.51 241	9.98 040	4	55
6	9.46 841	41	9.48 804	45	0.51 196	9.98 036	4	54
7	9.46 882	41	9.48 849	45	0.51 151	9.98 032	3	53
8	9.46 923	41	9.48 894	45	0.51 106	9.98 029	4	52
9	9.46 964	41	9.48 939	45	0.51 061	9.98 025	4	51
10	9.47 005	40	9.48 984	45	0.51 016	9.98 021	4	50
11	9.47 045	41	9.49 029	44	0.50 971	9.98 017	4	49
12	9.47 086	41	9.49 073	45	0.50 927	9.98 013	4	48
13	9.47 127	41	9.49 118	45	0.50 882	9.98 009	4	47
14	9.47 168	41	9.49 163	44	0.50 837	9.98 005	4	46
15	9.47 209	40	9.49 207	45	0.50 793	9.98 001	4	45
16	9.47 249	41	9.49 252	44	0.50 748	9.97 997	4	44
17	9.47 290	40	9.49 296	45	0.50 704	9.97 993	4	43
18	9.47 330	41	9.49 341	44	0.50 659	9.97 989	3	42
19	9.47 371	40	9.49 385	45	0.50 615	9.97 986	4	41
20	9.47 411	41	9.49 430	44	0.50 570	9.97 982	4	40
21	9.47 452	40	9.49 474	45	0.50 526	9.97 978	4	39
22	9.47 492	41	9.49 519	44	0.50 481	9.97 974	4	38
23	9.47 533	40	9.49 563	44	0.50 437	9.97 970	4	37
24	9.47 573	40	9.49 607	45	0.50 393	9.97 966	4	36
25	9.47 613	41	9.49 652	44	0.50 348	9.97 962	4	35
26	9.47 654	40	9.49 696	44	0.50 304	9.97 958	4	34
27	9.47 694	40	9.49 740	44	0.50 260	9.97 954	4	33
28	9.47 734	40	9.49 784	44	0.50 216	9.97 950	4	32
29	9.47 774	40	9.49 828	44	0.50 172	9.97 946	4	31
30	9.47 814	40	9.49 872	44	0.50 128	9.97 942	4	30
31	9.47 854	40	9.49 916	44	0.50 084	9.97 938	4	29
32	9.47 894	40	9.49 960	44	0.50 040	9.97 934	4	28
33	9.47 934	40	9.50 004	44	0.49 996	9.97 930	4	27
34	9.47 974	40	9.50 048	44	0.49 952	9.97 926	4	26
35	9.48 014	40	9.50 092	44	0.49 908	9.97 922	4	25
36	9.48 054	40	9.50 136	44	0.49 864	9.97 918	4	24
37	9.48 094	39	9.50 180	43	0.49 820	9.97 914	4	23
38	9.48 133	40	9.50 223	44	0.49 777	9.97 910	4	22
39	9.48 173	40	9.50 267	44	0.49 733	9.97 906	4	21
40	9.48 213	39	9.50 311	44	0.49 689	9.97 902	4	20
41	9.48 252	40	9.50 355	43	0.49 645	9.97 898	4	19
42	9.48 292	40	9.50 398	44	0.49 602	9.97 894	4	18
43	9.48 332	39	9.50 442	43	0.49 558	9.97 890	4	17
44	9.48 371	40	9.50 485	44	0.49 515	9.97 886	4	16
45	9.48 411	39	9.50 529	43	0.49 471	9.97 882	4	15
46	9.48 450	40	9.50 572	44	0.49 428	9.97 878	4	14
47	9.48 490	39	9.50 616	43	0.49 384	9.97 874	4	13
48	9.48 529	39	9.50 659	44	0.49 341	9.97 870	4	12
49	9.48 568	39	9.50 703	43	0.49 297	9.97 866	5	11
50	9.48 607	40	9.50 746	43	0.49 254	9.97 861	4	10
51	9.48 647	39	9.50 789	44	0.49 211	9.97 857	4	9
52	9.48 686	39	9.50 833	43	0.49 167	9.97 853	4	8
53	9.48 725	39	9.50 876	43	0.49 124	9.97 849	4	7
54	9.48 764	39	9.50 919	43	0.49 081	9.97 845	4	6
55	9.48 803	39	9.50 962	43	0.49 038	9.97 841	4	5
56	9.48 842	39	9.51 005	43	0.48 995	9.97 837	4	4
57	9.48 881	39	9.51 048	44	0.48 952	9.97 833	4	3
58	9.48 920	39	9.51 092	43	0.48 908	9.97 829	4	2
59	9.48 959	39	9.51 135	43	0.48 865	9.97 825	4	1
60	9.48 998		9.51 178		0.48 822	9.97 821		0
	Log Cos	d	Log Cot	c d	Log Tan	Log Sin	d	′

Prop. Parts

	45	44	43
1	4.5	4.4	4.3
2	9.0	8.8	8.6
3	13.5	13.2	12.9
4	18.0	17.6	17.2
5	22.5	22.0	21.5
6	27.0	26.4	25.8
7	31.5	30.8	30.1
8	36.0	35.2	34.4
9	40.5	39.6	38.7

	42	41	40
1	4.2	4.1	4.0
2	8.4	8.2	8.0
3	12.6	12.3	12.0
4	16.8	16.4	16.0
5	21.0	20.5	20.0
6	25.2	24.6	24.0
7	29.4	28.7	28.0
8	33.6	32.8	32.0
9	37.8	36.9	36.0

	39	5
1	3.9	0.5
2	7.8	1.0
3	11.7	1.5
4	15.6	2.0
5	19.5	2.5
6	23.4	3.0
7	27.3	3.5
8	31.2	4.0
9	35.1	4.5

	4	3
1	0.4	0.3
2	0.8	0.6
3	1.2	0.9
4	1.6	1.2
5	2.0	1.5
6	2.4	1.8
7	2.8	2.1
8	3.2	2.4
9	3.6	2.7

Prop. Parts

72°

Table II

18°

′	Log Sin	d	Log Tan	c d	Log Cot	Log Cos	d	
0	9.48 998	39	9.51 178	43	0.48 822	9.97 821	4	60
1	9.49 037	39	9.51 221	43	0.48 779	9.97 817	5	59
2	9.49 076	39	9.51 264	42	0.48 736	9.97 812	4	58
3	9.49 115	38	9.51 306	43	0.48 694	9.97 808	4	57
4	9.49 153	39	9.51 349	43	0.48 651	9.97 804	4	56
5	9.49 192	39	9.51 392	43	0.48 608	9.97 800	4	55
6	9.49 231	38	9.51 435	43	0.48 565	9.97 796	4	54
7	9.49 269	39	9.51 478	42	0.48 522	9.97 792	4	53
8	9.49 308	39	9.51 520	43	0.48 480	9.97 788	4	52
9	9.49 347	38	9.51 563	43	0.48 437	9.97 784	5	51
10	9.49 385	39	9.51 606	42	0.48 394	9.97 779	4	50
11	9.49 424	38	9.51 648	43	0.48 352	9.97 775	4	49
12	9.49 462	38	9.51 691	43	0.48 309	9.97 771	4	48
13	9.49 500	39	9.51 734	42	0.48 266	9.97 767	4	47
14	9.49 539	38	9.51 776	43	0.48 224	9.97 763	4	46
15	9.49 577	38	9.51 819	42	0.48 181	9.97 759	5	45
16	9.49 615	39	9.51 861	42	0.48 139	9.97 754	4	44
17	9.49 654	38	9.51 903	43	0.48 097	9.97 750	4	43
18	9.49 692	38	9.51 946	42	0.48 054	9.97 746	4	42
19	9.49 730	38	9.51 988	43	0.48 012	9.97 742	4	41
20	9.49 768	38	9.52 031	42	0.47 969	9.97 738	4	40
21	9.49 806	38	9.52 073	42	0.47 927	9.97 734	5	39
22	9.49 844	38	9.52 115	42	0.47 885	9.97 729	4	38
23	9.49 882	38	9.52 157	43	0.47 843	9.97 725	4	37
24	9.49 920	38	9.52 200	42	0.47 800	9.97 721	4	36
25	9.49 958	38	9.52 242	42	0.47 758	9.97 717	4	35
26	9.49 996	38	9.52 284	42	0.47 716	9.97 713	5	34
27	9.50 034	38	9.52 326	42	0.47 674	9.97 708	4	33
28	9.50 072	38	9.52 368	42	0.47 632	9.97 704	4	32
29	9.50 110	38	9.52 410	42	0.47 590	9.97 700	4	31
30	9.50 148	37	9.52 452	42	0.47 548	9.97 696	5	30
31	9.50 185	38	9.52 494	42	0.47 506	9.97 691	4	29
32	9.50 223	38	9.52 536	42	0.47 464	9.97 687	4	28
33	9.50 261	37	9.52 578	42	0.47 422	9.97 683	4	27
34	9.50 298	38	9.52 620	41	0.47 380	9.97 679	5	26
35	9.50 336	38	9.52 661	42	0.47 339	9.97 674	4	25
36	9.50 374	37	9.52 703	42	0.47 297	9.97 670	4	24
37	9.50 411	38	9.52 745	42	0.47 255	9.97 666	4	23
38	9.50 449	37	9.52 787	42	0.47 213	9.97 662	5	22
39	9.50 486	37	9.52 829	41	0.47 171	9.97 657	4	21
40	9.50 523	38	9.52 870	42	0.47 130	9.97 653	4	20
41	9.50 561	37	9.52 912	41	0.47 088	9.97 649	4	19
42	9.50 598	37	9.52 953	42	0.47 047	9.97 645	5	18
43	9.50 635	38	9.52 995	42	0.47 005	9.97 640	4	17
44	9.50 673	37	9.53 037	41	0.46 963	9.97 636	4	16
45	9.50 710	37	9.53 078	42	0.46 922	9.97 632	4	15
46	9.50 747	37	9.53 120	41	0.46 880	9.97 628	5	14
47	9.50 784	37	9.53 161	41	0.46 839	9.97 623	4	13
48	9.50 821	37	9.53 202	42	0.46 798	9.97 619	4	12
49	9.50 858	38	9.53 244	41	0.46 756	9.97 615	5	11
50	9.50 896	37	9.53 285	42	0.46 715	9.97 610	4	10
51	9.50 933	37	9.53 327	41	0.46 673	9.97 606	4	9
52	9.50 970	37	9.53 368	41	0.46 632	9.97 602	5	8
53	9.51 007	36	9.53 409	41	0.46 591	9.97 597	4	7
54	9.51 043	37	9.53 450	42	0.46 550	9.97 593	4	6
55	9.51 080	37	9.53 492	41	0.46 508	9.97 589	5	5
56	9.51 117	37	9.53 533	41	0.46 467	9.97 584	4	4
57	9.51 154	37	9.53 574	41	0.46 426	9.97 580	4	3
58	9.51 191	36	9.53 615	41	0.46 385	9.97 576	5	2
59	9.51 227	37	9.53 656	41	0.46 344	9.97 571	4	1
60	9.51 264		9.53 697		0.46 303	9.97 567		0
	Log Cos	d	Log Cot	c d	Log Tan	Log Sin	d	′

Prop. Parts

	43	42	41
1	4.3	4.2	4.1
2	8.6	8.4	8.2
3	12.9	12.6	12.3
4	17.2	16.8	16.4
5	21.5	21.0	20.5
6	25.8	25.2	24.6
7	30.1	29.4	28.7
8	34.4	33.6	32.8
9	38.7	37.8	36.9

	39	38	37
1	3.9	3.8	3.7
2	7.8	7.6	7.4
3	11.7	11.4	11.1
4	15.6	15.2	14.8
5	19.5	19.0	18.5
6	23.4	22.8	22.2
7	27.3	26.6	25.9
8	31.2	30.4	29.6
9	35.1	34.2	33.3

	36	5	4
1	3.6	0.5	0.4
2	7.2	1.0	0.8
3	10.8	1.5	1.2
4	14.4	2.0	1.6
5	18.0	2.5	2.0
6	21.6	3.0	2.4
7	25.2	3.5	2.8
8	28.8	4.0	3.2
9	32.4	4.5	3.6

Prop. Parts

19°

'	Log Sin	d	Log Tan	c d	Log Cot	Log Cos	d	
0	9.51 264	37	9.53 697	41	0.46 303	9.97 567	4	60
1	9.51 301	37	9.53 738	41	0.46 262	9.97 563	5	59
2	9.51 338	36	9.53 779	41	0.46 221	9.97 558	4	58
3	9.51 374	37	9.53 820	41	0.46 180	9.97 554	4	57
4	9.51 411	36	9.53 861	41	0.46 139	9.97 550	5	56
5	9.51 447	37	9.53 902	41	0.46 098	9.97 545	4	55
6	9.51 484	36	9.53 943	41	0.46 057	9.97 541	5	54
7	9.51 520	37	9.53 984	41	0.46 016	9.97 536	4	53
8	9.51 557	36	9.54 025	40	0.45 975	9.97 532	4	52
9	9.51 593	36	9.54 065	41	0.45 935	9.97 528	5	51
10	9.51 629	37	9.54 106	41	0.45 894	9.97 523	4	50
11	9.51 666	36	9.54 147	40	0.45 853	9.97 519	4	49
12	9.51 702	36	9.54 187	41	0.45 813	9.97 515	5	48
13	9.51 738	36	9.54 228	41	0.45 772	9.97 510	4	47
14	9.51 774	37	9.54 269	40	0.45 731	9.97 506	5	46
15	9.51 811	36	9.54 309	41	0.45 691	9.97 501	4	45
16	9.51 847	36	9.54 350	40	0.45 650	9.97 497	5	44
17	9.51 883	36	9.54 390	41	0.45 610	9.97 492	4	43
18	9.51 919	36	9.54 431	40	0.45 569	9.97 488	4	42
19	9.51 955	36	9.54 471	41	0.45 529	9.97 484	5	41
20	9.51 991	36	9.54 512	40	0.45 488	9.97 479	4	40
21	9.52 027	36	9.54 552	41	0.45 448	9.97 475	5	39
22	9.52 063	36	9.54 593	40	0.45 407	9.97 470	4	38
23	9.52 099	36	9.54 633	40	0.45 367	9.97 466	5	37
24	9.52 135	36	9.54 673	41	0.45 327	9.97 461	4	36
25	9.52 171	36	9.54 714	40	0.45 286	9.97 457	4	35
26	9.52 207	35	9.54 754	40	0.45 246	9.97 453	5	34
27	9.52 242	36	9.54 794	41	0.45 206	9.97 448	4	33
28	9.52 278	36	9.54 835	40	0.45 165	9.97 444	5	32
29	9.52 314	36	9.54 875	40	0.45 125	9.97 439	4	31
30	9.52 350	35	9.54 915	40	0.45 085	9.97 435	5	30
31	9.52 385	36	9.54 955	40	0.45 045	9.97 430	4	29
32	9.52 421	35	9.54 995	40	0.45 005	9.97 426	5	28
33	9.52 456	36	9.55 035	40	0.44 965	9.97 421	4	27
34	9.52 492	35	9.55 075	40	0.44 925	9.97 417	5	26
35	9.52 527	36	9.55 115	40	0.44 885	9.97 412	4	25
36	9.52 563	35	9.55 155	40	0.44 845	9.97 408	5	24
37	9.52 598	36	9.55 195	40	0.44 805	9.97 403	4	23
38	9.52 634	35	9.55 235	40	0.44 765	9.97 399	5	22
39	9.52 669	36	9.55 275	40	0.44 725	9.97 394	4	21
40	9.52 705	35	9.55 315	40	0.44 685	9.97 390	5	20
41	9.52 740	35	9.55 355	40	0.44 645	9.97 385	4	19
42	9.52 775	36	9.55 395	39	0.44 605	9.97 381	5	18
43	9.52 811	35	9.55 434	40	0.44 566	9.97 376	4	17
44	9.52 846	35	9.55 474	40	0.44 526	9.97 372	5	16
45	9.52 881	35	9.55 514	40	0.44 486	9.97 367	4	15
46	9.52 916	35	9.55 554	39	0.44 446	9.97 363	5	14
47	9.52 951	35	9.55 593	40	0.44 407	9.97 358	5	13
48	9.52 986	35	9.55 633	40	0.44 367	9.97 353	4	12
49	9.53 021	35	9.55 673	39	0.44 327	9.97 349	5	11
50	9.53 056	36	9.55 712	40	0.44 288	9.97 344	4	10
51	9.53 092	34	9.55 752	39	0.44 248	9.97 340	5	9
52	9.53 126	35	9.55 791	40	0.44 209	9.97 335	4	8
53	9.53 161	35	9.55 831	39	0.44 169	9.97 331	5	7
54	9.53 196	35	9.55 870	40	0.44 130	9.97 326	4	6
55	9.53 231	35	9.55 910	39	0.44 090	9.97 322	5	5
56	9.53 266	35	9.55 949	40	0.44 051	9.97 317	5	4
57	9.53 301	35	9.55 989	39	0.44 011	9.97 312	4	3
58	9.53 336	34	9.56 028	39	0.43 972	9.97 308	5	2
59	9.53 370	35	9.56 067	40	0.43 933	9.97 303	4	1
60	9.53 405		9.56 107		0.43 893	9.97 299		0
	Log Cos	d	Log Cot	c d	Log Tan	Log Sin	d	'

Prop. Parts

	41	40	39
1	4.1	4.0	3.9
2	8.2	8.0	7.8
3	12.3	12.0	11.7
4	16.4	16.0	15.6
5	20.5	20.0	19.5
6	24.6	24.0	23.4
7	28.7	28.0	27.3
8	32.8	32.0	31.2
9	36.9	36.0	35.1

	37	36	35
1	3.7	3.6	3.5
2	7.4	7.2	7.0
3	11.1	10.8	10.5
4	14.8	14.4	14.0
5	18.5	18.0	17.5
6	22.2	21.6	21.0
7	25.9	25.2	24.5
8	29.6	28.8	28.0
9	33.3	32.4	31.5

	34	5	4
1	3.4	0.5	0.4
2	6.8	1.0	0.8
3	10.2	1.5	1.2
4	13.6	2.0	1.6
5	17.0	2.5	2.0
6	20.4	3.0	2.4
7	23.8	3.5	2.8
8	27.2	4.0	3.2
9	30.6	4.5	3.6

Prop. Parts

70°

Table II

20°

′	Log Sin	d	Log Tan	c d	Log Cot	Log Cos	d			Prop. Parts		
0	9.53 405	35	9.56 107	39	0.43 893	9.97 299	5	60				
1	9.53 440	35	9.56 146	39	0.43 854	9.97 294	5	59				
2	9.53 475	34	9.56 185	39	0.43 815	9.97 289	4	58				
3	9.53 509	35	9.56 224	40	0.43 776	9.97 285	5	57				
4	9.53 544	34	9.56 264	39	0.43 736	9.97 280	4	56				
5	9.53 578	35	9.56 303	39	0.43 697	9.97 276	5	55				
6	9.53 613	34	9.56 342	39	0.43 658	9.97 271	5	54				
7	9.53 647	35	9.56 381	39	0.43 619	9.97 266	4	53				
8	9.53 682	34	9.56 420	39	0.43 580	9.97 262	5	52				
9	9.53 716	35	9.56 459	39	0.43 541	9.97 257	5	51		**40**	**39**	**38**
10	9.53 751	34	9.56 498	39	0.43 502	9.97 252	4	50	1	4.0	3.9	3.8
11	9.53 785	34	9.56 537	39	0.43 463	9.97 248	5	49	2	8.0	7.8	7.6
12	9.53 819	35	9.56 576	39	0.43 424	9.97 243	5	48	3	12.0	11.7	11.4
13	9.53 854	34	9.56 615	39	0.43 385	9.97 238	4	47	4	16.0	15.6	15.2
14	9.53 888	34	9.56 654	39	0.43 346	9.97 234	5	46	5	20.0	19.5	19.0
15	9.53 922	35	9.56 693	39	0.43 307	9.97 229	5	45	6	24.0	23.4	22.8
16	9.53 957	34	9.56 732	39	0.43 268	9.97 224	4	44	7	28.0	27.3	26.6
17	9.53 991	34	9.56 771	39	0.43 229	9.97 220	5	43	8	32.0	31.2	30.4
18	9.54 025	34	9.56 810	39	0.43 190	9.97 215	5	42	9	36.0	35.1	34.2
19	9.54 059	34	9.56 849	38	0.43 151	9.97 210	4	41				
20	9.54 093	34	9.56 887	39	0.43 113	9.97 206	5	40				
21	9.54 127	34	9.56 926	39	0.43 074	9.97 201	5	39				
22	9.54 161	34	9.56 965	39	0.43 035	9.97 196	4	38				
23	9.54 195	34	9.57 004	38	0.42 996	9.97 192	5	37				
24	9.54 229	34	9.57 042	39	0.42 958	9.97 187	5	36				
25	9.54 263	34	9.57 081	39	0.42 919	9.97 182	4	35		**37**	**35**	**34**
26	9.54 297	34	9.57 120	38	0.42 880	9.97 178	5	34	1	3.7	3.5	3.4
27	9.54 331	34	9.57 158	39	0.42 842	9.97 173	5	33	2	7.4	7.0	6.8
28	9.54 365	34	9.57 197	38	0.42 803	9.97 168	5	32	3	11.1	10.5	10.2
29	9.54 399	34	9.57 235	39	0.42 765	9.97 163	4	31	4	14.8	14.0	13.6
30	9.54 433	33	9.57 274	38	0.42 726	9.97 159	5	30	5	18.5	17.5	17.0
31	9.54 466	34	9.57 312	39	0.42 688	9.97 154	5	29	6	22.2	21.0	20.4
32	9.54 500	34	9.57 351	38	0.42 649	9.97 149	4	28	7	25.9	24.5	23.8
33	9.54 534	33	9.57 389	39	0.42 611	9.97 145	5	27	8	29.6	28.0	27.2
34	9.54 567	34	9.57 428	38	0.42 572	9.97 140	5	26	9	33.3	31.5	30.6
35	9.54 601	34	9.57 466	38	0.42 534	9.97 135	5	25				
36	9.54 635	33	9.57 504	39	0.42 496	9.97 130	4	24				
37	9.54 668	34	9.57 543	38	0.42 457	9.97 126	5	23				
38	9.54 702	33	9.57 581	38	0.42 419	9.97 121	5	22				
39	9.54 735	34	9.57 619	39	0.42 381	9.97 116	5	21				
40	9.54 769	33	9.57 658	38	0.42 342	9.97 111	4	20				
41	9.54 802	34	9.57 696	38	0.42 304	9.97 107	5	19				
42	9.54 836	33	9.57 734	38	0.42 266	9.97 102	5	18		**33**	**5**	**4**
43	9.54 869	34	9.57 772	38	0.42 228	9.97 097	5	17	1	3.3	0.5	0.4
44	9.54 903	33	9.57 810	39	0.42 190	9.97 092	5	16	2	6.6	1.0	0.8
45	9.54 936	33	9.57 849	38	0.42 151	9.97 087	4	15	3	9.9	1.5	1.2
46	9.54 969	34	9.57 887	38	0.42 113	9.97 083	5	14	4	13.2	2.0	1.6
47	9.55 003	33	9.57 925	38	0.42 075	9.97 078	5	13	5	16.5	2.5	2.0
48	9.55 036	33	9.57 963	38	0.42 037	9.97 073	5	12	6	19.8	3.0	2.4
49	9.55 069	33	9.58 001	38	0.41 999	9.97 068	5	11	7	23.1	3.5	2.8
50	9.55 102	34	9.58 039	38	0.41 961	9.97 063	4	10	8	26.4	4.0	3.2
51	9.55 136	33	9.58 077	38	0.41 923	9.97 059	5	9	9	29.7	4.5	3.6
52	9.55 169	33	9.58 115	38	0.41 885	9.97 054	5	8				
53	9.55 202	33	9.58 153	38	0.41 847	9.97 049	5	7				
54	9.55 235	33	9.58 191	38	0.41 809	9.97 044	5	6				
55	9.55 268	33	9.58 229	38	0.41 771	9.97 039	4	5				
56	9.55 301	33	9.58 267	37	0.41 733	9.97 035	5	4				
57	9.55 334	33	9.58 304	38	0.41 696	9.97 030	5	3				
58	9.55 367	33	9.58 342	38	0.41 658	9.97 025	5	2				
59	9.55 400	33	9.58 380	38	0.41 620	9.97 020	5	1				
60	9.55 433		9.58 418		0.41 582	9.97 015		0				
	Log Cos	d	Log Cot	c d	Log Tan	Log Sin	d	′		Prop. Parts		

69°

21°

′	Log Sin	d	Log Tan	c d	Log Cot	Log Cos	d		Prop. Parts		
0	9.55 433	33	9.58 418	37	0.41 582	9.97 015	5	60			
1	9.55 466	33	9.58 455	38	0.41 545	9.97 010	5	59			
2	9.55 499	33	9.58 493	38	0.41 507	9.97 005	4	58			
3	9.55 532	32	9.58 531	38	0.41 469	9.97 001	5	57			
4	9.55 564	33	9.58 569	37	0.41 431	9.96 996	5	56			
5	9.55 597	33	9.58 606	38	0.41 394	9.96 991	5	55			
6	9.55 630	33	9.58 644	37	0.41 356	9.96 986	5	54			
7	9.55 663	32	9.58 681	38	0.41 319	9.96 981	5	53			
8	9.55 695	33	9.58 719	38	0.41 281	9.96 976	5	52		38	37	36
9	9.55 728	33	9.58 757	37	0.41 243	9.96 971	5	51			
10	9.55 761	32	9.58 794	38	0.41 206	9.96 966	4	50	1	3.8	3.7	3.6
11	9.55 793	33	9.58 832	37	0.41 168	9.96 962	5	49	2	7.6	7.4	7.2
12	9.55 826	32	9.58 869	38	0.41 131	9.96 957	5	48	3	11.4	11.1	10.8
13	9.55 858	33	9.58 907	37	0.41 093	9.96 952	5	47	4	15.2	14.8	14.4
14	9.55 891	32	9.58 944	37	0.41 056	9.96 947	5	46	5	19.0	18.5	18.0
15	9.55 923	33	9.58 981	38	0.41 019	9.96 942	5	45	6	22.8	22.2	21.6
16	9.55 956	32	9.59 019	37	0.40 981	9.96 937	5	44	7	26.6	25.9	25.2
17	9.55 988	33	9.59 056	38	0.40 944	9.96 932	5	43	8	30.4	29.6	28.8
18	9.56 021	32	9.59 094	37	0.40 906	9.96 927	5	42	9	34.2	33.3	32.4
19	9.56 053	32	9.59 131	37	0.40 869	9.96 922	5	41			
20	9.56 085	33	9.59 168	37	0.40 832	9.96 917	5	40			
21	9.56 118	32	9.59 205	38	0.40 795	9.96 912	5	39			
22	9.56 150	32	9.59 243	37	0.40 757	9.96 907	4	38			
23	9.56 182	33	9.59 280	37	0.40 720	9.96 903	5	37			
24	9.56 215	32	9.59 317	37	0.40 683	9.96 898	5	36			
25	9.56 247	32	9.59 354	37	0.40 646	9.96 893	5	35		33	32	31
26	9.56 279	32	9.59 391	38	0.40 609	9.96 888	5	34			
27	9.56 311	32	9.59 429	37	0.40 571	9.96 883	5	33	1	3.3	3.2	3.1
28	9.56 343	32	9.59 466	37	0.40 534	9.96 878	5	32	2	6.6	6.4	6.2
29	9.56 375	33	9.59 503	37	0.40 497	9.96 873	5	31	3	9.9	9.6	9.3
30	9.56 408	32	9.59 540	37	0.40 460	9.96 868	5	30	4	13.2	12.8	12.4
31	9.56 440	32	9.59 577	37	0.40 423	9.96 863	5	29	5	16.5	16.0	15.5
32	9.56 472	32	9.59 614	37	0.40 386	9.96 858	5	28	6	19.8	19.2	18.6
33	9.56 504	32	9.59 651	37	0.40 349	9.96 853	5	27	7	23.1	22.4	21.7
34	9.56 536	32	9.59 688	37	0.40 312	9.96 848	5	26	8	26.4	25.6	24.8
35	9.56 568	31	9.59 725	37	0.40 275	9.96 843	5	25	9	29.7	28.8	27.9
36	9.56 599	32	9.59 762	37	0.40 238	9.96 838	5	24			
37	9.56 631	32	9.59 799	36	0.40 201	9.96 833	5	23			
38	9.56 663	32	9.59 835	37	0.40 165	9.96 828	5	22			
39	9.56 695	32	9.59 872	37	0.40 128	9.96 823	5	21			
40	9.56 727	32	9.59 909	37	0.40 091	9.96 818	5	20			
41	9.56 759	31	9.59 946	37	0.40 054	9.96 813	5	19			
42	9.56 790	32	9.59 983	36	0.40 017	9.96 808	5	18		6	5	4
43	9.56 822	32	9.60 019	37	0.39 981	9.96 803	5	17			
44	9.56 854	32	9.60 056	37	0.39 944	9.96 798	5	16	1	0.6	0.5	0.4
45	9.56 886	31	9.60 093	37	0.39 907	9.96 793	5	15	2	1.2	1.0	0.8
46	9.56 917	32	9.60 130	36	0.39 870	9.96 788	5	14	3	1.8	1.5	1.2
47	9.56 949	31	9.60 166	37	0.39 834	9.96 783	5	13	4	2.4	2.0	1.6
48	9.56 980	32	9.60 203	37	0.39 797	9.96 778	6	12	5	3.0	2.5	2.0
49	9.57 012	32	9.60 240	36	0.39 760	9.96 772	5	11	6	3.6	3.0	2.4
50	9.57 044	31	9.60 276	37	0.39 724	9.96 767	5	10	7	4.2	3.5	2.8
51	9.57 075	32	9.60 313	36	0.39 687	9.96 762	5	9	8	4.8	4.0	3.2
52	9.57 107	31	9.60 349	37	0.39 651	9.96 757	5	8	9	5.4	4.5	3.6
53	9.57 138	31	9.60 386	36	0.39 614	9.96 752	5	7			
54	9.57 169	32	9.60 422	37	0.39 578	9.96 747	5	6			
55	9.57 201	31	9.60 459	36	0.39 541	9.96 742	5	5			
56	9.57 232	32	9.60 495	37	0.39 505	9.96 737	5	4			
57	9.57 264	31	9.60 532	36	0.39 468	9.96 732	5	3			
58	9.57 295	31	9.60 568	37	0.39 432	9.96 727	5	2			
59	9.57 326	32	9.60 605	36	0.39 395	9.96 722	5	1			
60	9.57 358		9.60 641		0.39 359	9.96 717		0			
	Log Cos	d	Log Cot	c d	Log Tan	Log Sin	d	′	Prop. Parts		

68°

22°

′	Log Sin	d	Log Tan	cd	Log Cot	Log Cos	d	
0	9.57 358	31	9.60 641	36	0.39 359	9.96 717	6	60
1	9.57 389	31	9.60 677	37	0.39 323	9.96 711	5	59
2	9.57 420	31	9.60 714	36	0.39 286	9.96 706	5	58
3	9.57 451	31	9.60 750	36	0.39 250	9.96 701	5	57
4	9.57 482	32	9.60 786	37	0.39 214	9.96 696	5	56
5	9.57 514	31	9.60 823	36	0.39 177	9.96 691	5	55
6	9.57 545	31	9.60 859	36	0.39 141	9.96 686	5	54
7	9.57 576	31	9.60 895	36	0.39 105	9.96 681	5	53
8	9.57 607	31	9.60 931	36	0.39 069	9.96 676	6	52
9	9.57 638	31	9.60 967	37	0.39 033	9.96 670	5	51
10	9.57 669	31	9.61 004	36	0.38 996	9.96 665	5	50
11	9.57 700	31	9.61 040	36	0.38 960	9.96 660	5	49
12	9.57 731	31	9.61 076	36	0.38 924	9.96 655	5	48
13	9.57 762	31	9.61 112	36	0.38 888	9.96 650	5	47
14	9.57 793	31	9.61 148	36	0.38 852	9.96 645	5	46
15	9.57 824	31	9.61 184	36	0.38 816	9.96 640	6	45
16	9.57 855	30	9.61 220	36	0.38 780	9.96 634	5	44
17	9.57 885	31	9.61 256	36	0.38 744	9.96 629	5	43
18	9.57 916	31	9.61 292	36	0.38 708	9.96 624	5	42
19	9.57 947	31	9.61 328	36	0.38 672	9.96 619	5	41
20	9.57 978	30	9.61 364	36	0.38 636	9.96 614	6	40
21	9.58 008	31	9.61 400	36	0.38 600	9.96 608	5	39
22	9.58 039	31	9.61 436	36	0.38 564	9.96 603	5	38
23	9.58 070	31	9.61 472	36	0.38 528	9.96 598	5	37
24	9.58 101	30	9.61 508	36	0.38 492	9.96 593	5	36
25	9.58 131	31	9.61 544	35	0.38 456	9.96 588	6	35
26	9.58 162	30	9.61 579	36	0.38 421	9.96 582	5	34
27	9.58 192	31	9.61 615	36	0.38 385	9.96 577	5	33
28	9.58 223	30	9.61 651	36	0.38 349	9.96 572	5	32
29	9.58 253	31	9.61 687	35	0.38 313	9.96 567	5	31
30	9.58 284	30	9.61 722	36	0.38 278	9.96 562	6	30
31	9.58 314	31	9.61 758	36	0.38 242	9.96 556	5	29
32	9.58 345	30	9.61 794	36	0.38 206	9.96 551	5	28
33	9.58 375	31	9.61 830	35	0.38 170	9.96 546	5	27
34	9.58 406	30	9.61 865	36	0.38 135	9.96 541	6	26
35	9.58 436	31	9.61 901	35	0.38 099	9.96 535	5	25
36	9.58 467	30	9.61 936	36	0.38 064	9.96 530	5	24
37	9.58 497	30	9.61 972	36	0.38 028	9.96 525	5	23
38	9.58 527	30	9.62 008	35	0.37 992	9.96 520	6	22
39	9.58 557	31	9.62 043	36	0.37 957	9.96 514	5	21
40	9.58 588	30	9.62 079	35	0.37 921	9.96 509	5	20
41	9.58 618	30	9.62 114	36	0.37 886	9.96 504	5	19
42	9.58 648	30	9.62 150	35	0.37 850	9.96 498	6	18
43	9.58 678	31	9.62 185	36	0.37 815	9.96 493	5	17
44	9.58 709	30	9.62 221	35	0.37 779	9.96 488	5	16
45	9.58 739	30	9.62 256	36	0.37 744	9.96 483	6	15
46	9.58 769	30	9.62 292	35	0.37 708	9.96 477	5	14
47	9.58 799	30	9.62 327	35	0.37 673	9.96 472	5	13
48	9.58 829	30	9.62 362	36	0.37 638	9.96 467	6	12
49	9.58 859	30	9.62 398	35	0.37 602	9.96 461	5	11
50	9.58 889	30	9.62 433	35	0.37 567	9.96 456	5	10
51	9.58 919	30	9.62 468	36	0.37 532	9.96 451	6	9
52	9.58 949	30	9.62 504	35	0.37 496	9.96 445	5	8
53	9.58 979	30	9.62 539	35	0.37 461	9.96 440	5	7
54	9.59 009	30	9.62 574	35	0.37 426	9.96 435	6	6
55	9.59 039	30	9.62 609	36	0.37 391	9.96 429	5	5
56	9.59 069	29	9.62 645	35	0.37 355	9.96 424	5	4
57	9.59 098	30	9.62 680	35	0.37 320	9.96 419	6	3
58	9.59 128	30	9.62 715	35	0.37 285	9.96 413	5	2
59	9.59 158	30	9.62 750	35	0.37 250	9.96 408	5	1
60	9.59 188		9.62 785		0.37 215	9.96 403		0
	Log Cos	d	Log Cot	cd	Log Tan	Log Sin	d	′

Prop. Parts

	37	36	35
1	3.7	3.6	3.5
2	7.4	7.2	7.0
3	11.1	10.8	10.5
4	14.8	14.4	14.0
5	18.5	18.0	17.5
6	22.2	21.6	21.0
7	25.9	25.2	24.5
8	29.6	28.8	28.0
9	33.3	32.4	31.5

	32	31	30
1	3.2	3.1	3.0
2	6.4	6.2	6.0
3	9.6	9.3	9.0
4	12.8	12.4	12.0
5	16.0	15.5	15.0
6	19.2	18.6	18.0
7	22.4	21.7	21.0
8	25.6	24.8	24.0
9	28.8	27.9	27.0

	29	6	5
1	2.9	0.6	0.5
2	5.8	1.2	1.0
3	8.7	1.8	1.5
4	11.6	2.4	2.0
5	14.5	3.0	2.5
6	17.4	3.6	3.0
7	20.3	4.2	3.5
8	23.2	4.8	4.0
9	26.1	5.4	4.5

67°

23°

′	Log Sin	d	Log Tan	cd	Log Cot	Log Cos	d		Prop. Parts
0	9.59 188	30	9.62 785	35	0.37 215	9.96 403	6	60	
1	9.59 218	29	9.62 820	35	0.37 180	9.96 397	5	59	
2	9.59 247	30	9.62 855	35	0.37 145	9.96 392	5	58	
3	9.59 277	30	9.62 890	36	0.37 110	9.96 387	6	57	
4	9.59 307	29	9.62 926	35	0.37 074	9.96 381	5	56	
5	9.59 336	30	9.62 961	35	0.37 039	9.96 376	6	55	
6	9.59 366	30	9.62 996	35	0.37 004	9.96 370	5	54	
7	9.59 396	29	9.63 031	35	0.36 969	9.96 365	5	53	
8	9.59 425	30	9.63 066	35	0.36 934	9.96 360	6	52	
9	9.59 455	29	9.63 101	34	0.36 899	9.96 354	5	51	
10	9.59 484	30	9.63 135	35	0.36 865	9.96 349	6	50	
11	9.59 514	29	9.63 170	35	0.36 830	9.96 343	5	49	
12	9.59 543	30	9.63 205	35	0.36 795	9.96 338	5	48	
13	9.59 573	29	9.63 240	35	0.36 760	9.96 333	6	47	
14	9.59 602	30	9.63 275	35	0.36 725	9.96 327	5	46	
15	9.59 632	29	9.63 310	35	0.36 690	9.96 322	6	45	
16	9.59 661	29	9.63 345	34	0.36 655	9.96 316	5	44	
17	9.59 690	30	9.63 379	35	0.36 621	9.96 311	6	43	
18	9.59 720	29	9.63 414	35	0.36 586	9.96 305	5	42	
19	9.59 749	29	9.63 449	35	0.36 551	9.96 300	6	41	
20	9.59 778	30	9.63 484	35	0.36 516	9.96 294	5	40	
21	9.59 808	29	9.63 519	34	0.36 481	9.96 289	5	39	
22	9.59 837	29	9.63 553	35	0.36 447	9.96 284	6	38	
23	9.59 866	29	9.63 588	35	0.36 412	9.96 278	5	37	
24	9.59 895	29	9.63 623	34	0.36 377	9.96 273	6	36	
25	9.59 924	30	9.63 657	35	0.36 343	9.96 267	5	35	
26	9.59 954	29	9.63 692	34	0.36 308	9.96 262	6	34	
27	9.59 983	29	9.63 726	35	0.36 274	9.96 256	5	33	
28	9.60 012	29	9.63 761	35	0.36 239	9.96 251	6	32	
29	9.60 041	29	9.63 796	34	0.36 204	9.96 245	5	31	
30	9.60 070	29	9.63 830	35	0.36 170	9.96 240	6	30	
31	9.60 099	29	9.63 865	34	0.36 135	9.96 234	5	29	
32	9.60 128	29	9.63 899	35	0.36 101	9.96 229	6	28	
33	9.60 157	29	9.63 934	34	0.36 066	9.96 223	5	27	
34	9.60 186	29	9.63 968	35	0.36 032	9.96 218	6	26	
35	9.60 215	29	9.64 003	34	0.35 997	9.96 212	5	25	
36	9.60 244	29	9.64 037	35	0.35 963	9.96 207	6	24	
37	9.60 273	29	9.64 072	34	0.35 928	9.96 201	5	23	
38	9.60 302	29	9.64 106	34	0.35 894	9.96 196	6	22	
39	9.60 331	28	9.64 140	35	0.35 860	9.96 190	5	21	
40	9.60 359	29	9.64 175	34	0.35 825	9.96 185	6	20	
41	9.60 388	29	9.64 209	34	0.35 791	9.96 179	5	19	
42	9.60 417	29	9.64 243	35	0.35 757	9.96 174	6	18	
43	9.60 446	28	9.64 278	34	0.35 722	9.96 168	6	17	
44	9.60 474	29	9.64 312	34	0.35 688	9.96 162	5	16	
45	9.60 503	29	9.64 346	35	0.35 654	9.96 157	6	15	
46	9.60 532	29	9.64 381	34	0.35 619	9.96 151	5	14	
47	9.60 561	28	9.64 415	34	0.35 585	9.96 146	6	13	
48	9.60 589	29	9.64 449	34	0.35 551	9.96 140	5	12	
49	9.60 618	28	9.64 483	34	0.35 517	9.96 135	6	11	
50	9.60 646	29	9.64 517	35	0.35 483	9.96 129	6	10	
51	9.60 675	29	9.64 552	34	0.35 448	9.96 123	5	9	
52	9.60 704	28	9.64 586	34	0.35 414	9.96 118	6	8	
53	9.60 732	29	9.64 620	34	0.35 380	9.96 112	5	7	
54	9.60 761	28	9.64 654	34	0.35 346	9.96 107	6	6	
55	9.60 789	29	9.64 688	34	0.35 312	9.96 101	6	5	
56	9.60 818	28	9.64 722	34	0.35 278	9.96 095	5	4	
57	9.60 846	29	9.64 756	34	0.35 244	9.96 090	6	3	
58	9.60 875	28	9.64 790	34	0.35 210	9.96 084	5	2	
59	9.60 903	28	9.64 824	34	0.35 176	9.96 079	6	1	
60	9.60 931		9.64 858		0.35 142	9.96 073		0	
	Log Cos	d	Log Cot	cd	Log Tan	Log Sin	d	′	Prop. Parts

Prop. Parts

	36	35	34
1	3.6	3.5	3.4
2	7.2	7.0	6.8
3	10.8	10.5	10.2
4	14.4	14.0	13.6
5	18.0	17.5	17.0
6	21.6	21.0	20.4
7	25.2	24.5	23.8
8	28.8	28.0	27.2
9	32.4	31.5	30.6

	30	29	28
1	3.0	2.9	2.8
2	6.0	5.8	5.6
3	9.0	8.7	8.4
4	12.0	11.6	11.2
5	15.0	14.5	14.0
6	18.0	17.4	16.8
7	21.0	20.3	19.6
8	24.0	23.2	22.4
9	27.0	26.1	25.2

	6	5
1	0.6	0.5
2	1.2	1.0
3	1.8	1.5
4	2.4	2.0
5	3.0	2.5
6	3.6	3.0
7	4.2	3.5
8	4.8	4.0
9	5.4	4.5

24°

′	Log Sin	d	Log Tan	cd	Log Cot	Log Cos	d		Prop. Parts		
0	9.60 931	29	9.64 858	34	0.35 142	9.96 073	6	60			
1	9.60 960	28	9.64 892	34	0.35 108	9.96 067	5	59			
2	9.60 988	28	9.64 926	34	0.35 074	9.96 062	6	58			
3	9.61 016	29	9.64 960	34	0.35 040	9.96 056	6	57			
4	9.61 045	28	9.64 994	34	0.35 006	9.96 050	5	56			
5	9.61 073	28	9.65 028	34	0.34 972	9.96 045	6	55			
6	9.61 101	28	9.65 062	34	0.34 938	9.96 039	5	54			
7	9.61 129	29	9.65 096	34	0.34 904	9.96 034	6	53			
8	9.61 158	28	9.65 130	34	0.34 870	9.96 028	6	52			
9	9.61 186	28	9.65 164	33	0.34 836	9.96 022	5	51	**34**	**33**	**29**
10	9.61 214	28	9.65 197	34	0.34 803	9.96 017	6	50	1 3.4	3.3	2.9
11	9.61 242	28	9.65 231	34	0.34 769	9.96 011	6	49	2 6.8	6.6	5.8
12	9.61 270	28	9.65 265	34	0.34 735	9.96 005	5	48	3 10.2	9.9	8.7
13	9.61 298	28	9.65 299	34	0.34 701	9.96 000	6	47	4 13.6	13.2	11.6
14	9.61 326	28	9.65 333	33	0.34 667	9.95 994	6	46	5 17.0	16.5	14.5
15	9.61 354	28	9.65 366	34	0.34 634	9.95 988	6	45	6 20.4	19.8	17.4
16	9.61 382	29	9.65 400	34	0.34 600	9.95 982	5	44	7 23.8	23.1	20.3
17	9.61 411	27	9.65 434	33	0.34 566	9.95 977	6	43	8 27.2	26.4	23.2
18	9.61 438	28	9.65 467	34	0.34 533	9.95 971	6	42	9 30.6	29.7	26.1
19	9.61 466	28	9.65 501	34	0.34 499	9.95 965	5	41			
20	9.61 494	28	9.65 535	33	0.34 465	9.95 960	6	40			
21	9.61 522	28	9.65 568	34	0.34 432	9.95 954	6	39			
22	9.61 550	28	9.65 602	34	0.34 398	9.95 948	6	38			
23	9.61 578	28	9.65 636	33	0.34 364	9.95 942	5	37			
24	9.61 606	28	9.65 669	34	0.34 331	9.95 937	6	36			
25	9.61 634	28	9.65 703	33	0.34 297	9.95 931	6	35		**28**	**27**
26	9.61 662	27	9.65 736	34	0.34 264	9.95 925	5	34	1	2.8	2.7
27	9.61 689	28	9.65 770	33	0.34 230	9.95 920	6	33	2	5.6	5.4
28	9.61 717	28	9.65 803	34	0.34 197	9.95 914	6	32	3	8.4	8.1
29	9.61 745	28	9.65 837	33	0.34 163	9.95 908	6	31	4	11.2	10.8
30	9.61 773	27	9.65 870	34	0.34 130	9.95 902	5	30	5	14.0	13.5
31	9.61 800	28	9.65 904	33	0.34 096	9.95 897	6	29	6	16.8	16.2
32	9.61 828	28	9.65 937	34	0.34 063	9.95 891	6	28	7	19.6	18.9
33	9.61 856	27	9.65 971	33	0.34 029	9.95 885	6	27	8	22.4	21.6
34	9.61 883	28	9.66 004	34	0.33 996	9.95 879	6	26	9	25.2	24.3
35	9.61 911	28	9.66 038	33	0.33 962	9.95 873	5	25			
36	9.61 939	27	9.66 071	33	0.33 929	9.95 868	6	24			
37	9.61 966	28	9.66 104	34	0.33 896	9.95 862	6	23			
38	9.61 994	27	9.66 138	33	0.33 862	9.95 856	6	22			
39	9.62 021	28	9.66 171	33	0.33 829	9.95 850	6	21			
40	9.62 049	27	9.66 204	34	0.33 796	9.95 844	5	20			
41	9.62 076	28	9.66 238	33	0.33 762	9.95 839	6	19			
42	9.62 104	27	9.66 271	33	0.33 729	9.95 833	6	18		**6**	**5**
43	9.62 131	28	9.66 304	33	0.33 696	9.95 827	6	17	1	0.6	0.5
44	9.62 159	27	9.66 337	34	0.33 663	9.95 821	6	16	2	1.2	1.0
45	9.62 186	28	9.66 371	33	0.33 629	9.95 815	5	15	3	1.8	1.5
46	9.62 214	27	9.66 404	33	0.33 596	9.95 810	6	14	4	2.4	2.0
47	9.62 241	27	9.66 437	33	0.33 563	9.95 804	6	13	5	3.0	2.5
48	9.62 268	28	9.66 470	33	0.33 530	9.95 798	6	12	6	3.6	3.0
49	9.62 296	27	9.66 503	34	0.33 497	9.95 792	6	11	7	4.2	3.5
50	9.62 323	27	9.66 537	33	0.33 463	9.95 786	6	10	8	4.8	4.0
51	9.62 350	27	9.66 570	33	0.33 430	9.95 780	5	9	9	5.4	4.5
52	9.62 377	28	9.66 603	33	0.33 397	9.95 775	6	8			
53	9.62 405	27	9.66 636	33	0.33 364	9.95 769	6	7			
54	9.62 432	27	9.66 669	33	0.33 331	9.95 763	6	6			
55	9.62 459	27	9.66 702	33	0.33 298	9.95 757	6	5			
56	9.62 486	27	9.66 735	33	0.33 265	9.95 751	6	4			
57	9.62 513	28	9.66 768	33	0.33 232	9.95 745	6	3			
58	9.62 541	27	9.66 801	33	0.33 199	9.95 739	6	2			
59	9.62 568	27	9.66 834	33	0.33 166	9.95 733	5	1			
60	9.62 595		9.66 867		0.33 133	9.95 728		0			
	Log Cos	d	Log Cot	cd	Log Tan	Log Sin	d	′	Prop. Parts		

65°

25°

′	Log Sin	d	Log Tan	c d	Log Cot	Log Cos	d		Prop. Parts
0	9.62 595	27	9.66 867	33	0.33 133	9.95 728	6	60	
1	9.62 622	27	9.66 900	33	0.33 100	9.95 722	6	59	
2	9.62 649	27	9.66 933	33	0.33 067	9.95 716	6	58	
3	9.62 676	27	9.66 966	33	0.33 034	9.95 710	6	57	
4	9.62 703	27	9.66 999	33	0.33 001	9.95 704	6	56	
5	9.62 730	27	9.67 032	33	0.32 968	9.95 698	6	55	
6	9.62 757	27	9.67 065	33	0.32 935	9.95 692	6	54	
7	9.62 784	27	9.67 098	33	0.32 902	9.95 686	6	53	
8	9.62 811	27	9.67 131	32	0.32 869	9.95 680	6	52	
9	9.62 838	27	9.67 163	33	0.32 837	9.95 674	6	51	
10	9.62 865	27	9.67 196	33	0.32 804	9.95 668	5	50	
11	9.62 892	26	9.67 229	33	0.32 771	9.95 663	6	49	
12	9.62 918	27	9.67 262	33	0.32 738	9.95 657	6	48	
13	9.62 945	27	9.67 295	32	0.32 705	9.95 651	6	47	
14	9.62 972	27	9.67 327	33	0.32 673	9.95 645	6	46	
15	9.62 999	27	9.67 360	33	0.32 640	9.95 639	6	45	
16	9.63 026	26	9.67 393	33	0.32 607	9.95 633	6	44	
17	9.63 052	27	9.67 426	32	0.32 574	9.95 627	6	43	
18	9.63 079	27	9.67 458	33	0.32 542	9.95 621	6	42	
19	9.63 106	27	9.67 491	33	0.32 509	9.95 615	6	41	
20	9.63 133	26	9.67 524	32	0.32 476	9.95 609	6	40	
21	9.63 159	27	9.67 556	33	0.32 444	9.95 603	6	39	
22	9.63 186	27	9.67 589	33	0.32 411	9.95 597	6	38	
23	9.63 213	26	9.67 622	32	0.32 378	9.95 591	6	37	
24	9.63 239	27	9.67 654	33	0.32 346	9.95 585	6	36	
25	9.63 266	26	9.67 687	32	0.32 313	9.95 579	6	35	
26	9.63 292	27	9.67 719	33	0.32 281	9.95 573	6	34	
27	9.63 319	26	9.67 752	33	0.32 248	9.95 567	6	33	
28	9.63 345	27	9.67 785	32	0.32 215	9.95 561	6	32	
29	9.63 372	26	9.67 817	33	0.32 183	9.95 555	6	31	
30	9.63 398	27	9.67 850	32	0.32 150	9.95 549	6	30	
31	9.63 425	26	9.67 882	33	0.32 118	9.95 543	6	29	
32	9.63 451	27	9.67 915	32	0.32 085	9.95 537	6	28	
33	9.63 478	26	9.67 947	33	0.32 053	9.95 531	6	27	
34	9.63 504	27	9.67 980	32	0.32 020	9.95 525	6	26	
35	9.63 531	26	9.68 012	32	0.31 988	9.95 519	6	25	
36	9.63 557	26	9.68 044	33	0.31 956	9.95 513	6	24	
37	9.63 583	27	9.68 077	32	0.31 923	9.95 507	7	23	
38	9.63 610	26	9.68 109	33	0.31 891	9.95 500	6	22	
39	9.63 636	26	9.68 142	32	0.31 858	9.95 494	6	21	
40	9.63 662	27	9.68 174	32	0.31 826	9.95 488	6	20	
41	9.63 689	26	9.68 206	33	0.31 794	9.95 482	6	19	
42	9.63 715	26	9.68 239	32	0.31 761	9.95 476	6	18	
43	9.63 741	26	9.68 271	32	0.31 729	9.95 470	6	17	
44	9.63 767	27	9.68 303	33	0.31 697	9.95 464	6	16	
45	9.63 794	26	9.68 336	32	0.31 664	9.95 458	6	15	
46	9.63 820	26	9.68 368	32	0.31 632	9.95 452	6	14	
47	9.63 846	26	9.68 400	32	0.31 600	9.95 446	6	13	
48	9.63 872	26	9.68 432	33	0.31 568	9.95 440	6	12	
49	9.63 898	26	9.68 465	32	0.31 535	9.95 434	7	11	
50	9.63 924	26	9.68 497	32	0.31 503	9.95 427	6	10	
51	9.63 950	26	9.68 529	32	0.31 471	9.95 421	6	9	
52	9.63 976	26	9.68 561	32	0.31 439	9.95 415	6	8	
53	9.64 002	26	9.68 593	33	0.31 407	9.95 409	6	7	
54	9.64 028	26	9.68 626	32	0.31 374	9.95 403	6	6	
55	9.64 054	26	9.68 658	32	0.31 342	9.95 397	6	5	
56	9.64 080	26	9.68 690	32	0.31 310	9.95 391	7	4	
57	9.64 106	26	9.68 722	32	0.31 278	9.95 384	6	3	
58	9.64 132	26	9.68 754	32	0.31 246	9.95 378	6	2	
59	9.64 158	26	9.68 786	32	0.31 214	9.95 372	6	1	
60	9.64 184		9.68 818		0.31 182	9.95 366		0	
	Log Cos	d	Log Cot	c d	Log Tan	Log Sin	d	′	Prop. Parts

Prop. Parts

	33	32	27
1	3.3	3.2	2.7
2	6.6	6.4	5.4
3	9.9	9.6	8.1
4	13.2	12.8	10.8
5	16.5	16.0	13.5
6	19.8	19.2	16.2
7	23.1	22.4	18.9
8	26.4	25.6	21.6
9	29.7	28.8	24.3

	26	7
1	2.6	0.7
2	5.2	1.4
3	7.8	2.1
4	10.4	2.8
5	13.0	3.5
6	15.6	4.2
7	18.2	4.9
8	20.8	5.6
9	23.4	6.3

	6	5
1	0.6	0.5
2	1.2	1.0
3	1.8	1.5
4	2.4	2.0
5	3.0	2.5
6	3.6	3.0
7	4.2	3.5
8	4.8	4.0
9	5.4	4.5

64°

Table II

26°

′	Log Sin	d	Log Tan	c d	Log Cot	Log Cos	d	
0	9.64 184	26	9.68 818	32	0.31 182	9.95 366	6	60
1	9.64 210	26	9.68 850	32	0.31 150	9.95 360	6	59
2	9.64 236	26	9.68 882	32	0.31 118	9.95 354	6	58
3	9.64 262	26	9.68 914	32	0.31 086	9.95 348	7	57
4	9.64 288	25	9.68 946	32	0.31 054	9.95 341	6	56
5	9.64 313	26	9.68 978	32	0.31 022	9.95 335	6	55
6	9.64 339	26	9.69 010	32	0.30 990	9.95 329	6	54
7	9.64 365	26	9.69 042	32	0.30 958	9.95 323	6	53
8	9.64 391	26	9.69 074	32	0.30 926	9.95 317	7	52
9	9.64 417	25	9.69 106	32	0.30 894	9.95 310	6	51
10	9.64 442	26	9.69 138	32	0.30 862	9.95 304	6	50
11	9.64 468	26	9.69 170	32	0.30 830	9.95 298	6	49
12	9.64 494	25	9.69 202	32	0.30 798	9.95 292	6	48
13	9.64 519	26	9.69 234	32	0.30 766	9.95 286	7	47
14	9.64 545	26	9.69 266	32	0.30 734	9.95 279	6	46
15	9.64 571	25	9.69 298	31	0.30 702	9.95 273	6	45
16	9.64 596	26	9.69 329	32	0.30 671	9.95 267	6	44
17	9.64 622	25	9.69 361	32	0.30 639	9.95 261	7	43
18	9.64 647	26	9.69 393	32	0.30 607	9.95 254	6	42
19	9.64 673	25	9.69 425	32	0.30 575	9.95 248	6	41
20	9.64 698	26	9.69 457	31	0.30 543	9.95 242	6	40
21	9.64 724	25	9.69 488	32	0.30 512	9.95 236	7	39
22	9.64 749	26	9.69 520	32	0.30 480	9.95 229	6	38
23	9.64 775	25	9.69 552	32	0.30 448	9.95 223	6	37
24	9.64 800	26	9.69 584	31	0.30 416	9.95 217	6	36
25	9.64 826	25	9.69 615	32	0.30 385	9.95 211	7	35
26	9.64 851	26	9.69 647	32	0.30 353	9.95 204	6	34
27	9.64 877	25	9.69 679	31	0.30 321	9.95 198	6	33
28	9.64 902	25	9.69 710	32	0.30 290	9.95 192	7	32
29	9.64 927	26	9.69 742	32	0.30 258	9.95 185	6	31
30	9.64 953	25	9.69 774	31	0.30 226	9.95 179	6	30
31	9.64 978	25	9.69 805	32	0.30 195	9.95 173	6	29
32	9.65 003	26	9.69 837	31	0.30 163	9.95 167	7	28
33	9.65 029	25	9.69 868	32	0.30 132	9.95 160	6	27
34	9.65 054	25	9.69 900	32	0.30 100	9.95 154	6	26
35	9.65 079	25	9.69 932	31	0.30 068	9.95 148	7	25
36	9.65 104	26	9.69 963	32	0.30 037	9.95 141	6	24
37	9.65 130	25	9.69 995	31	0.30 005	9.95 135	6	23
38	9.65 155	25	9.70 026	32	0.29 974	9.95 129	7	22
39	9.65 180	25	9.70 058	31	0.29 942	9.95 122	6	21
40	9.65 205	25	9.70 089	32	0.29 911	9.95 116	6	20
41	9.65 230	25	9.70 121	31	0.29 879	9.95 110	7	19
42	9.65 255	26	9.70 152	32	0.29 848	9.95 103	6	18
43	9.65 281	25	9.70 184	31	0.29 816	9.95 097	7	17
44	9.65 306	25	9.70 215	32	0.29 785	9.95 090	6	16
45	9.65 331	25	9.70 247	31	0.29 753	9.95 084	6	15
46	9.65 356	25	9.70 278	31	0.29 722	9.95 078	7	14
47	9.65 381	25	9.70 309	32	0.29 691	9.95 071	6	13
48	9.65 406	25	9.70 341	31	0.29 659	9.95 065	6	12
49	9.65 431	25	9.70 372	32	0.29 628	9.95 059	7	11
50	9.65 456	25	9.70 404	31	0.29 596	9.95 052	6	10
51	9.65 481	25	9.70 435	31	0.29 565	9.95 046	7	9
52	9.65 506	25	9.70 466	32	0.29 534	9.95 039	6	8
53	9.65 531	25	9.70 498	31	0.29 502	9.95 033	6	7
54	9.65 556	24	9.70 529	31	0.29 471	9.95 027	7	6
55	9.65 580	25	9.70 560	32	0.29 440	9.95 020	6	5
56	9.65 605	25	9.70 592	31	0.29 408	9.95 014	7	4
57	9.65 630	25	9.70 623	31	0.29 377	9.95 007	6	3
58	9.65 655	25	9.70 654	31	0.29 346	9.95 001	6	2
59	9.65 680	25	9.70 685	32	0.29 315	9.94 995	7	1
60	9.65 705		9.70 717		0.29 283	9.94 988		0
	Log Cos	d	Log Cot	c d	Log Tan	Log Sin	d	′

Prop. Parts

	32	31	26
1	3.2	3.1	2.6
2	6.4	6.2	5.2
3	9.6	9.3	7.8
4	12.8	12.4	10.4
5	16.0	15.5	13.0
6	19.2	18.6	15.6
7	22.4	21.7	18.2
8	25.6	24.8	20.8
9	28.8	27.9	23.4

	25	24
1	2.5	2.4
2	5.0	4.8
3	7.5	7.2
4	10.0	9.6
5	12.5	12.0
6	15.0	14.4
7	17.5	16.8
8	20.0	19.2
9	22.5	21.6

	7	6
1	0.7	0.6
2	1.4	1.2
3	2.1	1.8
4	2.8	2.4
5	3.5	3.0
6	4.2	3.6
7	4.9	4.2
8	5.6	4.8
9	6.3	5.4

Prop. Parts

63°

27°

′	Log Sin	d	Log Tan	c d	Log Cot	Log Cos	d		Prop. Parts
0	9.65 705	24	9.70 717	31	0.29 283	9.94 988	6	60	
1	9.65 729	25	9.70 748	31	0.29 252	9.94 982	7	59	
2	9.65 754	25	9.70 779	31	0.29 221	9.94 975	6	58	
3	9.65 779	25	9.70 810	31	0.29 190	9.94 969	7	57	
4	9.65 804	24	9.70 841	32	0.29 159	9.94 962	6	56	
5	9.65 828	25	9.70 873	31	0.29 127	9.94 956	7	55	
6	9.65 853	25	9.70 904	31	0.29 096	9.94 949	6	54	
7	9.65 878	24	9.70 935	31	0.29 065	9.94 943	7	53	
8	9.65 902	25	9.70 966	31	0.29 034	9.94 936	6	52	
9	9.65 927	25	9.70 997	31	0.29 003	9.94 930	7	51	**32 31 30**
10	9.65 952	24	9.71 028	31	0.28 972	9.94 923	6	50	1 3.2 3.1 3.0
11	9.65 976	25	9.71 059	31	0.28 941	9.94 917	6	49	2 6.4 6.2 6.0
12	9.66 001	24	9.71 090	31	0.28 910	9.94 911	7	48	3 9.6 9.3 9.0
13	9.66 025	25	9.71 121	32	0.28 879	9.94 904	6	47	4 12.8 12.4 12.0
14	9.66 050	25	9.71 153	31	0.28 847	9.94 898	7	46	5 16.0 15.5 15.0
15	9.66 075	24	9.71 184	31	0.28 816	9.94 891	6	45	6 19.2 18.6 18.0
16	9.66 099	25	9.71 215	31	0.28 785	9.94 885	7	44	7 22.4 21.7 21.0
17	9.66 124	24	9.71 246	31	0.28 754	9.94 878	7	43	8 25.6 24.8 24.0
18	9.66 148	25	9.71 277	31	0.28 723	9.94 871	6	42	9 28.8 27.9 27.0
19	9.66 173	24	9.71 308	31	0.28 692	9.94 865	7	41	
20	9.66 197	24	9.71 339	31	0.28 661	9.94 858	6	40	
21	9.66 221	25	9.71 370	31	0.28 630	9.94 852	7	39	
22	9.66 246	24	9.71 401	30	0.28 599	9.94 845	6	38	
23	9.66 270	25	9.71 431	31	0.28 569	9.94 839	7	37	
24	9.66 295	24	9.71 462	31	0.28 538	9.94 832	6	36	
25	9.66 319	24	9.71 493	31	0.28 507	9.94 826	7	35	**25 24 23**
26	9.66 343	25	9.71 524	31	0.28 476	9.94 819	6	34	1 2.5 2.4 2.3
27	9.66 368	24	9.71 555	31	0.28 445	9.94 813	7	33	2 5.0 4.8 4.6
28	9.66 392	24	9.71 586	31	0.28 414	9.94 806	7	32	3 7.5 7.2 6.9
29	9.66 416	25	9.71 617	31	0.28 383	9.94 799	6	31	4 10.0 9.6 9.2
30	9.66 441	24	9.71 648	31	0.28 352	9.94 793	7	30	5 12.5 12.0 11.5
31	9.66 465	24	9.71 679	30	0.28 321	9.94 786	6	29	6 15.0 14.4 13.8
32	9.66 489	24	9.71 709	31	0.28 291	9.94 780	7	28	7 17.5 16.8 16.1
33	9.66 513	24	9.71 740	31	0.28 260	9.94 773	6	27	8 20.0 19.2 18.4
34	9.66 537	25	9.71 771	31	0.28 229	9.94 767	7	26	9 22.5 21.6 20.7
35	9.66 562	24	9.71 802	31	0.28 198	9.94 760	7	25	
36	9.66 586	24	9.71 833	30	0.28 167	9.94 753	6	24	
37	9.66 610	24	9.71 863	31	0.28 137	9.94 747	7	23	
38	9.66 634	24	9.71 894	31	0.28 106	9.94 740	6	22	
39	9.66 658	24	9.71 925	30	0.28 075	9.94 734	7	21	
40	9.66 682	24	9.71 955	31	0.28 045	9.94 727	7	20	
41	9.66 706	25	9.71 986	31	0.28 014	9.94 720	6	19	
42	9.66 731	24	9.72 017	31	0.27 983	9.94 714	7	18	
43	9.66 755	24	9.72 048	30	0.27 952	9.94 707	7	17	**7 6**
44	9.66 779	24	9.72 078	31	0.27 922	9.94 700	6	16	1 0.7 0.6
45	9.66 803	24	9.72 109	31	0.27 891	9.94 694	7	15	2 1.4 1.2
46	9.66 827	24	9.72 140	30	0.27 860	9.94 687	7	14	3 2.1 1.8
47	9.66 851	24	9.72 170	31	0.27 830	9.94 680	6	13	4 2.8 2.4
48	9.66 875	24	9.72 201	30	0.27 799	9.94 674	7	12	5 3.5 3.0
49	9.66 899	23	9.72 231	31	0.27 769	9.94 667	7	11	6 4.2 3.6
50	9.66 922	24	9.72 262	31	0.27 738	9.94 660	6	10	7 4.9 4.2
51	9.66 946	24	9.72 293	30	0.27 707	9.94 654	7	9	8 5.6 4.8
52	9.66 970	24	9.72 323	31	0.27 677	9.94 647	7	8	9 6.3 5.4
53	9.66 994	24	9.72 354	30	0.27 646	9.94 640	6	7	
54	9.67 018	24	9.72 384	31	0.27 616	9.94 634	7	6	
55	9.67 042	24	9.72 415	30	0.27 585	9.94 627	7	5	
56	9.67 066	24	9.72 445	31	0.27 555	9.94 620	6	4	
57	9.67 090	23	9.72 476	30	0.27 524	9.94 614	7	3	
58	9.67 113	24	9.72 506	31	0.27 494	9.94 607	7	2	
59	9.67 137	24	9.72 537	30	0.27 463	9.94 600	7	1	
60	9.67 161		9.72 567		0.27 433	9.94 593		0	
	Log Cos	d	Log Cot	c d	Log Tan	Log Sin	d	′	Prop. Parts

62°

28°

′	Log Sin	d	Log Tan	c d	Log Cot	Log Cos	d		Prop. Parts
0	9.67 161	24	9.72 567	31	0.27 433	9.94 593	6	60	
1	9.67 185	23	9.72 598	30	0.27 402	9.94 587	7	59	
2	9.67 208	24	9.72 628	31	0.27 372	9.94 580	7	58	
3	9.67 232	24	9.72 659	30	0.27 341	9.94 573	6	57	
4	9.67 256	24	9.72 689	31	0.27 311	9.94 567	7	56	
5	9.67 280	23	9.72 720	30	0.27 280	9.94 560	7	55	
6	9.67 303	24	9.72 750	30	0.27 250	9.94 553	7	54	
7	9.67 327	23	9.72 780	31	0.27 220	9.94 546	6	53	
8	9.67 350	24	9.72 811	30	0.27 189	9.94 540	7	52	
9	9.67 374	24	9.72 841	31	0.27 159	9.94 533	7	51	
10	9.67 398	23	9.72 872	30	0.27 128	9.94 526	7	50	
11	9.67 421	24	9.72 902	30	0.27 098	9.94 519	6	49	
12	9.67 445	23	9.72 932	31	0.27 068	9.94 513	7	48	
13	9.67 468	24	9.72 963	30	0.27 037	9.94 506	7	47	
14	9.67 492	23	9.72 993	30	0.27 007	9.94 499	7	46	
15	9.67 515	24	9.73 023	31	0.26 977	9.94 492	7	45	
16	9.67 539	23	9.73 054	30	0.26 946	9.94 485	6	44	
17	9.67 562	24	9.73 084	30	0.26 916	9.94 479	7	43	
18	9.67 586	23	9.73 114	30	0.26 886	9.94 472	7	42	
19	9.67 609	24	9.73 144	31	0.26 856	9.94 465	7	41	
20	9.67 633	23	9.73 175	30	0.26 825	9.94 458	7	40	
21	9.67 656	24	9.73 205	30	0.26 795	9.94 451	6	39	
22	9.67 680	23	9.73 235	30	0.26 765	9.94 445	7	38	
23	9.67 703	23	9.73 265	30	0.26 735	9.94 438	7	37	
24	9.67 726	24	9.73 295	31	0.26 705	9.94 431	7	36	
25	9.67 750	23	9.73 326	30	0.26 674	9.94 424	7	35	
26	9.67 773	23	9.73 356	30	0.26 644	9.94 417	7	34	
27	9.67 796	24	9.73 386	30	0.26 614	9.94 410	6	33	
28	9.67 820	23	9.73 416	30	0.26 584	9.94 404	7	32	
29	9.67 843	23	9.73 446	30	0.26 554	9.94 397	7	31	
30	9.67 866	24	9.73 476	31	0.26 524	9.94 390	7	30	
31	9.67 890	23	9.73 507	30	0.26 493	9.94 383	7	29	
32	9.67 913	23	9.73 537	30	0.26 463	9.94 376	7	28	
33	9.67 936	23	9.73 567	30	0.26 433	9.94 369	7	27	
34	9.67 959	23	9.73 597	30	0.26 403	9.94 362	7	26	
35	9.67 982	24	9.73 627	30	0.26 373	9.94 355	6	25	
36	9.68 006	23	9.73 657	30	0.26 343	9.94 349	7	24	
37	9.68 029	23	9.73 687	30	0.26 313	9.94 342	7	23	
38	9.68 052	23	9.73 717	30	0.26 283	9.94 335	7	22	
39	9.68 075	23	9.73 747	30	0.26 253	9.94 328	7	21	
40	9.68 098	23	9.73 777	30	0.26 223	9.94 321	7	20	
41	9.68 121	23	9.73 807	30	0.26 193	9.94 314	7	19	
42	9.68 144	23	9.73 837	30	0.26 163	9.94 307	7	18	
43	9.68 167	23	9.73 867	30	0.26 133	9.94 300	7	17	
44	9.68 190	23	9.73 897	30	0.26 103	9.94 293	7	16	
45	9.68 213	24	9.73 927	30	0.26 073	9.94 286	7	15	
46	9.68 237	23	9.73 957	30	0.26 043	9.94 279	6	14	
47	9.68 260	23	9.73 987	30	0.26 013	9.94 273	7	13	
48	9.68 283	22	9.74 017	30	0.25 983	9.94 266	7	12	
49	9.68 305	23	9.74 047	30	0.25 953	9.94 259	7	11	
50	9.68 328	23	9.74 077	30	0.25 923	9.94 252	7	10	
51	9.68 351	23	9.74 107	30	0.25 893	9.94 245	7	9	
52	9.68 374	23	9.74 137	29	0.25 863	9.94 238	7	8	
53	9.68 397	23	9.74 166	30	0.25 834	9.94 231	7	7	
54	9.68 420	23	9.74 196	30	0.25 804	9.94 224	7	6	
55	9.68 443	23	9.74 226	30	0.25 774	9.94 217	7	5	
56	9.68 466	23	9.74 256	30	0.25 744	9.94 210	7	4	
57	9.68 489	23	9.74 286	30	0.25 714	9.94 203	7	3	
58	9.68 512	22	9.74 316	29	0.25 684	9.94 196	7	2	
59	9.68 534	23	9.74 345	30	0.25 655	9.94 189	7	1	
60	9.68 557		9.74 375		0.25 625	9.94 182		0	
	Log Cos	d	Log Cot	c d	Log Tan	Log Sin	d	′	Prop. Parts

Prop. Parts

	31	30	29
1	3.1	3.0	2.9
2	6.2	6.0	5.8
3	9.3	9.0	8.7
4	12.4	12.0	11.6
5	15.5	15.0	14.5
6	18.6	18.0	17.4
7	21.7	21.0	20.3
8	24.8	24.0	23.2
9	27.9	27.0	26.1

	24	23	22
1	2.4	2.3	2.2
2	4.8	4.6	4.4
3	7.2	6.9	6.6
4	9.6	9.2	8.8
5	12.0	11.5	11.0
6	14.4	13.8	13.2
7	16.8	16.1	15.4
8	19.2	18.4	17.6
9	21.6	20.7	19.8

	7	6
1	0.7	0.6
2	1.4	1.2
3	2.1	1.8
4	2.8	2.4
5	3.5	3.0
6	4.2	3.6
7	4.9	4.2
8	5.6	4.8
9	6.3	5.4

29°

′	Log Sin	d	Log Tan	c d	Log Cot	Log Cos	d			Prop. Parts		
0	9.68 557	23	9.74 375	30	0.25 625	9.94 182	7	60				
1	9.68 580	23	9.74 405	30	0.25 595	9.94 175	7	59				
2	9.68 603	22	9.74 435	30	0.25 565	9.94 168	7	58				
3	9.68 625	23	9.74 465	29	0.25 535	9.94 161	7	57				
4	9.68 648	23	9.74 494	30	0.25 506	9.94 154	7	56				
5	9.68 671	23	9.74 524	30	0.25 476	9.94 147	7	55				
6	9.68 694	22	9.74 554	29	0.25 446	9.94 140	7	54				
7	9.68 716	23	9.74 583	30	0.25 417	9.94 133	7	53				
8	9.68 739	23	9.74 613	30	0.25 387	9.94 126	7	52				
9	9.68 762	22	9.74 643	30	0.25 357	9.94 119	7	51				
10	9.68 784	23	9.74 673	29	0.25 327	9.94 112	7	50				
11	9.68 807	22	9.74 702	30	0.25 298	9.94 105	7	49				
12	9.68 829	23	9.74 732	30	0.25 268	9.94 098	8	48				
13	9.68 852	23	9.74 762	29	0.25 238	9.94 090	7	47		30	29	23
14	9.68 875	22	9.74 791	30	0.25 209	9.94 083	7	46				
15	9.68 897	23	9.74 821	30	0.25 179	9.94 076	7	45	1	3.0	2.9	2.3
16	9.68 920	22	9.74 851	29	0.25 149	9.94 069	7	44	2	6.0	5.8	4.6
17	9.68 942	23	9.74 880	30	0.25 120	9.94 062	7	43	3	9.0	8.7	6.9
18	9.68 965	22	9.74 910	29	0.25 090	9.94 055	7	42	4	12.0	11.6	9.2
19	9.68 987	23	9.74 939	30	0.25 061	9.94 048	7	41	5	15.0	14.5	11.5
20	9.69 010	22	9.74 969	29	0.25 031	9.94 041	7	40	6	18.0	17.4	13.8
21	9.69 032	23	9.74 998	30	0.25 002	9.94 034	7	39	7	21.0	20.3	16.1
22	9.69 055	22	9.75 028	30	0.24 972	9.94 027	7	38	8	24.0	23.2	18.4
23	9.69 077	23	9.75 058	29	0.24 942	9.94 020	8	37	9	27.0	26.1	20.7
24	9.69 100	22	9.75 087	30	0.24 913	9.94 012	7	36				
25	9.69 122	22	9.75 117	29	0.24 883	9.94 005	7	35				
26	9.69 144	23	9.75 146	30	0.24 854	9.93 998	7	34				
27	9.69 167	22	9.75 176	29	0.24 824	9.93 991	7	33				
28	9.69 189	23	9.75 205	30	0.24 795	9.93 984	7	32				
29	9.69 212	22	9.75 235	29	0.24 765	9.93 977	7	31				
30	9.69 234	22	9.75 264	30	0.24 736	9.93 970	7	30				
31	9.69 256	23	9.75 294	29	0.24 706	9.93 963	8	29				
32	9.69 279	22	9.75 323	30	0.24 677	9.93 955	7	28				
33	9.69 301	22	9.75 353	29	0.24 647	9.93 948	7	27				
34	9.69 323	22	9.75 382	29	0.24 618	9.93 941	7	26				
35	9.69 345	23	9.75 411	30	0.24 589	9.93 934	7	25				
36	9.69 368	22	9.75 441	29	0.24 559	9.93 927	7	24		22	8	7
37	9.69 390	22	9.75 470	30	0.24 530	9.93 920	8	23				
38	9.69 412	22	9.75 500	29	0.24 500	9.93 912	7	22	1	2.2	0.8	0.7
39	9.69 434	22	9.75 529	29	0.24 471	9.93 905	7	21	2	4.4	1.6	1.4
40	9.69 456	23	9.75 558	30	0.24 442	9.93 898	7	20	3	6.6	2.4	2.1
41	9.69 479	22	9.75 588	29	0.24 412	9.93 891	7	19	4	8.8	3.2	2.8
42	9.69 501	22	9.75 617	30	0.24 383	9.93 884	8	18	5	11.0	4.0	3.5
43	9.69 523	22	9.75 647	29	0.24 353	9.93 876	7	17	6	13.2	4.8	4.2
44	9.69 545	22	9.75 676	29	0.24 324	9.93 869	7	16	7	15.4	5.6	4.9
45	9.69 567	22	9.75 705	30	0.24 295	9.93 862	7	15	8	17.6	6.4	5.6
46	9.69 589	22	9.75 735	29	0.24 265	9.93 855	8	14	9	19.8	7.2	6.3
47	9.69 611	22	9.75 764	29	0.24 236	9.93 847	7	13				
48	9.69 633	22	9.75 793	29	0.24 207	9.93 840	7	12				
49	9.69 655	22	9.75 822	30	0.24 178	9.93 833	7	11				
50	9.69 677	22	9.75 852	29	0.24 148	9.93 826	7	10				
51	9.69 699	22	9.75 881	29	0.24 119	9.93 819	8	9				
52	9.69 721	22	9.75 910	29	0.24 090	9.93 811	7	8				
53	9.69 743	22	9.75 939	30	0.24 061	9.93 804	7	7				
54	9.69 765	22	9.75 969	29	0.24 031	9.93 797	8	6				
55	9.69 787	22	9.75 998	29	0.24 002	9.93 789	7	5				
56	9.69 809	22	9.76 027	29	0.23 973	9.93 782	7	4				
57	9.69 831	22	9.76 056	30	0.23 944	9.93 775	7	3				
58	9.69 853	22	9.76 086	29	0.23 914	9.93 768	8	2				
59	9.69 875	22	9.76 115	29	0.23 885	9.93 760	7	1				
60	9.69 897		9.76 144		0.23 856	9.93 753		0				
	Log Cos	d	Log Cot	c d	Log Tan	Log Sin	d	′		Prop. Parts		

60°

Table II

30°

′	Log Sin	d	Log Tan	c d	Log Cot	Log Cos	d		Prop. Parts			
0	9.69 897	22	9.76 144	29	0.23 856	9.93 753	7	60				
1	9.69 919	22	9.76 173	29	0.23 827	9.93 746	7	59				
2	9.69 941	22	9.76 202	29	0.23 798	9.93 738	8	58				
3	9.69 963	21	9.76 231	30	0.23 769	9.93 731	7	57				
4	9.69 984	22	9.76 261	29	0.23 739	9.93 724	7	56				
5	9.70 006	22	9.76 290	29	0.23 710	9.93 717	7	55				
6	9.70 028	22	9.76 319	29	0.23 681	9.93 709	8	54				
7	9.70 050	22	9.76 348	29	0.23 652	9.93 702	7	53				
8	9.70 072	21	9.76 377	29	0.23 623	9.93 695	7	52				
9	9.70 093	22	9.76 406	29	0.23 594	9.93 687	8	51				
10	9.70 115	22	9.76 435	29	0.23 565	9.93 680	7	50		30	29	28
11	9.70 137	22	9.76 464	29	0.23 536	9.93 673	7	49	1	3.0	2.9	2.8
12	9.70 159	21	9.76 493	29	0.23 507	9.93 665	7	48	2	6.0	5.8	5.6
13	9.70 180	22	9.76 522	29	0.23 478	9.93 658	8	47	3	9.0	8.7	8.4
14	9.70 202	22	9.76 551	29	0.23 449	9.93 650	7	46	4	12.0	11.6	11.2
15	9.70 224	21	9.76 580	29	0.23 420	9.93 643	7	45	5	15.0	14.5	14.0
16	9.70 245	22	9.76 609	30	0.23 391	9.93 636	7	44	6	18.0	17.4	16.8
17	9.70 267	21	9.76 639	29	0.23 361	9.93 628	7	43	7	21.0	20.3	19.6
18	9.70 288	22	9.76 668	29	0.23 332	9.93 621	7	42	8	24.0	23.2	22.4
19	9.70 310	22	9.76 697	28	0.23 303	9.93 614	8	41	9	27.0	26.1	25.2
20	9.70 332	21	9.76 725	29	0.23 275	9.93 606	7	40				
21	9.70 353	22	9.76 754	29	0.23 246	9.93 599	8	39				
22	9.70 375	21	9.76 783	29	0.23 217	9.93 591	7	38				
23	9.70 396	22	9.76 812	29	0.23 188	9.93 584	7	37				
24	9.70 418	21	9.76 841	29	0.23 159	9.93 577	8	36				
25	9.70 439	22	9.76 870	29	0.23 130	9.93 569	7	35		22	21	
26	9.70 461	21	9.76 899	29	0.23 101	9.93 562	8	34	1	2.2	2.1	
27	9.70 482	22	9.76 928	29	0.23 072	9.93 554	7	33	2	4.4	4.2	
28	9.70 504	21	9.76 957	29	0.23 043	9.93 547	8	32	3	6.6	6.3	
29	9.70 525	22	9.76 986	29	0.23 014	9.93 539	7	31	4	8.8	8.4	
30	9.70 547	21	9.77 015	29	0.22 985	9.93 532	7	30	5	11.0	10.5	
31	9.70 568	22	9.77 044	29	0.22 956	9.93 525	8	29	6	13.2	12.6	
32	9.70 590	21	9.77 073	28	0.22 927	9.93 517	7	28	7	15.4	14.7	
33	9.70 611	22	9.77 101	29	0.22 899	9.93 510	8	27	8	17.6	16.8	
34	9.70 633	21	9.77 130	29	0.22 870	9.93 502	7	26	9	19.8	18.9	
35	9.70 654	21	9.77 159	29	0.22 841	9.93 495	8	25				
36	9.70 675	22	9.77 188	29	0.22 812	9.93 487	7	24				
37	9.70 697	21	9.77 217	29	0.22 783	9.93 480	8	23				
38	9.70 718	21	9.77 246	28	0.22 754	9.93 472	7	22				
39	9.70 739	22	9.77 274	29	0.22 726	9.93 465	8	21				
40	9.70 761	21	9.77 303	29	0.22 697	9.93 457	7	20				
41	9.70 782	21	9.77 332	29	0.22 668	9.93 450	8	19				
42	9.70 803	21	9.77 361	29	0.22 639	9.93 442	7	18				
43	9.70 824	22	9.77 390	28	0.22 610	9.93 435	8	17		8	7	
44	9.70 846	21	9.77 418	29	0.22 582	9.93 427	7	16	1	0.8	0.7	
45	9.70 867	21	9.77 447	29	0.22 553	9.93 420	8	15	2	1.6	1.4	
46	9.70 888	21	9.77 476	29	0.22 524	9.93 412	7	14	3	2.4	2.1	
47	9.70 909	22	9.77 505	28	0.22 495	9.93 405	8	13	4	3.2	2.8	
48	9.70 931	21	9.77 533	29	0.22 467	9.93 397	7	12	5	4.0	3.5	
49	9.70 952	21	9.77 562	29	0.22 438	9.93 390	8	11	6	4.8	4.2	
50	9.70 973	21	9.77 591	28	0.22 409	9.93 382	7	10	7	5.6	4.9	
51	9.70 994	21	9.77 619	29	0.22 381	9.93 375	8	9	8	6.4	5.6	
52	9.71 015	21	9.77 648	29	0.22 352	9.93 367	7	8	9	7.2	6.3	
53	9.71 036	22	9.77 677	29	0.22 323	9.93 360	8	7				
54	9.71 058	21	9.77 706	28	0.22 294	9.93 352	8	6				
55	9.71 079	21	9.77 734	29	0.22 266	9.93 344	7	5				
56	9.71 100	21	9.77 763	28	0.22 237	9.93 337	8	4				
57	9.71 121	21	9.77 791	29	0.22 209	9.93 329	7	3				
58	9.71 142	21	9.77 820	29	0.22 180	9.93 322	8	2				
59	9.71 163	21	9.77 849	28	0.22 151	9.93 314	7	1				
60	9.71 184		9.77 877		0.22 123	9.93 307		0				
	Log Cos	d	Log Cot	c d	Log Tan	Log Sin	d	′	Prop. Parts			

59°

31°

′	Log Sin	d	Log Tan	cd	Log Cot	Log Cos	d	
0	9.71 184	21	9.77 877	29	0.22 123	9.93 307	8	60
1	9.71 205	21	9.77 906	29	0.22 094	9.93 299	8	59
2	9.71 226	21	9.77 935	28	0.22 065	9.93 291	7	58
3	9.71 247	21	9.77 963	29	0.22 037	9.93 284	8	57
4	9.71 268	21	9.77 992	28	0.22 008	9.93 276	7	56
5	9.71 289	21	9.78 020	29	0.21 980	9.93 269	8	55
6	9.71 310	21	9.78 049	28	0.21 951	9.93 261	8	54
7	9.71 331	21	9.78 077	29	0.21 923	9.93 253	7	53
8	9.71 352	21	9.78 106	29	0.21 894	9.93 246	8	52
9	9.71 373	20	9.78 135	28	0.21 865	9.93 238	8	51
10	9.71 393	21	9.78 163	29	0.21 837	9.93 230	7	50
11	9.71 414	21	9.78 192	28	0.21 808	9.93 223	8	49
12	9.71 435	21	9.78 220	29	0.21 780	9.93 215	8	48
13	9.71 456	21	9.78 249	28	0.21 751	9.93 207	7	47
14	9.71 477	21	9.78 277	29	0.21 723	9.93 200	8	46
15	9.71 498	21	9.78 306	28	0.21 694	9.93 192	8	45
16	9.71 519	20	9.78 334	29	0.21 666	9.93 184	7	44
17	9.71 539	21	9.78 363	28	0.21 637	9.93 177	8	43
18	9.71 560	21	9.78 391	28	0.21 609	9.93 169	8	42
19	9.71 581	21	9.78 419	29	0.21 581	9.93 161	7	41
20	9.71 602	20	9.78 448	28	0.21 552	9.93 154	8	40
21	9.71 622	21	9.78 476	29	0.21 524	9.93 146	8	39
22	9.71 643	21	9.78 505	28	0.21 495	9.93 138	7	38
23	9.71 664	21	9.78 533	29	0.21 467	9.93 131	8	37
24	9.71 685	20	9.78 562	28	0.21 438	9.93 123	8	36
25	9.71 705	21	9.78 590	28	0.21 410	9.93 115	7	35
26	9.71 726	21	9.78 618	29	0.21 382	9.93 108	8	34
27	9.71 747	20	9.78 647	28	0.21 353	9.93 100	8	33
28	9.71 767	21	9.78 675	29	0.21 325	9.93 092	8	32
29	9.71 788	21	9.78 704	28	0.21 296	9.93 084	7	31
30	9.71 809	20	9.78 732	28	0.21 268	9.93 077	8	30
31	9.71 829	21	9.78 760	29	0.21 240	9.93 069	8	29
32	9.71 850	20	9.78 789	28	0.21 211	9.93 061	8	28
33	9.71 870	21	9.78 817	28	0.21 183	9.93 053	7	27
34	9.71 891	20	9.78 845	29	0.21 155	9.93 046	8	26
35	9.71 911	21	9.78 874	28	0.21 126	9.93 038	8	25
36	9.71 932	20	9.78 902	28	0.21 098	9.93 030	8	24
37	9.71 952	21	9.78 930	29	0.21 070	9.93 022	8	23
38	9.71 973	21	9.78 959	28	0.21 041	9.93 014	7	22
39	9.71 994	20	9.78 987	28	0.21 013	9.93 007	8	21
40	9.72 014	20	9.79 015	28	0.20 985	9.92 999	8	20
41	9.72 034	21	9.79 043	29	0.20 957	9.92 991	8	19
42	9.72 055	20	9.79 072	28	0.20 928	9.92 983	7	18
43	9.72 075	21	9.79 100	28	0.20 900	9.92 976	8	17
44	9.72 096	20	9.79 128	28	0.20 872	9.92 968	8	16
45	9.72 116	21	9.79 156	29	0.20 844	9.92 960	8	15
46	9.72 137	20	9.79 185	28	0.20 815	9.92 952	8	14
47	9.72 157	20	9.79 213	28	0.20 787	9.92 944	8	13
48	9.72 177	21	9.79 241	28	0.20 759	9.92 936	7	12
49	9.72 198	20	9.79 269	28	0.20 731	9.92 929	8	11
50	9.72 218	20	9.79 297	29	0.20 703	9.92 921	8	10
51	9.72 238	21	9.79 326	28	0.20 674	9.92 913	8	9
52	9.72 259	20	9.79 354	28	0.20 646	9.92 905	8	8
53	9.72 279	20	9.79 382	28	0.20 618	9.92 897	8	7
54	9.72 299	21	9.79 410	28	0.20 590	9.92 889	8	6
55	9.72 320	20	9.79 438	28	0.20 562	9.92 881	7	5
56	9.72 340	20	9.79 466	29	0.20 534	9.92 874	8	4
57	9.72 360	21	9.79 495	28	0.20 505	9.92 866	8	3
58	9.72 381	20	9.79 523	28	0.20 477	9.92 858	8	2
59	9.72 401	20	9.79 551	28	0.20 449	9.92 850	8	1
60	9.72 421		9.79 579		0.20 421	9.92 842		0
	Log Cos	d	Log Cot	cd	Log Tan	Log Sin	d	′

Prop. Parts

	29	28	21
1	2.9	2.8	2.1
2	5.8	5.6	4.2
3	8.7	8.4	6.3
4	11.6	11.2	8.4
5	14.5	14.0	10.5
6	17.4	16.8	12.6
7	20.3	19.6	14.7
8	23.2	22.4	16.8
9	26.1	25.2	18.9

	20	8	7
1	2.0	0.8	0.7
2	4.0	1.6	1.4
3	6.0	2.4	2.1
4	8.0	3.2	2.8
5	10.0	4.0	3.5
6	12.0	4.8	4.2
7	14.0	5.6	4.9
8	16.0	6.4	5.6
9	18.0	7.2	6.3

58°

32°

′	Log Sin	d	Log Tan	c d	Log Cot	Log Cos	d		Prop. Parts
0	9.72 421	20	9.79 579	28	0.20 421	9.92 842	8	60	
1	9.72 441	20	9.79 607	28	0.20 393	9.92 834	8	59	
2	9.72 461	21	9.79 635	28	0.20 365	9.92 826	8	58	
3	9.72 482	20	9.79 663	28	0.20 337	9.92 818	8	57	
4	9.72 502	20	9.79 691	28	0.20 309	9.92 810	7	56	
5	9.72 522	20	9.79 719	28	0.20 281	9.92 803	8	55	
6	9.72 542	20	9.79 747	29	0.20 253	9.92 795	8	54	
7	9.72 562	20	9.79 776	28	0.20 224	9.92 787	8	53	
8	9.72 582	20	9.79 804	28	0.20 196	9.92 779	8	52	
9	9.72 602	20	9.79 832	28	0.20 168	9.92 771	8	51	
10	9.72 622	21	9.79 860	28	0.20 140	9.92 763	8	50	
11	9.72 643	20	9.79 888	28	0.20 112	9.92 755	8	49	
12	9.72 663	20	9.79 916	28	0.20 084	9.92 747	8	48	
13	9.72 683	20	9.79 944	28	0.20 056	9.92 739	8	47	
14	9.72 703	20	9.79 972	28	0.20 028	9.92 731	8	46	
15	9.72 723	20	9.80 000	28	0.20 000	9.92 723	8	45	
16	9.72 743	20	9.80 028	28	0.19 972	9.92 715	8	44	
17	9.72 763	20	9.80 056	28	0.19 944	9.92 707	8	43	
18	9.72 783	20	9.80 084	28	0.19 916	9.92 699	8	42	
19	9.72 803	20	9.80 112	28	0.19 888	9.92 691	8	41	
20	9.72 823	20	9.80 140	28	0.19 860	9.92 683	8	40	
21	9.72 843	20	9.80 168	27	0.19 832	9.92 675	8	39	
22	9.72 863	20	9.80 195	28	0.19 805	9.92 667	8	38	
23	9.72 883	19	9.80 223	28	0.19 777	9.92 659	8	37	
24	9.72 902	20	9.80 251	28	0.19 749	9.92 651	8	36	
25	9.72 922	20	9.80 279	28	0.19 721	9.92 643	8	35	
26	9.72 942	20	9.80 307	28	0.19 693	9.92 635	8	34	
27	9.72 962	20	9.80 335	28	0.19 665	9.92 627	8	33	
28	9.72 982	20	9.80 363	28	0.19 637	9.92 619	8	32	
29	9.73 002	20	9.80 391	28	0.19 609	9.92 611	8	31	
30	9.73 022	19	9.80 419	28	0.19 581	9.92 603	8	30	
31	9.73 041	20	9.80 447	27	0.19 553	9.92 595	8	29	
32	9.73 061	20	9.80 474	28	0.19 526	9.92 587	8	28	
33	9.73 081	20	9.80 502	28	0.19 498	9.92 579	8	27	
34	9.73 101	20	9.80 530	28	0.19 470	9.92 571	8	26	
35	9.73 121	19	9.80 558	28	0.19 442	9.92 563	8	25	
36	9.73 140	20	9.80 586	28	0.19 414	9.92 555	9	24	
37	9.73 160	20	9.80 614	28	0.19 386	9.92 546	8	23	
38	9.73 180	20	9.80 642	27	0.19 358	9.92 538	8	22	
39	9.73 200	19	9.80 669	28	0.19 331	9.92 530	8	21	
40	9.73 219	20	9.80 697	28	0.19 303	9.92 522	8	20	
41	9.73 239	20	9.80 725	28	0.19 275	9.92 514	8	19	
42	9.73 259	19	9.80 753	28	0.19 247	9.92 506	8	18	
43	9.73 278	20	9.80 781	27	0.19 219	9.92 498	8	17	
44	9.73 298	20	9.80 808	28	0.19 192	9.92 490	8	16	
45	9.73 318	19	9.80 836	28	0.19 164	9.92 482	9	15	
46	9.73 337	20	9.80 864	28	0.19 136	9.92 473	8	14	
47	9.73 357	20	9.80 892	27	0.19 108	9.92 465	8	13	
48	9.73 377	19	9.80 919	28	0.19 081	9.92 457	8	12	
49	9.73 396	20	9.80 947	28	0.19 053	9.92 449	8	11	
50	9.73 416	19	9.80 975	28	0.19 025	9.92 441	8	10	
51	9.73 435	20	9.81 003	27	0.18 997	9.92 433	8	9	
52	9.73 455	19	9.81 030	28	0.18 970	9.92 425	9	8	
53	9.73 474	20	9.81 058	28	0.18 942	9.92 416	8	7	
54	9.73 494	19	9.81 086	27	0.18 914	9.92 408	8	6	
55	9.73 513	20	9.81 113	28	0.18 887	9.92 400	8	5	
56	9.73 533	19	9.81 141	28	0.18 859	9.92 392	8	4	
57	9.73 552	20	9.81 169	27	0.18 831	9.92 384	8	3	
58	9.73 572	19	9.81 196	28	0.18 804	9.92 376	9	2	
59	9.73 591	20	9.81 224	28	0.18 776	9.92 367	8	1	
60	9.73 611		9.81 252		0.18 748	9.92 359		0	
	Log Cos	d	Log Cot	c d	Log Tan	Log Sin	d	′	Prop. Parts

Prop. Parts

	29	28	27
1	2.9	2.8	2.7
2	5.8	5.6	5.4
3	8.7	8.4	8.1
4	11.6	11.2	10.8
5	14.5	14.0	13.5
6	17.4	16.8	16.2
7	20.3	19.6	18.9
8	23.2	22.4	21.6
9	26.1	25.2	24.3

	21	20	19
1	2.1	2.0	1.9
2	4.2	4.0	3.8
3	6.3	6.0	5.7
4	8.4	8.0	7.6
5	10.5	10.0	9.5
6	12.6	12.0	11.4
7	14.7	14.0	13.3
8	16.8	16.0	15.2
9	18.9	18.0	17.1

	9	8	7
1	0.9	0.8	0.7
2	1.8	1.6	1.4
3	2.7	2.4	2.1
4	3.6	3.2	2.8
5	4.5	4.0	3.5
6	5.4	4.8	4.2
7	6.3	5.6	4.9
8	7.2	6.4	5.6
9	8.1	7.2	6.3

57°

33°

′	Log Sin	d	Log Tan	c d	Log Cot	Log Cos	d	′
0	9.73 611	19	9.81 252	27	0.18 748	9.92 359	8	60
1	9.73 630	20	9.81 279	28	0.18 721	9.92 351	8	59
2	9.73 650	19	9.81 307	28	0.18 693	9.92 343	8	58
3	9.73 669	20	9.81 335	27	0.18 665	9.92 335	9	57
4	9.73 689	19	9.81 362	28	0.18 638	9.92 326	8	56
5	9.73 708	19	9.81 390	28	0.18 610	9.92 318	8	55
6	9.73 727	20	9.81 418	27	0.18 582	9.92 310	8	54
7	9.73 747	19	9.81 445	28	0.18 555	9.92 302	9	53
8	9.73 766	19	9.81 473	27	0.18 527	9.92 293	8	52
9	9.73 785	20	9.81 500	28	0.18 500	9.92 285	8	51
10	9.73 805	19	9.81 528	28	0.18 472	9.92 277	8	50
11	9.73 824	19	9.81 556	27	0.18 444	9.92 269	9	49
12	9.73 843	20	9.81 583	28	0.18 417	9.92 260	8	48
13	9.73 863	19	9.81 611	27	0.18 389	9.92 252	8	47
14	9.73 882	19	9.81 638	28	0.18 362	9.92 244	9	46
15	9.73 901	20	9.81 666	27	0.18 334	9.92 235	8	45
16	9.73 921	19	9.81 693	28	0.18 307	9.92 227	8	44
17	9.73 940	19	9.81 721	27	0.18 279	9.92 219	8	43
18	9.73 959	19	9.81 748	28	0.18 252	9.92 211	9	42
19	9.73 978	19	9.81 776	27	0.18 224	9.92 202	8	41
20	9.73 997	20	9.81 803	28	0.18 197	9.92 194	8	40
21	9.74 017	19	9.81 831	27	0.18 169	9.92 186	9	39
22	9.74 036	19	9.81 858	28	0.18 142	9.92 177	8	38
23	9.74 055	19	9.81 886	27	0.18 114	9.92 169	8	37
24	9.74 074	19	9.81 913	28	0.18 087	9.92 161	9	36
25	9.74 093	20	9.81 941	27	0.18 059	9.92 152	8	35
26	9.74 113	19	9.81 968	28	0.18 032	9.92 144	8	34
27	9.74 132	19	9.81 996	27	0.18 004	9.92 136	9	33
28	9.74 151	19	9.82 023	28	0.17 977	9.92 127	8	32
29	9.74 170	19	9.82 051	27	0.17 949	9.92 119	8	31
30	9.74 189	19	9.82 078	28	0.17 922	9.92 111	9	30
31	9.74 208	19	9.82 106	27	0.17 894	9.92 102	8	29
32	9.74 227	19	9.82 133	28	0.17 867	9.92 094	8	28
33	9.74 246	19	9.82 161	27	0.17 839	9.92 086	9	27
34	9.74 265	19	9.82 188	27	0.17 812	9.92 077	8	26
35	9.74 284	19	9.82 215	28	0.17 785	9.92 069	9	25
36	9.74 303	19	9.82 243	27	0.17 757	9.92 060	8	24
37	9.74 322	19	9.82 270	28	0.17 730	9.92 052	8	23
38	9.74 341	19	9.82 298	27	0.17 702	9.92 044	9	22
39	9.74 360	19	9.82 325	27	0.17 675	9.92 035	8	21
40	9.74 379	19	9.82 352	28	0.17 648	9.92 027	9	20
41	9.74 398	19	9.82 380	27	0.17 620	9.92 018	8	19
42	9.74 417	19	9.82 407	28	0.17 593	9.92 010	8	18
43	9.74 436	19	9.82 435	27	0.17 565	9.92 002	9	17
44	9.74 455	19	9.82 462	27	0.17 538	9.91 993	8	16
45	9.74 474	19	9.82 489	28	0.17 511	9.91 985	9	15
46	9.74 493	19	9.82 517	27	0.17 483	9.91 976	8	14
47	9.74 512	19	9.82 544	27	0.17 456	9.91 968	9	13
48	9.74 531	18	9.82 571	28	0.17 429	9.91 959	8	12
49	9.74 549	19	9.82 599	27	0.17 401	9.91 951	9	11
50	9.74 568	19	9.82 626	27	0.17 374	9.91 942	8	10
51	9.74 587	19	9.82 653	28	0.17 347	9.91 934	9	9
52	9.74 606	19	9.82 681	27	0.17 319	9.91 925	8	8
53	9.74 625	19	9.82 708	27	0.17 292	9.91 917	9	7
54	9.74 644	18	9.82 735	27	0.17 265	9.91 908	8	6
55	9.74 662	19	9.82 762	28	0.17 238	9.91 900	9	5
56	9.74 681	19	9.82 790	27	0.17 210	9.91 891	8	4
57	9.74 700	19	9.82 817	27	0.17 183	9.91 883	8	3
58	9.74 719	18	9.82 844	27	0.17 156	9.91 874	8	2
59	9.74 737	19	9.82 871	28	0.17 129	9.91 866	9	1
60	9.74 756		9.82 899		0.17 101	9.91 857		0
	Log Cos	d	Log Cot	c d	Log Tan	Log Sin	d	′

Prop. Parts

	28	27	20
1	2.8	2.7	2.0
2	5.6	5.4	4.0
3	8.4	8.1	6.0
4	11.2	10.8	8.0
5	14.0	13.5	10.0
6	16.8	16.2	12.0
7	19.6	18.9	14.0
8	22.4	21.6	16.0
9	25.2	24.3	18.0

	19	18
1	1.9	1.8
2	3.8	3.6
3	5.7	5.4
4	7.6	7.2
5	9.5	9.0
6	11.4	10.8
7	13.3	12.6
8	15.2	14.4
9	17.1	16.2

	9	8
1	0.9	0.8
2	1.8	1.6
3	2.7	2.4
4	3.6	3.2
5	4.5	4.0
6	5.4	4.8
7	6.3	5.6
8	7.2	6.4
9	8.1	7.2

56°

34°

′	Log Sin	d	Log Tan	c d	Log Cot	Log Cos	d	′
0	9.74 756	19	9.82 899	27	0.17 101	9.91 857	8	60
1	9.74 775	19	9.82 926	27	0.17 074	9.91 849	9	59
2	9.74 794	18	9.82 953	27	0.17 047	9.91 840	8	58
3	9.74 812	19	9.82 980	28	0.17 020	9.91 832	9	57
4	9.74 831	19	9.83 008	27	0.16 992	9.91 823	8	56
5	9.74 850	18	9.83 035	27	0.16 965	9.91 815	9	55
6	9.74 868	19	9.83 062	27	0.16 938	9.91 806	8	54
7	9.74 887	19	9.83 089	28	0.16 911	9.91 798	9	53
8	9.74 906	18	9.83 117	27	0.16 883	9.91 789	8	52
9	9.74 924	19	9.83 144	27	0.16 856	9.91 781	9	51
10	9.74 943	18	9.83 171	27	0.16 829	9.91 772	9	50
11	9.74 961	19	9.83 198	27	0.16 802	9.91 763	8	49
12	9.74 980	19	9.83 225	27	0.16 775	9.91 755	9	48
13	9.74 999	18	9.83 252	28	0.16 748	9.91 746	8	47
14	9.75 017	19	9.83 280	27	0.16 720	9.91 738	9	46
15	9.75 036	18	9.83 307	27	0.16 693	9.91 729	9	45
16	9.75 054	19	9.83 334	27	0.16 666	9.91 720	8	44
17	9.75 073	18	9.83 361	27	0.16 639	9.91 712	9	43
18	9.75 091	19	9.83 388	27	0.16 612	9.91 703	8	42
19	9.75 110	18	9.83 415	27	0.16 585	9.91 695	9	41
20	9.75 128	19	9.83 442	28	0.16 558	9.91 686	9	40
21	9.75 147	18	9.83 470	27	0.16 530	9.91 677	8	39
22	9.75 165	19	9.83 497	27	0.16 503	9.91 669	9	38
23	9.75 184	18	9.83 524	27	0.16 476	9.91 660	9	37
24	9.75 202	19	9.83 551	27	0.16 449	9.91 651	8	36
25	9.75 221	18	9.83 578	27	0.16 422	9.91 643	9	35
26	9.75 239	19	9.83 605	27	0.16 395	9.91 634	9	34
27	9.75 258	18	9.83 632	27	0.16 368	9.91 625	8	33
28	9.75 276	18	9.83 659	27	0.16 341	9.91 617	9	32
29	9.75 294	19	9.83 686	27	0.16 314	9.91 608	9	31
30	9.75 313	18	9.83 713	27	0.16 287	9.91 599	8	30
31	9.75 331	19	9.83 740	28	0.16 260	9.91 591	9	29
32	9.75 350	18	9.83 768	27	0.16 232	9.91 582	9	28
33	9.75 368	18	9.83 795	27	0.16 205	9.91 573	8	27
34	9.75 386	19	9.83 822	27	0.16 178	9.91 565	9	26
35	9.75 405	18	9.83 849	27	0.16 151	9.91 556	9	25
36	9.75 423	18	9.83 876	27	0.16 124	9.91 547	9	24
37	9.75 441	18	9.83 903	27	0.16 097	9.91 538	8	23
38	9.75 459	19	9.83 930	27	0.16 070	9.91 530	9	22
39	9.75 478	18	9.83 957	27	0.16 043	9.91 521	9	21
40	9.75 496	18	9.83 984	27	0.16 016	9.91 512	8	20
41	9.75 514	19	9.84 011	27	0.15 989	9.91 504	9	19
42	9.75 533	18	9.84 038	27	0.15 962	9.91 495	9	18
43	9.75 551	18	9.84 065	27	0.15 935	9.91 486	9	17
44	9.75 569	18	9.84 092	27	0.15 908	9.91 477	8	16
45	9.75 587	18	9.84 119	27	0.15 881	9.91 469	9	15
46	9.75 605	19	9.84 146	27	0.15 854	9.91 460	9	14
47	9.75 624	18	9.84 173	27	0.15 827	9.91 451	9	13
48	9.75 642	18	9.84 200	27	0.15 800	9.91 442	9	12
49	9.75 660	18	9.84 227	27	0.15 773	9.91 433	8	11
50	9.75 678	18	9.84 254	26	0.15 746	9.91 425	9	10
51	9.75 696	18	9.84 280	27	0.15 720	9.91 416	9	9
52	9.75 714	19	9.84 307	27	0.15 693	9.91 407	9	8
53	9.75 733	18	9.84 334	27	0.15 666	9.91 398	9	7
54	9.75 751	18	9.84 361	27	0.15 639	9.91 389	8	6
55	9.75 769	18	9.84 388	27	0.15 612	9.91 381	9	5
56	9.75 787	18	9.84 415	27	0.15 585	9.91 372	9	4
57	9.75 805	18	9.84 442	27	0.15 558	9.91 363	9	3
58	9.75 823	18	9.84 469	27	0.15 531	9.91 354	9	2
59	9.75 841	18	9.84 496	27	0.15 504	9.91 345	9	1
60	9.75 859		9.84 523		0.15 477	9.91 336		0
	Log Cos	d	Log Cot	c d	Log Tan	Log Sin	d	′

Prop. Parts

	28	27	26
1	2.8	2.7	2.6
2	5.6	5.4	5.2
3	8.4	8.1	7.8
4	11.2	10.8	10.4
5	14.0	13.5	13.0
6	16.8	16.2	15.6
7	19.6	18.9	18.2
8	22.4	21.6	20.8
9	25.2	24.3	23.4

	19	18
1	1.9	1.8
2	3.8	3.6
3	5.7	5.4
4	7.6	7.2
5	9.5	9.0
6	11.4	10.8
7	13.3	12.6
8	15.2	14.4
9	17.1	16.2

	9	8
1	0.9	0.8
2	1.8	1.6
3	2.7	2.4
4	3.6	3.2
5	4.5	4.0
6	5.4	4.8
7	6.3	5.6
8	7.2	6.4
9	8.1	7.2

55°

35°

′	Log Sin	d	Log Tan	c d	Log Cot	Log Cos	d	
0	9.75 859	18	9.84 523	27	0.15 477	9.91 336	8	60
1	9.75 877	18	9.84 550	26	0.15 450	9.91 328	9	59
2	9.75 895	18	9.84 576	27	0.15 424	9.91 319	9	58
3	9.75 913	18	9.84 603	27	0.15 397	9.91 310	9	57
4	9.75 931	18	9.84 630	27	0.15 370	9.91 301	9	56
5	9.75 949	18	9.84 657	27	0.15 343	9.91 292	9	55
6	9.75 967	18	9.84 684	27	0.15 316	9.91 283	9	54
7	9.75 985	18	9.84 711	27	0.15 289	9.91 274	8	53
8	9.76 003	18	9.84 738	26	0.15 262	9.91 266	9	52
9	9.76 021	18	9.84 764	27	0.15 236	9.91 257	9	51
10	9.76 039	18	9.84 791	27	0.15 209	9.91 248	9	50
11	9.76 057	18	9.84 818	27	0.15 182	9.91 239	9	49
12	9.76 075	18	9.84 845	27	0.15 155	9.91 230	9	48
13	9.76 093	18	9.84 872	27	0.15 128	9.91 221	9	47
14	9.76 111	18	9.84 899	26	0.15 101	9.91 212	9	46
15	9.76 129	17	9.84 925	27	0.15 075	9.91 203	9	45
16	9.76 146	18	9.84 952	27	0.15 048	9.91 194	9	44
17	9.76 164	18	9.84 979	27	0.15 021	9.91 185	9	43
18	9.76 182	18	9.85 006	27	0.14 994	9.91 176	9	42
19	9.76 200	18	9.85 033	26	0.14 967	9.91 167	9	41
20	9.76 218	18	9.85 059	27	0.14 941	9.91 158	9	40
21	9.76 236	17	9.85 086	27	0.14 914	9.91 149	8	39
22	9.76 253	18	9.85 113	27	0.14 887	9.91 141	9	38
23	9.76 271	18	9.85 140	26	0.14 860	9.91 132	9	37
24	9.76 289	18	9.85 166	27	0.14 834	9.91 123	9	36
25	9.76 307	17	9.85 193	27	0.14 807	9.91 114	9	35
26	9.76 324	18	9.85 220	27	0.14 780	9.91 105	9	34
27	9.76 342	18	9.85 247	26	0.14 753	9.91 096	9	33
28	9.76 360	18	9.85 273	27	0.14 727	9.91 087	9	32
29	9.76 378	17	9.85 300	27	0.14 700	9.91 078	9	31
30	9.76 395	18	9.85 327	27	0.14 673	9.91 069	9	30
31	9.76 413	18	9.85 354	26	0.14 646	9.91 060	9	29
32	9.76 431	17	9.85 380	27	0.14 620	9.91 051	9	28
33	9.76 448	18	9.85 407	27	0.14 593	9.91 042	9	27
34	9.76 466	18	9.85 434	26	0.14 566	9.91 033	10	26
35	9.76 484	17	9.85 460	27	0.14 540	9.91 023	9	25
36	9.76 501	18	9.85 487	27	0.14 513	9.91 014	9	24
37	9.76 519	18	9.85 514	26	0.14 486	9.91 005	9	23
38	9.76 537	17	9.85 540	27	0.14 460	9.90 996	9	22
39	9.76 554	18	9.85 567	27	0.14 433	9.90 987	9	21
40	9.76 572	18	9.85 594	26	0.14 406	9.90 978	9	20
41	9.76 590	17	9.85 620	27	0.14 380	9.90 969	9	19
42	9.76 607	18	9.85 647	27	0.14 353	9.90 960	9	18
43	9.76 625	17	9.85 674	26	0.14 326	9.90 951	9	17
44	9.76 642	18	9.85 700	27	0.14 300	9.90 942	9	16
45	9.76 660	17	9.85 727	27	0.14 273	9.90 933	9	15
46	9.76 677	18	9.85 754	26	0.14 246	9.90 924	9	14
47	9.76 695	17	9.85 780	27	0.14 220	9.90 915	9	13
48	9.76 712	18	9.85 807	27	0.14 193	9.90 906	10	12
49	9.76 730	17	9.85 834	26	0.14 166	9.90 896	9	11
50	9.76 747	18	9.85 860	27	0.14 140	9.90 887	9	10
51	9.76 765	17	9.85 887	26	0.14 113	9.90 878	9	9
52	9.76 782	18	9.85 913	27	0.14 087	9.90 869	9	8
53	9.76 800	17	9.85 940	27	0.14 060	9.90 860	9	7
54	9.76 817	18	9.85 967	26	0.14 033	9.90 851	9	6
55	9.76 835	17	9.85 993	27	0.14 007	9.90 842	10	5
56	9.76 852	18	9.86 020	26	0.13 980	9.90 832	9	4
57	9.76 870	17	9.86 046	27	0.13 954	9.90 823	9	3
58	9.76 887	17	9.86 073	27	0.13 927	9.90 814	9	2
59	9.76 904	18	9.86 100	26	0.13 900	9.90 805	9	1
60	9.76 922		9.86 126		0.13 874	9.90 796		0
	Log Cos	d	Log Cot	c d	Log Tan	Log Sin	d	′

Prop. Parts

	27	26	18
1	2.7	2.6	1.8
2	5.4	5.2	3.6
3	8.1	7.8	5.4
4	10.8	10.4	7.2
5	13.5	13.0	9.0
6	16.2	15.6	10.8
7	18.9	18.2	12.6
8	21.6	20.8	14.4
9	24.3	23.4	16.2

	17	10
1	1.7	1.0
2	3.4	2.0
3	5.1	3.0
4	6.8	4.0
5	8.5	5.0
6	10.2	6.0
7	11.9	7.0
8	13.6	8.0
9	15.3	9.0

	9	8
1	0.9	0.8
2	1.8	1.6
3	2.7	2.4
4	3.6	3.2
5	4.5	4.0
6	5.4	4.8
7	6.3	5.6
8	7.2	6.4
9	8.1	7.2

Prop. Parts

54°

36°

′	Log Sin	d	Log Tan	c d	Log Cot	Log Cos	d		Prop. Parts
0	9.76 922	17	9.86 126	27	0.13 874	9.90 796	9	60	
1	9.76 939	18	9.86 153	26	0.13 847	9.90 787	10	59	
2	9.76 957	17	9.86 179	27	0.13 821	9.90 777	9	58	
3	9.76 974	17	9.86 206	26	0.13 794	9.90 768	9	57	
4	9.76 991	18	9.86 232	27	0.13 768	9.90 759	9	56	
5	9.77 009	17	9.86 259	26	0.13 741	9.90 750	9	55	
6	9.77 026	17	9.86 285	27	0.13 715	9.90 741	10	54	
7	9.77 043	18	9.86 312	26	0.13 688	9.90 731	9	53	
8	9.77 061	17	9.86 338	27	0.13 662	9.90 722	9	52	
9	9.77 078	17	9.86 365	27	0.13 635	9.90 713	9	51	

	27	26	18
1	2.7	2.6	1.8
2	5.4	5.2	3.6
3	8.1	7.8	5.4
4	10.8	10.4	7.2
5	13.5	13.0	9.0
6	16.2	15.6	10.8
7	18.9	18.2	12.6
8	21.6	20.8	14.4
9	24.3	23.4	16.2

′	Log Sin	d	Log Tan	c d	Log Cot	Log Cos	d	
10	9.77 095	17	9.86 392	26	0.13 608	9.90 704	10	50
11	9.77 112	18	9.86 418	27	0.13 582	9.90 694	9	49
12	9.77 130	17	9.86 445	26	0.13 555	9.90 685	9	48
13	9.77 147	17	9.86 471	27	0.13 529	9.90 676	9	47
14	9.77 164	17	9.86 498	26	0.13 502	9.90 667	10	46
15	9.77 181	18	9.86 524	27	0.13 476	9.90 657	9	45
16	9.77 199	17	9.86 551	26	0.13 449	9.90 648	9	44
17	9.77 216	17	9.86 577	26	0.13 423	9.90 639	9	43
18	9.77 233	17	9.86 603	27	0.13 397	9.90 630	10	42
19	9.77 250	18	9.86 630	26	0.13 370	9.90 620	9	41
20	9.77 268	17	9.86 656	27	0.13 344	9.90 611	9	40
21	9.77 285	17	9.86 683	26	0.13 317	9.90 602	10	39
22	9.77 302	17	9.86 709	27	0.13 291	9.90 592	9	38
23	9.77 319	17	9.86 736	26	0.13 264	9.90 583	9	37
24	9.77 336	17	9.86 762	27	0.13 238	9.90 574	9	36
25	9.77 353	17	9.86 789	26	0.13 211	9.90 565	10	35
26	9.77 370	17	9.86 815	27	0.13 185	9.90 555	9	34
27	9.77 387	18	9.86 842	26	0.13 158	9.90 546	9	33
28	9.77 405	17	9.86 868	26	0.13 132	9.90 537	10	32
29	9.77 422	17	9.86 894	27	0.13 106	9.90 527	9	31

	17	16
1	1.7	1.6
2	3.4	3.2
3	5.1	4.8
4	6.8	6.4
5	8.5	8.0
6	10.2	9.6
7	11.9	11.2
8	13.6	12.8
9	15.3	14.4

′	Log Sin	d	Log Tan	c d	Log Cot	Log Cos	d	
30	9.77 439	17	9.86 921	26	0.13 079	9.90 518	9	30
31	9.77 456	17	9.86 947	27	0.13 053	9.90 509	10	29
32	9.77 473	17	9.86 974	26	0.13 026	9.90 499	9	28
33	9.77 490	17	9.87 000	27	0.13 000	9.90 490	10	27
34	9.77 507	17	9.87 027	26	0.12 973	9.90 480	9	26
35	9.77 524	17	9.87 053	26	0.12 947	9.90 471	9	25
36	9.77 541	17	9.87 079	27	0.12 921	9.90 462	10	24
37	9.77 558	17	9.87 106	26	0.12 894	9.90 452	9	23
38	9.77 575	17	9.87 132	26	0.12 868	9.90 443	9	22
39	9.77 592	17	9.87 158	27	0.12 842	9.90 434	10	21
40	9.77 609	17	9.87 185	26	0.12 815	9.90 424	9	20
41	9.77 626	17	9.87 211	27	0.12 789	9.90 415	10	19
42	9.77 643	17	9.87 238	26	0.12 762	9.90 405	9	18
43	9.77 660	17	9.87 264	26	0.12 736	9.90 396	10	17
44	9.77 677	17	9.87 290	27	0.12 710	9.90 386	9	16
45	9.77 694	17	9.87 317	26	0.12 683	9.90 377	9	15
46	9.77 711	17	9.87 343	26	0.12 657	9.90 368	10	14
47	9.77 728	16	9.87 369	27	0.12 631	9.90 358	9	13
48	9.77 744	17	9.87 396	26	0.12 604	9.90 349	10	12
49	9.77 761	17	9.87 422	26	0.12 578	9.90 339	10	11

	10	9
1	1.0	0.9
2	2.0	1.8
3	3.0	2.7
4	4.0	3.6
5	5.0	4.5
6	6.0	5.4
7	7.0	6.3
8	8.0	7.2
9	9.0	8.1

′	Log Sin	d	Log Tan	c d	Log Cot	Log Cos	d	
50	9.77 778	17	9.87 448	27	0.12 552	9.90 330	10	10
51	9.77 795	17	9.87 475	26	0.12 525	9.90 320	9	9
52	9.77 812	17	9.87 501	26	0.12 499	9.90 311	10	8
53	9.77 829	17	9.87 527	27	0.12 473	9.90 301	9	7
54	9.77 846	16	9.87 554	26	0.12 446	9.90 292	10	6
55	9.77 862	17	9.87 580	26	0.12 420	9.90 282	9	5
56	9.77 879	17	9.87 606	27	0.12 394	9.90 273	10	4
57	9.77 896	17	9.87 633	26	0.12 367	9.90 263	9	3
58	9.77 913	17	9.87 659	26	0.12 341	9.90 254	10	2
59	9.77 930	16	9.87 685	26	0.12 315	9.90 244	9	1
60	9.77 946		9.87 711		0.12 289	9.90 235		0
	Log Cos	d	Log Cot	c d	Log Tan	Log Sin	d	′	Prop. Parts

53°

37°

′	Log Sin	d	Log Tan	c d	Log Cot	Log Cos	d		Prop. Parts		
0	9.77 946	17	9.87 711	27	0.12 289	9.90 235	10	**60**			
1	9.77 963	17	9.87 738	26	0.12 262	9.90 225	9	59			
2	9.77 980	17	9.87 764	26	0.12 236	9.90 216	10	58			
3	9.77 997	16	9.87 790	27	0.12 210	9.90 206	9	57			
4	9.78 013	17	9.87 817	26	0.12 183	9.90 197	10	56			
5	9.78 030	17	9.87 843	26	0.12 157	9.90 187	9	**55**			
6	9.78 047	16	9.87 869	26	0.12 131	9.90 178	10	54			
7	9.78 063	17	9.87 895	27	0.12 105	9.90 168	9	53			
8	9.78 080	17	9.87 922	26	0.12 078	9.90 159	10	52			
9	9.78 097	16	9.87 948	26	0.12 052	9.90 149	10	51			
10	9.78 113	17	9.87 974	26	0.12 026	9.90 139	9	**50**			
11	9.78 130	17	9.88 000	27	0.12 000	9.90 130	10	49			
12	9.78 147	16	9.88 027	26	0.11 973	9.90 120	9	48			
13	9.78 163	17	9.88 053	26	0.11 947	9.90 111	10	47			
14	9.78 180	17	9.88 079	26	0.11 921	9.90 101	10	46			
15	9.78 197	16	9.88 105	26	0.11 895	9.90 091	9	**45**	**27**	**26**	**17**
16	9.78 213	17	9.88 131	27	0.11 869	9.90 082	10	44	1 2.7	2.6	1.7
17	9.78 230	16	9.88 158	26	0.11 842	9.90 072	9	43	2 5.4	5.2	3.4
18	9.78 246	17	9.88 184	26	0.11 816	9.90 063	10	42	3 8.1	7.8	5.1
19	9.78 263	17	9.88 210	26	0.11 790	9.90 053	10	41	4 10.8	10.4	6.8
20	9.78 280	16	9.88 236	26	0.11 764	9.90 043	9	**40**	5 13.5	13.0	8.5
21	9.78 296	17	9.88 262	27	0.11 738	9.90 034	10	39	6 16.2	15.6	10.2
22	9.78 313	16	9.88 289	26	0.11 711	9.90 024	10	38	7 18.9	18.2	11.9
23	9.78 329	17	9.88 315	26	0.11 685	9.90 014	9	37	8 21.6	20.8	13.6
24	9.78 346	16	9.88 341	26	0.11 659	9.90 005	10	36	9 24.3	23.4	15.3
25	9.78 362	17	9.88 367	26	0.11 633	9.89 995	10	**35**			
26	9.78 379	16	9.88 393	27	0.11 607	9.89 985	9	34			
27	9.78 395	17	9.88 420	26	0.11 580	9.89 976	10	33			
28	9.78 412	16	9.88 446	26	0.11 554	9.89 966	10	32			
29	9.78 428	17	9.88 472	26	0.11 528	9.89 956	9	31			
30	9.78 445	16	9.88 498	26	0.11 502	9.89 947	10	**30**			
31	9.78 461	17	9.88 524	26	0.11 476	9.89 937	10	29			
32	9.78 478	16	9.88 550	27	0.11 450	9.89 927	10	28			
33	9.78 494	16	9.88 577	26	0.11 423	9.89 918	10	27			
34	9.78 510	17	9.88 603	26	0.11 397	9.89 908	10	26			
35	9.78 527	16	9.88 629	26	0.11 371	9.89 898	10	**25**			
36	9.78 543	17	9.88 655	26	0.11 345	9.89 888	9	24	**16**	**10**	**9**
37	9.78 560	16	9.88 681	26	0.11 319	9.89 879	10	23	1 1.6	1.0	0.9
38	9.78 576	16	9.88 707	26	0.11 293	9.89 869	10	22	2 3.2	2.0	1.8
39	9.78 592	17	9.88 733	26	0.11 267	9.89 859	10	21	3 4.8	3.0	2.7
40	9.78 609	16	9.88 759	27	0.11 241	9.89 849	9	**20**	4 6.4	4.0	3.6
41	9.78 625	17	9.88 786	26	0.11 214	9.89 840	10	19	5 8.0	5.0	4.5
42	9.78 642	16	9.88 812	26	0.11 188	9.89 830	10	18	6 9.6	6.0	5.4
43	9.78 658	16	9.88 838	26	0.11 162	9.89 820	10	17	7 11.2	7.0	6.3
44	9.78 674	17	9.88 864	26	0.11 136	9.89 810	9	16	8 12.8	8.0	7.2
45	9.78 691	16	9.88 890	26	0.11 110	9.89 801	10	**15**	9 14.4	9.0	8.1
46	9.78 707	16	9.88 916	26	0.11 084	9.89 791	10	14			
47	9.78 723	16	9.88 942	26	0.11 058	9.89 781	10	13			
48	9.78 739	17	9.88 968	26	0.11 032	9.89 771	10	12			
49	9.78 756	16	9.88 994	26	0.11 006	9.89 761	9	11			
50	9.78 772	16	9.89 020	26	0.10 980	9.89 752	10	**10**			
51	9.78 788	17	9.89 046	27	0.10 954	9.89 742	10	9			
52	9.78 805	16	9.89 073	26	0.10 927	9.89 732	10	8			
53	9.78 821	16	9.89 099	26	0.10 901	9.89 722	10	7			
54	9.78 837	16	9.89 125	26	0.10 875	9.89 712	10	6			
55	9.78 853	16	9.89 151	26	0.10 849	9.89 702	9	**5**			
56	9.78 869	17	9.89 177	26	0.10 823	9.89 693	10	4			
57	9.78 886	16	9.89 203	26	0.10 797	9.89 683	10	3			
58	9.78 902	16	9.89 229	26	0.10 771	9.89 673	10	2			
59	9.78 918	16	9.89 255	26	0.10 745	9.89 663	10	1			
60	9.78 934		9.89 281		0.10 719	9.89 653		**0**			
	Log Cos	d	Log Cot	c d	Log Tan	Log Sin	d	′	Prop. Parts		

52°

Table II

38°

′	Log Sin	d	Log Tan	c d	Log Cot	Log Cos	d		Prop. Parts		
0	9.78 934	16	9.89 281	26	0.10 719	9.89 653	10	**60**			
1	9.78 950	17	9.89 307	26	0.10 693	9.89 643	10	59			
2	9.78 967	16	9.89 333	26	0.10 667	9.89 633	9	58			
3	9.78 983	16	9.89 359	26	0.10 641	9.89 624	10	57			
4	9.78 999	16	9.89 385	26	0.10 615	9.89 614	10	56			
5	9.79 015	16	9.89 411	26	0.10 589	9.89 604	10	**55**			
6	9.79 031	16	9.89 437	26	0.10 563	9.89 594	10	54			
7	9.79 047	16	9.89 463	26	0.10 537	9.89 584	10	53			
8	9.79 063	16	9.89 489	26	0.10 511	9.89 574	10	52			
9	9.79 079	16	9.89 515	26	0.10 485	9.89 564	10	51		**26**	**25**	**17**
10	9.79 095	16	9.89 541	26	0.10 459	9.89 554	10	**50**	1	2.6	2.5	1.7
11	9.79 111	17	9.89 567	26	0.10 433	9.89 544	10	49	2	5.2	5.0	3.4
12	9.79 128	16	9.89 593	26	0.10 407	9.89 534	10	48	3	7.8	7.5	5.1
13	9.79 144	16	9.89 619	26	0.10 381	9.89 524	10	47	4	10.4	10.0	6.8
14	9.79 160	16	9.89 645	26	0.10 355	9.89 514	10	46	5	13.0	12.5	8.5
15	9.79 176	16	9.89 671	26	0.10 329	9.89 504	9	**45**	6	15.6	15.0	10.2
16	9.79 192	16	9.89 697	26	0.10 303	9.89 495	10	44	7	18.2	17.5	11.9
17	9.79 208	16	9.89 723	26	0.10 277	9.89 485	10	43	8	20.8	20.0	13.6
18	9.79 224	16	9.89 749	26	0.10 251	9.89 475	10	42	9	23.4	22.5	15.3
19	9.79 240	16	9.89 775	26	0.10 225	9.89 465	10	41				
20	9.79 256	16	9.89 801	26	0.10 199	9.89 455	10	**40**				
21	9.79 272	16	9.89 827	26	0.10 173	9.89 445	10	39				
22	9.79 288	16	9.89 853	26	0.10 147	9.89 435	10	38				
23	9.79 304	15	9.89 879	26	0.10 121	9.89 425	10	37				
24	9.79 319	16	9.89 905	26	0.10 095	9.89 415	10	36				
25	9.79 335	16	9.89 931	26	0.10 069	9.89 405	10	**35**		**16**	**15**	**11**
26	9.79 351	16	9.89 957	26	0.10 043	9.89 395	10	34				
27	9.79 367	16	9.89 983	26	0.10 017	9.89 385	10	33	1	1.6	1.5	1.1
28	9.79 383	16	9.90 009	26	0.09 991	9.89 375	11	32	2	3.2	3.0	2.2
29	9.79 399	16	9.90 035	26	0.09 965	9.89 364	10	31	3	4.8	4.5	3.3
30	9.79 415	16	9.90 061	25	0.09 939	9.89 354	10	**30**	4	6.4	6.0	4.4
31	9.79 431	16	9.90 086	26	0.09 914	9.89 344	10	29	5	8.0	7.5	5.5
32	9.79 447	16	9.90 112	26	0.09 888	9.89 334	10	28	6	9.6	9.0	6.6
33	9.79 463	15	9.90 138	26	0.09 862	9.89 324	10	27	7	11.2	10.5	7.7
34	9.79 478	16	9.90 164	26	0.09 836	9.89 314	10	26	8	12.8	12.0	8.8
35	9.79 494	16	9.90 190	26	0.09 810	9.89 304	10	**25**	9	14.4	13.5	9.9
36	9.79 510	16	9.90 216	26	0.09 784	9.89 294	10	24				
37	9.79 526	16	9.90 242	26	0.09 758	9.89 284	10	23				
38	9.79 542	16	9.90 268	26	0.09 732	9.89 274	10	22				
39	9.79 558	15	9.90 294	26	0.09 706	9.89 264	10	21				
40	9.79 573	16	9.90 320	26	0.09 680	9.89 254	10	**20**				
41	9.79 589	16	9.90 346	25	0.09 654	9.89 244	11	19				
42	9.79 605	16	9.90 371	26	0.09 629	9.89 233	10	18		**10**	**9**	
43	9.79 621	15	9.90 397	26	0.09 603	9.89 223	10	17				
44	9.79 636	16	9.90 423	26	0.09 577	9.89 213	10	16	1	1.0	0.9	
45	9.79 652	16	9.90 449	26	0.09 551	9.89 203	10	**15**	2	2.0	1.8	
46	9.79 668	16	9.90 475	26	0.09 525	9.89 193	10	14	3	3.0	2.7	
47	9.79 684	15	9.90 501	26	0.09 499	9.89 183	10	13	4	4.0	3.6	
48	9.79 699	16	9.90 527	26	0.09 473	9.89 173	11	12	5	5.0	4.5	
49	9.79 715	16	9.90 553	25	0.09 447	9.89 162	10	11	6	6.0	5.4	
50	9.79 731	15	9.90 578	26	0.09 422	9.89 152	10	**10**	7	7.0	6.3	
51	9.79 746	16	9.90 604	26	0.09 396	9.89 142	10	9	8	8.0	7.2	
52	9.79 762	16	9.90 630	26	0.09 370	9.89 132	10	8	9	9.0	8.1	
53	9.79 778	15	9.90 656	26	0.09 344	9.89 122	10	7				
54	9.79 793	16	9.90 682	26	0.09 318	9.89 112	11	6				
55	9.79 809	16	9.90 708	26	0.09 292	9.89 101	10	**5**				
56	9.79 825	15	9.90 734	25	0.09 266	9.89 091	10	4				
57	9.79 840	16	9.90 759	26	0.09 241	9.89 081	10	3				
58	9.79 856	16	9.90 785	26	0.09 215	9.89 071	11	2				
59	9.79 872	15	9.90 811	26	0.09 189	9.89 060	10	1				
60	9.79 887		9.90 837		0.09 163	9.89 050		**0**				
	Log Cos	d	Log Cot	c d	Log Tan	Log Sin	d	′		Prop. Parts		

51°

39°

'	Log Sin	d	Log Tan	c d	Log Cot	Log Cos	d		Prop. Parts		
0	9.79 887	16	9.90 837	26	0.09 163	9.89 050	10	60			
1	9.79 903	15	9.90 863	26	0.09 137	9.89 040	10	59			
2	9.79 918	16	9.90 889	25	0.09 111	9.89 030	10	58			
3	9.79 934	16	9.90 914	26	0.09 086	9.89 020	11	57			
4	9.79 950	15	9.90 940	26	0.09 060	9.89 009	10	56			
5	9.79 965	16	9.90 966	26	0.09 034	9.88 999	10	55			
6	9.79 981	15	9.90 992	26	0.09 008	9.88 989	11	54			
7	9.79 996	16	9.91 018	25	0.08 982	9.88 978	10	53			
8	9.80 012	15	9.91 043	26	0.08 957	9.88 968	10	52			
9	9.80 027	16	9.91 069	26	0.08 931	9.88 958	10	51			
10	9.80 043	15	9.91 095	26	0.08 905	9.88 948	11	50			
11	9.80 058	16	9.91 121	26	0.08 879	9.88 937	10	49			
12	9.80 074	15	9.91 147	25	0.08 853	9.88 927	10	48			
13	9.80 089	16	9.91 172	26	0.08 828	9.88 917	11	47			
14	9.80 105	15	9.91 198	26	0.08 802	9.88 906	10	46	**26**	**25**	**16**
15	9.80 120	16	9.91 224	26	0.08 776	9.88 896	10	45	1 2.6	2.5	1.6
16	9.80 136	15	9.91 250	26	0.08 750	9.88 886	11	44	2 5.2	5.0	3.2
17	9.80 151	15	9.91 276	25	0.08 724	9.88 875	10	43	3 7.8	7.5	4.8
18	9.80 166	16	9.91 301	26	0.08 699	9.88 865	10	42	4 10.4	10.0	6.4
19	9.80 182	15	9.91 327	26	0.08 673	9.88 855	11	41	5 13.0	12.5	8.0
20	9.80 197	16	9.91 353	26	0.08 647	9.88 844	10	40	6 15.6	15.0	9.6
21	9.80 213	15	9.91 379	25	0.08 621	9.88 834	10	39	7 18.2	17.5	11.2
22	9.80 228	16	9.91 404	26	0.08 596	9.88 824	11	38	8 20.8	20.0	12.8
23	9.80 244	15	9.91 430	26	0.08 570	9.88 813	10	37	9 23.4	22.5	14.4
24	9.80 259	15	9.91 456	26	0.08 544	9.88 803	10	36			
25	9.80 274	16	9.91 482	25	0.08 518	9.88 793	11	35			
26	9.80 290	15	9.91 507	26	0.08 493	9.88 782	10	34			
27	9.80 305	15	9.91 533	26	0.08 467	9.88 772	11	33			
28	9.80 320	16	9.91 559	26	0.08 441	9.88 761	10	32			
29	9.80 336	15	9.91 585	25	0.08 415	9.88 751	10	31			
30	9.80 351	15	9.91 610	26	0.08 390	9.88 741	11	30			
31	9.80 366	16	9.91 636	26	0.08 364	9.88 730	10	29			
32	9.80 382	15	9.91 662	26	0.08 338	9.88 720	11	28			
33	9.80 397	15	9.91 688	25	0.08 312	9.88 709	10	27			
34	9.80 412	16	9.91 713	26	0.08 287	9.88 699	11	26			
35	9.80 428	15	9.91 739	26	0.08 261	9.88 688	10	25			
36	9.80 443	15	9.91 765	26	0.08 235	9.88 678	10	24			
37	9.80 458	15	9.91 791	25	0.08 209	9.88 668	11	23	**15**	**11**	**10**
38	9.80 473	16	9.91 816	26	0.08 184	9.88 657	10	22	1 1.5	1.1	1.0
39	9.80 489	15	9.91 842	26	0.08 158	9.88 647	11	21	2 3.0	2.2	2.0
40	9.80 504	15	9.91 868	25	0.08 132	9.88 636	10	20	3 4.5	3.3	3.0
41	9.80 519	15	9.91 893	26	0.08 107	9.88 626	11	19	4 6.0	4.4	4.0
42	9.80 534	16	9.91 919	26	0.08 081	9.88 615	10	18	5 7.5	5.5	5.0
43	9.80 550	15	9.91 945	26	0.08 055	9.88 605	10	17	6 9.0	6.6	6.0
44	9.80 565	15	9.91 971	25	0.08 029	9.88 594	10	16	7 10.5	7.7	7.0
45	9.80 580	15	9.91 996	26	0.08 004	9.88 584	11	15	8 12.0	8.8	8.0
46	9.80 595	15	9.92 022	26	0.07 978	9.88 573	10	14	9 13.5	9.9	9.0
47	9.80 610	15	9.92 048	25	0.07 952	9.88 563	11	13			
48	9.80 625	16	9.92 073	26	0.07 927	9.88 552	10	12			
49	9.80 641	15	9.92 099	26	0.07 901	9.88 542	11	11			
50	9.80 656	15	9.92 125	25	0.07 875	9.88 531	10	10			
51	9.80 671	15	9.92 150	26	0.07 850	9.88 521	11	9			
52	9.80 686	15	9.92 176	26	0.07 824	9.88 510	11	8			
53	9.80 701	15	9.92 202	25	0.07 798	9.88 499	10	7			
54	9.80 716	15	9.92 227	26	0.07 773	9.88 489	11	6			
55	9.80 731	15	9.92 253	26	0.07 747	9.88 478	10	5			
56	9.80 746	16	9.92 279	25	0.07 721	9.88 468	11	4			
57	9.80 762	15	9.92 304	26	0.07 696	9.88 457	10	3			
58	9.80 777	15	9.92 330	26	0.07 670	9.88 447	11	2			
59	9.80 792	15	9.92 356	25	0.07 644	9.88 436	11	1			
60	9.80 807		9.92 381		0.07 619	9.88 425		0			
	Log Cos	d	Log Cot	c d	Log Tan	Log Sin	d	'	Prop. Parts		

50°

40°

'	Log Sin	d	Log Tan	c d	Log Cot	Log Cos	d	
0	9.80 807	15	9.92 381	26	0.07 619	9.88 425	10	60
1	9.80 822	15	9.92 407	26	0.07 593	9.88 415	11	59
2	9.80 837	15	9.92 433	25	0.07 567	9.88 404	10	58
3	9.80 852	15	9.92 458	26	0.07 542	9.88 394	11	57
4	9.80 867	15	9.92 484	26	0.07 516	9.88 383	11	56
5	9.80 882	15	9.92 510	25	0.07 490	9.88 372	10	55
6	9.80 897	15	9.92 535	26	0.07 465	9.88 362	11	54
7	9.80 912	15	9.92 561	26	0.07 439	9.88 351	11	53
8	9.80 927	15	9.92 587	25	9.07 413	9.88 340	10	52
9	9.80 942	15	9.92 612	26	0.07 388	9.88 330	11	51
10	9.80 957	15	9.92 638	25	0.07 362	9.88 319	11	50
11	9.80 972	15	9.92 663	26	0.07 337	9.88 308	10	49
12	9.80 987	15	9.92 689	26	0.07 311	9.88 298	11	48
13	9.81 002	15	9.92 715	25	0.07 285	9.88 287	11	47
14	9.81 017	15	9.92 740	26	0.07 260	9.88 276	10	46
15	9.81 032	15	9.92 766	26	0.07 234	9.88 266	11	45
16	9.81 047	14	9.92 792	25	0.07 208	9.88 255	11	44
17	9.81 061	15	9.92 817	26	0.07 183	9.88 244	10	43
18	9.81 076	15	9.92 843	25	0.07 157	9.88 234	11	42
19	9.81 091	15	9.92 868	26	0.07 132	9.88 223	11	41
20	9.81 106	15	9.92 894	26	0.07 106	9.88 212	11	40
21	9.81 121	15	9.92 920	25	0.07 080	9.88 201	10	39
22	9.81 136	15	9.92 945	26	0.07 055	9.88 191	11	38
23	9.81 151	15	9.92 971	25	0.07 029	9.88 180	11	37
24	9.81 166	14	9.92 996	26	0.07 004	9.88 169	11	36
25	9.81 180	15	9.93 022	26	0.06 978	9.88 158	10	35
26	9.81 195	15	9.93 048	25	0.06 952	9.88 148	11	34
27	9.81 210	15	9.93 073	26	0.06 927	9.88 137	11	33
28	9.81 225	15	9.93 099	25	0.06 901	9.88 126	11	32
29	9.81 240	14	9.93 124	26	0.06 876	9.88 115	10	31
30	9.81 254	15	9.93 150	25	0.06 850	9.88 105	11	30
31	9.81 269	15	9.93 175	26	0.06 825	9.88 094	11	29
32	9.81 284	15	9.93 201	26	0.06 799	9.88 083	11	28
33	9.81 299	15	9.93 227	25	0.06 773	9.88 072	11	27
34	9.81 314	14	9.93 252	26	0.06 748	9.88 061	10	26
35	9.81 328	15	9.93 278	25	0.06 722	9.88 051	11	25
36	9.81 343	15	9.93 303	26	0.06 697	9.88 040	11	24
37	9.81 358	14	9.93 329	25	0.06 671	9.88 029	11	23
38	9.81 372	15	9.93 354	26	0.06 646	9.88 018	11	22
39	9.81 387	15	9.93 380	26	0.06 620	9.88 007	11	21
40	9.81 402	15	9.93 406	25	0.06 594	9.87 996	11	20
41	9.81 417	14	9.93 431	26	0.06 569	9.87 985	10	19
42	9.81 431	15	9.93 457	25	0.06 543	9.87 975	11	18
43	9.81 446	15	9.93 482	26	0.06 518	9.87 964	11	17
44	9.81 461	14	9.93 508	25	0.06 492	9.87 953	11	16
45	9.81 475	15	9.93 533	26	0.06 467	9.87 942	11	15
46	9.81 490	15	9.93 559	25	0.06 441	9.87 931	11	14
47	9.81 505	14	9.93 584	26	0.06 416	9.87 920	11	13
48	9.81 519	15	9.93 610	26	0.06 390	9.87 909	11	12
49	9.81 534	15	9.93 636	25	0.06 364	9.87 898	11	11
50	9.81 549	14	9.93 661	26	0.06 339	9.87 887	10	10
51	9.81 563	15	9.93 687	25	0.06 313	9.87 877	11	9
52	9.81 578	14	9.93 712	26	0.06 288	9.87 866	11	8
53	9.81 592	15	9.93 738	25	0.06 262	9.87 855	11	7
54	9.81 607	15	9.93 763	26	0.06 237	9.87 844	11	6
55	9.81 622	14	9.93 789	25	0.06 211	9.87 833	11	5
56	9.81 636	15	9.93 814	26	0.06 186	9.87 822	11	4
57	9.81 651	14	9.93 840	25	0.06 160	9.87 811	11	3
58	9.81 665	15	9.93 865	26	0.06 135	9.87 800	11	2
59	9.81 680	14	9.93 891	25	0.06 109	9.87 789	11	1
60	9.81 694		9.93 916		0.06 084	9.87 778		0
	Log Cos	d	Log Cot	c d	Log Tan	Log Sin	d	'

Prop. Parts

	26	25	15
1	2.6	2.5	1.5
2	5.2	5.0	3.0
3	7.8	7.5	4.5
4	10.4	10.0	6.0
5	13.0	12.5	7.5
6	15.6	15.0	9.0
7	18.2	17.5	10.5
8	20.8	20.0	12.0
9	23.4	22.5	13.5

	14	11	10
1	1.4	1.1	1.0
2	2.8	2.2	2.0
3	4.2	3.3	3.0
4	5.6	4.4	4.0
5	7.0	5.5	5.0
6	8.4	6.6	6.0
7	9.8	7.7	7.0
8	11.2	8.8	8.0
9	12.6	9.9	9.0

49°

41°

′	Log Sin	d	Log Tan	c d	Log Cot	Log Cos	d		Prop. Parts		
0	9.81 694	15	9.93 916	26	0.06 084	9.87 778	11	60			
1	9.81 709	14	9.93 942	25	0.06 058	9.87 767	11	59			
2	9.81 723	15	9.93 967	26	0.06 033	9.87 756	11	58			
3	9.81 738	14	9.93 993	25	0.06 007	9.87 745	11	57			
4	9.81 752	15	9.94 018	26	0.05 982	9.87 734	11	56			
5	9.81 767	14	9.94 044	25	0.05 956	9.87 723	11	55			
6	9.81 781	15	9.94 069	26	0.05 931	9.87 712	11	54			
7	9.81 796	14	9.94 095	25	0.05 905	9.87 701	11	53			
8	9.81 810	15	9.94 120	26	0.05 880	9.87 690	11	52			
9	9.81 825	14	9.94 146	25	0.05 854	9.87 679	11	51			
10	9.81 839	15	9.94 171	26	0.05 829	9.87 668	11	50			
11	9.81 854	14	9.94 197	25	0.05 803	9.87 657	11	49			
12	9.81 868	14	9.94 222	26	0.05 778	9.87 646	11	48			
13	9.81 882	15	9.94 248	25	0.05 752	9.87 635	11	47			
14	9.81 897	14	9.94 273	26	0.05 727	9.87 624	11	46		26	25	15
15	9.81 911	15	9.94 299	25	0.05 701	9.87 613	12	45	1	2.6	2.5	1.5
16	9.81 926	14	9.94 324	26	0.05 676	9.87 601	11	44	2	5.2	5.0	3.0
17	9.81 940	15	9.94 350	25	0.05 650	9.87 590	11	43	3	7.8	7.5	4.5
18	9.81 955	14	9.94 375	26	0.05 625	9.87 579	11	42	4	10.4	10.0	6.0
19	9.81 969	14	9.94 401	25	0.05 599	9.87 568	11	41	5	13.0	12.5	7.5
20	9.81 983	15	9.94 426	26	0.05 574	9.87 557	11	40	6	15.6	15.0	9.0
21	9.81 998	14	9.94 452	25	0.05 548	9.87 546	11	39	7	18.2	17.5	10.5
22	9.82 012	14	9.94 477	26	0.05 523	9.87 535	11	38	8	20.8	20.0	12.0
23	9.82 026	15	9.94 503	25	0.05 497	9.87 524	11	37	9	23.4	22.5	13.5
24	9.82 041	14	9.94 528	26	0.05 472	9.87 513	12	36				
25	9.82 055	14	9.94 554	25	0.05 446	9.87 501	11	35				
26	9.82 069	15	9.94 579	25	0.05 421	9.87 490	11	34				
27	9.82 084	14	9.94 604	26	0.05 396	9.87 479	11	33				
28	9.82 098	14	9.94 630	25	0.05 370	9.87 468	11	32				
29	9.82 112	14	9.94 655	26	0.05 345	9.87 457	11	31				
30	9.82 126	15	9.94 681	25	0.05 319	9.87 446	12	30				
31	9.82 141	14	9.94 706	26	0.05 294	9.87 434	11	29				
32	9.82 155	14	9.94 732	25	0.05 268	9.87 423	11	28				
33	9.82 169	15	9.94 757	26	0.05 243	9.87 412	11	27				
34	9.82 184	14	9.94 783	25	0.05 217	9.87 401	11	26				
35	9.82 198	14	9.94 808	26	0.05 192	9.87 390	12	25				
36	9.82 212	14	9.94 834	25	0.05 166	9.87 378	11	24				
37	9.82 226	14	9.94 859	25	0.05 141	9.87 367	11	23		14	12	11
38	9.82 240	15	9.94 884	26	0.05 116	9.87 356	11	22	1	1.4	1.2	1.1
39	9.82 255	14	9.94 910	25	0.05 090	9.87 345	11	21	2	2.8	2.4	2.2
40	9.82 269	14	9.94 935	26	0.05 065	9.87 334	12	20	3	4.2	3.6	3.3
41	9.82 283	14	9.94 961	25	0.05 039	9.87 322	11	19	4	5.6	4.8	4.4
42	9.82 297	14	9.94 986	26	0.05 014	9.87 311	11	18	5	7.0	6.0	5.5
43	9.82 311	15	9.95 012	25	0.04 988	9.87 300	12	17	6	8.4	7.2	6.6
44	9.82 326	14	9.95 037	25	0.04 963	9.87 288	11	16	7	9.8	8.4	7.7
45	9.82 340	14	9.95 062	26	0.04 938	9.87 277	11	15	8	11.2	9.6	8.8
46	9.82 354	14	9.95 088	25	0.04 912	9.87 266	11	14	9	12.6	10.8	9.9
47	9.82 368	14	9.95 113	26	0.04 887	9.87 255	12	13				
48	9.82 382	14	9.95 139	25	0.04 861	9.87 243	11	12				
49	9.82 396	14	9.95 164	26	0.04 836	9.87 232	11	11				
50	9.82 410	14	9.95 190	25	0.04 810	9.87 221	12	10				
51	9.82 424	15	9.95 215	25	0.04 785	9.87 209	11	9				
52	9.82 439	14	9.95 240	26	0.04 760	9.87 198	11	8				
53	9.82 453	14	9.95 266	25	0.04 734	9.87 187	12	7				
54	9.82 467	14	9.95 291	26	0.04 709	9.87 175	11	6				
55	9.82 481	14	9.95 317	25	0.04 683	9.87 164	11	5				
56	9.82 495	14	9.95 342	26	0.04 658	9.87 153	12	4				
57	9.82 509	14	9.95 368	25	0.04 632	9.87 141	11	3				
58	9.82 523	14	9.95 393	25	0.04 607	9.87 130	11	2				
59	9.82 537	14	9.95 418	26	0.04 582	9.87 119	12	1				
60	9.82 551		9.95 444		0.04 556	9.87 107		0				
	Log Cos	d	Log Cot	c d	Log Tan	Log Sin	d	′	Prop. Parts		

48°

Table II

42°

′	Log Sin	d	Log Tan	c d	Log Cot	Log Cos	d		Prop. Parts
0	9.82 551	14	9.95 444	25	0.04 556	9.87 107	11	60	
1	9.82 565	14	9.95 469	26	0.04 531	9.87 096	11	59	
2	9.82 579	14	9.95 495	25	0.04 505	9.87 085	12	58	
3	9.82 593	14	9.95 520	25	0.04 480	9.87 073	11	57	
4	9.82 607	14	9.95 545	26	0.04 455	9.87 062	12	56	
5	9.82 621	14	9.95 571	25	0.04 429	9.87 050	11	55	
6	9.82 635	14	9.95 596	26	0.04 404	9.87 039	11	54	
7	9.82 649	14	9.95 622	25	0.04 378	9.87 028	12	53	
8	9.82 663	14	9.95 647	25	0.04 353	9.87 016	11	52	
9	9.82 677	14	9.95 672	26	0.04 328	9.87 005	12	51	
10	9.82 691	14	9.95 698	25	0.04 302	9.86 993	11	50	
11	9.82 705	14	9.95 723	25	0.04 277	9.86 982	12	49	
12	9.82 719	14	9.95 748	26	0.04 252	9.86 970	11	48	
13	9.82 733	14	9.95 774	25	0.04 226	9.86 959	12	47	
14	9.82 747	14	9.95 799	26	0.04 201	9.86 947	11	46	
15	9.82 761	14	9.95 825	25	0.04 175	9.86 936	12	45	
16	9.82 775	13	9.95 850	25	0.04 150	9.86 924	11	44	
17	9.82 788	14	9.95 875	26	0.04 125	9.86 913	11	43	
18	9.82 802	14	9.95 901	25	0.04 099	9.86 902	12	42	
19	9.82 816	14	9.95 926	26	0.04 074	9.86 890	11	41	
20	9.82 830	14	9.95 952	25	0.04 048	9.86 879	12	40	
21	9.82 844	14	9.95 977	25	0.04 023	9.86 867	12	39	
22	9.82 858	14	9.96 002	26	0.03 998	9.86 855	11	38	
23	9.82 872	13	9.96 028	25	0.03 972	9.86 844	12	37	
24	9.82 885	14	9.96 053	25	0.03 947	9.86 832	11	36	
25	9.82 899	14	9.96 078	26	0.03 922	9.86 821	12	35	
26	9.82 913	14	9.96 104	25	0.03 896	9.86 809	11	34	
27	9.82 927	14	9.96 129	26	0.03 871	9.86 798	12	33	
28	9.82 941	14	9.96 155	25	0.03 845	9.86 786	11	32	
29	9.82 955	13	9.96 180	25	0.03 820	9.86 775	12	31	
30	9.82 968	14	9.96 205	26	0.03 795	9.86 763	11	30	
31	9.82 982	14	9.96 231	25	0.03 769	9.86 752	12	29	
32	9.82 996	14	9.96 256	25	0.03 744	9.86 740	12	28	
33	9.83 010	13	9.96 281	26	0.03 719	9.86 728	11	27	
34	9.83 023	14	9.96 307	25	0.03 693	9.86 717	12	26	
35	9.83 037	14	9.96 332	25	0.03 668	9.86 705	11	25	
36	9.83 051	14	9.96 357	26	0.03 643	9.86 694	12	24	
37	9.83 065	13	9.96 383	25	0.03 617	9.86 682	12	23	
38	9.83 078	14	9.96 408	25	0.03 592	9.86 670	11	22	
39	9.83 092	14	9.96 433	26	0.03 567	9.86 659	12	21	
40	9.83 106	14	9.96 459	25	0.03 541	9.86 647	12	20	
41	9.83 120	13	9.96 484	26	0.03 516	9.86 635	11	19	
42	9.83 133	14	9.96 510	25	0.03 490	9.86 624	12	18	
43	9.83 147	14	9.96 535	25	0.03 465	9.86 612	12	17	
44	9.83 161	13	9.96 560	26	0.03 440	9.86 600	11	16	
45	9.83 174	14	9.96 586	25	0.03 414	9.86 589	12	15	
46	9.83 188	14	9.96 611	25	0.03 389	9.86 577	12	14	
47	9.83 202	13	9.96 636	26	0.03 364	9.86 565	11	13	
48	9.83 215	14	9.96 662	25	0.03 338	9.86 554	12	12	
49	9.83 229	13	9.96 687	25	0.03 313	9.86 542	12	11	
50	9.83 242	14	9.96 712	26	0.03 288	9.86 530	12	10	
51	9.83 256	14	9.96 738	25	0.03 262	9.86 518	11	9	
52	9.83 270	13	9.96 763	25	0.03 237	9.86 507	12	8	
53	9.83 283	14	9.96 788	26	0.03 212	9.86 495	12	7	
54	9.83 297	13	9.96 814	25	0.03 186	9.86 483	11	6	
55	9.83 310	14	9.96 839	25	0.03 161	9.86 472	12	5	
56	9.83 324	14	9.96 864	26	0.03 136	9.86 460	12	4	
57	9.83 338	13	9.96 890	25	0.03 110	9.86 448	12	3	
58	9.83 351	14	9.96 915	25	0.03 085	9.86 436	11	2	
59	9.83 365	13	9.96 940	26	0.03 060	9.86 425	12	1	
60	9.83 378		9.96 966		0.03 034	9.86 413		0	
	Log Cos	d	Log Cot	c d	Log Tan	Log Sin	d	′	Prop. Parts

Prop. Parts

	26	25	14
1	2.6	2.5	1.4
2	5.2	5.0	2.8
3	7.8	7.5	4.2
4	10.4	10.0	5.6
5	13.0	12.5	7.0
6	15.6	15.0	8.4
7	18.2	17.5	9.8
8	20.8	20.0	11.2
9	23.4	22.5	12.6

	13	12	11
1	1.3	1.2	1.1
2	2.6	2.4	2.2
3	3.9	3.6	3.3
4	5.2	4.8	4.4
5	6.5	6.0	5.5
6	7.8	7.2	6.6
7	9.1	8.4	7.7
8	10.4	9.6	8.8
9	11.7	10.8	9.9

47°

43°

′	Log Sin	d	Log Tan	c d	Log Cot	Log Cos	d		Prop. Parts		
0	9.83 378	14	9.96 966	25	0.03 034	9.86 413	12	60			
1	9.83 392	13	9.96 991	25	0.03 009	9.86 401	12	59			
2	9.83 405	14	9.97 016	26	0.02 984	9.86 389	12	58			
3	9.83 419	13	9.97 042	25	0.02 958	9.86 377	11	57			
4	9.83 432	14	9.97 067	25	0.02 933	9.86 366	12	56			
5	9.83 446	13	9.97 092	26	0.02 908	9.86 354	12	55			
6	9.83 459	14	9.97 118	25	0.02 882	9.86 342	12	54			
7	9.83 473	13	9.97 143	25	0.02 857	9.86 330	12	53			
8	9.83 486	14	9.97 168	25	0.02 832	9.86 318	12	52			
9	9.83 500	13	9.97 193	26	0.02 807	9.86 306	11	51			
10	9.83 513	14	9.97 219	25	0.02 781	9.86 295	12	50			
11	9.83 527	13	9.97 244	25	0.02 756	9.86 283	12	49			
12	9.83 540	14	9.97 269	26	0.02 731	9.86 271	12	48			
13	9.83 554	13	9.97 295	25	0.02 705	9.86 259	12	47		26	25	14
14	9.83 567	14	9.97 320	25	0.02 680	9.86 247	12	46			
15	9.83 581	13	9.97 345	26	0.02 655	9.86 235	12	45	1	2.6	2.5	1.4
16	9.83 594	14	9.97 371	25	0.02 629	9.86 223	12	44	2	5.2	5.0	2.8
17	9.83 608	13	9.97 396	25	0.02 604	9.86 211	11	43	3	7.8	7.5	4.2
18	9.83 621	13	9.97 421	26	0.02 579	9.86 200	12	42	4	10.4	10.0	5.6
19	9.83 634	14	9.97 447	25	0.02 553	9.86 188	12	41	5	13.0	12.5	7.0
20	9.83 648	13	9.97 472	25	0.02 528	9.86 176	12	40	6	15.6	15.0	8.4
21	9.83 661	13	9.97 497	26	0.02 503	9.86 164	12	39	7	18.2	17.5	9.8
22	9.83 674	14	9.97 523	25	0.02 477	9.86 152	12	38	8	20.8	20.0	11.2
23	9.83 688	13	9.97 548	25	0.02 452	9.86 140	12	37	9	23.4	22.5	12.6
24	9.83 701	14	9.97 573	25	0.02 427	9.86 128	12	36			
25	9.83 715	13	9.97 598	26	0.02 402	9.86 116	12	35			
26	9.83 728	13	9.97 624	25	0.02 376	9.86 104	12	34			
27	9.83 741	14	9.97 649	25	0.02 351	9.86 092	12	33			
28	9.83 755	13	9.97 674	26	0.02 326	9.86 080	12	32			
29	9.83 768	13	9.97 700	25	0.02 300	9.86 068	12	31			
30	9.83 781	14	9.97 725	25	0.02 275	9.86 056	12	30			
31	9.83 795	13	9.97 750	26	0.02 250	9.86 044	12	29			
32	9.83 808	13	9.97 776	25	0.02 224	9.86 032	12	28			
33	9.83 821	13	9.97 801	25	0.02 199	9.86 020	12	27			
34	9.83 834	14	9.97 826	25	0.02 174	9.86 008	12	26			
35	9.83 848	13	9.97 851	26	0.02 149	9.85 996	12	25			
36	9.83 861	13	9.97 877	25	0.02 123	9.85 984	12	24			
37	9.83 874	13	9.97 902	25	0.02 098	9.85 972	12	23		13	12	11
38	9.83 887	14	9.97 927	26	0.02 073	9.85 960	12	22			
39	9.83 901	13	9.97 953	25	0.02 047	9.85 948	12	21	1	1.3	1.2	1.1
40	9.83 914	13	9.97 978	25	0.02 022	9.85 936	12	20	2	2.6	2.4	2.2
41	9.83 927	13	9.98 003	26	0.01 997	9.85 924	12	19	3	3.9	3.6	3.3
42	9.83 940	14	9.98 029	25	0.01 971	9.85 912	12	18	4	5.2	4.8	4.4
43	9.83 954	13	9.98 054	25	0.01 946	9.85 900	12	17	5	6.5	6.0	5.5
44	9.83 967	13	9.98 079	25	0.01 921	9.85 888	12	16	6	7.8	7.2	6.6
45	9.83 980	13	9.98 104	26	0.01 896	9.85 876	12	15	7	9.1	8.4	7.7
46	9.83 993	13	9.98 130	25	0.01 870	9.85 864	13	14	8	10.4	9.6	8.8
47	9.84 006	14	9.98 155	25	0.01 845	9.85 851	12	13	9	11.7	10.8	9.9
48	9.84 020	13	9.98 180	26	0.01 820	9.85 839	12	12			
49	9.84 033	13	9.98 206	25	0.01 794	9.85 827	12	11			
50	9.84 046	13	9.98 231	25	0.01 769	9.85 815	12	10			
51	9.84 059	13	9.98 256	25	0.01 744	9.85 803	12	9			
52	9.84 072	13	9.98 281	26	0.01 719	9.85 791	12	8			
53	9.84 085	13	9.98 307	25	0.01 693	9.85 779	13	7			
54	9.84 098	14	9.98 332	25	0.01 668	9.85 766	12	6			
55	9.84 112	13	9.98 357	26	0.01 643	9.85 754	12	5			
56	9.84 125	13	9.98 383	25	0.01 617	9.85 742	12	4			
57	9.84 138	13	9.98 408	25	0.01 592	9.85 730	12	3			
58	9.84 151	13	9.98 433	25	0.01 567	9.85 718	12	2			
59	9.84 164	13	9.98 458	26	0.01 542	9.85 706	13	1			
60	9.84 177		9.98 484		0.01 516	9.85 693		0			
	Log Cos	d	Log Cot	c d	Log Tan	Log Sin	d	′	Prop. Parts		

46°

Table II

44°

′	Log Sin	d	Log Tan	c d	Log Cot	Log Cos	d		Prop. Parts
0	9.84 177	13	9.98 484	25	0.01 516	9.85 693	12	60	
1	9.84 190	13	9.98 509	25	0.01 491	9.85 681	12	59	
2	9.84 203	13	9.98 534	26	0.01 466	9.85 669	12	58	
3	9.84 216	13	9.98 560	25	0.01 440	9.85 657	12	57	
4	9.84 229	13	9.98 585	25	0.01 415	9.85 645	13	56	
5	9.84 242	13	9.98 610	25	0.01 390	9.85 632	12	55	
6	9.84 255	14	9.98 635	26	0.01 365	9.85 620	12	54	
7	9.84 269	13	9.98 661	25	0.01 339	9.85 608	12	53	
8	9.84 282	13	9.98 686	25	0.01 314	9.85 596	13	52	
9	9.84 295	13	9.98 711	26	0.01 289	9.85 583	12	51	
10	9.84 308	13	9.98 737	25	0.01 263	9.85 571	12	50	
11	9.84 321	13	9.98 762	25	0.01 238	9.85 559	12	49	
12	9.84 334	13	9.98 787	25	0.01 213	9.85 547	13	48	
13	9.84 347	13	9.98 812	26	0.01 188	9.85 534	12	47	
14	9.84 360	13	9.98 838	25	0.01 162	9.85 522	12	46	
15	9.84 373	12	9.98 863	25	0.01 137	9.85 510	13	45	
16	9.84 385	13	9.98 888	25	0.01 112	9.85 497	12	44	
17	9.84 398	13	9.98 913	26	0.01 087	9.85 485	12	43	
18	9.84 411	13	9.98 939	25	0.01 061	9.85 473	13	42	
19	9.84 424	13	9.98 964	25	0.01 036	9.85 460	12	41	
20	9.84 437	13	9.98 989	26	0.01 011	9.85 448	12	40	
21	9.84 450	13	9.99 015	25	0.00 985	9.85 436	13	39	
22	9.84 463	13	9.99 040	25	0.00 960	9.85 423	12	38	
23	9.84 476	13	9.99 065	25	0.00 935	9.85 411	12	37	
24	9.84 489	13	9.99 090	26	0.00 910	9.85 399	13	36	
25	9.84 502	13	9.99 116	25	0.00 884	9.85 386	12	35	
26	9.84 515	13	9.99 141	25	0.00 859	9.85 374	13	34	
27	9.84 528	12	9.99 166	25	0.00 834	9.85 361	12	33	
28	9.84 540	13	9.99 191	26	0.00 809	9.85 349	12	32	
29	9.84 553	13	9.99 217	25	0.00 783	9.85 337	13	31	
30	9.84 566	13	9.99 242	25	0.00 758	9.85 324	12	30	
31	9.84 579	13	9.99 267	26	0.00 733	9.85 312	13	29	
32	9.84 592	13	9.99 293	25	0.00 707	9.85 299	12	28	
33	9.84 605	13	9.99 318	25	0.00 682	9.85 287	13	27	
34	9.84 618	12	9.99 343	25	0.00 657	9.85 274	12	26	
35	9.84 630	13	9.99 368	26	0.00 632	9.85 262	12	25	
36	9.84 643	13	9.99 394	25	0.00 606	9.85 250	13	24	
37	9.84 656	13	9.99 419	25	0.00 581	9.85 237	12	23	
38	9.84 669	13	9.99 444	25	0.00 556	9.85 225	13	22	
39	9.84 682	12	9.99 469	26	0.00 531	9.85 212	12	21	
40	9.84 694	13	9.99 495	25	0.00 505	9.85 200	13	20	
41	9.84 707	13	9.99 520	25	0.00 480	9.85 187	12	19	
42	9.84 720	13	9.99 545	25	0.00 455	9.85 175	13	18	
43	9.84 733	12	9.99 570	26	0.00 430	9.85 162	12	17	
44	9.84 745	13	9.99 596	25	0.00 404	9.85 150	13	16	
45	9.84 758	13	9.99 621	25	0.00 379	9.85 137	12	15	
46	9.84 771	13	9.99 646	26	0.00 354	9.85 125	13	14	
47	9.84 784	12	9.99 672	25	0.00 328	9.85 112	12	13	
48	9.84 796	13	9.99 697	25	0.00 303	9.85 100	13	12	
49	9.84 809	13	9.99 722	25	0.00 278	9.85 087	13	11	
50	9.84 822	13	9.99 747	26	0.00 253	9.85 074	12	10	
51	9.84 835	12	9.99 773	25	0.00 227	9.85 062	13	9	
52	9.84 847	13	9.99 798	25	0.00 202	9.85 049	12	8	
53	9.84 860	13	9.99 823	25	0.00 177	9.85 037	13	7	
54	9.84 873	12	9.99 848	26	0.00 152	9.85 024	12	6	
55	9.84 885	13	9.99 874	25	0.00 126	9.85 012	13	5	
56	9.84 898	13	9.99 899	25	0.00 101	9.84 999	13	4	
57	9.84 911	12	9.99 924	25	0.00 076	9.84 986	12	3	
58	9.84 923	13	9.99 949	26	0.00 051	9.84 974	13	2	
59	9.84 936	13	9.99 975	25	0.00 025	9.84 961	12	1	
60	9.84 949		0.00 000		0.00 000	9.84 949		0	
	Log Cos	d	Log Cot	c d	Log Tan	Log Sin	d	′	Prop. Parts

Prop. Parts

	26	25	14
1	2.6	2.5	1.4
2	5.2	5.0	2.8
3	7.8	7.5	4.2
4	10.4	10.0	5.6
5	13.0	12.5	7.0
6	15.6	15.0	8.4
7	18.2	17.5	9.8
8	20.8	20.0	11.2
9	23.4	22.5	12.6

	13	12
1	1.3	1.2
2	2.6	2.4
3	3.9	3.6
4	5.2	4.8
5	6.5	6.0
6	7.8	7.2
7	9.1	8.4
8	10.4	9.6
9	11.7	10.8

45°

Table III

Values of the trigonometric functions

to five decimal places

0°

′	Sin	Tan	Cot	Cos	
0	.00000	.00000	———	1.0000	60
1	029	029	3437.7	000	59
2	058	058	1718.9	000	58
3	087	087	1145.9	000	57
4	116	116	859.44	000	56
5	.00145	.00145	687.55	1.0000	55
6	175	175	572.96	000	54
7	204	204	491.11	000	53
8	233	233	429.72	000	52
9	262	262	381.97	000	51
10	.00291	.00291	343.77	1.0000	50
11	320	320	312.52	.99999	49
12	349	349	286.48	999	48
13	378	378	264.44	999	47
14	407	407	245.55	999	46
15	.00436	.00436	229.18	.99999	45
16	465	465	214.86	999	44
17	495	495	202.22	999	43
18	524	524	190.98	999	42
19	553	553	180.93	998	41
20	.00582	.00582	171.89	.99998	40
21	611	611	163.70	998	39
22	640	640	156.26	998	38
23	669	669	149.47	998	37
24	698	698	143.24	998	36
25	.00727	.00727	137.51	.99997	35
26	756	756	132.22	997	34
27	785	785	127.32	997	33
28	814	815	122.77	997	32
29	844	844	118.54	996	31
30	.00873	.00873	114.59	.99996	30
31	902	902	110.89	996	29
32	931	931	107.43	996	28
33	960	960	104.17	995	27
34	.00989	.00989	101.11	995	26
35	.01018	.01018	98.218	.99995	25
36	047	047	95.489	995	24
37	076	076	92.908	994	23
38	105	105	90.463	994	22
39	134	135	88.144	994	21
40	.01164	.01164	85.940	.99993	20
41	193	193	83.844	993	19
42	222	222	81.847	993	18
43	251	251	79.943	992	17
44	280	280	78.126	992	16
45	.01309	.01309	76.390	.99991	15
46	338	338	74.729	991	14
47	367	367	73.139	991	13
48	396	396	71.615	990	12
49	425	425	70.153	990	11
50	.01454	.01455	68.750	.99989	10
51	483	484	67.402	989	9
52	513	513	66.105	989	8
53	542	542	64.858	988	7
54	571	571	63.657	988	6
55	.01600	.01600	62.499	.99987	5
56	629	629	61.383	987	4
57	658	658	60.306	986	3
58	687	687	59.266	986	2
59	716	.716	58.261	985	1
60	.01745	.01746	57.290	.99985	0
	Cos	Cot	Tan	Sin	′

89°

1°

′	Sin	Tan	Cot	Cos	
0	.01745	.01746	57.290	.99985	60
1	774	775	56.351	984	59
2	803	804	55.442	984	58
3	832	833	54.561	983	57
4	862	862	53.709	983	56
5	.01891	.01891	52.882	.99982	55
6	920	920	52.081	982	54
7	949	949	51.303	981	53
8	.01978	.01978	50.549	980	52
9	.02007	.02007	49.816	980	51
10	.02036	.02036	49.104	.99979	50
11	065	066	48.412	979	49
12	094	095	47.740	978	48
13	123	124	47.085	977	47
14	152	153	46.449	977	46
15	.02181	.02182	45.829	.99976	45
16	211	211	45.226	976	44
17	240	240	44.639	975	43
18	269	269	44.066	974	42
19	298	298	43.508	974	41
20	.02327	.02328	42.964	.99973	40
21	356	357	42.433	972	39
22	385	386	41.916	972	38
23	414	415	41.411	971	37
24	443	444	40.917	970	36
25	.02472	.02473	40.436	.99969	35
26	501	502	39.965	969	34
27	530	531	39.506	968	33
28	560	560	39.057	967	32
29	589	589	38.618	966	31
30	.02618	.02619	38.188	.99966	30
31	647	648	37.769	965	29
32	676	677	37.358	964	28
33	705	706	36.956	963	27
34	734	735	36.563	963	26
35	.02763	.02764	36.178	.99962	25
36	792	793	35.801	961	24
37	821	822	35.431	960	23
38	850	851	35.070	959	22
39	879	881	34.715	959	21
40	.02908	.02910	34.368	.99958	20
41	938	939	34.027	957	19
42	967	968	33.694	956	18
43	.02996	.02997	33.366	955	17
44	.03025	.03026	33.045	954	16
45	.03054	.03055	32.730	.99953	15
46	083	084	32.421	952	14
47	112	114	32.118	952	13
48	141	143	31.821	951	12
49	170	172	31.528	950	11
50	.03199	.03201	31.242	.99949	10
51	228	230	30.960	948	9
52	257	259	30.683	947	8
53	286	288	30.412	946	7
54	316	317	30.145	945	6
55	.03345	.03346	29.882	.99944	5
56	374	376	29.624	943	4
57	403	405	29.371	942	3
58	432	434	29.122	941	2
59	461	463	28.877	940	1
60	.03490	.03492	28.636	.99939	0
	Cos	Cot	Tan	Sin	′

88°

2°

′	Sin	Tan	Cot	Cos	
0	.03490	.03492	28.636	.99939	60
1	519	521	.399	938	59
2	548	550	28.166	937	58
3	577	579	27.937	936	57
4	606	609	.712	935	56
5	.03635	.03638	27.490	.99934	55
6	664	667	.271	933	54
7	693	696	27.057	932	53
8	723	725	26.845	931	52
9	752	754	.637	930	51
10	.03781	.03783	26.432	.99929	50
11	810	812	.230	927	49
12	839	842	26.031	926	48
13	868	871	25.835	925	47
14	897	900	.642	924	46
15	.03926	.03929	25.452	.99923	45
16	955	958	.264	922	44
17	.03984	.03987	25.080	921	43
18	.04013	.04016	24.898	919	42
19	042	046	.719	918	41
20	.04071	.04075	24.542	.99917	40
21	100	104	.368	916	39
22	129	133	.196	915	38
23	159	162	24.026	913	37
24	188	191	23.859	912	36
25	.04217	.04220	23.695	.99911	35
26	246	250	.532	910	34
27	275	279	.372	909	33
28	304	308	.214	907	32
29	333	337	23.058	906	31
30	.04362	.04366	22.904	.99905	30
31	391	395	.752	904	29
32	420	424	.602	902	28
33	449	454	.454	901	27
34	478	483	.308	900	26
35	.04507	.04512	22.164	.99898	25
36	536	541	22.022	897	24
37	565	570	21.881	896	23
38	594	599	.743	894	22
39	623	628	.606	893	21
40	.04653	.04658	21.470	.99892	20
41	682	687	.337	890	19
42	711	716	.205	889	18
43	740	745	21.075	888	17
44	769	774	20.946	886	16
45	.04798	.04803	20.819	.99885	15
46	827	833	.693	883	14
47	856	862	.569	882	13
48	885	891	.446	881	12
49	914	920	.325	879	11
50	.04943	.04949	20.206	.99878	10
51	.04972	.04978	20.087	876	9
52	.05001	.05007	19.970	875	8
53	030	037	.855	873	7
54	059	066	.740	872	6
55	.05088	.05095	19.627	.99870	5
56	117	124	.516	869	4
57	146	153	.405	867	3
58	175	182	.296	866	2
59	205	212	.188	864	1
60	.05234	.05241	19.081	.99863	0
	Cos	Cot	Tan	Sin	′

87°

3°

′	Sin	Tan	Cot	Cos	
0	.05234	.05241	19.081	.99863	60
1	263	270	18.976	861	59
2	292	299	.871	860	58
3	321	328	.768	858	57
4	350	357	.666	857	56
5	.05379	.05387	18.564	.99855	55
6	408	416	.464	854	54
7	437	445	.366	852	53
8	466	474	.268	851	52
9	495	503	.171	849	51
10	.05524	.05533	18.075	.99847	50
11	553	562	17.980	846	49
12	582	591	.886	844	48
13	611	620	.793	842	47
14	640	649	.702	841	46
15	.05669	.05678	17.611	.99839	45
16	698	708	.521	838	44
17	727	737	.431	836	43
18	756	766	.343	834	42
19	785	795	.256	833	41
20	.05814	.05824	17.169	.99831	40
21	844	854	17.084	829	39
22	873	883	16.999	827	38
23	902	912	.915	826	37
24	931	941	.832	824	36
25	.05960	.05970	16.750	.99822	35
26	.05989	.05999	.668	821	34
27	.06018	.06029	.587	819	33
28	047	058	.507	817	32
29	076	087	.428	815	31
30	.06105	.06116	16.350	.99813	30
31	134	145	.272	812	29
32	163	175	.195	810	28
33	192	204	.119	808	27
34	221	233	16.043	806	26
35	.06250	.06262	15.969	.99804	25
36	279	291	.895	803	24
37	308	321	.821	801	23
38	337	350	.748	799	22
39	366	379	.676	797	21
40	.06395	.06408	15.605	.99795	20
41	424	438	.534	793	19
42	453	467	.464	792	18
43	482	496	.394	790	17
44	511	525	.325	788	16
45	.06540	.06554	15.257	.99786	15
46	569	584	.189	784	14
47	598	613	.122	782	13
48	627	642	15.056	780	12
49	656	671	14.990	778	11
50	.06685	.06700	14.924	.99776	10
51	714	730	.860	774	9
52	743	759	.795	772	8
53	773	788	.732	770	7
54	802	817	.669	768	6
55	.06831	.06847	14.606	.99766	5
56	860	876	.544	764	4
57	889	905	.482	762	3
58	918	934	.421	760	2
59	947	963	.361	758	1
60	.06976	.06993	14.301	.99756	0
	Cos	Cot	Tan	Sin	′

86°

Table III

4°

′	Sin	Tan	Cot	Cos	
0	.06976	.06993	14.301	.99756	60
1	.07005	.07022	.241	754	59
2	034	051	.182	752	58
3	063	080	.124	750	57
4	092	110	.065	748	56
5	.07121	.07139	14.008	.99746	55
6	150	168	13.951	744	54
7	179	197	.894	742	53
8	208	227	.838	740	52
9	237	256	.782	738	51
10	.07266	.07285	13.727	.99736	50
11	295	314	.672	734	49
12	324	344	.617	731	48
13	353	373	.563	729	47
14	382	402	.510	727	46
15	.07411	.07431	13.457	.99725	45
16	440	461	.404	723	44
17	469	490	.352	721	43
18	498	519	.300	719	42
19	527	548	.248	716	41
20	.07556	.07578	13.197	.99714	40
21	585	607	.146	712	39
22	614	636	.096	710	38
23	643	665	13.046	708	37
24	672	695	12.996	705	36
25	.07701	.07724	12.947	.99703	35
26	730	753	.898	701	34
27	759	782	.850	699	33
28	788	812	.801	696	32
29	817	841	.754	694	31
30	.07846	.07870	12.706	.99692	30
31	875	899	.659	689	29
32	904	929	.612	687	28
33	933	958	.566	685	27
34	962	.07987	.520	683	26
35	.07991	.08017	12.474	.99680	25
36	.08020	046	.429	678	24
37	049	075	.384	676	23
38	078	104	.339	673	22
39	107	134	.295	671	21
40	.08136	.08163	12.251	.99668	20
41	165	192	.207	666	19
42	194	221	.163	664	18
43	223	251	.120	661	17
44	252	280	.077	659	16
45	.08281	.08309	12.035	.99657	15
46	310	339	11.992	654	14
47	339	368	.950	652	13
48	368	397	.909	649	12
49	397	427	.867	647	11
50	.08426	.08456	11.826	.99644	10
51	455	485	.785	642	9
52	484	514	.745	639	8
53	513	544	.705	637	7
54	542	573	.664	635	6
55	.08571	.08602	11.625	.99632	5
56	600	632	.585	630	4
57	629	661	.546	627	3
58	658	690	.507	625	2
59	687	720	.468	622	1
60	.08716	.08749	11.430	.99619	0
	Cos	Cot	Tan	Sin	′

85°

5°

′	Sin	Tan	Cot	Cos	
0	.08716	.08749	11.430	.99619	60
1	745	778	.392	617	59
2	774	807	.354	614	58
3	803	837	.316	612	57
4	831	866	.279	609	56
5	.08860	.08895	11.242	.99607	55
6	889	925	.205	604	54
7	918	954	.168	602	53
8	947	.08983	.132	599	52
9	.08976	.09013	.095	596	51
10	.09005	.09042	11.059	.99594	50
11	034	071	11.024	591	49
12	063	101	10.988	588	48
13	092	130	.953	586	47
14	121	159	.918	583	46
15	.09150	.09189	10.883	.99580	45
16	179	218	.848	578	44
17	208	247	.814	575	43
18	237	277	.780	572	42
19	266	306	.746	570	41
20	.09295	.09335	10.712	.99567	40
21	324	365	.678	564	39
22	353	394	.645	562	38
23	382	423	.612	559	37
24	411	453	.579	556	36
25	.09440	.09482	10.546	.99553	35
26	469	511	.514	551	34
27	498	541	.481	548	33
28	527	570	.449	545	32
29	556	600	.417	542	31
30	.09585	.09629	10.385	.99540	30
31	614	658	.354	537	29
32	642	688	.322	534	28
33	671	717	.291	531	27
34	700	746	.260	528	26
35	.09729	.09776	10.229	.99526	25
36	758	805	.199	523	24
37	787	834	.168	520	23
38	816	864	.138	517	22
39	845	893	.108	514	21
40	.09874	.09923	10.078	.99511	20
41	903	952	.048	508	19
42	932	.09981	10.019	506	18
43	961	.10011	9.9893	503	17
44	.09990	040	.9601	500	16
45	.10019	.10069	9.9310	.99497	15
46	048	099	.9021	494	14
47	077	128	.8734	491	13
48	106	158	.8448	488	12
49	135	187	.8164	485	11
50	.10164	.10216	9.7882	.99482	10
51	192	246	.7601	479	9
52	221	275	.7322	476	8
53	250	305	.7044	473	7
54	279	334	.6768	470	6
55	.10308	.10363	9.6493	.99467	5
56	337	393	.6220	464	4
57	366	422	.5949	461	3
58	395	452	.5679	458	2
59	424	481	.5411	455	1
60	.10453	.10510	9.5144	.99452	0
	Cos	Cot	Tan	Sin	′

84°

6°

′	Sin	Tan	Cot	Cos	
0	.10453	.10510	9.5144	.99452	60
1	482	540	.4878	449	59
2	511	569	.4614	446	58
3	540	599	.4352	443	57
4	569	628	.4090	440	56
5	.10597	.10657	9.3831	.99437	55
6	626	687	.3572	434	54
7	655	716	.3315	431	53
8	684	746	.3060	428	52
9	713	775	.2806	424	51
10	.10742	.10805	9.2553	.99421	50
11	771	834	.2302	418	49
12	800	863	.2052	415	48
13	829	893	.1803	412	47
14	858	922	.1555	409	46
15	.10887	.10952	9.1309	.99406	45
16	916	.10981	.1065	402	44
17	945	.11011	.0821	399	43
18	.10973	040	.0579	396	42
19	.11002	070	.0338	393	41
20	.11031	.11099	9.0098	.99390	40
21	060	128	8.9860	386	39
22	089	158	.9623	383	38
23	118	187	.9387	380	37
24	147	217	.9152	377	36
25	.11176	.11246	8.8919	.99374	35
26	205	276	.8686	370	34
27	234	305	.8455	367	33
28	263	335	.8225	364	32
29	291	364	.7996	360	31
30	.11320	.11394	8.7769	.99357	30
31	349	423	.7542	354	29
32	378	452	.7317	351	28
33	407	482	.7093	347	27
34	436	511	.6870	344	26
35	.11465	.11541	8.6648	.99341	25
36	494	570	.6427	337	24
37	523	600	.6208	334	23
38	552	629	.5989	331	22
39	580	659	.5772	327	21
40	.11609	.11688	8.5555	.99324	20
41	638	718	.5340	320	19
42	667	747	.5126	317	18
43	696	777	.4913	314	17
44	725	806	.4701	310	16
45	.11754	.11836	8.4490	.99307	15
46	783	865	.4280	303	14
47	812	895	.4071	300	13
48	840	924	.3863	297	12
49	869	954	.3656	293	11
50	.11898	.11983	8.3450	.99290	10
51	927	.12013	.3245	286	9
52	956	042	.3041	283	8
53	.11985	072	.2838	279	7
54	.12014	101	.2636	276	6
55	.12043	.12131	8.2434	.99272	5
56	071	160	.2234	269	4
57	100	190	.2035	265	3
58	129	219	.1837	262	2
59	158	249	.1640	258	1
60	.12187	.12278	8.1443	.99255	0
	Cos	Cot	Tan	Sin	′

83°

7°

′	Sin	Tan	Cot	Cos	
0	.12187	.12278	8.1443	.99255	60
1	216	308	.1248	251	59
2	245	338	.1054	248	58
3	274	367	.0860	244	57
4	302	397	.0667	240	56
5	.12331	.12426	8.0476	.99237	55
6	360	456	.0285	233	54
7	389	485	8.0095	230	53
8	418	515	7.9906	226	52
9	447	544	.9718	222	51
10	.12476	.12574	7.9530	.99219	50
11	504	603	.9344	215	49
12	533	633	.9158	211	48
13	562	662	.8973	208	47
14	591	692	.8789	204	46
15	.12620	.12722	7.8606	.99200	45
16	649	751	.8424	197	44
17	678	781	.8243	193	43
18	706	810	.8062	189	42
19	735	840	.7882	186	41
20	.12764	.12869	7.7704	.99182	40
21	793	899	.7525	178	39
22	822	929	.7348	175	38
23	851	958	.7171	171	37
24	880	.12988	.6996	167	36
25	.12908	.13017	7.6821	.99163	35
26	937	047	.6647	160	34
27	966	076	.6473	156	33
28	.12995	106	.6301	152	32
29	.13024	136	.6129	148	31
30	.13053	.13165	7.5958	.99144	30
31	081	195	.5787	141	29
32	110	224	.5618	137	28
33	139	254	.5449	133	27
34	168	284	.5281	129	26
35	.13197	.13313	7.5113	.99125	25
36	226	343	.4947	122	24
37	254	372	.4781	118	23
38	283	402	.4615	114	22
39	312	432	.4451	110	21
40	.13341	.13461	7.4287	.99106	20
41	370	491	.4124	102	19
42	399	521	.3962	098	18
43	427	550	.3800	094	17
44	456	580	.3639	091	16
45	.13485	.13609	7.3479	.99087	15
46	514	639	.3319	083	14
47	543	669	.3160	079	13
48	572	698	.3002	075	12
49	600	728	.2844	071	11
50	.13629	.13758	7.2687	.99067	10
51	658	787	.2531	063	9
52	687	817	.2375	059	8
53	716	846	.2220	055	7
54	744	876	.2066	051	6
55	.13773	.13906	7.1912	.99047	5
56	802	935	.1759	043	4
57	831	965	.1607	039	3
58	860	.13995	.1455	035	2
59	889	.14024	.1304	031	1
60	.13917	.14054	7.1154	.99027	0
	Cos	Cot	Tan	Sin	′

82°

Table III

8°

′	Sin	Tan	Cot	Cos	
0	.13917	.14054	7.1154	.99027	60
1	946	084	.1004	023	59
2	.13975	113	.0855	019	58
3	.14004	143	.0706	015	57
4	033	173	.0558	011	56
5	.14061	.14202	7.0410	.99006	55
6	090	232	.0264	.99002	54
7	119	262	7.0117	.98998	53
8	148	291	6.9972	994	52
9	177	321	.9827	990	51
10	.14205	.14351	6.9682	.98986	50
11	234	381	.9538	982	49
12	263	410	.9395	978	48
13	292	440	.9252	973	47
14	320	470	.9110	969	46
15	.14349	.14499	6.8969	.98965	45
16	378	529	.8828	961	44
17	407	559	.8687	957	43
18	436	588	.8548	953	42
19	464	618	.8408	948	41
20	.14493	.14648	6.8269	.98944	40
21	522	678	.8131	940	39
22	551	707	.7994	936	38
23	580	737	.7856	931	37
24	608	767	.7720	927	36
25	.14637	.14796	6.7584	.98923	35
26	666	826	.7448	919	34
27	695	856	.7313	914	33
28	723	886	.7179	910	32
29	752	915	.7045	906	31
30	.14781	.14945	6.6912	.98902	30
31	810	.14975	.6779	897	29
32	838	.15005	.6646	893	28
33	867	034	.6514	889	27
34	896	064	.6383	884	26
35	.14925	.15094	6.6252	.98880	25
36	954	124	.6122	876	24
37	.14982	153	.5992	871	23
38	.15011	183	.5863	867	22
39	040	213	.5734	863	21
40	.15069	.15243	6.5606	.98858	20
41	097	272	.5478	854	19
42	126	302	.5350	849	18
43	155	332	.5223	845	17
44	184	362	.5097	841	16
45	.15212	.15391	6.4971	.98836	15
46	241	421	.4846	832	14
47	270	451	.4721	827	13
48	299	481	.4596	823	12
49	327	511	.4472	818	11
50	.15356	.15540	6.4348	.98814	10
51	385	570	.4225	809	9
52	414	600	.4103	805	8
53	442	630	.3980	800	7
54	471	660	.3859	796	6
55	.15500	.15689	6.3737	.98791	5
56	529	719	.3617	787	4
57	557	749	.3496	782	3
58	586	779	.3376	778	2
59	615	809	.3257	773	1
60	.15643	.15838	6.3138	.98769	0
	Cos	Cot	Tan	Sin	′

81°

9°

′	Sin	Tan	Cot	Cos	
0	.15643	.15838	6.3138	.98769	60
1	672	868	.3019	764	59
2	701	898	.2901	760	58
3	730	928	.2783	755	57
4	758	958	.2666	751	56
5	.15787	.15988	6.2549	.98746	55
6	816	.16017	.2432	741	54
7	845	047	.2316	737	53
8	873	077	.2200	732	52
9	902	107	.2085	728	51
10	.15931	.16137	6.1970	.98723	50
11	959	167	.1856	718	49
12	.15988	196	.1742	714	48
13	.16017	226	.1628	709	47
14	046	256	.1515	704	46
15	.16074	.16286	6.1402	.98700	45
16	103	316	.1290	695	44
17	132	346	.1178	690	43
18	160	376	.1066	686	42
19	189	405	.0955	681	41
20	.16218	.16435	6.0844	.98676	40
21	246	465	.0734	671	39
22	275	495	.0624	667	38
23	304	525	.0514	662	37
24	333	555	.0405	657	36
25	.16361	.16585	6.0296	.98652	35
26	390	615	.0188	648	34
27	419	645	6.0080	643	33
28	447	674	5.9972	638	32
29	476	704	.9865	633	31
30	.13505	.16734	5.9758	.98629	30
31	533	764	.9651	624	29
32	562	794	.9545	619	28
33	591	824	.9439	614	27
34	620	854	.9333	609	26
35	.16648	.16884	5.9228	.98604	25
36	677	914	.9124	600	24
37	706	944	.9019	595	23
38	734	.16974	.8915	590	22
39	763	.17004	.8811	585	21
40	.16792	.17033	5.8708	.98580	20
41	820	063	.8605	575	19
42	849	093	.8502	570	18
43	878	123	.8400	565	17
44	906	153	.8298	561	16
45	.16935	.17183	5.8197	.98556	15
46	964	213	.8095	551	14
47	.16992	243	.7994	546	13
48	.17021	273	.7894	541	12
49	050	303	.7794	536	11
50	.17078	.17333	5.7694	.98531	10
51	107	363	.7594	526	9
52	136	393	.7495	521	8
53	164	423	.7396	516	7
54	193	453	.7297	511	6
55	.17222	.17483	5.7199	.98506	5
56	250	513	.7101	501	4
57	279	543	.7004	496	3
58	308	573	.6906	491	2
59	336	603	.6809	486	1
60	.17365	.17633	5.6713	.98481	0
	Cos	Cot	Tan	Sin	′

80°

Natural functions
73

10°

′	Sin	Tan	Cot	Cos	
0	.17365	.17633	5.6713	.98481	60
1	393	663	.6617	476	59
2	422	693	.6521	471	58
3	451	723	.6425	466	57
4	479	753	.6329	461	56
5	.17508	.17783	5.6234	.98455	55
6	537	813	.6140	450	54
7	565	843	.6045	445	53
8	594	873	.5951	440	52
9	623	903	.5857	435	51
10	.17651	.17933	5.5764	.98430	50
11	680	963	.5671	425	49
12	708	.17993	.5578	420	48
13	737	.18023	.5485	414	47
14	766	053	.5393	409	46
15	.17794	.18083	5.5301	.98404	45
16	823	113	.5209	399	44
17	852	143	.5118	394	43
18	880	173	.5026	389	42
19	909	203	.4936	383	41
20	.17937	.18233	5.4845	.98378	40
21	966	263	.4755	373	39
22	.17995	293	.4665	368	38
23	.18023	323	.4575	362	37
24	052	353	.4486	357	36
25	.18081	.18384	5.4397	.98352	35
26	109	414	.4308	347	34
27	138	444	.4219	341	33
28	166	474	.4131	336	32
29	195	504	.4043	331	31
30	.18224	.18534	5.3955	.98325	30
31	252	564	.3868	320	29
32	281	594	.3781	315	28
33	309	624	.3694	310	27
34	338	654	.3607	304	26
35	.18367	.18684	5.3521	.98299	25
36	395	714	.3435	294	24
37	424	745	.3349	288	23
38	452	775	.3263	283	22
39	481	805	.3178	277	21
40	.18509	.18835	5.3093	.98272	20
41	538	865	.3008	267	19
42	567	895	.2924	261	18
43	595	925	.2839	256	17
44	624	955	.2755	250	16
45	.18652	.18986	5.2672	.98245	15
46	681	.19016	.2588	240	14
47	710	046	.2505	234	13
48	738	076	.2422	229	12
49	767	106	.2339	223	11
50	.18795	.19136	5.2257	.98218	10
51	824	166	.2174	212	9
52	852	197	.2092	207	8
53	881	227	.2011	201	7
54	910	257	.1929	196	6
55	.18938	.19287	5.1848	.98190	5
56	967	317	.1767	185	4
57	.18995	347	.1686	179	3
58	.19024	378	.1606	174	2
59	052	408	.1526	168	1
60	.19081	.19438	5.1446	.98163	0
	Cos	Cot	Tan	Sin	′

79°

11°

′	Sin	Tan	Cot	Cos	
0	.19081	.19438	5.1446	.98163	60
1	109	468	.1366	157	59
2	138	498	.1286	152	58
3	167	529	.1207	146	57
4	195	559	.1128	140	56
5	.19224	.19589	5.1049	.98135	55
6	252	619	.0970	129	54
7	281	649	.0892	124	53
8	309	680	.0814	118	52
9	338	710	.0736	112	51
10	.19366	.19740	5.0658	.98107	50
11	395	770	.0581	101	49
12	423	801	.0504	096	48
13	452	831	.0427	090	47
14	481	861	.0350	084	46
15	.19509	.19891	5.0273	.98079	45
16	538	921	.0197	073	44
17	566	952	.0121	067	43
18	595	.19982	5.0045	061	42
19	623	.20012	4.9969	056	41
20	.19652	.20042	4.9894	.98050	40
21	680	073	.9819	044	39
22	709	103	.9744	039	38
23	737	133	.9669	033	37
24	766	164	.9594	027	36
25	.19794	.20194	4.9520	.98021	35
26	823	224	.9446	016	34
27	851	254	.9372	010	33
28	880	285	.9298	.98004	32
29	908	315	.9225	.97998	31
30	.19937	.20345	4.9152	.97992	30
31	965	376	.9078	987	29
32	.19994	406	.9006	981	28
33	.20022	436	.8933	975	27
34	051	466	.8860	969	26
35	.20079	.20497	4.8788	.97963	25
36	108	527	.8716	958	24
37	136	557	.8644	952	23
38	165	588	.8573	946	22
39	193	618	.8501	940	21
40	.20222	.20648	4.8430	.97934	20
41	250	679	.8359	928	19
42	279	709	.8288	922	18
43	307	739	.8218	916	17
44	336	770	.8147	910	16
45	.20364	.20800	4.8077	.97905	15
46	393	830	.8007	899	14
47	421	861	.7937	893	13
48	450	891	.7867	887	12
49	478	921	.7798	881	11
50	.20507	.20952	4.7729	.97875	10
51	535	.20982	.7659	869	9
52	563	.21013	.7591	863	8
53	592	043	.7522	857	7
54	620	073	.7453	851	6
55	.20649	.21104	4.7385	.97845	5
56	677	134	.7317	839	4
57	706	164	.7249	833	3
58	734	195	.7181	827	2
59	763	225	.7114	821	1
60	.20791	.21256	4.7046	.97815	0
	Cos	Cot	Tan	Sin	′

78°

Table III

12°

′	Sin	Tan	Cot	Cos	
0	.20791	.21256	4.7046	.97815	60
1	820	286	.6979	809	59
2	848	316	.6912	803	58
3	877	347	.6845	797	57
4	905	377	.6779	791	56
5	.20933	.21408	4.6712	.97784	55
6	962	438	.6646	778	54
7	.20990	469	.6580	772	53
8	.21019	499	.6514	766	52
9	047	529	.6448	760	51
10	.21076	.21560	4.6382	.97754	50
11	104	590	.6317	748	49
12	132	621	.6252	742	48
13	161	651	.6187	735	47
14	189	682	.6122	729	46
15	.21218	.21712	4.6057	.97723	45
16	246	743	.5993	717	44
17	275	773	.5928	711	43
18	303	804	.5864	705	42
19	331	834	.5800	698	41
20	.21360	.21864	4.5736	.97692	40
21	388	895	.5673	686	39
22	417	925	.5609	680	38
23	445	956	.5546	673	37
24	474	.21986	.5483	667	36
25	.21502	.22017	4.5420	.97661	35
26	530	047	.5357	655	34
27	559	078	.5294	648	33
28	587	108	.5232	642	32
29	616	139	.5169	636	31
30	.21644	.22169	4.5107	.97630	30
31	672	200	.5045	623	29
32	701	231	.4983	617	28
33	729	261	.4922	611	27
34	758	292	.4860	604	26
35	.21786	.22322	4.4799	.97598	25
36	814	353	.4737	592	24
37	843	383	.4676	585	23
38	871	414	.4615	579	22
39	899	444	.4555	573	21
40	.21928	.22475	4.4494	.97566	20
41	956	505	.4434	560	19
42	.21985	536	.4373	553	18
43	.22013	567	.4313	547	17
44	041	597	.4253	541	16
45	.22070	.22628	4.4194	.97534	15
46	098	658	.4134	528	14
47	126	689	.4075	521	13
48	155	719	.4015	515	12
49	183	750	.3956	508	11
50	.22212	.22781	4.3897	.97502	10
51	240	811	.3838	496	9
52	268	842	.3779	489	8
53	297	872	.3721	483	7
54	325	903	.3662	476	6
55	.22353	.22934	4.3604	.97470	5
56	382	964	.3546	463	4
57	410	.22995	.3488	457	3
58	438	.23026	.3430	450	2
59	467	056	.3372	444	1
60	.22495	.23087	4.3315	.97437	0
	Cos	Cot	Tan	Sin	′

77°

13°

′	Sin	Tan	Cot	Cos	
0	.22495	.23087	4.3315	.97437	60
1	523	117	.3257	430	59
2	552	148	.3200	424	58
3	580	179	.3143	417	57
4	608	209	.3086	411	56
5	.22637	.23240	4.3029	.97404	55
6	665	271	.2972	398	54
7	693	301	.2916	391	53
8	722	332	.2859	384	52
9	750	363	.2803	378	51
10	.22778	.23393	4.2747	.97371	50
11	807	424	.2691	365	49
12	835	455	.2635	358	48
13	863	485	.2580	351	47
14	892	516	.2524	345	46
15	.22920	.23547	4.2468	.97338	45
16	948	578	.2413	331	44
17	.22977	608	.2358	325	43
18	.23005	639	.2303	318	42
19	033	670	.2248	311	41
20	.23062	.23700	4.2193	.97304	40
21	090	731	.2139	298	39
22	118	762	.2084	291	38
23	146	793	.2030	284	37
24	175	823	.1976	278	36
25	.23203	.23854	4.1922	.97271	35
26	231	885	.1868	264	34
27	260	916	.1814	257	33
28	288	946	.1760	251	32
29	316	.23977	.1706	244	31
30	.23345	.24008	4.1653	.97237	30
31	373	039	.1600	230	29
32	401	069	.1547	223	28
33	429	100	.1493	217	27
34	458	131	.1441	210	26
35	.23486	.24162	4.1388	.97203	25
36	514	193	.1335	196	24
37	542	223	.1282	189	23
38	571	254	.1230	182	22
39	599	285	.1178	176	21
40	.23627	.24316	4.1126	.97169	20
41	656	347	.1074	162	19
42	684	377	.1022	155	18
43	712	408	.0970	148	17
44	740	439	.0918	141	16
45	.23769	.24470	4.0867	.97134	15
46	797	501	.0815	127	14
47	825	532	.0764	120	13
48	853	562	.0713	113	12
49	882	593	.0662	106	11
50	.23910	.24624	4.0611	.97100	10
51	938	655	.0560	093	9
52	966	686	.0509	086	8
53	.23995	717	.0459	079	7
54	.24023	747	.0408	072	6
55	.24051	.24778	4.0358	.97065	5
56	079	809	.0308	058	4
57	108	840	.0257	051	3
58	136	871	.0207	044	2
59	164	902	.0158	037	1
60	.24192	.24933	4.0108	.97030	0
	Cos	Cot	Tan	Sin	′

76°

14°

′	Sin	Tan	Cot	Cos	
0	.24192	.24933	4.0108	.97030	60
1	220	964	.0058	023	59
2	249	.24995	4.0009	015	58
3	277	.25026	3.9959	008	57
4	305	056	.9910	.97001	56
5	.24333	.25087	3.9861	.96994	55
6	362	118	.9812	987	54
7	390	149	.9763	980	53
8	418	180	.9714	973	52
9	446	211	.9665	966	51
10	.24474	.25242	3.9617	.96959	50
11	503	273	.9568	952	49
12	531	304	.9520	945	48
13	559	335	.9471	937	47
14	587	366	.9423	930	46
15	.24615	.25397	3.9375	.96923	45
16	644	428	.9327	916	44
17	672	459	.9279	909	43
18	700	490	.9232	902	42
19	728	521	.9184	894	41
20	.24756	.25552	3.9136	.96887	40
21	784	583	.9089	880	39
22	813	614	.9042	873	38
23	841	645	.8995	866	37
24	869	676	.8947	858	36
25	.24897	.25707	3.8900	.96851	35
26	925	738	.8854	844	34
27	954	769	.8807	837	33
28	.24982	800	.8760	829	32
29	.25010	831	.8714	822	31
30	.25038	.25862	3.8667	.96815	30
31	066	893	.8621	807	29
32	094	924	.8575	800	28
33	122	955	.8528	793	27
34	151	.25986	.8482	786	26
35	.25179	.26017	3.8436	.96778	25
36	207	048	.8391	771	24
37	235	079	.8345	764	23
38	263	110	.8299	756	22
39	291	141	.8254	749	21
40	.25320	.26172	3.8208	.96742	20
41	348	203	.8163	734	19
42	376	235	.8118	727	18
43	404	266	.8073	719	17
44	432	297	.8028	712	16
45	.25460	.26328	3.7983	.96705	15
46	488	359	.7938	697	14
47	516	390	.7893	690	13
48	545	421	.7848	682	12
49	573	452	.7804	675	11
50	.25601	.26483	3.7760	.96667	10
51	629	515	.7715	660	9
52	657	546	.7671	653	8
53	685	577	.7627	645	7
54	713	608	.7583	638	6
55	.25741	.26639	3.7539	.96630	5
56	769	670	.7495	623	4
57	798	701	.7451	615	3
58	826	733	.7408	608	2
59	854	764	.7364	600	1
60	.25882	.26795	3.7321	.96593	0
	Cos	Cot	Tan	Sin	′

75°

15°

′	Sin	Tan	Cot	Cos	
0	.25882	.26795	3.7321	.96593	60
1	910	826	.7277	• 585	59
2	938	857	.7234	578	58
3	966	888	.7191	570	57
4	.25994	920	.7148	562	56
5	.26022	.26951	3.7105	.96555	55
6	050	.26982	.7062	547	54
7	079	.27013	.7019	540	53
8	107	044	.6976	532	52
9	135	076	.6933	524	51
10	.26163	.27107	3.6891	.96517	50
11	191	138	.6848	509	49
12	219	169	.6806	502	48
13	247	201	.6764	494	47
14	275	232	.6722	486	46
15	.26303	.27263	3.6680	.96479	45
16	331	294	.6638	471	44
17	359	326	.6596	463	43
18	387	357	.6554	456	42
19	415	388	.6512	448	41
20	.26443	.27419	3.6470	.96440	40
21	471	451	.6429	433	39
22	500	482	.6387	425	38
23	528	513	.6346	417	37
24	556	545	.6305	410	36
25	.26584	.27576	3.6264	.96402	35
26	612	607	.6222	394	34
27	640	638	.6181	386	33
28	668	670	.6140	379	32
29	696	701	.6100	371	31
30	.26724	.27732	3.6059	.96363	30
31	752	764	.6018	355	29
32	780	795	.5978	347	28
33	808	826	.5937	340	27
34	836	858	.5897	332	26
35	.26864	.27889	3.5856	.96324	25
36	892	921	.5816	316	24
37	920	952	.5776	308	23
38	948	.27983	.5736	301	22
39	.26976	.28015	.5696	293	21
40	.27004	.28046	3.5656	.96285	20
41	032	077	.5616	277	19
42	060	109	.5576	269	18
43	088	140	.5536	261	17
44	116	172	.5497	253	16
45	.27144	.28203	3.5457	.96246	15
46	172	234	.5418	238	14
47	200	266	.5379	230	13
48	228	297	.5339	222	12
49	256	329	.5300	214	11
50	.27284	.28360	3.5261	.96206	10
51	312	391	.5222	198	9
52	340	423	.5183	190	8
53	368	454	.5144	182	7
54	396	486	.5105	174	6
55	.27424	.28517	3.5067	.96166	5
56	452	549	.5028	158	4
57	480	580	.4989	150	3
58	508	612	.4951	142	2
59	536	643	.4912	134	1
60	.27564	.28675	3.4874	.96126	0
	Cos	Cot	Tan	Sin	′

74°

Table III

16°

'	Sin	Tan	Cot	Cos	
0	.27564	.28675	3.4874	.96126	60
1	592	706	.4836	118	59
2	620	738	.4798	110	58
3	648	769	.4760	102	57
4	676	801	.4722	094	56
5	.27704	.28832	3.4684	.96086	55
6	731	864	.4646	078	54
7	759	895	.4608	070	53
8	787	927	.4570	062	52
9	815	958	.4533	054	51
10	.27843	.28990	3.4495	.96046	50
11	871	.29021	.4458	037	49
12	899	053	.4420	029	48
13	927	084	.4383	021	47
14	955	116	.4346	013	46
15	.27983	.29147	3.4308	.96005	45
16	.28011	179	.4271	.95997	44
17	039	210	.4234	989	43
18	067	242	.4197	981	42
19	095	274	.4160	972	41
20	.28123	.29305	3.4124	.95964	40
21	150	337	.4087	956	39
22	178	368	.4050	948	38
23	206	400	.4014	940	37
24	234	432	.3977	931	36
25	.28262	.29463	3.3941	.95923	35
26	290	495	.3904	915	34
27	318	526	.3868	907	33
28	346	558	.3832	898	32
29	374	590	.3796	890	31
30	.28402	.29621	3.3759	.95882	30
31	429	653	.3723	874	29
32	457	685	.3687	865	28
33	485	716	.3652	857	27
34	513	748	.3616	849	26
35	.28541	.29780	3.3580	.95841	25
36	569	811	.3544	832	24
37	597	843	.3509	824	23
38	625	875	.3473	816	22
39	652	906	.3438	807	21
40	.28680	.29938	3.3402	.95799	20
41	708	.29970	.3367	791	19
42	736	.30001	.3332	782	18
43	764	033	.3297	774	17
44	792	065	.3261	766	16
45	.28820	.30097	3.3226	.95757	15
46	847	128	.3191	749	14
47	875	160	.3156	740	13
48	903	192	.3122	732	12
49	931	224	.3087	724	11
50	.28959	.30255	3.3052	.95715	10
51	.28987	287	.3017	707	9
52	.29015	319	.2983	698	8
53	042	351	.2948	690	7
54	070	382	.2914	681	6
55	.29098	.30414	3.2879	.95673	5
56	126	446	.2845	664	4
57	154	478	.2811	656	3
58	182	509	.2777	647	2
59	209	541	.2743	639	1
60	.29237	.30573	3.2709	.95630	0
	Cos	Cot	Tan	Sin	'

73°

17°

'	Sin	Tan	Cot	Cos	
0	.29237	.30573	3.2709	.95630	60
1	265	605	.2675	622	59
2	293	637	.2641	613	58
3	321	669	.2607	605	57
4	348	700	.2573	596	56
5	.29376	.30732	3.2539	.95588	55
6	404	764	.2506	579	54
7	432	796	.2472	571	53
8	460	828	.2438	562	52
9	487	860	.2405	554	51
10	.29515	.30891	3.2371	.95545	50
11	543	923	.2338	536	49
12	571	955	.2305	528	48
13	599	.30987	.2272	519	47
14	626	.31019	.2238	511	46
15	.29654	.31051	3.2205	.95502	45
16	682	083	.2172	493	44
17	710	115	.2139	485	43
18	737	147	.2106	476	42
19	765	178	.2073	467	41
20	.29793	.31210	3.2041	.95459	40
21	821	242	.2008	450	39
22	849	274	.1975	441	38
23	876	306	.1943	433	37
24	904	338	.1910	424	36
25	.29932	.31370	3.1878	.95415	35
26	960	402	.1845	407	34
27	.29987	434	.1813	398	33
28	.30015	466	.1780	389	32
29	043	498	.1748	380	31
30	.30071	.31530	3.1716	.95372	30
31	098	562	.1684	363	29
32	126	594	.1652	354	28
33	154	626	.1620	345	27
34	182	658	.1588	337	26
35	.30209	.31690	3.1556	.95328	25
36	237	722	.1524	319	24
37	265	754	.1492	310	23
38	292	786	.1460	301	22
39	320	818	.1429	293	21
40	.30348	.31850	3.1397	.95284	20
41	376	882	.1366	275	19
42	403	914	.1334	266	18
43	431	946	.1303	257	17
44	459	.31978	.1271	248	16
45	.30486	.32010	3.1240	.95240	15
46	514	042	.1209	231	14
47	542	074	.1178	222	13
48	570	106	.1146	213	12
49	597	139	.1115	204	11
50	.30625	.32171	3.1084	.95195	10
51	653	203	.1053	186	9
52	680	235	.1022	177	8
53	708	267	.0991	168	7
54	736	299	.0961	159	6
55	.30763	.32331	3.0930	.95150	5
56	791	363	.0899	142	4
57	819	396	.0868	133	3
58	846	428	.0838	124	2
59	874	460	.0807	115	1
60	.30902	.32492	3.0777	.95106	0
	Cos	Cot	Tan	Sin	'

72°

18°

′	Sin	Tan	Cot	Cos	
0	.30902	.32492	3.0777	.95106	60
1	929	524	.0746	097	59
2	957	556	.0716	088	58
3	.30985	588	.0686	079	57
4	.31012	621	.0655	070	56
5	.31040	.32653	3.0625	.95061	55
6	068	685	.0595	052	54
7	095	717	.0565	043	53
8	123	749	.0535	033	52
9	151	782	.0505	024	51
10	.31178	.32814	3.0475	.95015	50
11	206	846	.0445	.95006	49
12	233	878	.0415	.94997	48
13	261	911	.0385	988	47
14	289	943	.0356	979	46
15	.31316	.32975	3.0326	.94970	45
16	344	.33007	.0296	961	44
17	372	040	.0267	952	43
18	399	072	.0237	943	42
19	427	104	.0208	933	41
20	.31454	.33136	3.0178	.94924	40
21	482	169	.0149	915	39
22	510	201	.0120	906	38
23	537	233	.0090	897	37
24	565	266	.0061	888	36
25	.31593	.33298	3.0032	.94878	35
26	620	330	3.0003	869	34
27	648	363	2.9974	860	33
28	675	395	.9945	851	32
29	703	427	.9916	842	31
30	.31730	.33460	2.9887	.94832	30
31	758	492	.9858	823	29
32	786	524	.9829	814	28
33	813	557	.9800	805	27
34	841	589	.9772	795	26
35	.31868	.33621	2.9743	.94786	25
36	896	654	.9714	777	24
37	923	686	.9686	768	23
38	951	718	.9657	758	22
39	.31979	751	.9629	749	21
40	.32006	.33783	2.9600	.94740	20
41	034	816	.9572	730	19
42	061	848	.9544	721	18
43	089	881	.9515	712	17
44	116	913	.9487	702	16
45	.32144	.33945	2.9459	.94693	15
46	171	.33978	.9431	684	14
47	199	.34010	.9403	674	13
48	227	043	.9375	665	12
49	254	075	.9347	656	11
50	.32282	.34108	2.9319	.94646	10
51	309	140	.9291	637	9
52	337	173	.9263	627	8
53	364	205	.9235	618	7
54	392	238	.9208	609	6
55	.32419	.34270	2.9180	.94599	5
56	447	303	.9152	590	4
57	474	335	.9125	580	3
58	502	368	.9097	571	2
59	529	400	.9070	561	1
60	.32557	.34433	2.9042	.94552	0
	Cos	Cot	Tan	Sin	′

71°

19°

′	Sin	Tan	Cot	Cos	
0	.32557	.34433	2.9042	.94552	60
1	584	465	.9015	542	59
2	612	498	.8987	533	58
3	639	530	.8960	523	57
4	667	563	.8933	514	56
5	.32694	.34596	2.8905	.94504	55
6	722	628	.8878	495	54
7	749	661	.8851	485	53
8	777	693	.8824	476	52
9	804	726	.8797	466	51
10	.32832	.34758	2.8770	.94457	50
11	859	791	.8743	447	49
12	887	824	.8716	438	48
13	914	856	.8689	428	47
14	942	889	.8662	418	46
15	.32969	.34922	2.8636	.94409	45
16	.32997	954	.8609	399	44
17	.33024	.34987	.8582	390	43
18	051	.35020	.8556	380	42
19	079	052	.8529	370	41
20	.33106	.35085	2.8502	.94361	40
21	134	118	.8476	351	39
22	161	150	.8449	342	38
23	189	183	.8423	332	37
24	216	216	.8397	322	36
25	.33244	.35248	2.8370	.94313	35
26	271	281	.8344	303	34
27	298	314	.8318	293	33
28	326	346	.8291	284	32
29	353	379	.8265	274	31
30	.33381	.35412	2.8239	.94264	30
31	408	445	.8213	254	29
32	436	477	.8187	245	28
33	463	510	.8161	235	27
34	490	543	.8135	225	26
35	.33518	.35576	2.8109	.94215	25
36	545	608	.8083	206	24
37	573	641	.8057	196	23
38	600	674	.8032	186	22
39	627	707	.8006	176	21
40	.33655	.35740	2.7980	.94167	20
41	682	772	.7955	157	19
42	710	805	.7929	147	18
43	737	838	.7903	137	17
44	764	871	.7878	127	16
45	.33792	.35904	2.7852	.94118	15
46	819	937	.7827	108	14
47	846	.35969	.7801	098	13
48	874	.36002	.7776	088	12
49	901	035	.7751	078	11
50	.33929	.36068	2.7725	.94068	10
51	956	101	.7700	058	9
52	.33983	134	.7675	049	8
53	.34011	167	.7650	039	7
54	038	199	.7625	029	6
55	.34065	.36232	2.7600	.94019	5
56	093	265	.7575	.94009	4
57	120	298	.7550	.93999	3
58	147	331	.7525	989	2
59	175	364	.7500	979	1
60	.34202	.36397	2.7475	.93969	0
	Cos	Cot	Tan	Sin	′

70°

20°

′	Sin	Tan	Cot	Cos	
0	.34202	.36397	2.7475	.93969	60
1	229	430	.7450	959	59
2	257	463	.7425	949	58
3	284	496	.7400	939	57
4	311	529	.7376	929	56
5	.34339	.36562	2.7351	.93919	55
6	366	595	.7326	909	54
7	393	628	.7302	899	53
8	421	661	.7277	889	52
9	448	694	.7253	879	51
10	.34475	.36727	2.7228	.93869	50
11	503	760	.7204	859	49
12	530	793	.7179	849	48
13	557	826	.7155	839	47
14	584	859	.7130	829	46
15	.34612	.36892	2.7106	.93819	45
16	639	925	.7082	809	44
17	666	958	.7058	799	43
18	694	.36991	.7034	789	42
19	721	.37024	.7009	779	41
20	.34748	.37057	2.6985	.93769	40
21	775	090	.6961	759	39
22	803	123	.6937	748	38
23	830	157	.6913	738	37
24	857	190	.6889	728	36
25	.34884	.37223	2.6865	.93718	35
26	912	256	.6841	708	34
27	939	289	.6818	698	33
28	966	322	.6794	688	32
29	.34993	355	.6770	677	31
30	.35021	.37388	2.6746	.93667	30
31	048	422	.6723	657	29
32	075	455	.6699	647	28
33	102	488	.6675	637	27
34	130	521	.6652	626	26
35	.35157	.37554	2.6628	.93616	25
36	184	588	.6605	606	24
37	211	621	.6581	596	23
38	239	654	.6558	585	22
39	266	687	.6534	575	21
40	.35293	.37720	2.6511	.93565	20
41	320	754	.6488	555	19
42	347	787	.6464	544	18
43	375	820	.6441	534	17
44	402	853	.6418	524	16
45	.35429	.37887	2.6395	.93514	15
46	456	920	.6371	503	14
47	484	953	.6348	493	13
48	511	.37986	.6325	483	12
49	538	.38020	.6302	472	11
50	.35565	.38053	2.6279	.93462	10
51	592	086	.6256	452	9
52	619	120	.6233	441	8
53	647	153	.6210	431	7
54	674	186	.6187	420	6
55	.35701	.38220	2.6165	.93410	5
56	728	253	.6142	400	4
57	755	286	.6119	389	3
58	782	320	.6096	379	2
59	810	353	.6074	368	1
60	.35837	.38386	2.6051	.93358	0
	Cos	Cot	Tan	Sin	′

69°

21°

′	Sin	Tan	Cot	Cos	
0	.35837	.38386	2.6051	.93358	60
1	864	420	.6028	348	59
2	891	453	.6006	337	58
3	918	487	.5983	327	57
4	945	520	.5961	316	56
5	.35973	.38553	2.5938	.93306	55
6	.36000	587	.5916	295	54
7	027	620	.5893	285	53
8	054	654	.5871	274	52
9	081	687	.5848	264	51
10	.36108	.38721	2.5826	.93253	50
11	135	754	.5804	243	49
12	162	787	.5782	232	48
13	190	821	.5759	222	47
14	217	854	.5737	211	46
15	.36244	.38888	2.5715	.93201	45
16	271	921	.5693	190	44
17	298	955	.5671	180	43
18	325	.38988	.5649	169	42
19	352	.39022	.5627	159	41
20	.36379	.39055	2.5605	.93148	40
21	406	089	.5583	137	39
22	434	122	.5561	127	38
23	461	156	.5539	116	37
24	488	190	.5517	106	36
25	.36515	.39223	2.5495	.93095	35
26	542	257	.5473	084	34
27	569	290	.5452	074	33
28	596	324	.5430	063	32
29	623	357	.5408	052	31
30	.36650	.39391	2.5386	.93042	30
31	677	425	.5365	031	29
32	704	458	.5343	020	28
33	731	492	.5322	.93010	27
34	758	526	.5300	.92999	26
35	.36785	.39559	2.5279	.92988	25
36	812	593	.5257	978	24
37	839	626	.5236	967	23
38	867	660	.5214	956	22
39	894	694	.5193	945	21
40	.36921	.39727	2.5172	.92935	20
41	948	761	.5150	924	19
42	.36975	795	.5129	913	18
43	.37002	829	.5108	902	17
44	029	862	.5086	892	16
45	.37056	.39896	2.5065	.92881	15
46	083	930	.5044	870	14
47	110	963	.5023	859	13
48	137	.39997	.5002	849	12
49	164	.40031	.4981	838	11
50	.37191	.40065	2.4960	.92827	10
51	218	098	.4939	816	9
52	245	132	.4918	805	8
53	272	166	.4897	794	7
54	299	200	.4876	784	6
55	.37326	.40234	2.4855	.92773	5
56	353	267	.4834	762	4
57	380	301	.4813	751	3
58	407	335	.4792	740	2
59	434	369	.4772	729	1
60	.37461	.40403	2.4751	.92718	0
	Cos	Cot	Tan	Sin	′

68°

22°

'	Sin	Tan	Cot	Cos	
0	.37461	.40403	2.4751	.92718	60
1	488	436	.4730	707	59
2	515	470	.4709	697	58
3	542	504	.4689	686	57
4	569	538	.4668	675	56
5	.37595	.40572	2.4648	.92664	55
6	622	606	.4627	653	54
7	649	640	.4606	642	53
8	676	674	.4586	631	52
9	703	707	.4566	620	51
10	.37730	.40741	2.4545	.92609	50
11	757	775	.4525	598	49
12	784	809	.4504	587	48
13	811	843	.4484	576	47
14	838	877	.4464	565	46
15	.37865	.40911	2.4443	.92554	45
16	892	945	.4423	543	44
17	919	.40979	.4403	532	43
18	946	.41013	.4383	521	42
19	973	047	.4362	510	41
20	.37999	.41081	2.4342	.92499	40
21	.38026	115	.4322	488	39
22	053	149	.4302	477	38
23	080	183	.4282	466	37
24	107	217	.4262	455	36
25	.38134	.41251	2.4242	.92444	35
26	161	285	.4222	432	34
27	188	319	.4202	421	33
28	215	353	.4182	410	32
29	241	387	.4162	399	31
30	.38268	.41421	2.4142	.92388	30
31	295	455	.4122	377	29
32	322	490	.4102	366	28
33	349	524	.4083	355	27
34	376	558	.4063	343	26
35	.38403	.41592	2.4043	.92332	25
36	430	626	.4023	321	24
37	456	660	.4004	310	23
38	483	694	.3984	299	22
39	510	728	.3964	287	21
40	.38537	.41763	2.3945	.92276	20
41	564	797	.3925	265	19
42	591	831	.3906	254	18
43	617	865	.3886	243	17
44	644	899	.3867	231	16
45	.38671	.41933	2.3847	.92220	15
46	698	.41968	.3828	209	14
47	725	.42002	.3808	198	13
48	752	036	.3789	186	12
49	778	070	.3770	175	11
50	.38805	.42105	2.3750	.92164	10
51	832	139	.3731	152	9
52	859	173	.3712	141	8
53	886	207	.3693	130	7
54	912	242	.3673	119	6
55	.38939	.42276	2.3654	.92107	5
56	966	310	.3635	096	4
57	.38993	345	.3616	085	3
58	.39020	379	.3597	073	2
59	046	413	.3578	062	1
60	.39073	.42447	2.3559	.92050	0
	Cos	Cot	Tan	Sin	'

67°

23°

'	Sin	Tan	Cot	Cos	
0	.39073	.42447	2.3559	.92050	60
1	100	482	.3539	039	59
2	127	516	.3520	028	58
3	153	551	.3501	016	57
4	180	585	.3483	.92005	56
5	.39207	.42619	2.3464	.91994	55
6	234	654	.3445	982	54
7	260	688	.3426	971	53
8	287	722	.3407	959	52
9	314	757	.3388	948	51
10	.39341	.42791	2.3369	.91936	50
11	367	826	.3351	925	49
12	394	860	.3332	914	48
13	421	894	.3313	902	47
14	448	929	.3294	891	46
15	.39474	.42963	2.3276	.91879	45
16	501	.42998	.3257	868	44
17	528	.43032	.3238	856	43
18	555	067	.3220	845	42
19	581	101	.3201	833	41
20	.39608	.43136	2.3183	.91822	40
21	635	170	.3164	810	39
22	661	205	.3146	799	38
23	688	239	.3127	787	37
24	715	274	.3109	775	36
25	.39741	.43308	2.3090	.91764	35
26	768	343	.3072	752	34
27	795	378	.3053	741	33
28	822	412	.3035	729	32
29	848	447	.3017	718	31
30	.39875	.43481	2.2998	.91706	30
31	902	516	.2980	694	29
32	928	550	.2962	683	28
33	955	585	.2944	671	27
34	.39982	620	.2925	660	26
35	.40008	.43654	2.2907	.91648	25
36	035	689	.2889	636	24
37	062	724	.2871	625	23
38	088	758	.2853	613	22
39	115	793	.2835	601	21
40	.40141	.43828	2.2817	.91590	20
41	168	862	.2799	578	19
42	195	897	.2781	566	18
43	221	932	.2763	555	17
44	248	.43966	.2745	543	16
45	.40275	.44001	2.2727	.91531	15
46	301	036	.2709	519	14
47	328	071	.2691	508	13
48	355	105	.2673	496	12
49	381	140	.2655	484	11
50	.40408	.44175	2.2637	.91472	10
51	434	210	.2620	461	9
52	461	244	.2602	449	8
53	488	279	.2584	437	7
54	514	314	.2566	425	6
55	.40541	.44349	2.2549	.91414	5
56	567	384	.2531	402	4
57	594	418	.2513	390	3
58	621	453	.2496	378	2
59	647	488	.2478	366	1
60	.40674	.44523	2.2460	.91355	0
	Cos	Cot	Tan	Sin	'

66°

Table III

24°

'	Sin	Tan	Cot	Cos	
0	.40674	.44523	2.2460	.91355	60
1	700	558	.2443	343	59
2	727	593	.2425	331	58
3	753	627	.2408	319	57
4	780	662	.2390	307	56
5	.40806	.44697	2.2373	.91295	55
6	833	732	.2355	283	54
7	860	767	.2338	272	53
8	886	802	.2320	260	52
9	913	837	.2303	248	51
10	.40939	.44872	2.2286	.91236	50
11	966	907	.2268	224	49
12	.40992	942	.2251	212	48
13	.41019	.44977	.2234	200	47
14	045	.45012	.2216	188	46
15	.41072	.45047	2.2199	.91176	45
16	098	082	.2182	164	44
17	125	117	.2165	152	43
18	151	152	.2148	140	42
19	178	187	.2130	128	41
20	.41204	.45222	2.2113	.91116	40
21	231	257	.2096	104	39
22	257	292	.2079	092	38
23	284	327	.2062	080	37
24	310	362	.2045	068	36
25	.41337	.45397	2.2028	.91056	35
26	363	432	.2011	044	34
27	390	467	.1994	032	33
28	416	502	.1977	020	32
29	443	538	.1960	.91008	31
30	.41469	.45573	2.1943	.90996	30
31	496	608	.1926	984	29
32	522	643	.1909	972	28
33	549	678	.1892	960	27
34	575	713	.1876	948	26
35	.41602	.45748	2.1859	.90936	25
36	628	784	.1842	924	24
37	655	819	.1825	911	23
38	681	854	.1808	899	22
39	707	889	.1792	887	21
40	.41734	.45924	2.1775	.90875	20
41	760	960	.1758	863	19
42	787	.45995	.1742	851	18
43	813	.46030	.1725	839	17
44	840	065	.1708	826	16
45	.41866	.46101	2.1692	.90814	15
46	892	136	.1675	802	14
47	919	171	.1659	790	13
48	945	206	.1642	778	12
49	972	242	.1625	766	11
50	.41998	.46277	2.1609	.90753	10
51	.42024	312	.1592	741	9
52	051	348	.1576	729	8
53	077	383	.1560	717	7
54	104	418	.1543	704	6
55	.42130	.46454	2.1527	.90692	5
56	156	489	.1510	680	4
57	183	525	.1494	668	3
58	209	560	.1478	655	2
59	235	595	.1461	643	1
60	.42262	.46631	2.1445	.90631	0
	Cos	Cot	Tan	Sin	'

65°

25°

'	Sin	Tan	Cot	Cos	
0	.42262	.46631	2.1445	.90631	60
1	288	666	.1429	618	59
2	315	702	.1413	606	58
3	341	737	.1396	594	57
4	367	772	.1380	582	56
5	.42394	.46808	2.1364	.90569	55
6	420	843	.1348	557	54
7	446	879	.1332	545	53
8	473	914	.1315	532	52
9	499	950	.1299	520	51
10	.42525	.46985	2.1283	.90507	50
11	552	.47021	.1267	495	49
12	578	056	.1251	483	48
13	604	092	.1235	470	47
14	631	128	.1219	458	46
15	.42657	.47163	2.1203	.90446	45
16	683	199	.1187	433	44
17	709	234	.1171	421	43
18	736	270	.1155	408	42
19	762	305	.1139	396	41
20	.42788	.47341	2.1123	.90383	40
21	815	377	.1107	371	39
22	841	412	.1092	358	38
23	867	448	.1076	346	37
24	894	483	.1060	334	36
25	.42920	.47519	2.1044	.90321	35
26	946	555	.1028	309	34
27	972	590	.1013	296	33
28	.42999	626	.0997	284	32
29	.43025	662	.0981	271	31
30	.43051	.47698	2.0965	.90259	30
31	077	733	.0950	246	29
32	104	769	.0934	233	28
33	130	805	.0918	221	27
34	156	840	.0903	208	26
35	.43182	.47876	2.0887	.90196	25
36	209	912	.0872	183	24
37	235	948	.0856	171	23
38	261	.47984	.0840	158	22
39	287	.48019	.0825	146	21
40	.43313	.48055	2.0809	.90133	20
41	340	091	.0794	120	19
42	366	127	.0778	108	18
43	392	163	.0763	095	17
44	418	198	.0748	082	16
45	.43445	.48234	2.0732	.90070	15
46	471	270	.0717	057	14
47	497	306	.0701	045	13
48	523	342	.0686	032	12
49	549	378	.0671	019	11
50	.43575	.48414	2.0655	.90007	10
51	602	450	.0640	.89994	9
52	628	486	.0625	981	8
53	654	521	.0609	968	7
54	680	557	.0594	956	6
55	.43706	.48593	2.0579	.89943	5
56	733	629	.0564	930	4
57	759	665	.0549	918	3
58	785	701	.0533	905	2
59	811	737	.0518	892	1
60	.43837	.48773	2.0503	.89879	0
	Cos	Cot	Tan	Sin	'

64°

26° | 27°

′	Sin	Tan	Cot	Cos		′	Sin	Tan	Cot	Cos	
0	.43837	.48773	2.0503	.89879	60	0	.45399	.50953	1.9626	.89101	60
1	863	809	.0488	867	59	1	425	.50989	.9612	087	59
2	889	845	.0473	854	58	2	451	.51026	.9598	074	58
3	916	881	.0458	841	57	3	477	063	.9584	061	57
4	942	917	.0443	828	56	4	503	099	.9570	048	56
5	.43968	.48953	2.0428	.89816	55	5	.45529	.51136	1.9556	.89035	55
6	.43994	.48989	.0413	803	54	6	554	173	.9542	021	54
7	.44020	.49026	.0398	790	53	7	580	209	.9528	.89008	53
8	046	062	.0383	777	52	8	606	246	.9514	.88995	52
9	072	098	.0368	764	51	9	632	283	.9500	981	51
10	.44098	.49134	2.0353	.89752	50	10	.45658	.51319	1.9486	.88968	50
11	124	170	.0338	739	49	11	684	356	.9472	955	49
12	151	206	.0323	726	48	12	710	393	.9458	942	48
13	177	242	.0308	713	47	13	736	430	.9444	928	47
14	203	278	.0293	700	46	14	762	467	.9430	915	46
15	.44229	.49315	2.0278	.89687	45	15	.45787	.51503	1.9416	.88902	45
16	255	351	.0263	674	44	16	813	540	.9402	888	44
17	281	387	.0248	662	43	17	839	577	.9388	875	43
18	307	423	.0233	649	42	18	865	614	.9375	862	42
19	333	459	.0219	636	41	19	891	651	.9361	848	41
20	.44359	.49495	2.0204	.89623	40	20	.45917	.51688	1.9347	.88835	40
21	385	532	.0189	610	39	21	942	724	.9333	822	39
22	411	568	.0174	597	38	22	968	761	.9319	808	38
23	437	604	.0160	584	37	23	.45994	798	.9306	795	37
24	464	640	.0145	571	36	24	.46020	835	.9292	782	36
25	.44490	.49677	2.0130	.89558	35	25	.46046	.51872	1.9278	.88768	35
26	516	713	.0115	545	34	26	072	909	.9265	755	34
27	542	749	.0101	532	33	27	097	946	.9251	741	33
28	568	786	.0086	519	32	28	123	.51983	.9237	728	32
29	594	822	.0072	506	31	29	149	.52020	.9223	715	31
30	.44620	.49858	2.0057	.89493	30	30	.46175	.52057	1.9210	.88701	30
31	646	894	.0042	480	29	31	201	094	.9196	688	29
32	672	931	.0028	467	28	32	226	131	.9183	674	28
33	698	.49967	2.0013	454	27	33	252	168	.9169	661	27
34	724	.50004	1.9999	441	26	34	278	205	.9155	647	26
35	.44750	.50040	1.9984	.89428	25	35	.46304	.52242	1.9142	.88634	25
36	776	076	.9970	415	24	36	330	279	.9128	620	24
37	802	113	.9955	402	23	37	355	316	.9115	607	23
38	828	149	.9941	389	22	38	381	353	.9101	593	22
39	854	185	.9926	376	21	39	407	390	.9088	580	21
40	.44880	.50222	1.9912	.89363	20	40	.46433	.52427	1.9074	.88566	20
41	906	258	.9897	350	19	41	458	464	.9061	553	19
42	932	295	.9883	337	18	42	484	501	.9047	539	18
43	958	331	.9868	324	17	43	510	538	.9034	526	17
44	.44984	368	.9854	311	16	44	536	575	.9020	512	16
45	.45010	.50404	1.9840	.89298	15	45	.46561	.52613	1.9007	.88499	15
46	036	441	.9825	285	14	46	587	650	.8993	485	14
47	062	477	.9811	272	13	47	613	687	.8980	472	13
48	088	514	.9797	259	12	48	639	724	.8967	458	12
49	114	550	.9782	245	11	49	664	761	.8953	445	11
50	.45140	.50587	1.9768	.89232	10	50	.46690	.52798	1.8940	.88431	10
51	166	623	.9754	219	9	51	716	836	.8927	417	9
52	192	660	.9740	206	8	52	742	873	.8913	404	8
53	218	696	.9725	193	7	53	767	910	.8900	390	7
54	243	733	.9711	180	6	54	793	947	.8887	377	6
55	.45269	.50769	1.9697	.89167	5	55	.46819	.52985	1.8873	.88363	5
56	295	806	.9683	153	4	56	844	.53022	.8860	349	4
57	321	843	.9669	140	3	57	870	059	.8847	336	3
58	347	879	.9654	127	2	58	896	096	.8834	322	2
59	373	916	.9640	114	1	59	921	134	.8820	308	1
60	.45399	.50953	1.9626	.89101	0	60	.46947	.53171	1.8807	.88295	0
	Cos	Cot	Tan	Sin	′		Cos	Cot	Tan	Sin	′

63° | 62°

28°

′	Sin	Tan	Cot	Cos	
0	.46947	.53171	1.8807	.88295	60
1	973	208	.8794	281	59
2	.46999	246	.8781	267	58
3	.47024	283	.8768	254	57
4	050	320	.8755	240	56
5	.47076	.53358	1.8741	.88226	55
6	101	395	.8728	213	54
7	127	432	.8715	199	53
8	153	470	.8702	185	52
9	178	507	.8689	172	51
10	.47204	.53545	1.8676	.88158	50
11	229	582	.8663	144	49
12	255	620	.8650	130	48
13	281	657	.8637	117	47
14	306	694	.8624	103	46
15	.47332	.53732	1.8611	.88089	45
16	358	769	.8598	075	44
17	383	807	.8585	062	43
18	409	844	.8572	048	42
19	434	882	.8559	034	41
20	.47460	.53920	1.8546	.88020	40
21	486	957	.8533	.88006	39
22	511	.53995	.8520	.87993	38
23	537	.54032	.8507	979	37
24	562	070	.8495	965	36
25	.47588	.54107	1.8482	.87951	35
26	614	145	.8469	937	34
27	639	183	.8456	923	33
28	665	220	.8443	909	32
29	690	258	.8430	896	31
30	.47716	.54296	1.8418	.87882	30
31	741	333	.8405	868	29
32	767	371	.8392	854	28
33	793	409	.8379	840	27
34	818	446	.8367	826	26
35	.47844	.54484	1.8354	.87812	25
36	869	522	.8341	798	24
37	895	560	.8329	784	23
38	920	597	.8316	770	22
39	946	635	.8303	756	21
40	.47971	.54673	1.8291	.87743	20
41	.47997	711	.8278	729	19
42	.48022	748	.8265	715	18
43	048	786	.8253	701	17
44	073	824	.8240	687	16
45	.48099	.54862	1.8228	.87673	15
46	124	900	.8215	659	14
47	150	938	.8202	645	13
48	175	.54975	.8190	631	12
49	201	.55013	.8177	617	11
50	.48226	.55051	1.8165	.87603	10
51	252	089	.8152	589	9
52	277	127	.8140	575	8
53	303	165	.8127	561	7
54	328	203	.8115	546	6
55	.48354	.55241	1.8103	.87532	5
56	379	279	.8090	518	4
57	405	317	.8078	504	3
58	430	355	.8065	490	2
59	456	393	.8053	476	1
60	.48481	.55431	1.8040	.87462	0
	Cos	Cot	Tan	Sin	′

61°

29°

′	Sin	Tan	Cot	Cos	
0	.48481	.55431	1.8040	.87462	60
1	506	469	.8028	448	59
2	532	507	.8016	434	58
3	557	545	.8003	420	57
4	583	583	.7991	406	56
5	.48608	.55621	1.7979	.87391	55
6	634	659	.7966	377	54
7	659	697	.7954	363	53
8	684	736	.7942	349	52
9	710	774	.7930	335	51
10	.48735	.55812	1.7917	.87321	50
11	761	850	.7905	306	49
12	786	888	.7893	292	48
13	811	926	.7881	278	47
14	837	.55964	.7868	264	46
15	.48862	.56003	1.7856	.87250	45
16	888	041	.7844	235	44
17	913	079	.7832	221	43
18	938	117	.7820	207	42
19	964	156	.7808	193	41
20	.48989	.56194	1.7796	.87178	40
21	.49014	232	.7783	164	39
22	040	270	.7771	150	38
23	065	209	.7759	136	37
24	090	347	.7747	121	36
25	.49116	.56385	1.7735	.87107	35
26	141	424	.7723	093	34
27	166	462	.7711	079	33
28	192	501	.7699	064	32
29	217	539	.7687	050	31
30	.49242	.56577	1.7675	.87036	30
31	268	616	.7663	021	29
32	293	654	.7651	.87007	28
33	318	693	.7639	.86993	27
34	344	731	.7627	978	26
35	.49369	.56769	1.7615	.86964	25
36	394	808	.7603	949	24
37	419	846	.7591	935	23
38	445	885	.7579	921	22
39	470	923	.7567	906	21
40	.49495	.56962	1.7556	.86892	20
41	521	.57000	.7544	878	19
42	546	039	.7532	863	18
43	571	078	.7520	849	17
44	596	116	.7508	834	16
45	.49622	.57155	1.7496	.86820	15
46	647	193	.7485	805	14
47	672	232	.7473	791	13
48	697	271	.7461	777	12
49	723	309	.7449	762	11
50	.49748	.57348	1.7437	.86748	10
51	773	386	.7426	733	9
52	798	425	.7414	719	8
53	824	464	.7402	704	7
54	849	503	.7391	690	6
55	.49874	.57541	1.7379	.86675	5
56	899	580	.7367	661	4
57	924	619	.7355	646	3
58	950	657	.7344	632	2
59	.49975	696	.7332	617	1
60	.50000	.57735	1.7321	.86603	0
	Cos	Cot	Tan	Sin	′

60°

30°

'	Sin	Tan	Cot	Cos	
0	.50000	.57735	1.7321	.86603	60
1	025	774	.7309	588	59
2	050	813	.7297	573	58
3	076	851	.7286	559	57
4	101	890	.7274	544	56
5	.50126	.57929	1.7262	.86530	55
6	151	.57968	.7251	515	54
7	176	.58007	.7239	501	53
8	201	046	.7228	486	52
9	227	085	.7216	471	51
10	.50252	.58124	1.7205	.86457	50
11	277	162	.7193	442	49
12	302	201	.7182	427	48
13	327	240	.7170	413	47
14	352	279	.7159	398	46
15	.50377	.58318	1.7147	.86384	45
16	403	357	.7136	369	44
17	428	396	.7124	354	43
18	453	435	.7113	340	42
19	478	474	.7102	325	41
20	.50503	.58513	1.7090	.86310	40
21	528	552	.7079	295	39
22	553	591	.7067	281	38
23	578	631	.7056	266	37
24	603	670	.7045	251	36
25	.50628	.58709	1.7033	.86237	35
26	654	748	.7022	222	34
27	679	787	.7011	207	33
28	704	826	.6999	192	32
29	729	865	.6988	178	31
30	.50754	.58905	1.6977	.86163	30
31	779	944	.6965	148	29
32	804	.58983	.6954	133	28
33	829	.59022	.6943	119	27
34	854	061	.6932	104	26
35	.50879	.59101	1.6920	.86089	25
36	904	140	.6909	074	24
37	929	179	.6898	059	23
38	954	218	.6887	045	22
39	.50979	258	.6875	030	21
40	.51004	.59297	1.6864	.86015	20
41	029	336	.6853	.86000	19
42	054	376	.6842	.85985	18
43	079	415	.6831	970	17
44	104	454	.6820	956	16
45	.51129	.59494	1.6808	.85941	15
46	154	533	.6797	926	14
47	179	573	.6786	911	13
48	204	612	.6775	896	12
49	229	651	.6764	881	11
50	.51254	.59691	1.6753	.85866	10
51	279	730	.6742	851	9
52	304	770	.6731	836	8
53	329	809	.6720	821	7
54	354	849	.6709	806	6
55	.51379	.59888	1.6698	.85792	5
56	404	928	.6687	777	4
57	429	.59967	.6676	762	3
58	454	.60007	.6665	747	2
59	479	046	.6654	732	1
60	.51504	.60086	1.6643	.85717	0
	Cos	Cot	Tan	Sin	'

31°

'	Sin	Tan	Cot	Cos	
0	.51504	.60086	1.6643	.85717	60
1	529	126	.6632	702	59
2	554	165	.6621	687	58
3	579	205	.6610	672	57
4	604	245	.6599	657	56
5	.51628	.60284	1.6588	.85642	55
6	653	324	.6577	627	54
7	678	364	.6566	612	53
8	703	403	.6555	597	52
9	728	443	.6545	582	51
10	.51753	.60483	1.6534	.85567	50
11	778	522	.6523	551	49
12	803	562	.6512	536	48
13	828	602	.6501	521	47
14	852	642	.6490	506	46
15	.51877	.60681	1.6479	.85491	45
16	902	721	.6469	476	44
17	927	761	.6458	461	43
18	952	801	.6447	446	42
19	.51977	841	.6436	431	41
20	.52002	.60881	1.6426	.85416	40
21	026	921	.6415	401	39
22	051	.60960	.6404	385	38
23	076	.61000	.6393	370	37
24	101	040	.6383	355	36
25	.52126	.61080	1.6372	.85340	35
26	151	120	.6361	325	34
27	175	160	.6351	310	33
28	200	200	.6340	294	32
29	225	240	.6329	279	31
30	.52250	.61280	1.6319	.85264	30
31	275	320	.6308	249	29
32	299	360	.6297	234	28
33	324	400	.6287	218	27
34	349	440	.6276	203	26
35	.52374	.61480	1.6265	.85188	25
36	399	520	.6255	173	24
37	423	561	.6244	157	23
38	448	601	.6234	142	22
39	473	641	.6223	127	21
40	.52498	.61681	1.6212	.85112	20
41	522	721	.6202	096	19
42	547	761	.6191	081	18
43	572	801	.6181	066	17
44	597	842	.6170	051	16
45	.52621	.61882	1.6160	.85035	15
46	646	922	.6149	020	14
47	671	.61962	.6139	.85005	13
48	696	.62003	.6128	.84989	12
49	720	043	.6118	974	11
50	.52745	.62083	1.6107	.84959	10
51	770	124	.6097	943	9
52	794	164	.6087	928	8
53	819	204	.6076	913	7
54	844	245	.6066	897	6
55	.52869	.62285	1.6055	.84882	5
56	893	325	.6045	866	4
57	918	366	.6034	851	3
58	943	406	.6024	836	2
59	967	446	.6014	820	1
60	.52992	.62487	1.6003	.84805	0
	Cos	Cot	Tan	Sin	'

59°   58°

32°

′	Sin	Tan	Cot	Cos	
0	.52992	.62487	1.6003	.84805	60
1	.53017	527	.5993	789	59
2	041	568	.5983	774	58
3	066	608	.5972	759	57
4	091	649	.5962	743	56
5	.53115	.62689	1.5952	.84728	55
6	140	730	.5941	712	54
7	164	770	.5931	697	53
8	189	811	.5921	681	52
9	214	852	.5911	666	51
10	.53238	.62892	1.5900	.84650	50
11	263	933	.5890	635	49
12	288	.62973	.5880	619	48
13	312	.63014	.5869	604	47
14	337	055	.5859	588	46
15	.53361	.63095	1.5849	.84573	45
16	386	136	.5839	557	44
17	411	177	.5829	542	43
18	435	217	.5818	526	42
19	460	258	.5808	511	41
20	.53484	.63299	1.5798	.84495	40
21	509	340	.5788	480	39
22	534	380	.5778	464	38
23	558	421	.5768	448	37
24	583	462	.5757	433	36
25	.53607	.63503	1.5747	.84417	35
26	632	544	.5737	402	34
27	656	584	.5727	386	33
28	681	625	.5717	370	32
29	705	666	.5707	355	31
30	.53730	.63707	1.5697	.84339	30
31	754	748	.5687	324	29
32	779	789	.5677	308	28
33	804	830	.5667	292	27
34	828	871	.5657	277	26
35	.53853	.63912	1.5647	.84261	25
36	877	953	.5637	245	24
37	902	.63994	.5627	230	23
38	926	.64035	.5617	214	22
39	951	076	.5607	198	21
40	.53975	.64117	1.5597	.84182	20
41	.54000	158	.5587	167	19
42	024	199	.5577	151	18
43	049	240	.5567	135	17
44	073	281	.5557	120	16
45	.54097	.64322	1.5547	.84104	15
46	122	363	.5537	088	14
47	146	404	.5527	072	13
48	171	446	.5517	057	12
49	195	487	.5507	041	11
50	.54220	.64528	1.5497	.84025	10
51	244	569	.5487	.84009	9
52	269	610	.5477	.83994	8
53	293	652	.5468	978	7
54	317	693	.5458	962	6
55	.54342	.64734	1.5448	.83946	5
56	366	775	.5438	930	4
57	391	817	.5428	915	3
58	415	858	.5418	899	2
59	440	899	.5408	883	1
60	.54464	.64941	1.5399	.83867	0
	Cos	Cot	Tan	Sin	′

57°

33°

′	Sin	Tan	Cot	Cos	
0	.54464	.64941	1.5399	.83867	60
1	488	.64982	′.5389	851	59
2	513	.65024	.5379	835	58
3	537	065	.5369	819	57
4	561	106	.5359	804	56
5	.54586	.65148	1.5350	.83788	55
6	610	189	.5340	772	54
7	635	231	.5330	756	53
8	659	272	.5320	740	52
9	683	314	.5311	724	51
10	.54708	.65355	1.5301	.83708	50
11	732	397	.5291	692	49
12	756	438	.5282	676	48
13	781	480	.5272	660	47
14	805	521	.5262	645	46
15	.54829	.65563	1.5253	.83629	45
16	854	604	.5243	613	44
17	878	646	.5233	597	43
18	902	688	.5224	581	42
19	927	729	.5214	565	41
20	.54951	.65771	1.5204	.83549	40
21	975	813	′.5195	533	39
22	.54999	854	.5185	517	38
23	.55024	896	.5175	501	37
24	048	938	.5166	485	36
25	.55072	.65980	1.5156	.83469	35
26	097	.66021	.5147	453	34
27	121	063	.5137	437	33
28	145	105	.5127	421	32
29	169	147	.5118	405	31
30	.55194	.66189	1.5108	.83389	30
31	218	230	.5099	373	29
32	242	272	.5089	356	28
33	266	314	.5080	340	27
34	291	356	.5070	324	26
35	.55315	.66398	1.5061	.83308	25
36	339	440	.5051	292	24
37	363	482	.5042	276	23
38	388	524	.5032	260	22
39	412	566	.5023	244	21
40	.55436	.66608	1.5013	.83228	20
41	460	650	.5004	212	19
42	484	692	.4994	195	18
43	509	734	.4985	179	17
44	533	776	.4975	163	16
45	.55557	.66818	1.4966	.83147	15
46	581	860	.4957	131	14
47	605	902	.4947	115	13
48	630	944	.4938	098	12
49	654	.66986	.4928	082	11
50	.55678	.67028	1.4919	.83066	10
51	702	071	.4910	050	9
52	726	113	.4900	034	8
53	750	155	.4891	017	7
54	775	197	.4882	.83001	6
55	.55799	.67239	1.4872	.82985	5
56	823	282	.4863	969	4
57	847	324	.4854	953	3
58	871	366	.4844	936	2
59	895	409	.4835	920	1
60	.55919	.67451	1.4826	.82904	0
	Cos	Cot	Tan	Sin	′

56°

34°

′	Sin	Tan	Cot	Cos	
0	.55919	.67451	1.4826	.82904	60
1	943	493	.4816	887	59
2	968	536	.4807	871	58
3	.55992	578	.4798	855	57
4	.56016	620	.4788	839	56
5	.56040	.67663	1.4779	.82822	55
6	064	705	.4770	806	54
7	088	748	.4761	790	53
8	112	790	.4751	773	52
9	136	832	.4742	757	51
10	.56160	.67875	1.4733	.82741	50
11	184	917	.4724	724	49
12	208	.67960	.4715	708	48
13	232	.68002	.4705	692	47
14	256	045	.4696	675	46
15	.56280	.68088	1.4687	.82659	45
16	305	130	.4678	643	44
17	329	173	.4669	626	43
18	353	215	.4659	610	42
19	377	258	.4650	593	41
20	.56401	.68301	1.4641	.82577	40
21	425	343	.4632	561	39
22	449	386	.4623	544	38
23	473	429	.4614	528	37
24	497	471	.4605	511	36
25	.56521	.68514	1.4596	.82495	35
26	545	557	.4586	478	34
27	569	600	.4577	462	33
28	593	642	.4568	446	32
29	617	685	.4559	429	31
30	.56641	.68728	1.4550	.82413	30
31	665	771	.4541	396	29
32	689	814	.4532	380	28
33	713	857	.4523	363	27
34	736	900	.4514	347	26
35	.56760	.68942	1.4505	.82330	25
36	784	.68985	.4496	314	24
37	808	.69028	.4487	297	23
38	832	071	.4478	281	22
39	856	114	.4469	264	21
40	.56880	.69157	1.4460	.82248	20
41	904	200	.4451	231	19
42	928	243	.4442	214	18
43	952	286	.4433	198	17
44	.56976	329	.4424	181	16
45	.57000	.69372	1.4425	.82165	15
46	024	416	.4406	148	14
47	047	459	.4397	132	13
48	071	502	.4388	115	12
49	095	545	.4379	098	11
50	.57119	.69588	1.4370	.82082	10
51	143	631	.4361	065	9
52	167	675	.4352	048	8
53	191	718	.4344	032	7
54	215	761	.4335	.82015	6
55	.57238	.69804	1.4326	.81999	5
56	262	847	.4317	982	4
57	286	891	.4308	965	3
58	310	934	.4299	949	2
59	334	.69977	.4290	932	1
60	.57358	.70021	1.4281	.81915	0
	Cos	Cot	Tan	Sin	′

55°

35°

′	Sin	Tan	Cot	Cos	
0	.57358	.70021	1.4281	.81915	60
1	381	064	.4273	899	59
2	405	107	.4264	882	58
3	429	151	.4255	865	57
4	453	194	.4246	848	56
5	.57477	.70238	1.4237	.81832	55
6	501	281	.4229	815	54
7	524	325	.4220	798	53
8	548	368	.4211	782	52
9	572	412	.4202	765	51
10	.57596	.70455	1.4193	.81748	50
11	619	499	.4185	731	49
12	643	542	.4176	714	48
13	667	586	.4167	698	47
14	691	629	.4158	681	46
15	.57715	.70673	1.4150	.81664	45
16	738	717	.4141	647	44
17	762	760	.4132	631	43
18	786	804	.4124	614	42
19	810	848	.4115	597	41
20	.57833	.70891	1.4106	.81580	40
21	857	935	.4097	563	39
22	881	.70979	.4089	546	38
23	904	.71023	.4080	530	37
24	928	066	.4071	513	36
25	.57952	.71110	1.4063	.81496	35
26	976	154	.4054	479	34
27	.57999	198	.4045	462	33
28	.58023	242	.4037	445	32
29	047	285	.4028	428	31
30	.58070	.71329	1.4019	.81412	30
31	094	373	.4011	395	29
32	118	417	.4002	378	28
33	141	461	.3994	361	27
34	165	505	.3985	344	26
35	.58189	.71549	1.3976	.81327	25
36	212	593	.3968	310	24
37	236	637	.3959	293	23
38	260	681	.3951	276	22
39	283	725	.3942	259	21
40	.58307	.71769	1.3934	.81242	20
41	330	813	.3925	225	19
42	354	857	.3916	208	18
43	378	901	.3908	191	17
44	401	946	.3899	174	16
45	.58425	.71990	1.3891	.81157	15
46	449	.72034	.3882	140	14
47	472	078	.3874	123	13
48	496	122	.3865	106	12
49	519	167	.3857	089	11
50	.58543	.72211	1.3848	.81072	10
51	567	255	.3840	055	9
52	590	299	.3831	038	8
53	614	344	.3823	021	7
54	637	388	.3814	.81004	6
55	.58661	.72432	1.3806	.80987	5
56	684	477	.3798	970	4
57	708	521	.3789	953	3
58	731	565	.3781	936	2
59	755	610	.3772	919	1
60	.58779	.72654	1.3764	.80902	0
	Cos	Cot	Tan	Sin	′

54°

Table III

36°

′	Sin	Tan	Cot	Cos	
0	.58779	.72654	1.3764	.80902	60
1	802	699	.3755	885	59
2	826	743	.3747	867	58
3	849	788	.3.39	850	57
4	873	832	.3730	833	56
5	.58896	.72877	1.3722	.80816	55
6	920	921	.3713	799	54
7	943	.72966	.3705	782	53
8	967	.73010	.3697	765	52
9	.58990	055	.3688	748	51
10	.59014	.73100	1.3680	.80730	50
11	037	144	.3672	713	49
12	061	189	.3663	696	48
13	084	234	.3655	679	47
14	108	278	.3647	662	46
15	.59131	.73323	1.3638	.80644	45
16	154	368	.3630	627	44
17	178	413	.3622	610	43
18	201	457	.3613	593	42
19	225	502	.3605	576	41
20	.59248	.73547	1.3597	.80558	40
21	272	592	.3588	541	39
22	295	637	.3580	524	38
23	318	681	.3572	507	37
24	342	726	.3564	489	36
25	.59365	.73771	1.3555	.80472	35
26	389	816	.3547	455	34
27	412	861	.3539	438	33
28	436	906	.3531	420	32
29	459	951	.3522	403	31
30	.59482	.73996	1.3514	.80386	30
31	506	.74041	.3506	368	29
32	529	086	.3498	351	28
33	552	131	.3490	334	27
34	576	176	.3481	316	26
35	.59599	.74221	1.3473	.80299	25
36	622	267	.3465	282	24
37	646	312	.3457	264	23
38	669	357	.3449	247	22
39	693	402	.3440	230	21
40	.59716	.74447	1.3432	.80212	20
41	739	492	.3424	195	19
42	763	538	.3416	178	18
43	786	583	.3408	160	17
44	809	628	.3400	143	16
45	.59832	.74674	1.3392	.80125	15
46	856	719	.3384	108	14
47	879	764	.3375	091	13
48	902	810	.3367	073	12
49	926	855	.3359	056	11
50	.59949	.74900	1.3351	.80038	10
51	972	946	.3343	021	9
52	.59995	.74991	.3335	.80003	8
53	.60019	.75037	.3327	.79986	7
54	042	082	.3319	968	6
55	.60065	.75128	1.3311	.79951	5
56	089	173	.3303	934	4
57	112	219	.3295	916	3
58	135	264	.3287	899	2
59	158	310	.3278	881	1
60	.60182	.75355	1.3270	.79864	0
	Cos	Cot	Tan	Sin	′

53°

37°

′	Sin	Tan	Cot	Cos	
0	.60182	.75355	1.3270	.79864	60
1	205	401	.3262	846	59
2	228	447	.3254	829	58
3	251	492	.3246	811	57
4	274	538	.3238	793	56
5	.60298	.75584	1.3230	.79776	55
6	321	629	.3222	758	54
7	344	675	.3214	741	53
8	367	721	.3206	723	52
9	390	767	.3198	706	51
10	.60414	.75812	1.3190	.79688	50
11	437	858	.3182	671	49
12	460	904	.3175	653	48
13	483	950	.3167	635	47
14	506	.75996	.3159	618	46
15	.60529	.76042	1.3151	.79600	45
16	553	088	.3143	583	44
17	576	134	.3135	565	43
18	599	180	.3127	547	42
19	622	226	.3119	530	41
20	.60645	.76272	1.3111	.79512	40
21	668	318	.3103	494	39
22	691	364	.3095	477	38
23	714	410	.3087	459	37
24	738	456	.3079	441	36
25	.60761	.76502	1.3072	.79424	35
26	784	548	.3064	406	34
27	807	594	.3056	388	33
28	830	640	.3048	371	32
29	853	686	.3040	353	31
30	.60876	.76733	1.3032	.79335	30
31	899	779	.3024	318	29
32	922	825	.3017	300	28
33	945	871	.3009	282	27
34	968	918	.3001	264	26
35	.60991	.76964	1.2993	.79247	25
36	.61015	.77010	.2985	229	24
37	038	057	.2977	211	23
38	061	103	.2970	193	22
39	084	149	.2962	176	21
40	.61107	.77196	1.2954	.79158	20
41	130	242	.2946	140	19
42	153	289	.2938	122	18
43	176	335	.2931	105	17
44	199	382	.2923	087	16
45	.61222	.77428	1.2915	.79069	15
46	245	475	.2907	051	14
47	268	521	.2900	033	13
48	291	568	.2892	.79016	12
49	314	615	.2884	.78998	11
50	.61337	.77661	1.2876	.78980	10
51	360	708	.2869	962	9
52	383	754	.2861	944	8
53	406	801	.2853	926	7
54	429	848	.2846	908	6
55	.61451	.77895	1.2838	.78891	5
56	474	941	.2830	873	4
57	497	.77988	.2822	855	3
58	520	.78035	.2815	837	2
59	543	082	.2807	819	1
60	.61566	.78129	1.2799	.78801	0
	Cos	Cot	Tan	Sin	′

52°

38°

′	Sin	Tan	Cot	Cos	
0	.61566	.78129	1.2799	.78801	60
1	589	175	.2792	783	59
2	612	222	.2784	765	58
3	635	269	.2776	747	57
4	658	316	.2769	729	56
5	.61681	.78363	1.2761	.78711	55
6	704	410	.2753	694	54
7	726	457	.2746	676	53
8	749	504	.2738	658	52
9	772	551	.2731	640	51
10	.61795	.78598	1.2723	.78622	50
11	818	645	.2715	604	49
12	841	692	.2708	586	48
13	864	739	.2700	568	47
14	887	786	.2693	550	46
15	.61909	.78834	1.2685	.78532	45
16	932	881	.2677	514	44
17	955	928	.2670	496	43
18	.61978	.78975	.2662	478	42
19	.62001	.79022	.2655	460	41
20	.62024	.79070	1.2647	.78442	40
21	046	117	.2640	424	39
22	069	164	.2632	405	38
23	092	212	.2624	387	37
24	115	259	.2617	369	36
25	.62138	.79306	1.2609	.78351	35
26	160	354	.2602	333	34
27	183	401	.2594	315	33
28	206	449	.2587	297	32
29	229	496	.2579	279	31
30	.62251	.79544	1.2572	.78261	30
31	274	591	.2564	243	29
32	297	639	.2557	225	28
33	320	686	.2549	206	27
34	342	734	.2542	188	26
35	.62365	.79781	1.2534	.78170	25
36	388	829	.2527	152	24
37	411	877	.2519	134	23
38	433	924	.2512	116	22
39	456	.79972	.2504	098	21
40	.62479	.80020	1.2497	.78079	20
41	502	067	.2489	061	19
42	524	115	.2482	043	18
43	547	163	.2475	025	17
44	570	211	.2467	.78007	16
45	.62592	.80258	1.2460	.77988	15
46	615	306	.2452	970	14
47	638	354	.2445	952	13
48	660	402	.2437	934	12
49	683	450	.2430	916	11
50	.62706	.80498	1.2423	.77897	10
51	728	546	.2415	879	9
52	751	594	.2408	861	8
53	774	642	.2401	843	7
54	796	690	.2393	824	6
55	.62819	.80738	1.2386	.77806	5
56	842	786	.2378	788	4
57	864	834	.2371	769	3
58	887	882	.2364	751	2
59	909	930	.2356	733	1
60	.62932	.80978	1.2349	.77715	0
	Cos	Cot	Tan	Sin	′

51°

39°

′	Sin	Tan	Cot	Cos	
0	.62932	.80978	1.2349	.77715	60
1	955	.81027	.2342	696	59
2	.62977	075	.2334	678	58
3	.63000	123	.2327	660	57
4	022	171	.2320	641	56
5	.63045	.81220	1.2312	.77623	55
6	068	268	.2305	605	54
7	090	316	.2298	586	53
8	113	364	.2290	568	52
9	135	413	.2283	550	51
10	.63158	.81461	1.2276	.77531	50
11	180	510	.2268	513	49
12	203	558	.2261	494	48
13	225	606	.2254	476	47
14	248	655	.2247	458	46
15	.63271	.81703	1.2239	.77439	45
16	293	752	.2232	421	44
17	316	800	.2225	402	43
18	338	849	.2218	384	42
19	361	898	.2210	366	41
20	.63383	.81946	1.2203	.77347	40
21	406	.81995	.2196	329	39
22	428	.82044	.2189	310	38
23	451	092	.2181	292	37
24	473	141	.2174	273	36
25	.63496	.82190	1.2167	.77255	35
26	518	238	.2160	236	34
27	540	287	.2153	218	33
28	563	336	.2145	199	32
29	585	385	.2138	181	31
30	.63608	.82434	1.2131	.77162	30
31	630	483	.2124	144	29
32	653	531	.2117	125	28
33	675	580	.2109	107	27
34	698	629	.2102	088	26
35	.63720	.82678	1.2095	.77070	25
36	742	727	.2088	051	24
37	765	776	.2081	033	23
38	787	825	.2074	.77014	22
39	810	874	.2066	.76996	21
40	.63832	.82923	1.2059	.76977	20
41	854	.82972	.2052	959	19
42	877	.83022	.2045	940	18
43	899	071	.2038	921	17
44	922	120	.2031	903	16
45	.63944	.83169	1.2024	.76884	15
46	966	218	.2017	866	14
47	.63989	268	.2009	847	13
48	.64011	317	.2002	828	12
49	033	366	.1995	810	11
50	.64056	.83415	1.1988	.76791	10
51	078	465	.1981	772	9
52	100	514	.1974	754	8
53	123	564	.1967	735	7
54	145	613	.1960	717	6
55	.64167	.83662	1.1953	.76698	5
56	190	712	.1946	679	4
57	212	761	.1939	661	3
58	234	811	.1932	642	2
59	256	860	.1925	623	1
60	.64279	.83910	1.1918	.76604	0
	Cos	Cot	Tan	Sin	′

50°

Table III

40°

′	Sin	Tan	Cot	Cos	
0	.64279	.83910	1.1918	.76604	60
1	301	.83960	.1910	586	59
2	323	.84009	.1903	567	58
3	346	059	.1896	548	57
4	368	108	.1889	530	56
5	.64390	.84158	1.1882	.76511	55
6	412	208	.1875	492	54
7	435	258	.1868	473	53
8	457	307	.1861	455	52
9	479	357	.1854	436	51
10	.64501	.84407	1.1847	.76417	50
11	524	457	.1840	398	49
12	546	507	.1833	380	48
13	568	556	.1826	361	47
14	590	606	.1819	342	46
15	.64612	.84656	1.1812	.76323	45
16	635	706	.1806	304	44
17	657	756	.1799	286	43
18	679	806	.1792	267	42
19	701	856	.1785	248	41
20	.64723	.84906	1.1778	.76229	40
21	746	.84956	.1771	210	39
22	768	.85006	.1764	192	38
23	790	057	.1757	173	37
24	812	107	.1750	154	36
25	.64834	.85157	1.1743	.76135	35
26	856	207	.1736	116	34
27	878	257	.1729	097	33
28	901	308	.1722	078	32
29	923	358	.1715	059	31
30	.64945	.85408	1.1708	.76041	30
31	967	458	.1702	022	29
32	.64989	509	.1695	.76003	28
33	.65011	559	.1688	.75984	27
34	033	609	.1681	965	26
35	.65055	.85660	1.1674	.75946	25
36	077	710	.1667	927	24
37	100	761	.1660	908	23
38	122	811	.1653	889	22
39	144	862	.1647	870	21
40	.65166	.85912	1.1640	.75851	20
41	188	.85963	.1633	832	19
42	210	.86014	.1626	813	18
43	232	064	.1619	794	17
44	254	115	.1612	775	16
45	.65276	.86166	1.1606	.75756	15
46	298	216	.1599	738	14
47	320	267	.1592	719	13
48	342	318	.1585	700	12
49	364	368	.1578	680	11
50	.65386	.86419	1.1571	.75661	10
51	408	470	.1565	642	9
52	430	521	.1558	623	8
53	452	572	.1551	604	7
54	474	623	.1544	585	6
55	.65496	.86674	1.1538	.75566	5
56	518	725	.1531	547	4
57	540	776	.1524	528	3
58	562	827	.1517	509	2
59	584	878	.1510	490	1
60	.65606	.86929	1.1504	.75471	0
	Cos	Cot	Tan	Sin	′

49°

41°

′	Sin	Tan	Cot	Cos	
0	.65606	.86929	1.1504	.75471	60
1	628	.86980	.1497	452	59
2	650	.87031	.1490	433	58
3	672	082	.1483	414	57
4	694	133	.1477	395	56
5	.65716	.87184	1.1470	.75375	55
6	738	236	.1463	356	54
7	759	287	.1456	337	53
8	781	338	.1450	318	52
9	803	389	.1443	299	51
10	.65825	.87441	1.1436	.75280	50
11	847	492	.1430	261	49
12	869	543	.1423	241	48
13	891	595	.1416	222	47
14	913	646	.1410	203	46
15	.65935	.87698	1.1403	.75184	45
16	956	749	.1396	165	44
17	.65978	801	.1389	146	43
18	.66000	852	.1383	126	42
19	022	904	.1376	107	41
20	.66044	.87955	1.1369	.75088	40
21	066	.88007	.1363	069	39
22	088	059	.1356	050	38
23	109	110	.1349	030	37
24	131	162	.1343	.75011	36
25	.66153	.88214	1.1336	.74992	35
26	175	265	.1329	973	34
27	197	317	.1323	953	33
28	218	369	.1316	934	32
29	240	421	.1310	915	31
30	.66262	.88473	1.1303	.74896	30
31	284	524	.1296	876	29
32	306	576	.1290	857	28
33	327	628	.1283	838	27
34	349	680	.1276	818	26
35	.66371	.88732	1.1270	.74799	25
36	393	784	.1263	780	24
37	414	836	.1257	760	23
38	436	888	.1250	741	22
39	458	940	.1243	722	21
40	.66480	.88992	1.1237	.74703	20
41	501	.89045	.1230	683	19
42	523	097	.1224	664	18
43	545	149	.1217	644	17
44	566	201	.1211	625	16
45	.66588	.89253	1.1204	.74606	15
46	610	306	.1197	586	14
47	632	358	.1191	567	13
48	653	410	.1184	548	12
49	675	463	.1178	528	11
50	.66697	.89515	1.1171	.74509	10
51	718	567	.1165	489	9
52	740	620	.1158	470	8
53	762	672	.1152	451	7
54	783	725	.1145	431	6
55	.66805	.89777	1.1139	.74412	5
56	827	830	.1132	392	4
57	848	883	.1126	373	3
58	870	935	.1119	353	2
59	891	.89988	.1113	334	1
60	.66913	.90040	1.1106	.74314	0
	Cos	Cot	Tan	Sin	′

48°

42°

′	Sin	Tan	Cot	Cos	
0	.66913	.90040	1.1106	.74314	60
1	935	093	.1100	295	59
2	956	146	.1093	276	58
3	978	199	.1087	256	57
4	.66999	251	.1080	237	56
5	.67021	.90304	1.1074	.74217	55
6	043	357	.1067	198	54
7	064	410	.1061	178	53
8	086	463	.1054	159	52
9	107	516	.1048	139	51
10	.67129	.90569	1.1041	.74120	50
11	151	621	.1035	100	49
12	172	674	.1028	080	48
13	194	727	.1022	061	47
14	215	781	.1016	041	46
15	.67237	.90834	1.1009	.74022	45
16	258	887	.1003	.74002	44
17	280	940	.0996	.73983	43
18	301	.90993	.0990	963	42
19	323	.91046	.0983	944	41
20	.67344	.91099	1.0977	.73924	40
21	366	153	.0971	904	39
22	387	206	.0964	885	38
23	409	259	.0958	865	37
24	430	313	.0951	846	36
25	.67452	.91366	1.0945	.73826	35
26	473	419	.0939	806	34
27	495	473	.0932	787	33
28	516	526	.0926	767	32
29	538	580	.0919	747	31
30	.67559	.91633	1.0913	.73728	30
31	580	687	.0907	708	29
32	602	740	.0900	688	28
33	623	794	.0894	669	27
34	645	847	.0888	649	26
35	.67666	.91901	1.0881	.73629	25
36	688	.91955	.0875	610	24
37	709	.92008	.0869	590	23
38	730	062	.0862	570	22
39	752	116	.0856	551	21
40	.67773	.92170	1.0850	.73531	20
41	795	224	.0843	511	19
42	816	277	.0837	491	18
43	837	331	.0831	472	17
44	859	385	.0824	452	16
45	.67880	.92439	1.0818	.73432	15
46	901	493	.0812	413	14
47	923	547	.0805	393	13
48	944	601	.0799	373	12
49	965	655	.0793	353	11
50	.67987	.92709	1.0786	.73333	10
51	.68008	763	.0780	314	9
52	029	817	.0774	294	8
53	051	872	.0768	274	7
54	072	926	.0761	254	6
55	.68093	.92980	1.0755	.73234	5
56	115	.93034	.0749	215	4
57	136	088	.0742	195	3
58	157	143	.0736	175	2
59	179	197	.0730	155	1
60	.68200	.93252	1.0724	.73135	0
	Cos	Cot	Tan	Sin	′

47°

43°

′	Sin	Tan	Cot	Cos	
0	.68200	.93252	1.0724	.73135	60
1	221	306	.0717	116	59
2	242	360	.0711	096	58
3	264	415	.0705	076	57
4	285	469	.0699	056	56
5	.68306	.93524	1.0692	.73036	55
6	327	578	.0686	.73016	54
7	349	633	.0680	.72996	53
8	370	688	.0674	976	52
9	391	742	.0668	957	51
10	.68412	.93797	1.0661	.72937	50
11	434	852	.0655	917	49
12	455	906	.0649	897	48
13	476	.93961	.0643	877	47
14	497	.94016	.0637	857	46
15	.68518	.94071	1.0630	.72837	45
16	539	125	.0624	817	44
17	561	180	.0618	797	43
18	582	235	.0612	777	42
19	603	290	.0606	757	41
20	.68624	.94345	1.0599	.72737	40
21	645	400	.0593	717	39
22	666	455	.0587	697	38
23	688	510	.0581	677	37
24	709	565	.0575	657	36
25	.68730	.94620	1.0569	.72637	35
26	751	676	.0562	617	34
27	772	731	.0556	597	33
28	793	786	.0550	577	32
29	814	841	.0544	557	31
30	.68835	.94896	1.0538	.72537	30
31	857	.94952	.0532	517	29
32	878	.95007	.0526	497	28
33	899	062	.0519	477	27
34	920	118	.0513	457	26
35	.68941	.95173	1.0507	.72437	25
36	962	229	.0501	417	24
37	.68983	284	.0495	397	23
38	.69004	340	.0489	377	22
39	025	395	.0483	357	21
40	.69046	.95451	1.0477	.72337	20
41	067	506	.0470	317	19
42	088	562	.0464	297	18
43	109	618	.0458	277	17
44	130	673	.0452	257	16
45	.69151	.95729	1.0446	.72236	15
46	172	785	.0440	216	14
47	193	841	.0434	196	13
48	214	897	.0428	176	12
49	235	.95952	.0422	156	11
50	.69256	.96008	1.0416	.72136	10
51	277	064	.0410	116	9
52	298	120	.0404	095	8
53	319	176	.0398	075	7
54	340	232	.0392	055	6
55	.69361	.96288	1.0385	.72035	5
56	382	344	.0379	.72015	4
57	403	400	.0373	.71995	3
58	424	457	.0367	974	2
59	445	513	.0361	954	1
60	.69466	.96569	1.0355	.71934	0
	Cos	Cot	Tan	Sin	′

46°

Table III

44°

′	Sin	Tan	Cot	Cos	
0	.69466	.96569	1.0355	.71934	60
1	487	625	.0349	914	59
2	508	681	.0343	894	58
3	529	738	.0337	873	57
4	549	794	.0331	853	56
5	.69570	.96850	1.0325	.71833	55
6	591	907	.0319	813	54
7	612	.96963	.0313	792	53
8	633	.97020	.0307	772	52
9	654	076	.0301	752	51
10	.69675	.97133	1.0295	.71732	50
11	696	189	.0289	711	49
12	717	246	.0283	691	48
13	737	302	.0277	671	47
14	758	359	.0271	650	46
15	.69779	.97416	1.0265	.71630	45
16	800	472	.0259	610	44
17	821	529	.0253	590	43
18	842	586	.0247	569	42
19	862	643	.0241	549	41
20	.69883	.97700	1.0235	.71529	40
21	904	756	.0230	508	39
22	925	813	.0224	488	38
23	946	870	.0218	468	37
24	966	927	.0212	447	36
25	.69987	.97984	1.0206	.71427	35
26	.70008	.98041	.0200	407	34
27	029	098	.0194	386	33
28	049	155	.0188	366	32
29	070	213	.0182	345	31
30	.70091	.98270	1.0176	.71325	30
31	112	327	.0170	305	29
32	132	384	.0164	284	28
33	153	441	.0158	264	27
34	174	499	.0152	243	26
35	.70195	.98556	1.0147	.71223	25
36	215	613	.0141	203	24
37	236	671	.0135	182	23
38	257	728	.0129	162	22
39	277	786	.0123	141	21
40	.70298	.98843	1.0117	.71121	20
41	319	901	.0111	100	19
42	339	.98958	.0105	080	18
43	360	.99016	.0099	059	17
44	381	073	.0094	039	16
45	.70401	.99131	1.0088	.71019	15
46	422	189	.0082	.70998	14
47	443	247	.0076	978	13
48	463	304	.0070	957	12
49	484	362	.0064	937	11
50	.70505	.99420	1.0058	.70916	10
51	525	478	.0052	896	9
52	546	536	.0047	875	8
53	567	594	.0041	855	7
54	587	652	.0035	834	6
55	.70608	.99710	1.0029	.70813	5
56	628	768	.0023	793	4
57	649	826	.0017	772	3
58	670	884	.0012	752	2
59	690	.99942	.0006	731	1
60	.70711	1.0000	1.0000	.70711	0
	Cos	Cot	Tan	Sin	′

45°

Table IV

Natural values and logarithms of the trigonometric functions

to four decimal places

Table IV

0° — 9°

Degrees	Sine		Tangent		Cotangent		Cosine		
	Value	Log	Value	Log	Value	Log	Value	Log	
0° 00′	.0000	——	.0000	——	——	——	1.0000	0.0000	**90° 00′**
10	.0029	7.4637	.0029	7.4637	343.77	2.5363	1.0000	0.0000	50
20	.0058	7.7648	.0058	7.7648	171.89	2.2352	1.0000	0.0000	40
30	.0087	7.9408	.0087	7.9409	114.59	2.0591	1.0000	0.0000	30
40	.0116	8.0658	.0116	8.0658	85.940	1.9342	.9999	0.0000	20
50	.0145	8.1627	.0145	8.1627	68.750	1.8373	.9999	0.0000	10
1° 00′	.0175	8.2419	.0175	8.2419	57.290	1.7581	.9998	9.9999	**89° 00′**
10	.0204	8.3088	.0204	8.3089	49.104	1.6911	.9998	9.9999	50
20	.0233	8.3668	.0233	8.3669	42.964	1.6331	.9997	9.9999	40
30	.0262	8.4179	.0262	8.4181	38.188	1.5819	.9997	9.9999	30
40	.0291	8.4637	.0291	8.4638	34.368	1.5362	.9996	9.9998	20
50	.0320	8.5050	.0320	8.5053	31.242	1.4947	.9995	9.9998	10
2° 00′	.0349	8.5428	.0349	8.5431	28.636	1.4569	.9994	9.9997	**88° 00′**
10	.0378	8.5776	.0378	8.5779	26.432	1.4221	.9993	9.9997	50
20	.0407	8.6097	.0407	8.6101	24.542	1.3899	.9992	9.9996	40
30	.0436	8.6397	.0437	8.6401	22.904	1.3599	.9990	9.9996	30
40	.0465	8.6677	.0466	8.6682	21.470	1.3318	.9989	9.9995	20
50	.0494	8.6940	.0495	8.6945	20.206	1.3055	.9988	9.9995	10
3° 00′	.0523	8.7188	.0524	8.7194	19.081	1.2806	.9986	9.9994	**87° 00′**
10	.0552	8.7423	.0553	8.7429	18.075	1.2571	.9985	9.9993	50
20	.0581	8.7645	.0582	8.7652	17.169	1.2348	.9983	9.9993	40
30	.0610	8.7857	.0612	8.7865	16.350	1.2135	.9981	9.9992	30
40	.0640	8.8059	.0641	8.8067	15.605	1.1933	.9980	9.9991	20
50	.0669	8.8251	.0670	8.8261	14.924	1.1739	.9978	9.9990	10
4° 00′	.0698	8.8436	.0699	8.8446	14.301	1.1554	.9976	9.9989	**86° 00′**
10	.0727	8.8613	.0729	8.8624	13.727	1.1376	.9974	9.9989	50
20	.0756	8.8783	.0758	8.8795	13.197	1.1205	.9971	9.9988	40
30	.0785	8.8946	.0787	8.8960	12.706	1.1040	.9969	9.9987	30
40	.0814	8.9104	.0816	8.9118	12.251	1.0882	.9967	9.9986	20
50	.0843	8.9256	.0846	8.9272	11.826	1.0728	.9964	9.9985	10
5° 00′	.0872	8.9403	.0875	8.9420	11.430	1.0580	.9962	9.9983	**85° 00′**
10	.0901	8.9545	.0904	8.9563	11.059	1.0437	.9959	9.9982	50
20	.0929	8.9682	.0934	8.9701	10.712	1.0299	.9957	9.9981	40
30	.0958	8.9816	.0963	8.9836	10.385	1.0164	.9954	9.9980	30
40	.0987	8.9945	.0992	8.9966	10.078	1.0034	.9951	9.9979	20
50	.1016	9.0070	.1022	9.0093	9.7882	0.9907	.9948	9.9977	10
6° 00′	.1045	9.0192	.1051	9.0216	9.5144	0.9784	.9945	9.9976	**84° 00′**
10	.1074	9.0311	.1080	9.0336	9.2553	0.9664	.9942	9.9975	50
20	.1103	9.0426	.1110	9.0453	9.0098	0.9547	.9939	9.9973	40
30	.1132	9.0539	.1139	9.0567	8.7769	0.9433	.9936	9.9972	30
40	.1161	9.0648	.1169	9.0678	8.5555	0.9322	.9932	9.9971	20
50	.1190	9.0755	.1198	9.0786	8.3450	0.9214	.9929	9.9969	10
7° 00′	.1219	9.0859	.1228	9.0891	8.1443	0.9109	.9925	9.9968	**83° 00′**
10	.1248	9.0961	.1257	9.0995	7.9530	0.9005	.9922	9.9966	50
20	.1276	9.1060	.1287	9.1096	7.7704	0.8904	.9918	9.9964	40
30	.1305	9.1157	.1317	9.1194	7.5958	0.8806	.9914	9.9963	30
40	.1334	9.1252	.1346	9.1291	7.4287	0.8709	.9911	9.9961	20
50	.1363	9.1345	.1376	9.1385	7.2687	0.8615	.9907	9.9959	10
8° 00′	.1392	9.1436	.1405	9.1478	7.1154	0.8522	.9903	9.9958	**82° 00′**
10	.1421	9.1525	.1435	9.1569	6.9682	0.8431	.9899	9.9956	50
20	.1449	9.1612	.1465	9.1658	6.8269	0.8342	.9894	9.9954	40
30	.1478	9.1697	.1495	9.1745	6.6912	0.8255	.9890	9.9952	30
40	.1507	9.1781	.1524	9.1831	6.5606	0.8169	.9886	9.9950	20
50	.1536	9.1863	.1554	9.1915	6.4348	0.8085	.9881	9.9948	10
9° 00′	.1564	9.1943	.1584	9.1997	6.3138	0.8003	.9877	9.9946	**81° 00′**
	Value	Log	Value	Log	Value	Log	Value	Log	Degrees
	Cosine		Cotangent		Tangent		Sine		

81° — 90°

9° — 18°

Degrees	Sine Value	Sine Log	Tangent Value	Tangent Log	Cotangent Value	Cotangent Log	Cosine Value	Cosine Log	
9° 00′	.1564	9.1943	.1584	9.1997	6.3138	0.8003	.9877	9.9946	81° 00′
10	.1593	9.2022	.1614	9.2078	6.1970	0.7922	.9872	9.9944	50′
20	.1622	9.2100	.1644	9.2158	6.0844	0.7842	.9868	9.9942	40
30	.1650	9.2176	.1673	9.2236	5.9758	0.7764	.9863	9.9940	30
40	.1679	9.2251	.1703	9.2313	5.8708	0.7687	.9858	9.9938	20
50	.1708	9.2324	.1733	9.2389	5.7694	0.7611	.9853	9.9936	10
10° 00′	.1736	9.2397	.1763	9.2463	5.6713	0.7537	.9848	9.9934	80° 00′
10	.1765	9.2468	.1793	9.2536	5.5764	0.7464	.9843	9.9931	50
20	.1794	9.2538	.1823	9.2609	5.4845	0.7391	.9838	9.9929	40
30	.1822	9.2606	.1853	9.2680	5.3955	0.7320	.9833	9.9927	30
40	.1851	9.2674	.1883	9.2750	5.3093	0.7250	.9827	9.9924	20
50	.1880	9.2740	.1914	9.2819	5.2257	0.7181	.9822	9.9922	10
11° 00′	.1908	9.2806	.1944	9.2887	5.1446	0.7113	.9816	9.9919	79° 00′
10	.1937	9.2870	.1974	9.2953	5.0658	0.7047	.9811	9.9917	50
20	.1965	9.2934	.2004	9.3020	4.9894	0.6980	.9805	9.9914	40
30	.1994	9.2997	.2035	9.3085	4.9152	0.6915	.9799	9.9912	30
40	.2022	9.3058	.2065	9.3149	4.8430	0.6851	.9793	9.9909	20
50	.2051	9.3119	.2095	9.3212	4.7729	0.6788	.9787	9.9907	10
12° 00′	.2079	9.3179	.2126	9.3275	4.7046	0.6725	.9781	9.9904	78° 00′
10	.2108	9.3238	.2156	9.3336	4.6382	0.6664	.9775	9.9901	50
20	.2136	9.3296	.2186	9.3397	4.5736	0.6603	.9769	9.9899	40
30	.2164	9.3353	.2217	9.3458	4.5107	0.6542	.9763	9.9896	30
40	.2193	9.3410	.2247	9.3517	4.4494	0.6483	.9757	9.9893	20
50	.2221	9.3466	.2278	9.3576	4.3897	0.6424	.9750	9.9890	10
13° 00′	.2250	9.3521	.2309	9.3634	4.3315	0.6366	.9744	9.9887	77° 00′
10	.2278	9.3575	.2339	9.3691	4.2747	0.6309	.9737	9.9884	50
20	.2306	9.3629	.2370	9.3748	4.2193	0.6252	.9730	9.9881	40
30	.2334	9.3682	.2401	9.3804	4.1653	0.6196	.9724	9.9878	30
40	.2363	9.3734	.2432	9.3859	4.1126	0.6141	.9717	9.9875	20
50	.2391	9.3786	.2462	9.3914	4.0611	0.6086	.9710	9.9872	10
14° 00′	.2419	9.3837	.2493	9.3968	4.0108	0.6032	.9703	9.9869	76° 00′
10	.2447	9.3887	.2524	9.4021	3.9617	0.5979	.9696	9.9866	50
20	.2476	9.3937	.2555	9.4074	3.9136	0.5926	.9689	9.9863	40
30	.2504	9.3986	.2586	9.4127	3.8667	0.5873	.9681	9.9859	30
40	.2532	9.4035	.2617	9.4178	3.8208	0.5822	.9674	9.9856	20
50	.2560	9.4083	.2648	9.4230	3.7760	0.5770	.9667	9.9853	10
15° 00′	.2588	9.4130	.2679	9.4281	3.7321	0.5719	.9659	9.9849	75° 00′
10	.2616	9.4177	.2711	9.4331	3.6891	0.5669	.9652	9.9846	50
20	.2644	9.4223	.2742	9.4381	3.6470	0.5619	.9644	9.9843	40
30	.2672	9.4269	.2773	9.4430	3.6059	0.5570	.9636	9.9839	30
40	.2700	9.4314	.2805	9.4479	3.5656	0.5521	.9628	9.9836	20
50	.2728	9.4359	.2836	9.4527	3.5261	0.5473	.9621	9.9832	10
16° 00′	.2756	9.4403	.2867	9.4575	3.4874	0.5425	.9613	9.9828	74° 00′
10	.2784	9.4447	.2899	9.4622	3.4495	0.5378	.9605	9.9825	50
20	.2812	9.4491	.2931	9.4669	3.4124	0.5331	.9596	9.9821	40
30	.2840	9.4533	.2962	9.4716	3.3759	0.5284	.9588	9.9817	30
40	.2868	9.4576	.2994	9.4762	3.3402	0.5238	.9580	9.9814	20
50	.2896	9.4618	.3026	9.4808	3.3052	0.5192	.9572	9.9810	10
17° 00′	.2924	9.4659	.3057	9.4853	3.2709	0.5147	.9563	9.9806	73° 00′
10	.2952	9.4700	.3089	9.4898	3.2371	0.5102	.9555	9.9802	50
20	.2979	9.4741	.3121	9.4943	3.2041	0.5057	.9546	9.9798	40
30	.3007	9.4781	.3153	9.4987	3.1716	0.5013	.9537	9.9794	30
40	.3035	9.4821	.3185	9.5031	3.1397	0.4969	.9528	9.9790	20
50	.3062	9.4861	.3217	9.5075	3.1084	0.4925	.9520	9.9786	10
18° 00′	.3090	9.4900	.3249	9.5118	3.0777	0.4882	.9511	9.9782	72° 00′
	Value	Log	Value	Log	Value	Log	Value	Log	Degrees
	Cosine		Cotangent		Tangent		Sine		

72° — 81°

Table IV

18° — 27°

Degrees	Sine		Tangent		Cotangent		Cosine		
	Value	Log	Value	Log	Value	Log	Value	Log	
18° 00′	.3090	9.4900	.3249	9.5118	3.0777	0.4882	.9511	9.9782	72° 00′
10	.3118	9.4939	.3281	9.5161	3.0475	0.4839	.9502	9.9778	50
20	.3145	9.4977	.3314	9.5203	3.0178	0.4797	.9492	9.9774	40
30	.3173	9.5015	.3346	9.5245	2.9887	0.4755	.9483	9.9770	30
40	.3201	9.5052	.3378	9.5287	2.9600	0.4713	.9474	9.9765	20
50	.3228	9.5090	.3411	9.5329	2.9319	0.4671	.9465	9.9761	10
19° 00′	.3256	9.5126	.3443	9.5370	2.9042	0.4630	.9455	9.9757	71° 00′
10	.3283	9.5163	.3476	9.5411	2.8770	0.4589	.9446	9.9752	50
20	.3311	9.5199	.3508	9.5451	2.8502	0.4549	.9436	9.9748	40
30	.3338	9.5235	.3541	9.5491	2.8239	0.4509	.9426	9.9743	30
40	.3365	9.5270	.3574	9.5531	2.7980	0.4469	.9417	9.9739	20
50	.3393	9.5306	.3607	9.5571	2.7725	0.4429	.9407	9.9734	10
20° 00′	.3420	9.5341	.3640	9.5611	2.7475	0.4389	.9397	9.9730	70° 00′
10	.3448	9.5375	.3673	9.5650	2.7228	0.4350	.9387	9.9725	50
20	.3475	9.5409	.3706	9.5689	2.6985	0.4311	.9377	9.9721	40
30	.3502	9.5443	.3739	9.5727	2.6746	0.4273	.9367	9.9716	30
40	.3529	9.5477	.3772	9.5766	2.6511	0.4234	.9356	9.9711	20
50	.3557	9.5510	.3805	9.5804	2.6279	0.4196	.9346	9.9706	10
21° 00′	.3584	9.5543	.3839	9.5842	2.6051	0.4158	.9336	9.9702	69° 00′
10	.3611	9.5576	.3872	9.5879	2.5826	0.4121	.9325	9.9697	50
20	.3638	9.5609	.3906	9.5917	2.5605	0.4083	.9315	9.9692	40
30	.3665	9.5641	.3939	9.5954	2.5386	0.4046	.9304	9.9687	30
40	.3692	9.5673	.3973	9.5991	2.5172	0.4009	.9293	9.9682	20
50	.3719	9.5704	.4006	9.6028	2.4960	0.3972	.9283	9.9677	10
22° 00′	.3746	9.5736	.4040	9.6064	2.4751	0.3936	.9272	9.9672	68° 00′
10	.3773	9.5767	.4074	9.6100	2.4545	0.3900	.9261	9.9667	50
20	.3800	9.5798	.4108	9.6136	2.4342	0.3864	.9250	9.9661	40
30	.3827	9.5828	.4142	9.6172	2.4142	0.3828	.9239	9.9656	30
40	.3854	9.5859	.4176	9.6208	2.3945	0.3792	.9228	9.9651	20
50	.3881	9.5889	.4210	9.6243	2.3750	0.3757	.9216	9.9646	10
23° 00′	.3907	9.5919	.4245	9.6279	2.3559	0.3721	.9205	9.9640	67° 00′
10	.3934	9.5948	.4279	9.6314	2.3369	0.3686	.9194	9.9635	50
20	.3961	9.5978	.4314	9.6348	2.3183	0.3652	.9182	9.9629	40
30	.3987	9.6007	.4348	9.6383	2.2998	0.3617	.9171	9.9624	30
40	.4014	9.6036	.4383	9.6417	2.2817	0.3583	.9159	9.9618	20
50	.4041	9.6065	.4417	9.6452	2.2637	0.3548	.9147	9.9613	10
24° 00′	.4067	9.6093	.4452	9.6486	2.2460	0.3514	.9135	9.9607	66° 00′
10	.4094	9.6121	.4487	9.6520	2.2286	0.3480	.9124	9.9602	50
20	.4120	9.6149	.4522	9.6553	2.2113	0.3447	.9112	9.9596	40
30	.4147	9.6177	.4557	9.6587	2.1943	0.3413	.9100	9.9590	30
40	.4173	9.6205	.4592	9.6620	2.1775	0.3380	.9088	9.9584	20
50	.4200	9.6232	.4628	9.6654	2.1609	0.3346	.9075	9.9579	10
25° 00′	.4226	9.6259	.4663	9.6687	2.1445	0.3313	.9063	9.9573	65° 00′
10	.4253	9.6286	.4699	9.6720	2.1283	0.3280	.9051	9.9567	50
20	.4279	9.6313	.4734	9.6752	2.1123	0.3248	.9038	9.9561	40
30	.4305	9.6340	.4770	9.6785	2.0965	0.3215	.9026	9.9555	30
40	.4331	9.6366	.4806	9.6817	2.0809	0.3183	.9013	9.9549	20
50	.4358	9.6392	.4841	9.6850	2.0655	0.3150	.9001	9.9543	10
26° 00′	.4384	9.6418	.4877	9.6882	2.0503	0.3118	.8988	9.9537	64° 00′
10	.4410	9.6444	.4913	9.6914	2.0353	0.3086	.8975	9.9530	50
20	.4436	9.6470	.4950	9.6946	2.0204	0.3054	.8962	9.9524	40
30	.4462	9.6495	.4986	9.6977	2.0057	0.3023	.8949	9.9518	30
40	.4488	9.6521	.5022	9.7009	1.9912	0.2991	.8936	9.9512	20
50	.4514	9.6546	.5059	9.7040	1.9768	0.2960	.8923	9.9505	10
27° 00′	.4540	9.6570	.5095	9.7072	1.9626	0.2928	.8910	9.9499	63° 00′
	Value	Log	Value	Log	Value	Log	Value	Log	
	Cosine		Cotangent		Tangent		Sine		Degrees

63° — 72°

27° — 36°

Degrees	Sine		Tangent		Cotangent		Cosine		
	Value	Log	Value	Log	Value	Log	Value	Log	
27° 00′	.4540	9.6570	.5095	9.7072	1.9626	0.2928	.8910	9.9499	**63° 00′**
10	.4566	9.6595	.5132	9.7103	1.9486	0.2897	.8897	9.9492	50
20	.4592	9.6620	.5169	9.7134	1.9347	0.2866	.8884	9.9486	40
30	.4617	9.6644	.5206	9.7165	1.9210	0.2835	.8870	9.9479	30
40	.4643	9.6668	.5243	9.7196	1.9074	0.2804	.8857	9.9473	20
50	.4669	9.6692	.5280	9.7226	1.8940	0.2774	.8843	9.9466	10
28° 00′	.4695	9.6716	.5317	9.7257	1.8807	0.2743	.8829	9.9459	**62° 00′**
10	.4720	9.6740	.5354	9.7287	1.8676	0.2713	.8816	9.9453	50
20	.4746	9.6763	.5392	9.7317	1.8546	0.2683	.8802	9.9446	40
30	.4772	9.6787	.5430	9.7348	1.8418	0.2652	.8788	9.9439	30
40	.4797	9.6810	.5467	9.7378	1.8291	0.2622	.8774	9.9432	20
50	.4823	9.6833	.5505	9.7408	1.8165	0.2592	.8760	9.9425	10
29° 00′	.4848	9.6856	.5543	9.7438	1.8040	0.2562	.8746	9.9418	**61° 00′**
10	.4874	9.6878	.5581	9.7467	1.7917	0.2533	.8732	9.9411	50
20	.4899	9.6901	.5619	9.7497	1.7796	0.2503	.8718	9.9404	40
30	.4924	9.6923	.5658	9.7526	1.7675	0.2474	.8704	9.9397	30
40	.4950	9.6946	.5696	9.7556	1.7556	0.2444	.8689	9.9390	20
50	.4975	9.6968	.5735	9.7585	1.7437	0.2415	.8675	9.9383	10
30° 00′	.5000	9.6990	.5774	9.7614	1.7321	0.2386	.8660	9.9375	**60° 00′**
10	.5025	9.7012	.5812	9.7644	1.7205	0.2356	.8646	9.9368	50
20	.5050	9.7033	.5851	9.7673	1.7090	0.2327	.8631	9.9361	40
30	.5075	9.7055	.5890	9.7701	1.6977	0.2299	.8616	9.9353	30
40	.5100	9.7076	.5930	9.7730	1.6864	0.2270	.8601	9.9346	20
50	.5125	9.7097	.5969	9.7759	1.6753	0.2241	.8587	9.9338	10
31° 00′	.5150	9.7118	.6009	9.7788	1.6643	0.2212	.8572	9.9331	**59° 00′**
10	.5175	9.7139	.6048	9.7816	1.6534	0.2184	.8557	9.9323	50
20	.5200	9.7160	.6088	9.7845	1.6426	0.2155	.8542	9.9315	40
30	.5225	9.7181	.6128	9.7873	1.6319	0.2127	.8526	9.9308	30
40	.5250	9.7201	.6168	9.7902	1.6212	0.2098	.8511	9.9300	20
50	.5275	9.7222	.6208	9.7930	1.6107	0.2070	.8496	9.9292	10
32° 00′	.5299	9.7242	.6249	9.7958	1.6003	0.2042	.8480	9.9284	**58° 00′**
10	.5324	9.7262	.6289	9.7986	1.5900	0.2014	.8465	9.9276	50
20	.5348	9.7282	.6330	9.8014	1.5798	0.1986	.8450	9.9268	40
30	.5373	9.7302	.6371	9.8042	1.5697	0.1958	.8434	9.9260	30
40	.5398	9.7322	.6412	9.8070	1.5597	0.1930	.8418	9.9252	20
50	.5422	9.7342	.6453	9.8097	1.5497	0.1903	.8403	9.9244	10
33° 00′	.5446	9.7361	.6494	9.8125	1.5399	0.1875	.8387	9.9236	**57° 00′**
10	.5471	9.7380	.6536	9.8153	1.5301	0.1847	.8371	9.9228	50
20	.5495	9.7400	.6577	9.8180	1.5204	0.1820	.8355	9.9219	40
30	.5519	9.7419	.6619	9.8208	1.5108	0.1792	.8339	9.9211	30
40	.5544	9.7438	.6661	9.8235	1.5013	0.1765	.8323	9.9203	20
50	.5568	9.7457	.6703	9.8263	1.4919	0.1737	.8307	9.9194	10
34° 00′	.5592	9.7476	.6745	9.8290	1.4826	0.1710	.8290	9.9186	**56° 00′**
10	.5616	9.7494	.6787	9.8317	1.4733	0.1683	.8274	9.9177	50
20	.5640	9.7513	.6830	9.8344	1.4641	0.1656	.8258	9.9169	40
30	.5664	9.7531	.6873	9.8371	1.4550	0.1629	.8241	9.9160	30
40	.5688	9.7550	.6916	9.8398	1.4460	0.1602	.8225	9.9151	20
50	.5712	9.7568	.6959	9.8425	1.4370	0.1575	.8208	9.9142	10
35° 00′	.5736	9.7586	.7002	9.8452	1.4281	0.1548	.8192	9.9134	**55° 00′**
10	.5760	9.7604	.7046	9.8479	1.4193	0.1521	.8175	9.9125	50
20	.5783	9.7622	.7089	9.8506	1.4106	0.1494	.8158	9.9116	40
30	.5807	9.7640	.7133	9.8533	1.4019	0.1467	.8141	9.9107	30
40	.5831	9.7657	.7177	9.8559	1.3934	0.1441	.8124	9.9098	20
50	.5854	9.7675	.7221	9.8586	1.3848	0.1414	.8107	9.9089	10
36° 00′	.5878	9.7692	.7265	9.8613	1.3764	0.1387	.8090	9.9080	**54° 00′**
	Value	Log	Value	Log	Value	Log	Value	Log	**Degrees**
	Cosine		Cotangent		Tangent		Sine		

54° — 63°

Table IV

36° — 45°

Degrees	Sine		Tangent		Cotangent		Cosine		
	Value	Log	Value	Log	Value	Log	Value	Log	
36° 00′	.5878	9.7692	.7265	9.8613	1.3764	0.1387	.8090	9.9080	**54° 00′**
10	.5901	9.7710	.7310	9.8639	1.3680	0.1361	.8073	9.9070	50
20	.5925	9.7727	.7355	9.8666	1.3597	0.1334	.8056	9.9061	40
30	.5948	9.7744	.7400	9.8692	1.3514	0.1308	.8039	9.9052	30
40	.5972	9.7761	.7445	9.8718	1.3432	0.1282	.8021	9.9042	20
50	.5995	9.7778	.7490	9.8745	1.3351	0.1255	.8004	9.9033	10
37° 00′	.6018	9.7795	.7536	9.8771	1.3270	0.1229	.7986	9.9023	**53° 00′**
10	.6041	9.7811	.7581	9.8797	1.3190	0.1203	.7969	9.9014	50
20	.6065	9.7828	.7627	9.8824	1.3111	0.1176	.7951	9.9004	40
30	.6088	9.7844	.7673	9.8850	1.3032	0.1150	.7934	9.8995	30
40	.6111	9.7861	.7720	9.8876	1.2954	0.1124	.7916	9.8985	20
50	.6134	9.7877	.7766	9.8902	1.2876	0.1098	.7898	9.8975	10
38° 00′	.6157	9.7893	.7813	9.8928	1.2799	0.1072	.7880	9.8965	**52° 00′**
10	.6180	9.7910	.7860	9.8954	1.2723	0.1046	.7862	9.8955	50
20	.6202	9.7926	.7907	9.8980	1.2647	0.1020	.7844	9.8945	40
30	.6225	9.7941	.7954	9.9006	1.2572	0.0994	.7826	9.8935	30
40	.6248	9.7957	.8002	9.9032	1.2497	0.0968	.7808	9.8925	20
50	.6271	9.7973	.8050	9.9058	1.2423	0.0942	.7790	9.8915	10
39° 00′	.6293	9.7989	.8098	9.9084	1.2349	0.0916	.7771	9.8905	**51° 00′**
10	.6316	9.8004	.8146	9.9110	1.2276	0.0890	.7753	9.8895	50
20	.6338	9.8020	.8195	9.9135	1.2203	0.0865	.7735	9.8884	40
30	.6361	9.8035	.8243	9.9161	1.2131	0.0839	.7716	9.8874	30
40	.6383	9.8050	.8292	9.9187	1.2059	0.0813	.7698	9.8864	20
50	.6406	9.8066	.8342	9.9212	1.1988	0.0788	.7679	9.8853	10
40° 00′	.6428	9.8081	.8391	9.9238	1.1918	0.0762	.7660	9.8843	**50° 00′**
10	.6450	9.8096	.8441	9.9264	1.1847	0.0736	.7642	9.8832	50
20	.6472	9.8111	.8491	9.9289	1.1778	0.0711	.7623	9.8821	40
30	.6494	9.8125	.8541	9.9315	1.1708	0.0685	.7604	9.8810	30
40	.6517	9.8140	.8591	9.9341	1.1640	0.0659	.7585	9.8800	20
50	.6539	9.8155	.8642	9.9366	1.1571	0.0634	.7566	9.8789	10
41° 00′	.6561	9.8169	.8693	9.9392	1.1504	0.0608	.7547	9.8778	**49° 00′**
10	.6583	9.8184	.8744	9.9417	1.1436	0.0583	.7528	9.8767	50
20	.6604	9.8198	.8796	9.9443	1.1369	0.0557	.7509	9.8756	40
30	.6626	9.8213	.8847	9.9468	1.1303	0.0532	.7490	9.8745	30
40	.6648	9.8227	.8899	9.9494	1.1237	0.0506	.7470	9.8733	20
50	.6670	9.8241	.8952	9.9519	1.1171	0.0481	.7451	9.8722	10
42° 00′	.6691	9.8255	.9004	9.9544	1.1106	0.0456	.7431	9.8711	**48° 00′**
10	.6713	9.8269	.9057	9.9570	1.1041	0.0430	.7412	9.8699	50
20	.6734	9.8283	.9110	9.9595	1.0977	0.0405	.7392	9.8688	40
30	.6756	9.8297	.9163	9.9621	1.0913	0.0379	.7373	9.8676	30
40	.6777	9.8311	.9217	9.9646	1.0850	0.0354	.7353	9.8665	20
50	.6799	9.8324	.9271	9.9671	1.0786	0.0329	.7333	9.8653	10
43° 00′	.6820	9.8338	.9325	9.9697	1.0724	0.0303	.7314	9.8641	**47° 00′**
10	.6841	9.8351	.9380	9.9722	1.0661	0.0278	.7294	9.8629	50
20	.6862	9.8365	.9435	9.9747	1.0599	0.0253	.7274	9.8618	40
30	.6884	9.8378	.9490	9.9772	1.0538	0.0228	.7254	9.8606	30
40	.6905	9.8391	.9545	9.9798	1.0477	0.0202	.7234	9.8594	20
50	.6926	9.8405	.9601	9.9823	1.0416	0.0177	.7214	9.8582	10
44° 00′	.6947	9.8418	.9657	9.9848	1.0355	0.0152	.7193	9.8569	**46° 00′**
10	.6967	9.8431	.9713	9.9874	1.0295	0.0126	.7173	9.8557	50
20	.6988	9.8444	.9770	9.9899	1.0235	0.0101	.7153	9.8545	40
30	.7009	9.8457	.9827	9.9924	1.0176	0.0076	.7133	9.8532	30
40	.7030	9.8469	.9884	9.9949	1.0117	0.0051	.7112	9.8520	20
50	.7050	9.8482	.9942	9.9975	1.0058	0.0025	.7092	9.8507	10
45° 00′	.7071	9.8495	1.0000	0.0000	1.0000	0.0000	.7071	9.8495	**45° 00′**
	Value	Log	Value	Log	Value	Log	Value	Log	
	Cosine		Cotangent		Tangent		Sine		Degrees

45° — 54°

Table V

Common logarithms of numbers

to four decimal places

Four-Place Logarithms of Numbers

n	0	1	2	3	4	5	6	7	8	9
10	0000	0043	0086	0128	0170	0212	0253	0294	0334	0374
11	0414	0453	0492	0531	0569	0607	0645	0682	0719	0755
12	0792	0828	0864	0899	0934	0969	1004	1038	1072	1106
13	1139	1173	1206	1239	1271	1303	1335	1367	1399	1430
14	1461	1492	1523	1553	1584	1614	1644	1673	1703	1732
15	1761	1790	1818	1847	1875	1903	1931	1959	1987	2014
16	2041	2068	2095	2122	2148	2175	2201	2227	2253	2279
17	2304	2330	2355	2380	2405	2430	2455	2480	2504	2529
18	2553	2577	2601	2625	2648	2672	2695	2718	2742	2765
19	2788	2810	2833	2856	2878	2900	2923	2945	2967	2989
20	3010	3032	3054	3075	3096	3118	3139	3160	3181	3201
21	3222	3243	3263	3284	3304	3324	3345	3365	3385	3404
22	3424	3444	3464	3483	3502	3522	3541	3560	3579	3598
23	3617	3636	3655	3674	3692	3711	3729	3747	3766	3784
24	3802	3820	3838	3856	3874	3892	3909	3927	3945	3962
25	3979	3997	4014	4031	4048	4065	4082	4099	4116	4133
26	4150	4166	4183	4200	4216	4232	4249	4265	4281	4298
27	4314	4330	4346	4362	4378	4393	4409	4425	4440	4456
28	4472	4487	4502	4518	4533	4548	4564	4579	4594	4609
29	4624	4639	4654	4669	4683	4698	4713	4728	4742	4757
30	4771	4786	4800	4814	4829	4843	4857	4871	4886	4900
31	4914	4928	4942	4955	4969	4983	4997	5011	5024	5038
32	5051	5065	5079	5092	5105	5119	5132	5145	5159	5172
33	5185	5198	5211	5224	5237	5250	5263	5276	5289	5302
34	5315	5328	5340	5353	5366	5378	5391	5403	5416	5428
35	5441	5453	5465	5478	5490	5502	5514	5527	5539	5551
36	5563	5575	5587	5599	5611	5623	5635	5647	5658	5670
37	5682	5694	5705	5717	5729	5740	5752	5763	5775	5786
38	5798	5809	5821	5832	5843	5855	5866	5877	5888	5899
39	5911	5922	5933	5944	5955	5966	5977	5988	5999	6010
40	6021	6031	6042	6053	6064	6075	6085	6096	6107	6117
41	6128	6138	6149	6160	6170	7180	6191	6201	6212	6222
42	6232	6243	6253	6263	6274	6284	6294	6304	6314	6325
43	6335	6345	6355	6365	6375	6385	6395	6405	6415	6425
44	6435	6444	6454	6464	6474	6484	6493	6503	6513	6522
45	6532	6542	6551	6561	6571	6580	6590	6599	6609	6618
46	6628	6637	6646	6656	6665	6675	6684	6693	6702	6712
47	6721	6730	6739	6749	6758	6767	6776	6785	6794	6803
48	6812	6821	6830	6839	6848	6857	6866	6875	6884	6893
49	6902	6911	6920	6928	6937	6946	6955	6964	6972	6981
50	6990	6998	7007	7016	7024	7033	7042	7050	7059	7067
51	7076	7084	7093	7101	7110	7118	7126	7135	7143	7152
52	7160	7168	7177	7185	7193	7202	7210	7218	7226	7235
53	7243	7251	7259	7267	7275	7284	7292	7300	7308	7316
54	7324	7332	7340	7348	7356	7364	7372	7380	7388	7396
n	0	1	2	3	4	5	6	7	8	9

Prop. Parts

	43	42	41	40
1	4.3	4.2	4.1	4.0
2	8.6	8.4	8.2	8.0
3	12.9	12.6	12.3	12.0
4	17.2	16.8	16.4	16.0
5	21.5	21.0	20.5	20.0
6	25.8	25.2	24.6	24.0
7	30.1	29.4	28.7	28.0
8	34.4	33.6	32.8	32.0
9	38.7	37.8	36.9	36.0

	39	38	37	36
1	3.9	3.8	3.7	3.6
2	7.8	7.6	7.4	7.2
3	11.7	11.4	11.1	10.8
4	15.6	15.2	14.8	14.4
5	19.5	19.0	18.5	18.0
6	23.4	22.8	22.2	21.6
7	27.3	26.6	25.9	25.2
8	31.2	30.4	29.6	28.8
9	35.1	34.2	33.3	32.4

	35	34	33	32
1	3.5	3.4	3.3	3.2
2	7.0	6.8	6.6	6.4
3	10.5	10.2	9.9	9.6
4	14.0	13.6	13.2	12.8
5	17.5	17.0	16.5	16.0
6	21.0	20.4	19.8	19.2
7	24.5	23.8	23.1	22.4
8	28.0	27.2	26.4	25.6
9	31.5	30.6	29.7	28.8

	31	30	29	28
1	3.1	3.0	2.9	2.8
2	6.2	6.0	5.8	5.6
3	9.3	9.0	8.7	8.4
4	12.4	12.0	11.6	11.2
5	15.5	15.0	14.5	14.0
6	18.6	18.0	17.4	16.8
7	21.7	21.0	20.3	19.6
8	24.8	24.0	23.2	22.4
9	27.9	27.0	26.1	25.2

	27	26	25	24
1	2.7	2.6	2.5	2.4
2	5.4	5.2	5.0	4.8
3	8.1	7.8	7.5	7.2
4	10.8	10.4	10.0	9.6
5	13.5	13.0	12.5	12.0
6	16.2	15.6	15.0	14.4
7	18.9	18.2	17.5	16.8
8	21.6	20.8	20.0	19.2
9	24.3	23.4	22.5	21.6

Proportional Parts

	23	22	21	20
1	2.3	2.2	2.1	2.0
2	4.6	4.4	4.2	4.0
3	6.9	6.6	6.3	6.0
4	9.2	8.8	8.4	8.0
5	11.5	11.0	10.5	10.0
6	13.8	13.2	12.6	12.0
7	16.1	15.4	14.7	14.0
8	18.4	17.6	16.8	16.0
9	20.7	19.8	18.9	18.0

	19	18	17	16
1	1.9	1.8	1.7	1.6
2	3.8	3.6	3.4	3.2
3	5.7	5.4	5.1	4.8
4	7.6	7.2	6.8	6.4
5	9.5	9.0	8.5	8.0
6	11.4	10.8	10.2	9.6
7	13.3	12.6	11.9	11.2
8	15.2	14.4	13.6	12.8
9	17.1	16.2	15.3	14.4

	15	14	13	12
1	1.5	1.4	1.3	1.2
2	3.0	2.8	2.6	2.4
3	4.5	4.2	3.9	3.6
4	6.0	5.6	5.2	4.8
5	7.5	7.0	6.5	6.0
6	9.0	8.4	7.8	7.2
7	10.5	9.8	9.1	8.4
8	12.0	11.2	10.4	9.6
9	13.5	12.6	11.7	10.8

	11	10	9	8
1	1.1	1.0	0.9	0.8
2	2.2	2.0	1.8	1.6
3	3.3	3.0	2.7	2.4
4	4.4	4.0	3.6	3.2
5	5.5	5.0	4.5	4.0
6	6.6	6.0	5.4	4.8
7	7.7	7.0	6.3	5.6
8	8.8	8.0	7.2	6.4
9	9.9	9.0	8.1	7.2

	7	6	5	4
1	0.7	0.6	0.5	0.4
2	1.4	1.2	1.0	0.8
3	2.1	1.8	1.5	1.2
4	2.8	2.4	2.0	1.6
5	3.5	3.0	2.5	2.0
6	4.2	3.6	3.0	2.4
7	4.9	4.2	3.5	2.8
8	5.6	4.8	4.0	3.2
9	6.3	5.4	4.5	3.6

Main Table

n	0	1	2	3	4	5	6	7	8	9
55	7404	7412	7419	7427	7435	7443	7451	7459	7466	7474
56	7482	7490	7497	7505	7513	7520	7528	7536	7543	7551
57	7559	7566	7574	7582	7589	7597	7604	7612	7619	7627
58	7634	7642	7649	7657	7664	7672	7679	7686	7694	7701
59	7709	7716	7723	7731	7738	7745	7752	7760	7767	7774
60	7782	7789	7796	7803	7810	7818	7825	7832	7839	7846
61	7853	7860	7868	7875	7882	7889	7896	7903	7910	7917
62	7924	7931	7938	7945	7952	7959	7966	7973	7980	7987
63	7993	8000	8007	8014	8021	8028	8035	8041	8048	8055
64	8062	8069	8075	8082	8089	8096	8102	8109	8116	8122
65	8129	8136	8142	8149	8156	8162	8169	8176	8182	8189
66	8195	8202	8209	8215	8222	8228	8235	8241	8248	8254
67	8261	8267	8274	8280	8287	8293	8299	8306	8312	8319
68	8325	8331	8338	8344	8351	8357	8363	8370	8376	8382
69	8388	8395	8401	8407	8414	8420	8426	8432	8439	8445
70	8451	8457	8463	8470	8476	8482	8488	8494	8500	8506
71	8513	8519	8525	8531	8537	8543	8549	8555	8561	8567
72	8573	8579	8585	8591	8597	8603	8609	8615	8621	8627
73	8633	8639	8645	8651	8657	8663	8669	8675	8681	8686
74	8692	8698	8704	8710	8716	8722	8727	8733	8739	8745
75	8751	8756	8762	8768	8774	8779	8785	8791	8797	8802
76	8808	8814	8820	8825	8831	8837	8842	8848	8854	8859
77	8865	8871	8876	8882	8887	8893	8899	8904	8910	8915
78	8921	8927	8932	8938	8943	8949	8954	8960	8965	8971
79	8976	8982	8987	8993	8998	9004	9009	9015	9020	9025
80	9031	9036	9042	9047	9053	9058	9063	9069	9074	9079
81	9085	9090	9096	9101	9106	9112	9117	9122	9128	9133
82	9138	9143	9149	9154	9159	9165	9170	9175	9180	9186
83	9191	9196	9201	9206	9212	9217	9222	9227	9232	9238
84	9243	9248	9253	9258	9263	9269	9274	9279	9284	9289
85	9294	9299	9304	9309	9315	9320	9325	9330	9335	9340
86	9345	9350	9355	9360	9365	9370	9375	9380	9385	9390
87	9395	9400	9405	9410	9415	9420	9425	9430	9435	9440
88	9445	9450	9455	9460	9465	9469	9474	9479	9484	9489
89	9494	9499	9504	9509	9513	9518	9523	9528	9533	9538
90	9542	9547	9552	9557	9562	9566	9571	9576	9581	9586
91	9590	9595	9600	9605	9609	9614	9619	9624	9628	9633
92	9638	9643	9647	9652	9657	9661	9666	9671	9675	9680
93	9685	9689	9694	9699	9703	9708	9713	9717	9722	9727
94	9731	9736	9741	9745	9750	9754	9759	9763	9768	9773
95	9777	9782	9786	9791	9795	9800	9805	9809	9814	9818
96	9823	9827	9832	9836	9841	9845	9850	9854	9859	9863
97	9868	9872	9877	9881	9886	9890	9894	9899	9903	9908
98	9912	9917	9921	9926	9930	9934	9939	9943	9948	9952
99	9956	9961	9965	9969	9974	9978	9983	9987	9991	9996

Prop. Parts　　n　0　1　2　3　4　5　6　7　8　9

Table VI

Trigonometric functions–radian measures

Table VI

Rad.	Degrees	Sin	Cos	Tan
.00	0° 00.0'	.00000	1.0000	.00000
.01	0° 34.4'	.01000	.99995	.01000
.02	1° 08.8'	.02000	.99980	.02000
.03	1° 43.1'	.03000	.99955	.03001
.04	2° 17.5'	.03999	.99920	.04002
.05	2° 51.9'	.04998	.99875	.05004
.06	3° 26.3'	.05996	.99820	.06007
.07	4° 00.6'	.06994	.99755	.07011
.08	4° 35.0'	.07991	.99680	.08017
.09	5° 09.4'	.08988	.99595	.09024
.10	5° 43.8'	.09983	.99500	.10033
.11	6° 18.2'	.10978	.99396	.11045
.12	6° 52.5'	.11971	.99281	.12058
.13	7° 26.9'	.12963	.99156	.13074
.14	8° 01.3'	.13954	.99022	.14092
.15	8° 35.7'	.14944	.98877	.15114
.16	9° 10.0'	.15932	.98723	.16138
.17	9° 44.4'	.16918	.98558	.17166
.18	10° 18.8'	.17903	.98384	.18197
.19	10° 53.2'	.18886	.98200	.19232
.20	11° 27.5'	.19867	.98007	.20271
.21	12° 01.9'	.20846	.97803	.21314
.22	12° 36.3'	.21823	.97590	.22362
.23	13° 10.7'	.22798	.97367	.23414
.24	13° 45.1'	.23770	.97134	.24472
.25	14° 19.4'	.24740	.96891	.25534
.26	14° 53.8'	.25708	.96639	.26602
.27	15° 28.2'	.26673	.96377	.27676
.28	16° 02.6'	.27636	.96106	.28755
.29	16° 36.9'	.28595	.95824	.29841
.30	17° 11.3'	.29552	.95534	.30934
.31	17° 45.7'	.30506	.95233	.32033
.32	18° 20.1'	.31457	.94924	.33139
.33	18° 54.5'	.32404	.94604	.34252
.34	19° 28.8'	.33349	.94275	.35374
.35	20° 03.2'	.34290	.93937	.36503
.36	20° 37.6'	.35227	.93590	.37640
.37	21° 12.0'	.36162	.93233	.38786
.38	21° 46.3'	.37092	.92866	.39941
.39	22° 20.7'	.38019	.92491	.41106
.40	22° 55.1'	.38942	.92106	.42279
.41	23° 29.5'	.39861	.91712	.43463
.42	24° 03.9'	.40776	.91309	.44657
.43	24° 38.2'	.41687	.90897	.45862
.44	25° 12.6'	.42594	.90475	.47078
.45	25° 47.0'	.43497	.90045	.48305
.46	26° 21.4'	.44395	.89605	.49545
.47	26° 55.7'	.45289	.89157	.50795
.48	27° 30.1'	.46178	.88699	.52061
.49	28° 04.5'	.47063	.88233	.53339
.50	28° 38.9'	.47943	.87758	.54630
.51	29° 13.3'	.48818	.87274	.55936
.52	29° 47.6'	.49688	.86782	.57256
.53	30° 22.0'	.50553	.86281	.58592
.54	30° 56.4'	.51414	.85771	.59943
.55	31° 30.8'	.52269	.85252	.61311
.56	32° 05.1'	.53119	.84726	.62695
.57	32° 39.5'	.53963	.84190	.64097
.58	33° 13.9'	.54802	.83646	.65517
.59	33° 48.3'	.55636	.83094	.66956
.60	34° 22.6'	.56464	.82534	.68414

Rad.	Degrees	Sin	Cos	Tan
.60	34° 22.6'	.56464	.82534	.68414
.61	34° 57.0'	.57287	.81965	.69892
.62	35° 31.4'	.58104	.81388	.71391
.63	36° 05.8'	.58914	.80803	.72911
.64	36° 40.2'	.59720	.80210	.74454
.65	37° 14.5'	.60519	.79608	.76020
.66	37° 48.9'	.61312	.78999	.77610
.67	38° 23.3'	.62099	.78382	.79225
.68	38° 57.7'	.62879	.77757	.80866
.69	39° 32.0'	.63654	.77125	.82533
.70	40° 06.4'	.64422	.76484	.84229
.71	40° 40.8'	.65183	.75836	.85953
.72	41° 15.2'	.65938	.75181	.87707
.73	41° 49.6'	.66687	.74517	.89492
.74	42° 23.9'	.67429	.73847	.91309
.75	42° 58.3'	.68164	.73169	.93160
.76	43° 32.7'	.68892	.72484	.95055
.77	44° 07.1'	.69614	.71791	.96967
.78	44° 41.4'	.70328	.71091	.98926
.79	45° 15.8'	.71035	.70385	1.0092
.80	45° 50.2'	.71736	.69671	1.0296
.81	46° 24.6'	.72429	.68950	1.0505
.82	46° 59.0'	.73115	.68222	1.0717
.83	47° 33.3'	.73793	.67488	1.0934
.84	48° 07.7'	.74464	.66746	1.1156
.85	48° 42.1'	.75128	65998	1.1383
.86	49° 16.5'	.75784	.65244	1.1616
.87	49° 50.8'	.76433	.64483	1.1853
.88	50° 25.2'	.77074	.63715	1.2097
.89	50° 59.6'	.77707	.62941	1.2346
.90	51° 34.0'	.78333	.62161	1.2602
.91	52° 08.3'	.78950	.61375	1.2864
.92	52° 42.7'	.79560	.60582	1.3133
.93	53° 17.1'	.80162	.59783	1.3409
.94	53° 51.5'	.80756	.58979	1.3692
.95	54° 25.9'	.81342	.58168	1.3984
.96	55° 00.2'	.81919	.57352	1.4284
.97	55° 34.6'	.82489	.56530	1.4592
.98	56° 09.0'	.83050	.55702	1.4910
.99	56° 43.4'	.83603	.54869	1.5237
1.00	57° 17.7'	.84147	.54030	1.5574
1.01	57° 52.1'	.84683	.53186	1.5922
1.02	58° 26.5'	.85211	.52337	1.6281
1.03	59° 00.9'	.85730	.51482	1.6652
1.04	59° 35.3'	.86240	.50622	1.7036
1.05	60° 09.6'	.86742	.49757	1.7433
1.06	60° 44.0'	.87236	.48887	1.7844
1.07	61° 18.4'	.87720	.48012	1.8270
1.08	61° 52.8'	.88196	.47133	1.8712
1.09	62° 27.1'	.88663	.46249	1.9171
1.10	63° 01.5'	.89121	.45360	1.9648
1.11	63° 35.9'	.89570	.44466	2.0143
1.12	64° 10.3'	.90010	.43568	2.0660
1.13	64° 44.7'	.90441	.42666	2.1198
1.14	65° 19.0'	.90863	.41759	2.1759
1.15	65° 53.4'	.91276	.40849	2.2345
1.16	66° 27.8'	.91680	.39934	2.2958
1.17	67° 02.2'	.92075	.39015	2.3600
1.18	67° 36.5'	.92461	.38092	2.4273
1.19	68° 10.9'	.92837	.37166	2.4979
1.20	68° 45.3'	.93204	.36236	2.5722

Trigonometric Functions—Radian Measure (*Cont.*)

Rad.	Degrees	Sin	Cos	Tan
1.20	68° 45.3'	.93204	.36236	2.5722
1.21	69° 19.7'	.93562	.35302	2.6503
1.22	69° 54.1'	.93910	.34365	2.7328
1.23	70° 28.4'	.94249	.33424	2.8198
1.24	71° 02.8'	.94578	.32480	2.9119
1.25	71° 37.2'	.94898	.31532	3.0096
1.26	72° 11.6'	.95209	.30582	3.1133
1.27	72° 45.9'	.95510	.29628	3.2236
1.28	73° 20.3'	.95802	.28672	3.3413
1.29	73° 54.7'	.96084	.27712	3.4672
1.30	74° 29.1'	.96356	.26750	3.6021
1.31	75° 03.4'	.96618	.25785	3.7470
1.32	75° 37.8'	.96872	.24818	3.9033
1.33	76° 12.2'	.97115	.23848	4.0723
1.34	76° 46.6'	.97348	.22875	4.2556
1.35	77° 21.0'	.97572	.21901	4.4552
1.36	77° 55.3'	.97786	.20924	4.6734
1.37	78° 29.7'	.97991	.19945	4.9131
1.38	79° 04.1'	.98185	.18964	5.1774
1.39	79° 38.5'	.98370	.17981	5.4707
1.40	80° 12.8'	.98545	.16997	5.7979
1.40	80° 12.8'	.98545	.16997	5.7979
1.41	80° 47.2'	.98710	.16010	6.1654
1.42	81° 21.6'	.98865	.15023	6.5811
1.43	81° 56.0'	.99010	.14033	7.0555
1.44	82° 30.4'	.99146	.13042	7.6018
1.45	83° 04.7'	.99271	.12050	8.2381
1.46	83° 39.1'	.99387	.11057	8.9886
1.47	84° 13.5'	.99492	.10063	9.8874
1.48	84° 47.9'	.99588	.09067	10.983
1.49	85° 22.2'	.99674	.08071	12.350
1.50	85° 56.6'	.99749	.07074	14.101
1.51	86° 31.0'	.99815	.06076	16.428
1.52	87° 05.4'	.99871	.05077	19.670
1.53	87° 39.8'	.99917	.04079	24.498
1.54	88° 14.1'	.99953	.03079	32.461
1.55	88° 48.5'	.99978	.02079	48.078
1.56	89° 22.9'	.99994	.01080	92.621
1.57	89° 57.3'	1.0000	.00080	1255.8
1.58	90° 31.6'	.99996	−.00920	−108.65
1.59	91° 06.0'	.99982	−.01920	−52.067
1.60	91° 40.4'	.99957	−.02920	−34.233

Degrees in Radians

°		°		°		°		°		°	
0°	0.00000	15°	0.26180	30°	0.52360	45°	0.78540	60°	1.04720	75°	1.30900
1	0.01745	16	0.27925	31	0.54105	46	0.80285	61	1.06465	76	1.32645
2	0.03491	17	0.29671	32	0.55851	47	0.82030	62	1.08210	77	1.34390
3	0.05236	18	0.31416	33	0.57596	48	0.83776	63	1.09956	78	1.36136
4	0.06981	19	0.33161	34	0.59341	49	0.85521	64	1.11701	79	1.37881
5	0.08727	20	0.34907	35	0.61087	50	0.87266	65	1.13446	80	1.39626
6	0.10472	21	0.36652	36	0.62832	51	0.89012	66	1.15192	81	1.41372
7	0.12217	22	0.38397	37	0.64577	52	0.90757	67	1.16937	82	1.43117
8	0.13963	23	0.40143	38	0.66323	53	0.92502	68	1.18682	83	1.44862
9	0.15708	24	0.41888	39	0.68068	54	0.94248	69	1.20428	84	1.46608
10	0.17453	25	0.43633	40	0.69813	55	0.95993	70	1.22173	85	1.48353
11	0.19199	26	0.45379	41	0.71558	56	0.97738	71	1.23918	86	1.50098
12	0.20944	27	0.47124	42	0.73304	57	0.99484	72	1.25664	87	1.51844
13	0.22689	28	0.48869	43	0.75049	58	1.01229	73	1.27409	88	1.53589
14	0.24435	29	0.50615	44	0.76794	59	1.02974	74	1.29154	89	1.55334
15	0.26180	30	0.52360	45	0.78540	60	1.04720	75	1.30900	90	1.57080

Minutes in Radians

'		'		'		'		'		'	
0'	0.00000	10'	0.00291	20'	0.00582	30'	0.00873	40'	0.01164	50'	0.01454
1	0.00029	11	0.00320	21	0.00611	31	0.00902	41	0.01193	51	0.01484
2	0.00058	12	0.00349	22	0.00640	32	0.00931	42	0.01222	52	0.01513
3	0.00087	13	0.00378	23	0.00669	33	0.00960	43	0.01251	53	0.01542
4	0.00116	14	0.00407	24	0.00698	34	0.00989	44	0.01280	54	0.01571
5	0.00145	15	0.00436	25	0.00727	35	0.01018	45	0.01309	55	0.01600
6	0.00174	16	0.00465	26	0.00756	36	0.01047	46	0.01338	56	0.01629
7	0.00204	17	0.00495	27	0.00785	37	0.01076	47	0.01367	57	0.01658
8	0.00233	18	0.00524	28	0.00814	38	0.01105	48	0.01396	58	0.01687
9	0.00262	19	0.00553	29	0.00844	39	0.01134	49	0.01425	59	0.01716
10	0.00291	20	0.00582	30	0.00873	40	0.01164	50	0.01454	60	0.01745

Table VII

Squares, square roots, reciprocals

Table VII

n	n^2	$\sqrt{n}$	$\sqrt{10n}$	$1/n$	n	n^2	$\sqrt{n}$	$\sqrt{10n}$	$1/n$
1.00	1.0000	1.00000	3.16228	1.00000	**1.50**	2.2500	1.22474	3.87298	.666667
1.01	1.0201	1.00499	3.17805	.990099	1.51	2.2801	1.22882	3.88587	.662252
1.02	1.0404	1.00995	3.19374	.980392	1.52	2.3104	1.23288	3.89872	.657895
1.03	1.0609	1.01489	3.20936	.970874	1.53	2.3409	1.23693	3.91152	.653595
1.04	1.0816	1.01980	3.22490	.961538	1.54	2.3716	1.24097	3.92428	.649351
1.05	1.1025	1.02470	3.24037	.952381	1.55	2.4025	1.24499	3.93700	.645161
1.06	1.1236	1.02956	3.25576	.943396	1.56	2.4336	1.24900	3.94968	.641026
1.07	1.1449	1.03441	3.27109	.934579	1.57	2.4649	1.25300	3.96232	.636943
1.08	1.1664	1.03923	3.28634	.925926	1.58	2.4964	1.25698	3.97492	.632911
1.09	1.1881	1.04403	3.30151	.917431	1.59	2.5281	1.26095	3.98748	.628931
1.10	1.2100	1.04881	3.31662	.909091	**1.60**	2.5600	1.26491	4.00000	.625000
1.11	1.2321	1.05357	3.33167	.900901	1.61	2.5921	1.26886	4.01248	.621118
1.12	1.2544	1.05830	3.34664	.892857	1.62	2.6244	1.27279	4.02492	.617284
1.13	1.2769	1.06301	3.36155	.884956	1.63	2.6569	1.27671	4.03733	.613497
1.14	1.2996	1.06771	3.37639	.877193	1.64	2.6896	1.28062	4.04969	.609756
1.15	1.3225	1.07238	3.39116	.869565	1.65	2.7225	1.28452	4.06202	.606061
1.16	1.3456	1.07703	3.40588	.862069	1.66	2.7556	1.28841	4.07431	.602410
1.17	1.3689	1.08167	3.42053	.854701	1.67	2.7889	1.29228	4.08656	.598802
1.18	1.3924	1.08628	3.43511	.847458	1.68	2.8224	1.29615	4.09878	.595238
1.19	1.4161	1.09087	3.44964	.840336	1.69	2.8561	1.30000	4.11096	.591716
1.20	1.4400	1.09545	3.46410	.833333	**1.70**	2.8900	1.30384	4.12311	.588235
1.21	1.4641	1.10000	3.47851	.826446	1.71	2.9241	1.30767	4.13521	.584795
1.22	1.4884	1.10454	3.49285	.819672	1.72	2.9584	1.31149	4.14729	.581395
1.23	1.5129	1.10905	3.50714	.813008	1.73	2.9929	1.31529	4.15933	.578035
1.24	1.5376	1.11355	3.52136	.806452	1.74	3.0276	1.31909	4.17133	.574713
1.25	1.5625	1.11803	3.53553	.800000	1.75	3.0625	1.32288	4.18330	.571429
1.26	1.5876	1.12250	3.54965	.793651	1.76	3.0976	1.32665	4.19524	.568182
1.27	1.6129	1.12694	3.56371	.787402	1.77	3.1329	1.33041	4.20714	.564972
1.28	1.6384	1.13137	3.57771	.781250	1.78	3.1684	1.33417	4.21900	.561798
1.29	1.6641	1.13578	3.59166	.775194	1.79	3.2041	1.33791	4.23084	.558659
1.30	1.6900	1.14018	3.60555	.769231	**1.80**	3.2400	1.34164	4.24264	.555556
1.31	1.7161	1.14455	3.61939	.763359	1.81	3.2761	1.34536	4.25441	.552486
1.32	1.7424	1.14891	3.63318	.757576	1.82	3.3124	1.34907	4.26615	.549451
1.33	1.7689	1.15326	3.64692	.751880	1.83	3.3489	1.35277	4.27785	.546448
1.34	1.7956	1.15758	3.66060	.746269	1.84	3.3856	1.35647	4.28952	.543478
1.35	1.8225	1.16190	3.67423	.740741	1.85	3.4225	1.36015	4.30116	.540541
1.36	1.8496	1.16619	3.68782	.735294	1.86	3.4596	1.36382	4.31277	.537634
1.37	1.8769	1.17047	3.70135	.729927	1.87	3.4969	1.36748	4.32435	.534759
1.38	1.9044	1.17473	3.71484	.724638	1.88	3.5344	1.37113	4.33590	.531915
1.39	1.9321	1.17898	3.72827	.719424	1.89	3.5721	1.37477	4.34741	.529101
1.40	1.9600	1.18322	3.74166	.714286	**1.90**	3.6100	1.37840	4.35890	.526316
1.41	1.9881	1.18743	3.75500	.709220	1.91	3.6481	1.38203	4.37035	.523560
1.42	2.0164	1.19164	3.76829	.704225	1.92	3.6864	1.38564	4.38178	.520833
1.43	2.0449	1.19583	3.78153	.699301	1.93	3.7249	1.38924	4.39318	.518135
1.44	2.0736	1.20000	3.79473	.694444	1.94	3.7636	1.39284	4.40454	.515464
1.45	2.1025	1.20416	3.80789	.689655	1.95	3.8025	1.39642	4.41588	.512821
1.46	2.1316	1.20830	3.82099	.684932	1.96	3.8416	1.40000	4.42719	.510204
1.47	2.1609	1.21244	3.83406	.680272	1.97	3.8809	1.40357	4.43847	.507614
1.48	2.1904	1.21655	3.84708	.675676	1.98	3.9204	1.40712	4.44972	.505051
1.49	2.2201	1.22066	3.86005	.671141	1.99	3.9601	1.41067	4.46094	.502513
1.50	2.2500	1.22474	3.87298	.666667	**2.00**	4.0000	1.41421	4.47214	.500000
n	n^2	$\sqrt{n}$	$\sqrt{10n}$	$1/n$	n	n^2	$\sqrt{n}$	$\sqrt{10n}$	$1/n$

n	n^2	$\sqrt{n}$	$\sqrt{10n}$	$1/n$	n	n^2	$\sqrt{n}$	$\sqrt{10n}$	$1/n$
2.00	4.0000	1.41421	4.47214	.500000	2.50	6.2500	1.58114	5.00000	.400000
2.01	4.0401	1.41774	4.48330	.497512	2.51	6.3001	1.58430	5.00999	.398406
2.02	4.0804	1.42127	4.49444	.495050	2.52	6.3504	1.58745	5.01996	.396825
2.03	4.1209	1.42478	4.50555	.492611	2.53	6.4009	1.59060	5.02991	.395257
2.04	4.1616	1.42829	4.51664	.490196	2.54	6.4516	1.59374	5.03984	.393701
2.05	4.2025	1.43178	4.52769	.487805	2.55	6.5025	1.59687	5.04975	.392157
2.06	4.2436	1.43527	4.53872	.485437	2.56	6.5536	1.60000	5.05964	.390625
2.07	4.2849	1.43875	4.54973	.483092	2.57	6.6049	1.60312	5.06952	.389105
2.08	4.3264	1.44222	4.56070	.480769	2.58	6.6564	1.60624	5.07937	.387597
2.09	4.3681	1.44568	4.57165	.478469	2.59	6.7081	1.60935	5.08920	.386100
2.10	4.4100	1.44914	4.58258	.476190	2.60	6.7600	1.61245	5.09902	.384615
2.11	4.4521	1.45258	4.59347	.473934	2.61	6.8121	1.61555	5.10882	.383142
2.12	4.4944	1.45602	4.60435	.471698	2.62	6.8644	1.61864	5.11859	.381679
2.13	4.5369	1.45945	4.61519	.469434	2.63	6.9169	1.62173	5.12835	.380228
2.14	4.5796	1.46287	4.62601	.467290	2.64	6.9696	1.62481	5.13809	.378788
2.15	4.6225	1.46629	4.63681	.465116	2.65	7.0225	1.62788	5.14782	.377358
2.16	4.6656	1.46969	4.64758	.462963	2.66	7.0756	1.63095	5.15752	.375940
2.17	4.7089	1.47309	4.65833	.460829	2.67	7.1289	1.63401	5.16720	.374532
2.18	4.7524	1.47648	4.66905	.458716	2.68	7.1824	1.63707	5.17687	.373134
2.19	4.7961	1.47986	4.67974	.456621	2.69	7.2361	1.64012	5.18652	.371747
2.20	4.8400	1.48324	4.69042	.454545	2.70	7.2900	1.64317	5.19615	.370370
2.21	4.8841	1.48661	4.70106	.452489	2.71	7.3441	1.64621	5.20577	.369004
2.22	4.9284	1.48997	4.71169	.450450	2.72	7.3984	1.64924	5.21536	.367647
2.23	4.9729	1.49332	4.72229	.448430	2.73	7.4529	1.65227	5.22494	.366300
2.24	5.0176	1.49666	4.73286	.446429	2.74	7.5076	1.65529	5.23450	.364964
2.25	5.0625	1.50000	4.74342	.444444	2.75	7.5625	1.65831	5.24404	.363636
2.26	5.1076	1.50333	4.75395	.442478	2.76	7.6176	1.66132	5.25357	.362319
2.27	5.1529	1.50665	4.76445	.440529	2.77	7.6729	1.66433	5.26308	.361011
2.28	5.1984	1.50997	4.77493	.438596	2.78	7.7284	1.66733	5.27257	.359712
2.29	5.2441	1.51327	4.78539	.436681	2.79	7.7841	1.67033	5.28205	.358423
2.30	5.2900	1.51658	4.79583	.434783	2.80	7.8400	1.67332	5.29150	.357143
2.31	5.3361	1.51987	4.80625	.432900	2.81	7.8961	1.67631	5.30094	.355872
2.32	5.3824	1.52315	4.81664	.431034	2.82	7.9524	1.67929	5.31037	.354610
2.33	5.4289	1.52643	4.82701	.429185	2.83	8.0089	1.68226	5.31977	.353357
2.34	5.4756	1.52971	4.83735	.427350	2.84	8.0656	1.68523	5.32917	.352113
2.35	5.5225	1.53297	4.84768	.425532	2.85	8.1225	1.68819	5.33854	.350877
2.36	5.5696	1.53623	4.85798	.423729	2.86	8.1796	1.69115	5.34790	.349650
2.37	5.6169	1.53948	4.86826	.421941	2.87	8.2369	1.69411	5.35724	.348432
2.38	5.6644	1.54272	4.87852	.420168	2.88	8.2944	1.69706	5.36656	.347222
2.39	5.7121	1.54596	4.88876	.418410	2.89	8.3521	1.70000	5.37587	.346021
2.40	5.7600	1.54919	4.89898	.416667	2.90	8.4100	1.70294	5.38516	.344828
2.41	5.8081	1.55242	4.90918	.414938	2.91	8.4681	1.70587	5.39444	.343643
2.42	5.8564	1.55563	4.91935	.413223	2.92	8.5264	1.70880	5.40370	.342466
2.43	5.9049	1.55885	4.92950	.411523	2.93	8.5849	1.71172	5.41295	.341297
2.44	5.9536	1.56205	4.93964	.409836	2.94	8.6436	1.71464	5.42218	3.40136
2.45	6.0025	1.56525	4.94975	.408163	2.95	8.7025	1.71756	5.43139	.338983
2.46	6.0516	1.56844	4.95984	.406504	2.96	8.7616	1.72047	5.44059	.337838
2.47	6.1009	1.57162	4.96991	.404858	2.97	8.8209	1.72337	5.44977	.336700
2.48	6.1504	1.57480	4.97996	.403226	2.98	8.8804	1.72627	5.45894	.335570
2.49	6.2001	1.57797	4.98999	.401606	2.99	8.9401	1.72916	5.46809	.334448
2.50	6.2500	1.58114	5.00000	.400000	3.00	9.0000	1.73205	5.47723	.333333
n	n^2	$\sqrt{n}$	$\sqrt{10n}$	$1/n$	n	n^2	$\sqrt{n}$	$\sqrt{10n}$	$1/n$

Table VII

n	n^2	$\sqrt{n}$	$\sqrt{10n}$	$1/n$	n	n^2	$\sqrt{n}$	$\sqrt{10n}$	$1/n$
3.00	9.0000	1.73205	5.47723	.333333	3.50	12.2500	1.87083	5.91608	.285714
3.01	9.0601	1.73494	5.48635	.332226	3.51	12.3201	1.87350	5.92453	.284900
3.02	9.1204	1.73781	5.49545	.331126	3.52	12.3904	1.87617	5.93296	.284091
3.03	9.1809	1.74069	5.50454	.330033	3.53	12.4609	1.87883	5.94138	.283286
3.04	9.2416	1.74356	5.51362	.328947	3.54	12.5316	1.88149	5.94979	.282486
3.05	9.3025	1.74642	5.52268	.327869	3.55	12.6025	1.88414	5.95819	.281690
3.06	9.3636	1.74929	5.53173	.326797	3.56	12.6736	1.88680	5.96657	.280899
3.07	9.4249	1.75214	5.54076	.325733	3.57	12.7449	1.88944	5.97495	.280112
3.08	9.4864	1.75499	5.54977	.324675	3.58	12.8164	1.89209	5.98331	.279330
3.09	9.5481	1.75784	5.55878	.323625	3.59	12.8881	1.89473	5.99166	.278552
3.10	9.6100	1.76068	5.56776	.322581	3.60	12.9600	1.89737	6.00000	.277778
3.11	9.6721	1.76352	5.57674	.321543	3.61	13.0321	1.90000	6.00833	.277008
3.12	9.7344	1.76635	5.58570	.320513	3.62	13.1044	1.90263	6.01664	.276243
3.13	9.7969	1.76918	5.59464	.319489	3.63	13.1769	1.90526	6.02495	.275482
3.14	9.8596	1.77200	5.60357	.318471	3.64	13.2496	1.90788	6.03324	.274725
3.15	9.9225	1.77482	5.61249	.317460	3.65	13.3225	1.91050	6.04152	.273973
3.16	9.9856	1.77764	5.62139	.316456	3.66	13.3956	1.91311	6.04979	.273224
3.17	10.0489	1.78045	5.63028	.315457	3.67	13.4689	1.91572	6.05805	.272480
3.18	10.1124	1.78326	5.63915	.314465	3.68	13.5424	1.91833	6.06630	.271739
3.19	10.1761	1.78606	5.64801	.313480	3.69	13.6161	1.92094	6.07454	.271003
3.20	10.2400	1.78885	5.65685	.312500	3.70	13.6900	1.92354	6.08276	.270270
3.21	10.3041	1.79165	5.66569	.311526	3.71	13.7641	1.92614	6.09098	.269542
3.22	10.3684	1.79444	5.67450	.310559	3.72	13.8384	1.92873	6.09918	.268817
3.23	10.4329	1.79722	5.68331	.309598	3.73	13.9129	1.93132	6.10737	.268097
3.24	10.4976	1.80000	5.69210	.308642	3.74	13.9876	1.93391	6.11555	.267380
3.25	10.5625	1.80278	5.70088	.307692	3.75	14.0625	1.93649	6.12372	.266667
3.26	10.6276	1.80555	5.70964	.306748	3.76	14.1376	1.93907	6.13188	.265957
3.27	10.6929	1.80831	5.71839	.305810	3.77	14.2129	1.94165	6.14003	.265252
3.28	10.7584	1.81108	5.72713	.304878	3.78	14.2884	1.94422	6.14817	.264550
3.29	10.8241	1.81384	5.73585	.303951	3.79	14.3641	1.94679	6.15630	.263852
3.30	10.8900	1.81659	5.74456	.303030	3.80	14.4400	1.94936	6.16441	.263158
3.31	10.9561	1.81934	5.75326	.302115	3.81	14.5161	1.95192	6.17252	.262467
3.32	11.0224	1.82209	5.76194	.301205	3.82	14.5924	1.95448	6.18061	.261780
3.33	11.0889	1.82483	5.77062	.300300	3.83	14.6689	1.95704	6.18870	.261097
3.34	11.1556	1.82757	5.77927	.299401	3.84	14.7456	1.95959	6.19677	.260417
3.35	11.2225	1.83030	5.78792	.298507	3.85	14.8225	1.96214	6.20484	.259740
3.36	11.2896	1.83303	5.79655	.297619	3.86	14.8996	1.96469	6.21289	.259067
3.37	11.3569	1.83576	5.80517	.296736	3.87	14.9769	1.96723	6.22093	.258398
3.38	11.4244	1.83848	5.81378	.295858	3.88	15.0544	1.96977	6.22896	.257732
3.39	11.4921	1.84120	5.82237	.294985	3.89	15.1321	1.97231	6.23699	.257069
3.40	11.5600	1.84391	5.83095	.294118	3.90	15.2100	1.97484	6.24500	.256410
3.41	11.6281	1.84662	5.83952	.293255	3.91	15.2881	1.97737	6.25300	.255754
3.42	11.6964	1.84932	5.84808	.292398	3.92	15.3664	1.97990	6.26099	.255102
3.43	11.7649	1.85203	5.85662	.291545	3.93	15.4449	1.98242	6.26897	.254453
3.44	11.8336	1.85472	5.86515	.290698	3.94	15.5236	1.98494	6.27694	.253807
3.45	11.9025	1.85742	5.87367	.289855	3.95	15.6025	1.98746	6.28490	.253165
3.46	11.9716	1.86011	5.88218	.289017	3.96	15.6816	1.98997	6.29285	.252525
3.47	12.0409	1.86279	5.89067	.288184	3.97	15.7609	1.99249	6.30079	.251889
3.48	12.1104	1.86548	5.89915	.287356	3.98	15.8404	1.99499	6.30872	.251256
3.49	12.1801	1.86815	5.90762	.286533	3.99	15.9201	1.99750	6.31664	.250627
3.50	12.2500	1.87083	5.91608	.285714	4.00	16.0000	2.00000	6.32456	.250000
n	n^2	$\sqrt{n}$	$\sqrt{10n}$	$1/n$	n	n^2	$\sqrt{n}$	$\sqrt{10n}$	$1/n$

n	n^2	$\sqrt{n}$	$\sqrt{10n}$	$1/n$	n	n^2	$\sqrt{n}$	$\sqrt{10n}$	$1/n$
4.00	16.0000	2.00000	6.32456	.250000	**4.50**	20.2500	2.12132	6.70820	.222222
4.01	16.0801	2.00250	6.33246	.249377	4.51	20.3401	2.12368	6.71565	.221729
4.02	16.1604	2.00499	6.34035	.248756	4.52	20.4304	2.12603	6.72309	.221239
4.03	16.2409	2.00749	6.34823	.248139	4.53	20.5209	2.12838	6.73053	.220751
4.04	16.3216	2.00998	6.35610	.247525	4.54	20.6116	2.13073	6.73795	.220264
4.05	16.4025	2.01246	6.36396	.246914	4.55	20.7025	2.13307	6.74537	.219780
4.06	16.4836	2.01494	6.37181	.246305	4.56	20.7936	2.13542	6.75278	.219298
4.07	16.5649	2.01742	6.37966	.245700	4.57	20.8849	2.13776	6.76018	.218818
4.08	16.6464	2.01990	6.38749	.245098	4.58	20.9764	2.14009	6.76757	.218341
4.09	16.7281	2.02237	6.39531	.244499	4.59	21.0681	2.14243	6.77495	.217865
4.10	16.8100	2.02485	6.40312	.243902	**4.60**	21.1600	2.14476	6.78233	.217391
4.11	16.8921	2.02731	6.41093	.243309	4.61	21.2521	2.14709	6.78970	.216920
4.12	16.9744	2.02978	6.41872	.242718	4.62	21.3444	2.14942	6.79706	.216450
4.13	17.0569	2.03224	6.42651	.242131	4.63	21.4369	2.15174	6.80441	.215983
4.14	17.1396	2.03470	6.43428	.241546	4.64	21.5296	2.15407	6.81175	.215517
4.15	17.2225	2.03715	6.44205	.240964	4.65	21.6225	2.15639	6.81909	.215054
4.16	17.3056	2.03961	6.44981	.240385	4.66	21.7156	2.15870	6.82642	.214592
4.17	17.3889	2.04206	6.45755	.239808	4.67	21.8089	2.16102	6.83374	.214133
4.18	17.4724	2.04450	6.46529	.239234	4.68	21.9024	2.16333	6.84105	.213675
4.19	17.5561	2.04695	6.47302	.238663	4.69	21.9961	2.16564	6.84836	.213220
4.20	17.6400	2.04939	6.48074	.238095	**4.70**	22.0900	2.16795	6.85565	.212766
4.21	17.7241	2.05183	6.48845	.237530	4.71	22.1841	2.17025	6.86294	.212314
4.22	17.8084	2.05426	6.49615	.236967	4.72	22.2784	2.17256	6.87023	.211864
4.23	17.8929	2.05670	6.50384	.236407	4.73	22.3729	2.17486	6.87750	.211416
4.24	17.9776	2.05913	6.51153	.235849	4.74	22.4676	2.17715	6.88477	.210970
4.25	18.0625	2.06155	6.51920	.235294	4.75	22.5625	2.17945	6.89202	.210526
4.26	18.1476	2.06398	6.52687	.234742	4.76	22.6576	2.18174	6.89928	.210084
4.27	18.2329	2.06640	6.53452	.234192	4.77	22.7529	2.18403	6.90652	.209644
4.28	18.3184	2.06882	6.54217	.233645	4.78	22.8484	2.18632	6.91375	.209205
4.29	18.4041	2.07123	6.54981	.233100	4.79	22.9441	2.18861	6.92098	.208768
4.30	18.4900	2.07364	6.55744	.232558	**4.80**	23.0400	2.19089	6.92820	.208333
4.31	18.5761	2.07605	6.56506	.232019	4.81	23.1361	2.19317	6.93542	.207900
4.32	18.6624	2.07846	6.57267	.231481	4.82	23.2324	2.19545	6.94262	.207469
4.33	18.7489	2.08087	6.58027	.230947	4.83	23.3289	2.19773	6.94982	.207039
4.34	18.8356	2.08327	6.58787	.230415	4.84	23.4256	2.20000	6.95701	.206612
4.35	18.9225	2.08567	6.59545	.229885	4.85	23.5225	2.20227	6.96419	.206186
4.36	19.0096	2.08806	6.60303	.229358	4.86	23.6196	2.20454	6.97137	.205761
4.37	19.0969	2.09045	6.61060	.228833	4.87	23.7169	2.20681	6.97854	.205339
4.38	19.1844	2.09284	6.61816	.228311	4.88	23.8144	2.20907	6.98570	.204918
4.39	19.2721	2.09523	6.62571	.227790	4.89	23.9121	2.21133	6.99285	.204499
4.40	19.3600	2.09762	6.63325	.227273	**4.90**	24.0100	2.21359	7.00000	.204082
4.41	19.4481	2.10000	6.64078	.226757	4.91	24.1081	2.21585	7.00714	.203666
4.42	19.5364	2.10238	6.64831	.226244	4.92	24.2064	2.21811	7.01427	.203252
4.43	19.6249	2.10476	6.65582	.225734	4.93	24.3049	2.22036	7.02140	.202840
4.44	19.7136	2.10713	6.66333	.225225	4.94	24.4036	2.22261	7.02851	.202429
4.45	19.8025	2.10950	6.67083	.224719	4.95	24.5025	2.22486	7.03562	.202020
4.46	19.8916	2.11187	6.67832	.224215	4.96	24.6016	2.22711	7.04273	.201613
4.47	19.9809	2.11424	6.68581	.223714	4.97	24.7009	2.22935	7.04982	.201207
4.48	20.0704	2.11660	6.69328	.223214	4.98	24.8004	2.23159	7.05691	.200803
4.49	20.1601	2.11896	6.70075	.222717	4.99	24.9001	2.23383	7.06399	.200401
4.50	20.2500	2.12132	6.70820	.222222	**5.00**	25.0000	2.23607	7.07107	.200000
n	n^2	$\sqrt{n}$	$\sqrt{10n}$	$1/n$	n	n^2	$\sqrt{n}$	$\sqrt{10n}$	$1/n$

Table VII

n	n^2	$\sqrt{n}$	$\sqrt{10n}$	$1/n$	n	n^2	$\sqrt{n}$	$\sqrt{10n}$	$1/n$
5.00	25.0000	2.23607	7.07107	.200000	5.50	30.2500	2.34521	7.41620	.181818
5.01	25.1001	2.23830	7.07814	.199601	5.51	30.3601	2.34734	7.42294	.181488
5.02	25.2004	2.24054	7.08520	.199203	5.52	30.4704	2.34947	7.42967	.181159
5.03	25.3009	2.24277	7.09225	.198807	5.53	30.5809	2.35160	7.43640	.180832
5.04	25.4016	2.24499	7.09930	.198413	5.54	30.6916	2.35372	7.44312	.180505
5.05	25.5025	2.24722	7.10634	.198020	5.55	30.8025	2.35584	7.44983	.180180
5.06	25.6036	2.24944	7.11337	.197628	5.56	30.9136	2.35797	7.45654	.179856
5.07	25.7049	2.25167	7.12039	.197239	5.57	31.0249	2.36008	7.46324	.179533
5.08	25.8064	2.25389	7.12741	.196850	5.58	31.1364	2.36220	7.46994	.179211
5.09	25.9081	2.25610	7.13442	.196464	5.59	31.2481	2.36432	7.47663	.178891
5.10	26.0100	2.25832	7.14143	.196078	5.60	31.3600	2.36643	7.48331	.178571
5.11	26.1121	2.26053	7.14843	.195695	5.61	31.4721	2.36854	7.48999	.178253
5.12	26.2144	2.26274	7.15542	.195312	5.62	31.5844	2.37065	7.49667	.177936
5.13	26.3169	2.26495	7.16240	.194932	5.63	31.6969	2.37276	7.50333	.177620
5.14	26.4196	2.26716	7.16938	.194553	5.64	31.8096	2.37487	7.50999	.177305
5.15	26.5225	2.26936	7.17635	.194175	5.65	31.9225	2.37697	7.51665	.176991
5.16	26.6256	2.27156	7.18331	.193798	5.66	32.0356	2.37908	7.52330	.176678
5.17	26.7289	2.27376	7.19027	.193424	5.67	32.1489	2.38118	7.52994	.176367
5.18	26.8324	2.27596	7.19722	.193050	5.68	32.2624	2.38328	7.53658	.176056
5.19	26.9361	2.27816	7.20417	.192678	5.69	32.3761	2.38537	7.54321	.175747
5.20	27.0400	2.28035	7.21110	.192308	5.70	32.4900	2.38747	7.54983	.175439
5.21	27.1441	2.28254	7.21803	.191939	5.71	32.6041	2.38956	7.55645	.175131
5.22	27.2484	2.28473	7.22496	.191571	5.72	32.7184	2.39165	7.56307	.174825
5.23	27.3529	2.28692	7.23187	.191205	5.73	32.8329	2.39374	7.56968	.174520
5.24	27.4576	2.28910	7.23878	.190840	5.74	32.9476	2.39583	7.57628	.174216
5.25	27.5625	2.29129	7.24569	.190476	5.75	33.0625	2.39792	7.58288	.173913
5.26	27.6676	2.29347	7.25259	.190114	5.76	33.1776	2.40000	7.58947	.173611
5.27	27.7729	2.29565	7.25948	.189753	5.77	33.2929	2.40208	7.59605	.173310
5.28	27.8784	2.29783	7.26636	.189394	5.78	33.4084	2.40416	7.60263	.173010
5.29	27.9841	2.30000	7.27324	.189036	5.79	33.5241	2.40624	7.60920	.172712
5.30	28.0900	2.30217	7.28011	.188679	5.80	33.6400	2.40832	7.61577	.172414
5.31	28.1961	2.30434	7.28697	.188324	5.81	33.7561	2.41039	7.62234	.172117
5.32	28.3024	2.30651	7.29383	.187970	5.82	33.8724	2.41247	7.62889	.171821
5.33	28.4089	2.30868	7.30068	.187617	5.83	33.9889	2.41454	7.63544	.171527
5.34	28.5156	2.31084	7.30753	.187266	5.84	34.1056	2.41661	7.64199	.171233
5.35	28.6225	2.31301	7.31437	.186916	5.85	34.2225	2.41868	7.64853	.170940
5.36	28.7296	2.31517	7.32120	.186567	5.86	34.3396	2.42074	7.65506	.170649
5.37	28.8369	2.31733	7.32803	.186220	5.87	34.4569	2.42281	7.66159	.170358
5.38	28.9444	2.31948	7.33485	.185874	5.88	34.5744	2.42487	7.66812	.170068
5.39	29.0521	2.32164	7.34166	.185529	5.89	34.6921	2.42693	7.67463	.169779
5.40	29.1600	2.32379	7.34847	.185185	5.90	34.8100	2.42899	7.68115	.169492
5.41	29.2681	2.32594	7.35527	.184843	5.91	34.9281	2.43105	7.68765	.169205
5.42	29.3764	2.32809	7.36206	.184502	5.92	35.0464	2.43311	7.69415	.168919
5.43	29.4849	2.33024	7.36885	.184162	5.93	35.1649	2.43516	7.70065	.168634
5.44	29.5936	2.33238	7.37564	.183824	5.94	35.2836	2.43721	7.70714	.168350
5.45	29.7025	2.33452	7.38241	.183486	5.95	35.4025	2.43926	7.71362	.168067
5.46	29.8116	2.33666	7.38918	.183150	5.96	35.5216	2.44131	7.72010	.167785
5.47	29.9209	2.33880	7.39594	.182815	5.97	35.6409	2.44336	7.72658	.167504
5.48	30.0304	2.34094	7.40270	.182482	5.98	35.7604	2.44540	7.73305	.167224
5.49	30.1401	2.34307	7.40945	.182149	5.99	35.8801	2.44745	7.73951	.166945
5.50	30.2500	2.34521	7.41620	.181818	6.00	36.0000	2.44949	7.74597	.166667
n	n^2	$\sqrt{n}$	$\sqrt{10n}$	$1/n$	n	n^2	$\sqrt{n}$	$\sqrt{10n}$	$1/n$

n	n^2	$\sqrt{n}$	$\sqrt{10n}$	$1/n$
6.00	36.0000	2.44949	7.74597	.166667
6.01	36.1201	2.45153	7.75242	.166389
6.02	36.2404	2.45357	7.75887	.166113
6.03	36.3609	2.45561	7.76531	.165837
6.04	36.4816	2.45764	7.77174	.165563
6.05	36.6025	2.45967	7.77817	.165289
6.06	36.7236	2.46171	7.78460	.165017
6.07	36.8449	2.46374	7.79102	.164745
6.08	36.9664	2.46577	7.79744	.164474
6.09	37.0881	2.46779	7.80385	.164204
6.10	37.2100	2.46982	7.81025	.163934
6.11	37.3321	2.47184	7.81665	.163666
6.12	37.4544	2.47386	7.82304	.163399
6.13	37.5769	2.47588	7.82943	.163132
6.14	37.6996	2.47790	7.83582	.162866
6.15	37.8225	2.47992	7.84219	.162602
6.16	37.9456	2.48193	7.84857	.162338
6.17	38.0689	2.48395	7.85493	.162075
6.18	38.1924	2.48596	7.86130	.161812
6.19	38.3161	2.48797	7.86766	.161551
6.20	38.4400	2.48998	7.87401	.161290
6.21	38.5641	2.49199	7.88036	.161031
6.22	38.6884	2.49399	7.88670	.160772
6.23	38.8129	2.49600	7.89303	.160514
6.24	38.9376	2.49800	7.89937	.160256
6.25	39.0625	2.50000	7.90569	.160000
6.26	39.1876	2.50200	7.91202	.159744
6.27	39.3129	2.50400	7.91833	.159490
6.28	39.4384	2.50599	7.92465	.159236
6.29	39.5641	2.50799	7.93095	.158983
6.30	39.6900	2.50998	7.93725	.158730
6.31	39.8161	2.51197	7.94355	.158479
6.32	39.9424	2.51396	7.94984	.158228
6.33	40.0689	2.51595	7.95613	.157978
6.34	40.1956	2.51794	7.96241	.157729
6.35	40.3225	2.51992	7.96869	.157480
6 36	40.4496	2.52190	7.97496	.157233
6.37	40.5769	2.52389	7.98123	.156986
6.38	40.7044	2.52587	7.98749	.156740
6.39	40.8321	2.52784	7.99375	.156495
6.40	40.9600	2.52982	8.00000	.156250
6.41	41.0881	2.53180	8.00625	.156006
6.42	41.2164	2.53377	8.01249	.155763
6.43	41.3449	2.53574	8.01873	.155521
6.44	41.4736	2.53772	8.02496	.155280
6.45	41.6025	2.53969	8.03119	.155039
6.46	41.7316	2.54165	8.03741	.154799
6.47	41.8609	2.54362	8.04363	.154560
6.48	41.9904	2.54558	8.04984	.154321
6.49	42.1201	2.54755	8.05605	.154083
6.50	42.2500	2.54951	8.06226	.153846
6.50	42.2500	2.54951	8.06226	.153846
6.51	42.3801	2.55147	8.06846	.153610
6.52	42.5104	2.55343	8.07465	.153374
6.53	42.6409	2.55539	8.08084	.153139
6.54	42.7716	2.55734	8.08703	.152905
6.55	42.9025	2.55930	8.09321	.152672
6.56	43.0336	2.56125	8.09938	.152439
6.57	43.1649	2.56320	8.10555	.152207
6.58	43.2964	2.56515	8.11172	.151976
6 59	43.4281	2.56710	8.11788	.151745
6.60	43.5600	2.56905	8.12404	.151515
6.61	43.6921	2.57099	8.13019	.151286
6.62	43.8244	2.57294	8.13634	.151057
6.63	43.9569	2.57488	8.14248	.150830
6.64	44.0896	2.57682	8.14862	.150602
6.65	44.2225	2.57876	8.15475	.150376
6.66	44.3556	2.58070	8.16088	.150150
6.67	44.4889	2.58263	8.16701	.149925
6.68	44.6224	2.58457	8.17313	.149701
6.69	44.7561	2.58650	8.17924	.149477
6.70	44.8900	2.58844	8.18535	.149254
6.71	45.0241	2.59037	8.19146	.149031
6.72	45.1584	2.59230	8.19756	.148810
6.73	45.2929	2.59422	8.20366	.148588
6.74	45.4276	2.59615	8.20975	.148368
6.75	45.5625	2.59808	8.21584	.148148
6.76	45.6976	2.60000	8.22192	.147929
6.77	45.8329	2.60192	8.22800	.147710
6.78	45.9684	2.60384	8.23408	.147493
6.79	46.1041	2.60576	8.24015	.147275
6.80	46.2400	2.60768	8.24621	.147059
6.81	46.3761	2.60960	8.25227	.146843
6.82	46.5124	2.61151	8.25833	.146628
6.83	46.6489	2.61343	8.26438	.146413
6.84	46.7856	2.61534	8.27043	.146199
6.85	46.9225	2.61725	8.27647	.145985
6.86	47.0596	2.61916	8.28251	.145773
6.87	47.1969	2.62107	8.28855	.145560
6.88	47.3344	2.62298	8.29458	.145349
6.89	47.4721	2.62488	8.30060	.145138
6.90	47.6100	2.62679	8.30662	.144928
6.91	47.7481	2.62869	8.31264	.144718
6.92	47.8864	2.63059	8.31865	.144509
6.93	48.0249	2.63249	8.32466	.144300
6.94	48.1636	2.63439	8.33067	.144092
6.95	48.3025	2.63629	8.33667	.143885
6.96	48.4416	2.63818	8.34266	.143678
6.97	48.5809	2.64008	8.34865	.143472
6.98	48.7204	2.64197	8.35464	.143266
6.99	48.8601	2.64386	8.36062	.143062
7.00	49.0000	2.64575	8.36660	.142857

n	n^2	$\sqrt{n}$	$\sqrt{10n}$	$1/n$

Table VII

n	n^2	$\sqrt{n}$	$\sqrt{10n}$	$1/n$	n	n^2	$\sqrt{n}$	$\sqrt{10n}$	$1/n$
7.00	49.0000	2.64575	8.36660	.142857	**7.50**	56.2500	2.73861	8.66025	.133333
7.01	49.1401	2.64764	8.37257	.142653	7.51	56.4001	2.74044	8.66603	.133156
7.02	49.2804	2.64953	8.37854	.142450	7.52	56.5504	2.74226	8.67179	.132979
7.03	49.4209	2.65141	8.38451	.142248	7.53	56.7009	2.74408	8.67756	.132802
7.04	49.5616	2.65330	8.39047	.142045	7.54	56.8516	2.74591	8.68332	.132626
7.05	49.7025	2.65518	8.39643	.141844	7.55	57.0025	2.74773	8.68907	.132450
7.06	49.8436	2.65707	8.40238	.141643	7.56	57.1536	2.74955	8.69483	.132275
7.07	49.9849	2.65895	8.40833	.141443	7.57	57.3049	2.75136	8.70057	.132100
7.08	50.1264	2.66083	8.41427	.141243	7.58	57.4564	2.75318	8.70632	.131926
7.09	50.2681	2.66271	8.42021	.141044	7.59	57.6081	2.75500	8.71206	.131752
7.10	50.4100	2.66458	8.42615	.140845	**7.60**	57.7600	2.75681	8.71780	.131579
7.11	50.5521	2.66646	8.43208	.140647	7.61	57.9121	2.75862	8.72353	.131406
7.12	50.6944	2.66833	8.43801	.140449	7.62	58.0644	2.76043	8.72926	.131234
7.13	50.8369	2.67021	8.44393	.140252	7.63	58.2169	2.76225	8.73499	.131062
7.14	50.9796	2.67208	8.44985	.140056	7.64	58.3696	2.76405	8.74071	.130890
7.15	51.1225	2.67395	8.45577	.139860	7.65	58.5225	2.76586	8.74643	.130719
7.16	51.2656	2.67582	8.46168	.139665	7.66	58.6756	2.76767	8.75214	.130548
7.17	51.4089	2.67769	8.46759	.139470	7.67	58.8289	2.76948	8.75785	.130378
7.18	51.5524	2.67955	8.47349	.139276	7.68	58.9824	2.77128	8.76356	.130208
7.19	51.6961	2.68142	8.47939	.139082	7.69	59.1361	2.77308	8.76926	.130039
7.20	51.8400	2.68328	8.48528	.138889	**7.70**	59.2900	2.77489	8.77496	.129870
7.21	51.9841	2.68514	8.49117	.138696	7.71	59.4441	2.77669	8.78066	.129702
7.22	52.1284	2.68701	8.49706	.138504	7.72	59.5984	2.77849	8.78635	.129534
7.23	52.2729	2.68887	8.50294	.138313	7.73	59.7529	2.78029	8.79204	.129366
7.24	52.4176	2.69072	8.50882	.138122	7.74	59.9076	2.78209	8.79773	.129199
7.25	52.5625	2.69258	8.51469	.137931	7.75	60.0625	2.78388	8.80341	.129032
7.26	52.7076	2.69444	8.52056	.137741	7.76	60.2176	2.78568	8.80909	.128866
7.27	52.8529	2.69629	8.52643	.137552	7.77	60.3729	2.78747	8.81476	.128700
7.28	52.9984	2.69815	8.53229	.137363	7.78	60.5284	2.78927	8.82043	.128535
7.29	53.1441	2.70000	8.53815	.137174	7.79	60.6841	2.79106	8.82610	.128370
7.30	53.2900	2.70185	8.54400	.136986	**7.80**	60.8400	2.79285	8.83176	.128205
7.31	53.4361	2.70370	8.54985	.136799	7.81	60.9961	2.79464	8.83742	.128041
7.32	53.5824	2.70555	8.55570	.136612	7.82	61.1524	2.79643	8.84308	.127877
7.33	53.7289	2.70740	8.56154	.136426	7.83	61.3089	2.79821	8.84873	.127714
7.34	53.8756	2.70924	8.56738	.136240	7.84	61.4656	2.80000	8.85438	.127551
7.35	54.0225	2.71109	8.57321	.136054	7.85	61.6225	2.80179	8.86002	.127389
7.36	54.1696	2.71293	8.57904	.135870	7.86	61.7796	2.80357	8.86566	.127226
7.37	54.3169	2.71477	8.58487	.135685	7.87	61.9369	2.80535	8.87130	.127065
7.38	54.4644	2.71662	8.59069	.135501	7.88	62.0944	2.80713	8.87694	.126904
7.39	54.6121	2.71846	8.59651	.135318	7.89	62.2521	2.80891	8.88257	.126743
7.40	54.7600	2.72029	8.60233	.135135	**7.90**	62.4100	2.81069	8.88819	.126582
7.41	54.9081	2.72213	8.60814	.134953	7.91	62.5681	2.81247	8.89382	.126422
7.42	55.0564	2.72397	8.61394	.134771	7.92	62.7264	2.81425	8.89944	.126263
7.43	55.2049	2.72580	8.61974	.134590	7.93	62.8849	2.81603	8.90505	.126103
7.44	55.3536	2.72764	8.62554	.134409	7.94	63.0436	2.81780	8.91067	.125945
7.45	55.5025	2.72947	8.63134	.134228	7.95	63.2025	2.81957	8.91628	.125786
7.46	55.6516	2.73130	8.63713	.134048	7.96	63.3616	2.82135	8.92188	.125628
7.47	55.8009	2.73313	8.64292	.133869	7.97	63.5209	2.82312	8.92749	.125471
7.48	55.9504	2.73496	8.64870	.133690	7.98	63.6804	2.82489	8.93308	.125313
7.49	56.1001	2.73679	8.65448	.133511	7.99	63.8401	2.82666	8.93868	.125156
7.50	56.2500	2.73861	8.66025	.133333	**8.00**	64.0000	2.82843	8.94427	.125000
n	n^2	$\sqrt{n}$	$\sqrt{10n}$	$1/n$	n	n^2	$\sqrt{n}$	$\sqrt{10n}$	$1/n$

n	n^2	$\sqrt{n}$	$\sqrt{10n}$	$1/n$	n	n^2	$\sqrt{n}$	$\sqrt{10n}$	$1/n$
8.00	64.0000	2.82843	8.94427	.125000	8.50	72.2500	2.91548	9.21954	.117647
8.01	64.1601	2.83019	8.94986	.124844	8.51	72.4201	2.91719	9.22497	.117509
8.02	64.3204	2.83196	8.95545	.124688	8.52	72.5904	2.91890	9.23038	.117371
8.03	64.4809	2.83373	8.96103	.124533	8.53	72.7609	2.92062	9.23580	.117233
8.04	64.6416	2.83549	8.96660	.124378	8.54	72.9316	2.92233	9.24121	.117096
8.05	64.8025	2.83725	8.97218	.124224	8.55	73.1025	2.92404	9.24662	.116959
8.06	64.9636	2.83901	8.97775	.124069	8.56	73.2736	2.92575	9.25203	.116822
8.07	65.1249	2.84077	8.98332	.123916	8.57	73.4449	2.92746	9.25743	.116686
8.08	65.2864	2.84253	8.98888	.123762	8.58	73.6164	2.92916	9.26283	.116550
8.09	65.4481	2.84429	8.99444	.123609	8.59	73.7881	2.93087	9.26823	.116414
8.10	65.6100	2.84605	9.00000	.123457	8.60	73.9600	2.93258	9.27362	.116279
8.11	65.7721	2.84781	9.00555	.123305	8.61	74.1321	2.93428	9.27901	.116144
8.12	65.9344	2.84956	9.01110	.123153	8.62	74.3044	2.93598	9.28440	.116009
8.13	66.0969	2.85132	9.01665	.123001	8.63	74.4769	2.93769	9.28978	.115875
8.14	66.2596	2.85307	9.02219	.122850	8.64	74.6496	2.93939	9.29516	.115741
8.15	66.4225	2.85482	9.02774	.122699	8.65	74.8225	2.94109	9.30054	.115607
8.16	66.5856	2.85657	9.03327	.122549	8.66	74.9956	2.94279	9.30591	.115473
8.17	66.7489	2.85832	9.03881	.122399	8.67	75.1689	2.94449	9.31128	.115340
8.18	66.9124	2.86007	9.04434	.122249	8.68	75.3424	2.94618	9.31665	.115207
8.19	67.0761	2.86182	9.04986	.122100	8.69	75.5161	2.94788	9.32202	.115075
8.20	67.2400	2.86356	9.05539	.121951	8.70	75.6900	2.94958	9.32738	.114943
8.21	67.4041	2.86531	9.06091	.121803	8.71	75.8641	2.95127	9.33274	.114811
8.22	67.5684	2.86705	9.06642	.121655	8.72	76.0384	2.95296	9.33809	.114679
8.23	67.7329	2.86880	9.07193	.121507	8.73	76.2129	2.95466	9.34345	.114548
8.24	67.8976	2.87054	9.07744	.121359	8.74	76.3876	2.95635	9.34880	.114416
8.25	68.0625	2.87228	9.08295	.121212	8.75	76.5625	2.95804	9.35414	.114286
8.26	68.2276	2.87402	9.08845	.121065	8.76	76.7376	2.95973	9.35949	.114155
8.27	68.3929	2.87576	9.09395	.120919	8.77	76.9129	2.96142	9.36483	.114025
8.28	68.5584	2.87750	9.09945	.120773	8.78	77.0884	2.96311	9.37017	.113895
8.29	68.7241	2.87924	9.10494	.120627	8.79	77.2641	2.96479	9.37550	.113766
8.30	68.8900	2.88097	9,11043	.120482	8.80	77.4400	2.96648	9.38083	.113636
8.31	69.0561	2.88271	9.11592	.120337	8.81	77.6161	2.96816	9.38616	.113507
8.32	69.2224	2.88444	9.12140	.120192	8.82	77.7924	2.96985	9.39149	.113379
8.33	69.3889	2.88617	9.12688	.120048	8.83	77.9689	2.97153	9.39681	.113250
8.34	69.5556	2.88791	9.13236	.119904	8.84	78.1456	2.97321	9.40213	.113122
8.35	69.7225	2.88964	9.13783	.119760	8.85	78.3225	2.97489	9.40744	.112994
8.36	69.8896	2.89137	9.14330	.119617	8.86	78.4996	2.97658	9.41276	.112867
8.37	70.0569	2.89310	9.14877	.119474	8.87	78.6769	2.97825	9.41807	.112740
8.38	70.2244	2.89482	9.15423	.119332	8.88	78.8544	2.97993	9.42338	.112613
8.39	70.3921	2.89655	9.15969	.119190	8.89	79.0321	2.98161	9.42868	.112486
8.40	70.5600	2.89828	9.16515	.119048	8.90	79.2100	2.98329	9.43398	.112360
8.41	70.7281	2.90000	9.17061	.118906	8.91	79.3881	2.98496	9.43928	.112233
8.42	70.8964	2.90172	9.17606	.118765	8.92	79.5664	2.98664	9.44458	.112108
8.43	71.0649	2.90345	9.18150	.118624	8.93	79.7449	2.98831	9.44987	.111982
8.44	71.2336	2.90517	9.18695	.118483	8.94	79.9236	2.98998	9.45516	.111857
8.45	71.4025	2.90689	9.19239	.118343	8.95	80.1025	2.99166	9.46044	.111732
8.46	71.5716	2.90861	9.19783	.118203	8.96	80.2816	2.99333	9.46573	.111607
8.47	71.7409	2.91033	9.20326	.118064	8.97	80.4609	2.99500	9.47101	.111483
8.48	71.9104	2.91204	9.20869	.117925	8.98	80.6404	2.99666	9.47629	.111359
8.49	72.0801	2.91376	9.21412	.117786	8.99	80.8201	2.99833	9.48156	.111235
8.50	72.2500	2.91548	9.21954	.117647	9.00	81.0000	3.00000	9.48683	.111111
n	n^2	$\sqrt{n}$	$\sqrt{10n}$	$1/n$	n	n^2	$\sqrt{n}$	$\sqrt{10n}$	$1/n$

Table VII

n	n^2	$\sqrt{n}$	$\sqrt{10n}$	$1/n$	n	n^2	$\sqrt{n}$	$\sqrt{10n}$	$1/n$
9.00	81.0000	3.00000	9.48683	.111111	**9.50**	90.2500	3.08221	9.74679	.105263
9.01	81.1801	3.00167	9.49210	.110988	9.51	90.4401	3.08383	9.75192	.105152
9.02	81.3604	3.00333	9.49737	.110865	9.52	90.6304	3.08545	9.75705	.105042
9.03	81.5409	3.00500	9.50263	.110742	9.53	90.8209	3.08707	9.76217	.104932
9.04	81.7216	3.00666	9.50789	.110619	9.54	91.0116	3.08869	9.76729	.104822
9.05	81.9025	3.00832	9.51315	.110497	9.55	91.2025	3.09031	9.77241	.104712
9.06	82.0836	3.00998	9.51840	.110375	9.56	91.3936	3.09192	9.77753	.104603
9.07	82.2649	3.01164	9.52365	.110254	9.57	91.5849	3.09354	9.78264	.104493
9.08	82.4464	3.01330	9.52890	.110132	9.58	91.7764	3.09516	9.78775	.104384
9.09	82.6281	3.01496	9.53415	.110011	9.59	91.9681	3.09677	9.79285	.104275
9.10	82.8100	3.01662	9.53939	.109890	**9.60**	92.1600	3.09839	9.79796	.104167
9.11	82.9921	3.01828	9.54463	.109769	9.61	92.3521	3.10000	9.80306	.104058
9.12	83.1744	3.01993	9.54987	.109649	9.62	92.5444	3.10161	9.80816	.103950
9.13	83.3569	3.02159	9.55510	.109529	9.63	92.7369	3.10322	9.81326	.103842
9.14	83.5396	3.02324	9.56033	.109409	9.64	92.9296	3.10483	9.81835	.103734
9.15	83.7225	3.02490	9.56556	.109290	9.65	93.1225	3.10644	9.82344	.103627
9.16	83.9056	3.02655	9.57079	.109170	9.66	93.3156	3.10805	9.82853	.103520
9.17	84.0889	3.02820	9.57601	.109051	9 67	93.5089	3.10966	9.83362	.103413
9.18	84.2724	3.02985	9.58123	.108932	9.68	93.7024	3.11127	9.83870	.103306
9.19	84.4561	3.03150	9.58645	.108814	9.69	93.8961	3.11288	9.84378	.103199
9.20	84.6400	3.03315	9.59166	.108696	**9.70**	94.0900	3.11448	9.84886	.103093
9.21	84.8241	3.03480	9.59687	.108578	9.71	94.2841	3.11609	9.85393	.102987
9.22	85.0084	3.03645	9.60208	.108460	9.72	94.4784	3.11769	9.85901	.102881
9.23	85.1929	3.03809	9.60729	.108342	9.73	94.6729	3.11929	9.86408	.102775
9.24	85.3776	3.03974	9.61249	.108225	9.74	94.8676	3.12090	9.86914	.102669
9.25	85.5625	3.04138	9.61769	.108108	9.75	95.0625	3.12250	9.87421	.102564
9.26	85.7476	3.04302	9.62289	.107991	9.76	95.2576	3.12410	9.87927	.102459
9.27	85.9329	3.04467	9.62808	.107875	9.77	95.4529	3.12570	9.88433	.102354
9.28	86.1184	3.04631	9.63328	.107759	9.78	95.6484	3.12730	9.88939	.102249
9.29	86.3041	3.04795	9.63846	.107643	9.79	95.8441	3.12890	9.89444	.102145
9.30	86.4900	3.04959	9.64365	.107527	**9.80**	96.0400	3.13050	9.89949	.102041
9.31	86.6761	3.05123	9.64883	.107411	9.81	96.2361	3.13209	9.90454	.101937
9.32	86.8624	3.05287	9.65401	.107296	9.82	96.4324	3.13369	9.90959	.101833
9.33	87.0489	3.05450	9.65919	.107181	9.83	96.6289	3.13528	9.91464	.101729
9.34	87.2356	3.05614	9.66437	.107066	9.84	96.8256	3.13688	9.91968	.101626
9.35	87.4225	3.05778	9.66954	.106952	9.85	97.0225	3.13847	9.92472	.101523
9.36	87.6096	3.05941	9.67471	.106838	9.86	97.2196	3.14006	9.92975	.101420
9.37	87.7969	3.06105	9.67988	.106724	9.87	97.4169	3.14166	9.93479	.101317
9.38	87.9844	3.06268	9.68504	.106610	9.88	97.6144	3.14325	9.93982	.101215
9.39	88.1721	3.06431	9.69020	.106496	9.89	97.8121	3.14484	9.94485	.101112
9.40	88.3600	3.06594	9.69536	.106383	**9.90**	98.0100	3.14643	9.94987	.101010
9.41	88.5481	3.06757	9.70052	.106270	9.91	98.2081	3.14802	9.95490	.100908
9.42	88.7364	3.06920	9.70567	.106157	9.92	98.4064	3.14960	9.95992	.100806
9.43	88.9249	3.07083	9.71082	.106045	9.93	98.6049	3.15119	9.96494	.100705
9.44	89.1136	3.07246	9.71597	.105932	9.94	98.8036	3.15278	9.96995	.100604
9.45	89.3025	3.07409	9.72111	.105820	9.95	99.0025	3.15436	9.97497	.100503
9.46	89.4916	3.07571	9.72625	.105708	9.96	99.2016	3.15595	9.97998	.100402
9.47	89.6809	3.07734	9.73139	.105597	9.97	99.4009	3.15753	9.98499	.100301
9.48	89.8704	3.07896	9.73653	.105485	9.98	99.6004	3.15911	9.98999	.100200
9.49	90.0601	3.08058	9.74166	.105374	9.99	99.8001	3.16070	9.99500	.100100
9.50	90.2500	3.08221	9.74679	.105263	**10.00**	100.000	3.16228	10.0000	.100000
n	n^2	$\sqrt{n}$	$\sqrt{10n}$	$1/n$	n	n^2	$\sqrt{n}$	$\sqrt{10n}$	$1/n$

TRIGONOMETRIC IDENTITIES

(1) $\csc \theta = \dfrac{1}{\sin \theta}$. **(2)** $\sec \theta = \dfrac{1}{\cos \theta}$. **(3)** $\cot \theta = \dfrac{1}{\tan \theta}$.

(4) $\tan \theta = \dfrac{\sin \theta}{\cos \theta}$. **(5)** $\cot \theta = \dfrac{\cos \theta}{\sin \theta}$.

(6) $\sin^2 \theta + \cos^2 \theta = 1$. **(7)** $1 + \tan^2 \theta = \sec^2 \theta$. **(8)** $1 + \cot^2 \theta = \csc^2 \theta$.

(9) $\sin(-\theta) = -\sin \theta$. **(10)** $\cos(-\theta) = \cos \theta$. **(11)** $\tan(-\theta) = -\tan \theta$.

(12) $\cot(-\theta) = -\cot \theta$. **(13)** $\sec(-\theta) = \sec \theta$. **(14)** $\csc(-\theta) = -\csc \theta$.

(15) $\cos(A + B) = \cos A \cos B - \sin A \sin B$.

(16) $\cos(A - B) = \cos A \cos B + \sin A \sin B$.

(17) $\sin(A + B) = \sin A \cos B + \cos A \sin B$.

(18) $\sin(A - B) = \sin A \cos B - \cos A \sin B$.

(19) $\tan(A + B) = \dfrac{\tan A + \tan B}{1 - \tan A \tan B}$.

(20) $\tan(A - B) = \dfrac{\tan A - \tan B}{1 + \tan A \tan B}$.

(21) $\sin 2A = 2 \sin A \cos A$.

(22a) $\cos 2A = \cos^2 A - \sin^2 A$.

(22b) $\quad\quad\quad = 1 - 2 \sin^2 A$.

(22c) $\quad\quad\quad = 2 \cos^2 A - 1$.